Monoaminergic Modulation of Cortical Excitability

Kuei-Yuan Tseng
Marco Atzori

Editors

Monoaminergic Modulation of Cortical Excitability

 Springer

Dr. Kuei-Yuan Tseng
Chicago Medical School Dept.
Cellular and Molecular Pharmacology
Rosalind Franklin University of Medicine
The Chicago Medical School
North Chicago, IL 60064
USA
Kuei.tseng@rosalindfranklin.edu

Dr. Marco Atzori
School of Brain and Behavioral
Sciences Neuroscience Program, GR 41
School for Behavioral and Brain Sciences
University of Texas
Dallas, TX 75083
USA
marco.atzori@utdallas.edu

Library of Congress Control Number: 2007930761

ISBN 978-0-387-72254-2   e-ISBN 978-0-387-72256-6

Printed on acid-free paper.

9 8 7 6 5 4 3 2 1

springer.com

Preface

The discovery by Patricia Goldman-Rakic that dopamine plays important roles in the modulation of prefrontal cortex neuronal activity during working memory was a milestone in the field of monoamine research. Since then, a significant number of studies from cellular to system physiology have investigated the neural basis of monoamine modulation in the cortex, providing seminal insights into potential treatments for schizophrenia and other neuropsychiatry disorders. In the past few years, many studies have provided compelling evidence indicating that other monoaminergic systems such as norepinephrine and serotonin are also essential in regulating neocortical functions. *Monoaminergic Modulation of Cortical Excitability* serves an integrative and comprehensive comparison among the diverse and complex modulatory action of dopamine, noradrenaline and serotonin receptors in the cortex. This volume intends to provide a current view of how cortical excitability and synaptic plasticity is regulated by monoamines and how these modulations interact with other neurotransmitter systems during critical periods of postnatal development such as puberty and adolescence. We have organized this volume in several sections offering a broad spectrum of opinions on how the monoamines affect cortical function from a cellular/subcellular level to system. The complexity of these interactions is discussed in light of recent data showing how disruption of these systems dramatically affects memory formation and information processing in the cortex.

Although the research field investigating the cortical roles of monoamines has been thriving for many decades now, a large amount of work is still needed to unveil several unresolved problems discussed in this volume. By understanding the common and distinctive mechanism coupling dopamine, norepinephrine, and serotonin actions has the potential to generate new therapeutic strategies for a number of devastating neuropsychiatric disorders. We expect further progress at an even faster rate in the incoming years.

Kuei-Yuan Tseng and Marco Atzori
Editors

Contents

Section IV   Monoamines and Psychiatric Disorders

Chapter 1

The Discovery of Multiple Serotonin Receptor Subtypes: *A Lesson from Molecular Cloning to Functional Expression*

Milt Teitler
Center for Neuropharmacology and Neuroscience, Albany Medical College, Albany, NY
   12208, USA

Abstract. The field of serotonin (5-HT) receptor pharmacology has been at least as dramatically altered by the advent of molecular neurobiology and recombinant DNA techniques, as has any other neurotransmitter receptor field. The principal reason for this is the unforeseen multitude of genes expressing different types of 5-HT receptors. Classical pharmacological studies as well as radioligand-binding studies convinced workers in the field that there were multiple 5-HT receptors. The extent of that multiplicity was not generally foreseen. Thirteen 5-HT receptor genes have been cloned and functionally expressed. All 13 receptors were cloned between the years 1987 and 1995. Adding to the complexity of 5-HT receptors pharmacology are the splice variants and, in one case, mRNA editing isoforms, detected in the last decade. Bringing order to this rapid outpouring of information at this time is a very difficult task. In order to provide a concise and timely review, this article focuses on the strategies used to clone and express multiple 5-HT receptors. Unique properties of the various receptors are emphasized.

1.1 Introduction

Neurotransmitter receptors have been divided into two general superfamilies: the G-protein-coupled receptors (GPCRs) and ligand-gated ion channels. The GPCR superfamily includes an astonishing variety of receptors, all having in common the stimulation of a GTP-binding protein upon agonist stimulation of the receptor. These receptors have the well-known general molecular architecture of seven alpha helices spanning the lipid bilayer. The N-terminus is extracellular and the C-terminus is intracellular. There is a great deal of primary structural conservation in the transmembrane-spanning regions. This conservation of amino acid sequence resulted in the rapid cloning of members of this receptor family before and after the determination of the human genome. Originally, cDNA probes based on known sequences of GPCRs were used to "fish out" sequences of genes that were similar to the already cloned GPCR sequence. A variation of this theme was the use of known sequences to design polymerase chain reaction (PCR) primers to amplify gene sequences that contain similar stretches of nucleic acids but code for novel receptors. These strategies are generally referred to as cloning by sequence homology and have been highly successful. With the determination of the human genome, programs searching for sequences common to GPCRs have allowed "virtual cloning" of orphan GPCR, that is, receptors that are clearly members of the GPCR family but for which no known ligands or function are yet determined. Differences in amino acids comprising the ligand-binding pocket account for the wide pharmacological variety of this vast family of receptors.

The second receptor superfamily is the ligand-gated ion channel family. They are ion channels and are composed of multimeric complexes of polypeptides, which form the channel. The binding of an agonist to a binding site on the cytoplasmic face of the receptor induces the opening of the channel. The number of known receptors in this family is currently far fewer than for the GPCR family. This may reflect fewer genes coding for this type of receptor. However, it appears at least as likely that the relative paucity of known ligand-gated ion channel receptors may be due to the much higher degree of difficulty involved in cloning the genes for this type of receptor. The potential pharmacological and neurobiological diversity possible due to cellular variability in expression of the multiple subunits making up the multimeric complexes appears to allow a given receptor subtype of this family to express a wide range of pharmacological and neurobiological properties (Table 1.1).

Table 1.1 5-HT Receptor Families.

Receptor family	Subtypes	Primary signal transduction
5-HT1	5-HT1A, 1B, 1D, 1E, 1F	Adenylate cyclase inhibition
5-HT2	5-HT2A, 2B, 2C	Phospholipase C stimulation
5-HT3	5-HT3A, 3B	Cation channel
5-HT4	5-HT4 (a–d) (splice variants)	Adenylate cyclase stimulation
5-HT5	5-HT5A	GIRK?
5-HT6		Adenylate cyclase stimulation
5-HT7	5-HT7 (a–d) (splice variants)	Adenylate cyclase stimulation

1.2 The 5-HT1A Receptor: Cloning by Sequence Homology

In 1987, the detection of a cDNA that was capable of hybridizing (under reduced stringency) to a full-length cDNA probe for the beta-2-adrenergic receptor was reported (Kobilka et al., 1987). Although the nucleic acid sequence of the open reading frame (ORF; nucleic acid sequence coding for a protein) predicted a protein highly homologous to the beta-2-adrenergic receptor, the receptor protein when expressed in a mammalian cell displayed pharmacological properties inconsistent with an adrenergic receptor. 125I-cyanopindolol, a nonselective beta-adrenergic/5-HT1B receptor radioligand, labeled membranes from mammalian cells transfected with the mysterious cDNA (code named G21). However, the overall affinity for 125I-cyanopindolol was two orders of magnitude lower for the receptor in the transfected cells than for beta-adrenergic receptors in mammalian tissues or in cells transfected with well-characterized, cloned beta-adrenergic receptors. Furthermore, epinephrine and norepinephrine did not compete for the binding site with the affinities consistent with a beta-adrenergic receptor. In 1988, the clone coded for by G21 was identified as the 5-HT1A receptor (Fargin et al., 1988). 3H-8-OH-DPAT, a 5-HT1A specific radioligand, exhibited saturable and high-affinity binding in COS-7 cells transfected with a plasmid containing the G21 cDNA. This high-affinity binding was inhibited by a series of drugs: the affinities of these drugs closely matched the affinities of drugs in binding to the 5-HT1A receptor detected in membrane homogenates of mammalian brain tissues. Data from both the saturation analysis of the radioligand (Scatchard transformation) and from the competition curves indicated that the radioligand was labeling two states of the receptor, consistent with the ternary complex model of receptor G-protein interactions. The G21 clone has been characterized as intronless. This was the first cloning of a serotonin receptor gene and was important for several reasons. For the serotonin receptor community, recombinant DNA technology was clearly on the verge of contributing major insights into an already fairly complex pharmacological field, bringing many unique opportunities to examine the 5-HT1A receptor in isolated recombinant cells. Of more global importance was the realization that the gene sequence and physical structure of GPCRs are highly conserved. Therefore, the methods that led to the "accidental" cloning of the 5-HT1A receptor gene (the use of sequence homology to "fish out" a previously uncloned receptor cDNA) became a standard strategy for receptor cloning. The intronless nature of the 5-HT1A receptor is presumably the reason that no splice variants have been reported (Table 1.1).

1.3 Cloning, Functional Expression, and Molecular Relationships of the 5-HT1B Receptor

In 1989, a novel method for cloning new members of the GPCR family using PCR was published (Libert et al., 1989). Taking advantage of the high degree of sequence homology residing in the TM regions of the clone GPCRs, it was hypothesized that degenerate PCR primers could be devised and would serve as generic templates for amplifying GPCR sequences using Taq polymerase. Thus, any receptor sequence with sufficient homology to at least one pair of the degenerate primers would be amplified between regions TM3 and TM6. The hypothesis was proven correct when several cDNAs were amplified from a dog thyroid cDNA library. These cDNAs were radiolabeled and four full-length clones were obtained. Sequence analysis of the full-length clones revealed four previously unpublished cDNAs called RDC1, RDC4, RDC7, and RDC8. The RDC4 clone was found to be approximately 50% homologous to the 5-HT1A receptor. However, the specific pharmacological identity of the cloned cDNA was not determined.

Radioligand-binding studies had revealed the presence of highly expressed receptors in rodent and human brain displaying different pharmacological properties (Pedigo et al., 1981; Hoyer et al., 1988). These receptors were named 5-HT1B and 5-HT1D, respectively. It was noted that the brain distribution of these highly expressed receptor are similar, and it was proposed that they were species variants of the same receptor despite pronounced differences in affinities for many drugs (Waeber et al., 1989). Cloning of the cDNAs coding for these receptors proved that, in fact, the originally termed 5-HT1B and 5-HT1D-beta, are species variants of the same receptor (Libert et al., 1989; Hamblin and Metcalf, 1991; Maenhaut et al., 1991; Suryanarayana et al., 1991; Adham et al., 1992; Demchyshyn et al., 1992; Hamblin et al., 1992; Jin et al., 1992; Levy et al., 1992; Maroteaux et al., 1992; Metcalf et al., 1992; Oksenberg et al., 1992; Weinshank et al., 1992). These receptors were found to have almost identical amino acid sequences. The reason for the difference in affinity for drugs in the different species is due to a difference in position 355 (TMVII) in humans versus rodents. In humans, there is a threonine at position 355. This amino acid at this position has been well known to lower GPCR affinities for many drugs including many of the beta-adrenergic blockers. Rodents express an asparagine at position 355 producing a receptor with high affinity for many beta-adrenergic blockers. Thus, the human and rodent 5-HT1B receptors appear quite distinct based on drug affinities. Distribution studies strongly supported the argument that they are species variants of the same receptor. The 99% amino acid homology between the human and rodent receptors proves that these clones are species variants. The situation was complicated by the discovery of another 5-HT receptor clone expressing lower amino acid homology but nearly identical drug affinities to the human 5-HT1B receptor (see below). It is now accepted that the 5-HT1B receptor displays significantly different drug affinities depending on the species due to position 355. For several years the term "5-HT1D-beta" was used to describe the human 5-HT1B receptor and may cause confusion when reading the original literature describing the cloning and properties of the 5-HT1B and 5-HT1D receptors.

The most notable pharmacological property of the 5-HT1B receptor is that it is the most likely target for the triptan antimigraine drugs (Peroutka and McCarthy, 1989; Miller et al., 1992b). Radioligand-binding studies demonstrated that sumatriptan, the first triptan antimigraine drug, displayed selective high affinity for the 5-HT1B receptor, using brain tissue. Functional studies of isolated cerebral arteries demonstrated a potent vasoconstrictor activity of sumatriptan mediated by a 5-HT receptor that appears to have the properties of the 5-HT1B receptor. In the brain the 5-HT1B receptors are highly expressed in basal ganglia, striatum, and frontal cortex and appear to function as terminal autoreceptors.

1.4 Cloning and Functional Expression of the 5-HT1E Receptor

The 5-HT1 receptor subfamily originally was characterized as responding potently to the drug 5-CT (5-carboxyamidotryptamine) as well as being radiolabeled with 3H-5-HT (Bradley et al., 1986). However, in 1989 the detection of a high-affinity 3H-5-HT-binding site in human brain tissue homogenates with properties that appeared to confound this classification scheme was reported (Leonhardt et al., 1989). The site was radiolabeled with high affinity by 3H-5-HT in a saturable manner, and the binding was potently and specifically inhibited by guanyl nucleotides in a manner consistent with a G-protein-coupled 5-HT receptor. Thus, the site appeared to be a member of the 5-HT1 receptor subfamily. However, the drug 5-CT displayed very low affinity for the site. Therefore, this novel binding site was called "5-HT1E" in an attempt to stay within the major accepted system of nomenclature and emphasize the unique pharmacological properties of the receptor.

In 1992, the cloning of several novel cDNA clones displaying high sequence homology to previously reported 5-HT receptor sequences was published (Levy et al., 1992). They detected these clones using probes designed from the 5-HT1A and 5-HT2C receptors. A human genomic library was probed using a mixture of the radiolabeled oligonucleotides. A positive clone was eventually isolated that displayed 50% homology with the 5-HT1A receptor and was termed "S31." When this cDNA was transfected and expressed in mammalian cells, a serotonin-dependent inhibition of forskolin-stimulated adenylate cyclase activity was observed. The precise pharmacological nature of this receptor was not reported. Several studies soon appeared reporting the cloning and functional expression of the human cDNA clones with identical sequences to the S31 clone originally reported by Levy et al. (McAllister et al., 1992; Zgombick et al., 1992; Gudermann et al., 1993). The pharmacological nature of this receptor clone was more fully examined in these reports. All three groups noted that the low affinities of 5-CT, ergotamine, and sumatriptan (drugs with high affinities for 5-HT1B and 5-HT1D receptors) for the receptor expressed in recombinant cells indicated that the receptor cloned was the 5-HT1E receptor discovered through radioligand binding (Leonhardt et al., 1989). It is likely that another 5-HT receptor detected with radioligand-binding studies using rabbit brain was also the 5-HT1E receptor (Xiong and Nelson, 1989). Two aspects of the 5-HT1E receptor are especially intriguing. The first is that this receptor is expressed at high levels in many areas of the human brain (Leonhardt et al., 1989; Miller and Teitler, 1992) and the receptor is not detected in mice and rat brain tissue but is expressed in guinea pig brain (Bai et al., 2004). It seems unusual that no specific drugs for the human 5-HT1E receptor have been developed given the high expression level in human brain. This may be related to the difficulty in designing drugs with selectivity for the h5-HT1E receptor due to the nature of the binding site (Dukat et al., 2004).

1.5 Cloning and Functional Expression of the 5-HT1F Receptor

Three groups reported the cloning and functional expression of a mouse, rat, and human "5-HT1F" receptor (Amlaiky et al., 1992; Adham et al., 1993; Lovenberg et al., 1993b). The mouse cloned displayed the highest homology to the human 5-HT1E receptor, approximately 70%, and two groups have referred to this clone as a "5-HT1E-beta" or "5-HT1E-like" receptor (Amlaiky et al., 1992; Lovenberg et al., 1993b). When the receptor was functionally expressed in recombinant cells, the pharmacological profiles were similar to each other, but distinct from the pharmacological profile of 5-HT1E receptor. The major distinction was that while the first 5-HT1E receptor cloned displayed low affinities for 5-HT, sumatriptan and ergotamine, the mouse and rat "5-HT1E-like" and human "5-HT1F" receptor displayed low affinity for 5-CT and fairly high affinity for sumatriptan and ergotamine. Radioligand-binding studies predicted the existence of high-affinity 5-HT receptors possessing pharmacological properties distinct from the known 5-HT1 receptors (Weisberg and Teitler, 1992). The question remains as to whether there are distinct receptors or species variants of the same receptor. The mouse clone mRNA was reported to be detectable only in hippocampus using in situ hybridization in mouse brain tissue slices (Amlaiky et al., 1992). The human clone mRNA was detectable in hippocampus and cortical areas of rat brain (Adham et al., 1993). The presence of 5-HT1F mRNA in the trigeminal nucleus has focused attention on the possible role of 5-HT1F receptors in the action of triptan antimigraine drugs (Bhalla et al., 2002).

1.6 Cloning and Functional Expression of the 5-HT2A Receptor by Sequence Homology

By 1987 and 1988 it was becoming clear that the GPCRs had conserved the basic seven transmembrane-spanning structures throughout evolution and that the TM regions often contained regions of high amino acid homology. With the publication of the nucleic acid sequence of the 5-HT2C receptor, it was reasoned that the pharmacologically related 5-HT2A (originally termed "5-HT2") receptor cDNA should be retrievable using radiolabeled cDNA probes based on TM sequences from the 5-HT2C sequence (Pritchett et al., 1988). This strategy proved successful. Using oligonucleotides probes based on sequences from TM2 and TM3 of the 5-HT2C receptor, a cDNA was detected in a rat forebrain cDNA library that coded for a protein containing an ORF with 51% sequence homology to the 5-HT2C receptor. Much higher sequence homologies were noted in the TM regions with an essentially identical TM5 sequence. Julius et al. (1990), using sequence homology screening of a rat whole brain cDNA and genomic library, cloned a cDNA with essentially the same nucleic acid sequence. A slight difference in the reading frame due to an extra thymidine reported by Pritchett et al. resulted in some minor differences in predicted structure. Radioligand-binding studies using 3H-spiperone or 125I-LSD and membranes from mammalian cells transfected with the putative 5-HT2A cDNA revealed the pharmacological properties associated with the brain 5-HT2A receptors. Spiperone and ketanserin (high-affinity 5-HT2A antagonists) both displayed high affinity for the radiolabeled receptor in the transfected cells. Pritchett et al. (1988) demonstrated that mRNA synthesized in vitro from the putative 5-HT2A cDNA, when injected into oocytes, produced a chloral-dependent calcium flux when the cells were exposed to 5-HT. This effect was potently blocked by spiperone. Also, a 5-HT-induced rise in intracellular calcium was detected in mammalian cells transfected with the putative 5-HT2A cDNA; this effect was blocked by spiperone.

There is a great deal of interest in the role of 5-HT2A receptor in cognitive processing (Teitler et al., 2002). This interest is due to several observations. The psychotomimetic effects of the LSD-like hallucinogens appear to be initiated by stimulation of brain 5-HT2A receptors, indicating the status of 5-HT2A receptor signaling is important in cognition (Glennon et al., 1984; Titeler et al., 1985, 1988). Brain 5-HT2A receptors are highly expressed in cortical areas associated with cognitive processing (Appel et al., 1990). It has been proposed that the atypical antipsychotic drugs have less extrapyramidal side effects and are more effective in improving negative symptoms of schizophrenia than the typical antipsychotic drugs due to their interaction with brain 5-HT2A receptors (Roth et al., 1994). The role of brain 5-HT2A receptors in brain function and dysfunction is currently of great interest to neuroscientists.

1.7 Cloning and Functional Expression of the 5-HT2B Receptor Using Reverse Transcriptase PCR

5-HT potently contracts the isolated rat stomach fundus and this receptor-mediated phenomenon has been studied for decades (Vane, 1957). However, the identification of this receptor mediating this effect had been elusive. None of the radiolabeled and/or cloned 5-HT receptors possessed a pharmacological profile that matched the potencies of drugs in contracting the rat fundus. In 1992, two groups used similar strategies to clone a novel receptor from rat stomach fundus tissue. This receptor appears to be the receptor mediating the contractile effects of 5-HT in the fundus and appears to be a new member of the 5-HT2 receptor family (Foguet et al., 1992; Kursar et al., 1992; Wainscott et al., 1993). Foguet et al. (1992) and Kursar et al. (1992) used reverse transcriptase PCR (RT PCR) to make cDNA from rat fundus mRNA. In this procedure, reverse transcriptase is used to produce a cDNA library from mRNA. The cDNA is then amplified with PCR. The primers for the PCR amplification were based on sequence information from the 5-HT2A and 5-HT2C receptor sequences. The pharmacological data indicated that the fundus receptor would be most homologous to the 5-HT2C receptor. A cDNA was amplified, radiolabeled, and used to clone a full-length cDNA from a rat stomach fundus cDNA library that contained an ORF of 479 amino acids. This receptor displayed 70% and 68% homology with the rat 5 HT2C and 5-HT2A receptors, respectively and was originally named 5-HT2F. Interestingly, of 14 tissues studied using quantitative PCR to detect low levels of mRNA for the 5-HT2F receptor, only the fundus appeared to be expressing this receptor. The receptor expressed in recombinant cells displayed a pharmacological profile very similar to the pharmacological profile observed in the isolated organ preparation experiments. The receptor stimulates phosphoinositol turnover when activated by agonists in transfected mammalian cells. In $Xenopus$ oocyte expression systems the 5-HT2F receptor activates a Ca^{+2}-dependent Cl-conductance. The sequence homology, the pharmacological similarities, and the stimulation of the same second messenger systems as the 5-HT2A and 5-HT2C receptors, clearly indicates that this receptor belongs in the 5-HT2 receptor family. The 5-HT2F receptor was eventually renamed 5-HT2B by the serotonin nomenclature committee (Julius et al., 1990).

1.8 Cloning of the 5-HT2C Receptor by Functional Expression in $Xenopus$ Oocytes

In 1987 and 1988, two groups reported the cloning of a cDNA encoding the 5-HT1C receptor (later renamed the "5-HT2C" receptor) (Lubbert et al., 1987; Julius et al., 1988). Unlike the cloning of the 5-HT1A receptor, which occurred due to the high sequence homology between serotonin and adrenergic receptors coupled to G-proteins, the 5-HT2C receptor was cloned using a functional expression assay. In order to clone a gene by functional expression, two major factors need to be present: a system capable of translating exogenous mRNA into a functional protein and sufficiently abundant source of mRNA. *Xenopus* oocytes had been demonstrated to be very useful for the expression of exogenous mRNA coding for receptor proteins (Sumikawa et al., 1981). Injection of mRNA extracts (from various tissues) into *Xenopus* oocytes results in the expression of cytoplasmic or membrane-bound proteins with properties similar to proteins expressed in the tissue from which the mRNA originated. As a source of mRNA, Julius and co-workers used rat choroid plexus tissue. Based on radioligand-binding assays, this tissue has been shown to have a very high density of 5-HT2C receptors (Yagaloff and Hartig, 1985). Julius and coworkers used a sucrose density gradient centrifugation procedure to enrich choroid plexus mRNA fractions that produced a 5-HT-activated chloral-dependent calcium current. A further enrichment of the active mRNA was achieved by constructing an expression cDNA library from the mRNA in the active fractions and transcribing mRNA from this cDNA library. Sequential pools of the cDNA library were tested for activity in the oocytes: the positive pool was subdivided and the resulting sublibrary tested for activity in the oocytes. Sequential testing, subdividing, and retesting eventually produced a purified positive clone. The transmembrane (TM) domains were between 30% and 50% homologous to similar regions in the 5-HT1A receptor. Although the oocyte expression system did not permit an exhaustive pharmacological examination, the activity of 5-HT, the low potency of spiperone in blocking the response, and the coupling of the receptor to elevation of intracellular calcium strongly indicated that the receptor cloned was the rat 5-HT2C receptor. Later studies in which this cDNA was stably expressed in mammalian cell lines confirmed the 5-HT2C receptor nature of the protein expressed by this clone (Julius et al., 1989).

1.9 Cloning and Functional Expression of the 5-HT4 Receptor

The 5-HT4 receptor was discovered and well characterized using "classical" pharmacological methodologies (Dumuis et al., 1988a, b, 1989a, b,; Bockaert et al., 1992). A unique pharmacological profile that featured of benzamide agonists such as cisapride made identification of the 5-HT4 receptor possible prior to cloning of the 5-HT4 cDNA and expression in recombinant cells. Bockaert and coworkers described the pharmacological properties of the receptor using primary culture of mouse colliculi neurons and various isolated organs depolarization, and resultant neurotransmitter release (Dumuis et al., 1989b; Fagni et al., 1992). High-potency, high-affinity agonists and antagonists were developed. Radiolabeling of these compounds allowed the direct labeling of the receptors in various mammalian species revealed a wide-spread distribution throughout the body, with highest levels found in the limbic system of the brain (Waeber et al., 1993, 1994, 1996). The pharmaceutical industry was highly motivated to develop peripheral 5-HT4 agonists for treatment of irritable bowel syndrome (IBS). Cisapride is prescribed for IBS (Zgombick et al., 1992; Gudermann et al., 1993), although there is some controversy concerning its efficacy (Yagaloff and Hartig, 1985). Tegaserod, a putative 5-HT4 partial agonist, is currently prescribed for IBS (Julius et al., 1988). There is a sense that a 5-HT4 agonist may prove to have beneficial effects on brain dysfunction, especially cognitive processes, due to its limbic distribution (Bockaert et al., 2004; Conductier et al., 2006). However at this time no therapeutic 5-HT4 drug for brain dysfunction has been developed.

Cloning of mammalian variants of the 5-HT4 receptor began with the cloning of the rat 5-HT4 receptor (Gerald et al., 1995). Primers designed from highly conserved regions of TM2 and TM4 of other 5-HT receptors were used to amplify fragments of 5-HT4 receptor from rat brain tissue. These fragments were used to isolate the full-length rat 5-HT4 cDNA from a rat cDNA library. Two splice variants were found varying in the length of their C-terminus and were labeled 5-HT4s and 5-HT4L. Eventually the mouse and human 5-HT4 receptors were cloned and seven C-terminal splice variants have been reported (Claeysen et al., 1996, 1997, 1998; Blondel et al., 1998;). In addition, a second extracellular loop splice variant was found in human tissue (Bockaert et al., 2004). Studies of 5-HT4 receptor functionality had revealed differences in agonist efficacies, potencies, and desensitization previous to the cloning of the receptor (Dumuis et al., 1988a, b, 1989a, b; Bockaert et al., 1992). Thus, a great deal of attention has been focused on the relationship between the splice variant and the variations in properties of the 5-HT4 receptor in various tissues. No obvious pattern has been observed.

1.10 Cloning and Functional Expression of Mouse 5-HT5A and 5-HT5B Receptors

The strategy of using TM-based nucleic acid sequences to act as PCR primers has been the single most successful strategy in the cloning of novel 5-HT receptors. The procedure consists of designing degenerate 5′ and 3′ primers based on TM3 and TM6 from previously cloned 5-HT receptors. Mouse brain RNA was reverse transcribed (Plassat et al., 1992). This DNA was used as a template to amplify DNA between and including nucleic acid sequences capable of hybridizing the two primers. This strategy produced several bands of DNA of varying lengths. The bands were isolated and sequenced, and the sequences were analyzed for significant homology to known 5-HT receptors. One of the DNA was then radiolabeled and used to probe a mouse brain cDNA library in order to obtain a full-length clone, hopefully containing the full DNA sequence for the receptor. A 4-kb cDNA was isolated. The amino acid sequence deduced from the nucleic acid sequence of this cDNA predicted a seven transmembrane-spanning protein with all of the expected properties of a GPCR. The N-terminal was short and displayed little or no homology to known receptors. One consensus glycosylation site was present, which is a common observation. The TM regions showed a surprisingly low homology to cloned 5-HT receptors, the highest homology (37%) occurring with a cloned *Drosophila* 5-HT receptor (Saudou et al., 1992). Despite the low sequence homology, the receptor expressed in transfected mammalian cells was clearly serotonergic in nature. The expressed receptor was radiolabeled with 125I-LSD with high affinity and 5-HT displayed fairly high affinity for the radiolabeled receptor (200 nM). Computer-assisted analysis indicated that the receptor existed in agonist high- and low-affinity states, similar to radiolabeled

5-HT2A receptors in brain (Battaglia et al., 1984). The high-affinity state appeared to represent a small proportion of the total number of sites, which was also very reminiscent of the brain 5-HT2A receptor. Pharmacologically, this receptor fit the 5-HT1D receptor category as it displayed very high affinity for 5-CT. However, the potent 5-HT1D receptor agonist, sumatriptan, displayed low affinity for the receptor. In situ hybridization and Northern blots clearly indicated significant levels of expression in brain. Quantitative PCR (in which mRNA is reverse transcribed and amplified with PCR) indicated that this receptor may only be expressed in the CNS. All of the cloned and expressed 5-HT receptors discussed above have been functionally coupled to an intracellular second messenger system, either inhibition of adenylate cyclase or stimulation of intracellular calcium levels through chloride-dependent mechanisms (see above). This clone did not stimulate or inhibit adenylate cyclase, or alter intracellular calcium levels. The authors postulate that the receptor may be linked to a potassium channel or, alternatively, that the G-protein necessary for the receptor to demonstrate functionality is not present in these cells. Because of the unique amino acid sequence, the somewhat unique pharmacology, and the apparent difference in second messenger coupling, the authors chose to categorize this receptor clone as a 5-HT5 receptor.

Using essentially the same strategy and reagents, the same group soon reported another clone with properties quite similar to the 5-HT5 receptor (Matthes et al., 1993). This clone was 69% homologous to the previously cloned 5-HT5 receptor in the TM regions. The pharmacological profile of this clone, when expressed in mammalian cells, was essentially identical to the first 5-HT5 clone and unlike any of the other 5-HT receptor subfamilies. Also, this receptor was not found to mediate any effect on adenylate cyclase or inositol phosphate formation when stimulated by 5-HT. The first 5-HT5 receptor clone mRNA was found to be most abundant in the cortex followed by the hippocampus. This second "5-HT5-like" clone was found to be most highly transcribed in the habenula and the hippocampus. Chromosomal mapping in both mouse and human chromosomes localized the genes for the two receptors to different chromosomes. Because of the similar amino acid sequence in the TM regions and the very similar pharmacological profile, Plassat et al. (1992) have chosen to place both of these clones in the 5-HT5 receptor subfamily. The first clone has been termed the 5-HT5A receptor and the second clone the 5-HT5B receptor. It is interesting to note that radioligand-binding studies using 3H-5-CT detected a radioligand-binding site(s) with high affinity for 5-HT and 5-CT and low affinity for sumatriptan (Mahle et al., 1991). It is possible that the site(s) labeled with 3H-5-CT displaying low affinity for sumatriptan and the 5-HT5 receptors are the same. Further studies in brain and in recombinant cells will be needed to establish the validity of this hypothesis.

The human 5-HT5A receptor was cloned and found to produce abundant mRNA in human tissues and appeared to be coupled to an inwardly rectifying K^+ channel, at least in recombinant cells (Grailhe et al., 2001). The human 5-HTB clone was found to contain stop codons and did not express mRNA in human tissues. It appears as though the 5-HT5B receptor is a "pseudogene," at least in humans, that is, it has lost its function during evolution.

1.11 Cloning and Functional Expression of a 5-HT6 Receptor

Using a PCR strategy, a clone expressing a seven TM-spanning protein was found when transfected into mammalian cells displayed the pharmacological properties of a 5-HT receptor (Monsma et al., 1993). This receptor displays a 36–41% homology with other cloned 5-HT receptors, has high affinity for 125I-LSD and 5-HT displays a 150 nM affinity for the radiolabeled receptor when expressed in mammalian cells. Northern blot analysis detected mRNA for this clone in various brain regions, with the striatum displaying the highest levels of mRNA. The pharmacological profile of this clone is particularly interesting as several important psychotropic drugs (clozapine, amoxapine, amitryptiline) display high affinity for this site. This receptor was found to produce a stimulation of adenylate cyclase activity in transfected cells: to this point only the 5-HT4 receptor (Bockaert et al., 1992) had been shown to have a similar activity among the 5-HT receptor family. Pharmacologically, this receptor has attracted a great deal of interest because of its high affinity for clozapine, an atypical antipsychotic agent. This drug is considered to be more efficacious in many types of psychosis and produces fewer side effects than the more commonly used antipsychotic drugs, whose actions are generally attributed to dopamine receptor blockade. The mechanism of action of clozapine is controversial and the high affinity of clozapine for several 5-HT receptor subtypes has led several groups to postulate a serotonergic mechanism as responsible for the unique pharmacological profile of clozapine.

A splice variant of the full-size receptor has been detected (Olsen et al., 1999). This truncated version is coexpressed in various brain regions and does translocate to the plasma membrane. However, no functional activity has been observed for the splice variant.

1.12 Cloning and Functional Expression of a 5-HT7 Receptor

In 1993 several groups reported the cloning and functional expression of yet another 5-HT receptor linked to stimulation of adenylate cyclase activity (Bard et al., 1993; Lovenberg et al., 1993a, Meycrhof et al., 1993; Shen et al., 1993). This clone is unique and has a homology ranging from 39% to 53% with other reported 5-HT receptor sequences. Interestingly, one cloning strategy that resulted in the isolation of the clone, designated "5-HT7," involved probes designed from *Drosophila* receptors that were used to detect this clone in a human cDNA library: the highest degree of homology is with a *Drosophila* receptor. This receptor is the third 5-HT receptor to display a stimulation of adenylate cyclase activity (5-HT4 and 5-HT6 are the others, see above). Whereas the pharmacological profile of this receptor is somewhat "5-HT1D-like" in displaying high affinity for 5-CT and moderate affinity for sumatriptan, the low sequence homology and the stimulation of adenylate cyclase activity indicates that this clone probably deserves to be placed in its own receptor category. As with the 5-HT6 receptor, clozapine displays a high affinity for the 5-HT7 receptor and therefore has attracted a great deal of attention from the psychiatric community. Also, this receptor has been implicated in the regulation of mammalian circadian rhythms as the pharmacology of the receptor matches the pharmacology of drugs that advance circadian rhythms (Lovenberg et al., 1993a).

Four C-terminal splice variants of the 5-HT7 receptor have been detected, similar to the 5-HT4 receptor (Heidmann et al., 1997). Although studies have attempted to correlate a difference in signal transduction and the splice variant, little or no differences in properties between the variants have been reported.

Recently an intriguing property of human 5-HT7 receptors was reported (Smith et al., 2006). Short-term (30 min), low concentrations (5 nM) of the antipsychotic drug risperidone inactivate the human 5-HT7 receptor in vitro. 9-OH-Risperidone and methiothepin produce a similar inactivation while other antipsychotic drugs do not. The mechanism appears to be a pseudo-irreversible interaction of some drugs with the receptor. A sudden inactivation of human 5-HT7 receptors in vivo may produce some of the undesirable side effects associated with risperidone. This anomalous inactivation process is under investigation.

1.13 Functional Expression of the 5-HT3 Receptor: Cloning of a Ligand-Gated Ion Channel by Functional Expression in *Xenopus* Oocytes

The existence of the 5-HT3 receptor was originally reported in 1957. At that time it was termed the "5-HTM" receptor (Gaddum and Picarelli, 1957). In 1985, selective antagonists of this receptor were developed (Richardson et al., 1985). This receptor was clearly quite distinct from the other known 5-HT receptors in several respects. The pharmacological structures capable of potently blocking serotonergic responses mediated by this receptor displayed essentially no affinity for other 5-HT receptors. 5-HT3 receptor antagonists have proven to be the most effective antiemetic agents and are now routinely used in conjunction with chemotherapy to control emesis associated with these drugs (Cunningham et al., 1987). The responses mediated by this receptor were of an electrophysiological nature (i.e., altered cation permeability, responses in the micro second time scale, and a rapid desensitization of the response) (Neijt et al., 1989).

Radioligand-binding studies soon confirmed the suspicions raised by the electrophysiological studies: that the 5-HT3 receptor is a member of the ligand-gated ion channel family of receptors and is not a member of the seven transmembrane domain spanning G-protein-coupled family of receptor (Barnard, 1988). This difference in the nature of the 5-HT3 receptor had profound implications for designing strategies to clone the gene coding for the receptor. Ligand-gated ion channel receptors display very low sequence homology to each other and almost no sequence homology to the GPCRs. Therefore, strategies based on sequence homology that have proven so successful for cloning numerous GPCRs were not anticipated to result in the cloning of the gene for the 5-HT3 receptor. Thus, two alternative general cloning strategies were considered. One strategy would have been to purify the 5-HT3 receptor, obtain the amino acid sequence of a peptide from the purified material, and design a cDNA probe from the deduced nucleic acid sequence of the peptide. Several groups provided evidence that this approach was feasible because the receptor could be solubilized from several tissue sources in a functional state (Miquel et al., 1990; Miller et al.,

1992a). The second strategy was based on functional expression of the mRNA coding for the receptor in a heterologous system allowing the isolation of a cDNA coding for the receptor (or a subunit of the receptor). This strategy proved successful in cloning the 5-HT2C receptor (see above). The latter strategy proved successful and is discussed herein.

In order for a ligand-gated ion channel receptor to be cloned, several factors have to be present at the start. A source of mRNA coding for a functional receptor is necessary, as well as a system capable of incorporating and translating the mRNA into a functional receptor. The source of 5-HT3 mRNA was not a problem. Several neuroblastoma-glioma cell lines, originally created by Nirenberg and coworkers (Neijt et al., 1989) had been shown to express high levels of functional 5-HT3 receptors (Hoyer and Neijt, 1988). In approaching the cloning of the 5-HT3 receptor several major technical obstacles were facing Julius and coworkers as they initiated their efforts that resulted in the successful cloning of a 5-HT3 receptor subunit (Maricq et al., 1991). The receptor mediated a relatively small electrophysiological response that desensitized rapidly. In addition, it was well established that ligand-gated ion channels are composed of multimeric subunits that associate to form a pore and a binding site (Barnard, 1988). Thus, even the successful cloning of one subunit was necessary, but not necessary sufficient, for the formation of an active channel. However, previous cloning studies had demonstrated that the individual subunit of a given ligand-gated ion channel can associate to form an active channel. These recombinant ligand-gated receptors tend to have properties reminiscent of the receptor in vivo. Furthermore, once one subunit is cloned the other subunits are soon cloned by sequence homology because within the ligand-gated channel subunit family there is a high degree of sequence homology. Thus, it was anticipated that detecting functional receptors in a transfected cell presented formidable but surmountable technical problems.

Polyadenylated mRNA from NCB-20 cells (one of the neuroblastoma-glioma cell lines known to express high levels of 5-HT3 receptors) was injected into *Xenopus* oocytes and electrophysiological recordings were made in the absence and presence of 10 μM 5-HT. A very small signal was noted in the presence of 1 mM Ca^{++} and Mg^{++}. In order to enhance this signal, the mRNA was size-fractionated. Two consecutive fractions were found to produce increased activity. A further enhancement of the signal (which was an inward, depolarizing current) was achieved by reducing the divalent cation concentrations to 0.1 mM. Desensitization was minimized by rapid exchange of solutions in a small perfusion chamber. Specific 5-HT3 blockers reversed the serotonergic response. An mRNA expression library was constructed from the two fractions producing the signal. RNA fractions produced by this library were prepared and tested for activity. Serial dilution testing of the RNA for activity and repetition of this procedure several times eventually led to the isolation of a single clone expressing the 5-HT3 mRNA. The cDNA insert in this clone contained an ORF of 1,462 bp that coded for a 487 amino acid protein. The subunit cloned (termed the beta-subunit) was found to have 27% sequence homology to the nicotinic acetylcholine receptor from *Torpedo californica*, 22% homology to the alpha-1 subunit of the GABAA receptor, and 22% homology to the 48K subunit of the rat glycine receptor.

Hydropathy analysis of the 5-HT3 clone (in which computer programs predict the extracellular, membrane-spanning, and cytoplasmic domains of a protein based on the amino acid sequence) produced a hydropathy map similar to other ligand-gated ion channels. Four membrane-spanning domains were predicted surrounded by a large N-terminal region and containing a large intracellular loop connecting TM3 and TM4. Transient transfection of the 5-HT3 cDNA in COS-7 cells produced a receptor with radioligand-binding characteristics essentially identical to 5-HT3 receptors in brain homogenates (Hoyer and Neijt, 1988; Neijt et al., 1989). The homomeric 5-HT3 receptor demonstrated many, but not all, of the electrophysiological properties of the native 5-HT3 receptor. The differences were presumably due to the existence of other types of subunits not yet cloned. Although Northern blots of RNA from a variety of mouse tissues proved negative, PCR amplification of mRNA did produce signals from mouse cortex, brainstem, midbrain, spinal cord, and heart. Surprisingly, 5-HT3 mRNA was not detected in the intestine, even though the GI tract has been shown to be extremely responsive to 5-HT3 drugs.

A second 5-HT3 subunit was cloned in 1999 (Davies et al., 1999; Dubin et al., 1999; Hanna et al., 2000). The first cloned subunit is now termed 5-HT3A and the second subunit 5-HT3B. Coexpressing these subunits results in a recombinant 5-HT3 receptor with conductance properties that more closely mimic 5-HT3 receptors in vivo than homomeric 5-HT3 receptors. 5-HT3B receptor subunit mRNA is found wherever 5-HT3A mRNA is found in rodent and human tissues and it appears that the natural 5-HT3 receptor is composed of some stoichiometry of the two subunits. The cation conductance is far larger with both subunits in the receptor than in the homomeric receptor composed of either subunit.

1.14 Summary

At the time of the preparation of this review, thirteen 5-HT receptors have been cloned and functionally expressed in mammalian tissues and *Xenopus* oocyte expression systems. For the medicinal chemist, this unexpected plethora of mammalian 5-HT receptor subtypes presents both wonderful opportunities for novel drug development as well as formidable challenges in designing compounds with a high degree of specificity. In fact, the synthesis of a truly specific serotonergic drug may already be impossible because of the multiplicity of receptors with similar binding domains. The increased knowledge about the physical structure of the receptors and the ability to alter receptor structures in an orderly and precise fashion using recombinant DNA techniques will allow the medicinal chemist to test hypotheses concerning binding domains in a degree of detail heretofore unattainable. The advent of molecular neurobiology should have as great an impact on the field of medicinal chemistry as it has on the other biomedical sciences.

Acknowledgments

M.T. is supported by PHS grant MH56650.

References

Adham N, Kao H T, Schecter L E, Bard J, Olsen M, Urquhart D, Durkin M, Hartig P R, Weinshank R L and Branchek T A (1993) Cloning of Another Human Serotonin Receptor (5-HT1F): A Fifth 5-HT1 Receptor Subtype Coupled to the Inhibition of Adenylate Cyclase. Proc Natl Acad Sci U S A 90: pp 408–412.

Adham N, Romanienko P, Hartig P, Weinshank R L and Branchek T (1992) The Rat 5-Hydroxytryptamine1B Receptor is the Species Homologue of the Human 5-Hydroxytryptamine1D Beta Receptor. Mol Pharmacol 41: pp 1–7.

Amlaiky N, Ramboz S, Boschert U, Plassat J L and Hen R (1992) Isolation of a Mouse "5HT1E-Like" Serotonin Receptor Expressed Predominantly in Hippocampus. J Biol Chem 267: pp 19761–19764.

Appel N M, Mitchell W M, Garlick R K, Glennon R A, Teitler M and De Souza E B (1990) Autoradiographic Characterization of (+-)-1-(2,5-Dimethoxy-4-[125I] Iodophenyl)-2-Aminopropane ([125I]DOI) Binding to 5-HT2 and 5-HT1c Receptors in Rat Brain. J Pharmacol Exp Ther 255: pp 843–857.

Bai F, Yin T, Johnstone E M, Su C, Varga G, Little S P and Nelson D L (2004) Molecular Cloning and Pharmacological Characterization of the Guinea Pig 5-HT1E Receptor. Eur J Pharmacol 484: pp 127–139.

Bard J A, Zgombick J, Adham N, Vaysse P, Branchek T A and Weinshank R L (1993) Cloning of a Novel Human Serotonin Receptor (5-HT7) Positively Linked to Adenylate Cyclase. J Biol Chem 268: pp 23422–23426.

Barnard E A (1988) Molecular Neurobiology. Separating Receptor Subtypes From Their Shadows. Nature 335: pp 301–302.

Battaglia G, Shannon M and Titeler M (1984) Guanyl Nucleotide and Divalent Cation Regulation of Cortical S2 Serotonin Receptors. J Neurochem 43: pp 1213–1219.

Bhalla P, Sharma H S, Wurch T, Pauwels P J and Saxena P R (2002) Molecular Cloning and Expression of the Porcine Trigeminal Ganglion CDNA Encoding a 5-Ht(1F) Receptor. Eur J Pharmacol 436: pp 23–33.

Blondel O, Gastineau M, Dahmoune Y, Langlois M and Fischmeister R (1998) Cloning, Expression, and Pharmacology of Four Human 5-Hydroxytryptamine 4 Receptor Isoforms Produced by Alternative Splicing in the Carboxyl Terminus. J Neurochem 70: pp 2252–2261.

Bockaert J, Claeysen S, Compan V and Dumuis A (2004) 5-HT4 Receptors. Curr Drug Targets CNS Neurol Disord 3: pp 39–51.

Bockaert J, Fozard J R, Dumuis A and Clarke D E (1992) The 5-HT4 Receptor: A Place in the Sun. Trends Pharmacol Sci 13: pp 141–145.

Bradley P B, Engel G, Feniuk W, Fozard J R, Humphrey P P, Middlemiss D N, Mylecharane E J, Richardson B P and Saxena P R (1986) Proposals for the Classification and Nomenclature of Functional Receptors for 5-Hydroxytryptamine. Neuropharmacology 25: pp 563–576.

Claeysen S, Faye P, Sebben M, Lemaire S, Bockaert J and Dumuis A (1997) Cloning and Expression of Human 5-HT4S Receptors. Effect of Receptor Density on Their Coupling to Adenylyl Cyclase. Neuroreport 8: pp 3189–3196.

Claeysen S, Faye P, Sebben M, Taviaux S, Bockaert J and Dumuis A (1998) 5-HT4 Receptors: Cloning and Expression of New Splice Variants. Ann N Y Acad Sci 861: pp 49–56.

Claeysen S, Sebben M, Journot L, Bockaert J and Dumuis A (1996) Cloning, Expression and Pharmacology of the Mouse 5-HT(4L) Receptor. FEBS Lett 398: pp 19–25.

Conductier G, Dusticier N, Lucas G, Cote F, Debonnel G, Daszuta A, Dumuis A, Nieoullon A, Hen R, Bockaert J and Compan V (2006) Adaptive Changes in Serotonin Neurons of the Raphe Nuclei in 5-HT(4) Receptor Knock-Out Mouse. Eur J Neurosci 24: pp 1053–1062.

Cunningham D, Hawthorn J, Pople A, Gazet J C, Ford H T, Challoner T and Coombes R C (1987) Prevention of Emesis in Patients Receiving Cytotoxic Drugs by GR38032F, a Selective 5-HT3 Receptor Antagonist. Lancet 1: pp 1461–1463.

Davies PA, Pistis M, Hanna M C, Peters J A, Lambert J J, Hales T G and Kirkness E F (1999) The 5-HT3B Subunit Is a Major Determinant of Serotonin-Receptor Function. Nature 397: pp 359–363.

Demchyshyn L, Sunahara R K, Miller K, Teitler M, Hoffman B J, Kennedy J L, Seeman P, Van Tol H H and Niznik H B (1992) A Human Serotonin 1D Receptor Variant (5HT1D Beta) Encoded by an Intronless Gene on Chromosome 6. Proc Natl Acad Sci U S A 89: pp 5522–5526.

Dubin A E, Huvar R, D'Andrea M R, Pyati J, Zhu J Y, Joy K C, Wilson S J, Galindo J E, Glass C A, Luo L, Jackson M R, Lovenberg T W and Erlander M G (1999) The Pharmacological and Functional Characteristics of the Serotonin 5-HT(3A) Receptor Are Specifically Modified by a 5-HT(3B) Receptor Subunit. J Biol Chem 274: pp 30799–30810.

Dukat M, Smith C, Herrick-Davis K, Teitler M and Glennon R A (2004) Binding of Tryptamine Analogs at H5-HT1E Receptors: A Structure-Affinity Investigation. Bioorg Med Chem 12: pp 2545–2552.

Dumuis A, Bouhelal R, Sebben M and Bockaert J (1988a) A 5-HT Receptor in the Central Nervous System, Positively Coupled With Adenylate Cyclase, Is Antagonized by ICS 205 930. Eur J Pharmacol 146: pp 187–188.

Dumuis A, Bouhelal R, Sebben M, Cory R and Bockaert J (1988b) A Nonclassical 5-Hydroxytryptamine Receptor Positively Coupled With Adenylate Cyclase in the Central Nervous System. Mol Pharmacol 34: pp 880–887.

Dumuis A, Sebben M and Bockaert J (1989a) BRL 24924: A Potent Agonist at a Non-Classical 5-HT Receptor Positively Coupled With Adenylate Cyclase in Colliculi Neurons. Eur J Pharmacol 162: pp 381–384.

Dumuis A, Sebben M and Bockaert J (1989b) The Gastrointestinal Prokinetic Benzamide Derivatives Are Agonists at the Non-Classical 5-HT Receptor (5-HT4) Positively Coupled to Adenylate Cyclase in Neurons. Naunyn Schmiedebergs Arch Pharmacol 340: pp 403–410.

Fagni L, Dumuis A, Sebben M and Bockaert J (1992) The 5-HT4 Receptor Subtype Inhibits K+ Current in Colliculi Neurones Via Activation of a Cyclic AMP-Dependent Protein Kinase. Br J Pharmacol 105: pp 973–979.

Fargin A, Raymond J R, Lohse M J, Kobilka B K, Caron M G and Lefkowitz R J (1988) The Genomic Clone G-21 Which Resembles a Beta-Adrenergic Receptor Sequence Encodes the 5-HT1A Receptor. Nature 335: pp 358–360.

Foguet M, Hoyer D, Pardo L A, Parekh A, Kluxen F W, Kalkman H O, Stuhmer W and Lubbert H (1992) Cloning and Functional Characterization of the Rat Stomach Fundus Serotonin Receptor. EMBO J 11: pp 3481–3487.

Gaddum J H and Picarelli Z P (1957) Two Kinds of Tryptamine Receptor. Br J Pharmacol Chemother 12: pp 323–328.

Gerald C, Adham N, Kao H T, Olsen M A, Laz T M, Schechter L E, Bard J A, Vaysse P J, Hartig P R, Branchek T A, and Weinshanb R L (1995) The 5-HT4 Receptor: Molecular Cloning and Pharmacological Characterization of Two Splice Variants. EMBO J 14: pp 2806–2815.

Glennon R A, Titeler M and McKenney J D (1984) Evidence for 5-HT2 Involvement in the Mechanism of Action of Hallucinogenic Agents. Life Sci 35: pp 2505–2511.

Grailhe R, Grabtree G W and Hen R (2001) Human 5-HT(5) Receptors: the 5-HT(5A) Receptor Is Functional but the 5-HT(5B) Receptor Was Lost During Mammalian Evolution. Eur J Pharmacol 418: pp 157–167.

Gudermann T, Levy F O, Birnbaumer M, Birnbaumer L and Kaumann A J (1993) Human S31 Serotonin Receptor Clone Encodes a 5-Hydroxytryptamine1E-Like Serotonin Receptor. Mol Pharmacol 43: pp 412–418.

Hamblin M W and Metcalf M A (1991) Primary Structure and Functional Characterization of a Human 5-HT1D-Type Serotonin Receptor. Mol Pharmacol 40: pp 143–148.

Hamblin M W, Metcalf M A, McGuffin R W and Karpells S (1992) Molecular Cloning and Functional Characterization of a Human 5-HT1B Serotonin Receptor: A Homologue of the Rat 5-HT1B Receptor With 5-HT1D-Like Pharmacological Specificity. Biochem Biophys Res Commun 184: pp 752–759.

Hanna M C, Davies P A, Hales T G and Kirkness E F (2000) Evidence for Expression of Heteromeric Serotonin 5-HT(3) Receptors in Rodents. J Neurochem 75: pp 240–247.

Heidmann D E, Metcalf M A, Kohen R and Hamblin M W (1997) Four 5-Hydroxytryptamine7 (5-HT7) Receptor Isoforms in Human and Rat Produced by Alternative Splicing: Species Differences Due to Altered Intron-Exon Organization. J Neurochem 68: pp 1372–1381.

Hoyer D and Neijt H C (1988) Identification of Serotonin 5-HT3 Recognition Sites in Membranes of N1E-115 Neuroblastoma Cells by Radioligand Binding. Mol Pharmacol 33: pp 303–309.

Hoyer D, Waeber C, Pazos A, Probst A and Palacios J M (1988) Identification of a 5-HT1 Recognition Site in Human Brain Membranes Different From 5-HT1A, 5-HT1B and 5-HT1C Sites. Neurosci Lett 85: pp 357–362.

Jin H, Oksenberg D, Ashkenazi A, Peroutka S J, Duncan A M, Rozmahel R, Yang Y, Mengod G, Palacios J M and O'Dowd B F (1992) Characterization of the Human 5-Hydroxytryptamine1B Receptor. J Biol Chem 267: pp 5735–5738.

Julius D, Huang K N, Livelli T J, Axel R and Jessell T M (1990) The 5HT2 Receptor Defines a Family of Structurally Distinct but Functionally Conserved Serotonin Receptors. Proc Natl Acad Sci U S A 87: pp 928–932.

Julius D, Livelli T J, Jessell T M and Axel R (1989) Ectopic Expression of the Serotonin 1c Receptor and the Triggering of Malignant Transformation. Science 244: pp 1057–1062.

Julius D, MacDermott A B, Axel R and Jessell T M (1988) Molecular Characterization of a Functional CDNA Encoding the Serotonin 1c Receptor. Science 241: pp 558–564.

Kobilka B K, Frielle T, Collins S, Yang-Feng T, Kobilka T S, Francke U, Lefkowitz R J and Caron M G (1987) An Intronless Gene Encoding a Potential Member of the Family of Receptors Coupled to Guanine Nucleotide Regulatory Proteins. Nature 329: pp 75–79.

Kursar J D, Nelson D L, Wainscott D B, Cohen M L and Baez M (1992) Molecular Cloning, Functional Expression, and Pharmacological Characterization of a Novel Serotonin Receptor (5-Hydroxytryptamine2F) From Rat Stomach Fundus. Mol Pharmacol 42: pp 549–557.

Leonhardt S, Herrick-Davis K and Titeler M (1989) Detection of a Novel Serotonin Receptor Subtype (5-HT1E) in Human Brain: Interaction With a GTP-Binding Protein. J Neurochem 53: pp 465–471.

Levy FO, Gudermann T, Birnbaumer M, Kaumann A J and Birnbaumer L (1992) Molecular Cloning of a Human Gene (S31) Encoding a Novel Serotonin Receptor Mediating Inhibition of Adenylyl Cyclase. FEBS Lett 296: pp 201–206.

Libert F, Parmentier M, Lefort A, Dinsart C, Van Sande J, Maenhaut C, Simons M J, Dumont J E and Vassart G (1989) Selective Amplification and Cloning of Four New Members of the G Protein-Coupled Receptor Family. Science 244: pp 569–572.

Lovenberg T W, Baron B M, de Lecea L, Miller J D, Prosser R A, Rea M A, Foye P E, Racke M, Slone A L, Siegel B W (1993a) A Novel Adenylyl Cyclase-Activating Serotonin Receptor (5-HT7) Implicated in the Regulation of Mammalian Circadian Rhythms. Neuron 11: pp 449–458.

Lovenberg T W, Erlander M G, Baron B M, Racke M, Slone A L, Siegel B W, Craft C M, Burns J E, Danielson P E and Sutcliffe J G (1993b) Molecular Cloning and Functional Expression of 5-HT1E-Like Rat and Human 5-Hydroxytryptamine Receptor Genes. Proc Natl Acad Sci U S A 90: pp 2184–2188.

Lubbert H, Hoffman B J, Snutch T P, van Dyke T, Levine A J, Hartig P R, Lester H A and Davidson N (1987) CDNA Cloning of a Serotonin 5-HT1C Receptor by Electrophysiological Assays of MRNA-Injected *Xenopus* Oocytes. Proc Natl Acad Sci U S A 84: pp 4332–4336.

Maenhaut C, Van Sande J, Massart C, Dinsart C, Libert F, Monferini E, Giraldo E, Ladinsky H, Vassart G and Dumont J E (1991) The Orphan Receptor CDNA RDC4 Encodes a 5-HT1D Serotonin Receptor. Biochem Biophys Res Commun 180: pp 1460–1468.

Mahle C D, Nowak H P, Mattson R J, Hurt S D and Yocca F D (1991) [3H]5-Carboxamidotryptamine Labels Multiple High Affinity 5-HT1D-Like Sites in Guinea Pig Brain. Eur J Pharmacol 205: pp 323–324.

Maricq A V, Peterson A S, Brake A J, Myers R M and Julius D (1991) Primary Structure and Functional Expression of the 5HT3 Receptor, a Serotonin-Gated Ion Channel. Science 254: pp 432–437.

Maroteaux L, Saudou F, Amlaiky N, Boschert U, Plassat J L and Hen R (1992) Mouse 5HT1B Serotonin Receptor: Cloning, Functional Expression, and Localization in Motor Control Centers. Proc Natl Acad Sci U S A 89: pp 3020–3024.

Matthes H, Boschert U, Amlaiky N, Grailhe R, Plassat J L, Muscatelli F, Mattei M G and Hen R (1993) Mouse 5-Hydroxytryptamine5A and 5-Hydroxytryptamine5B Receptors Define a New Family of Serotonin Receptors: Cloning, Functional Expression, and Chromosomal Localization. Mol Pharmacol 43: pp 313–319.

McAllister G, Charlesworth A, Snodin C, Beer M S, Noble A J, Middlemiss D N, Iversen L L and Whiting P (1992) Molecular Cloning of a Serotonin Receptor From Human Brain (5HT1E): a Fifth 5HT1-Like Subtype. Proc Natl Acad Sci U S A 89: pp 5517–5521.

Metcalf M A, McGuffin R W and Hamblin M W (1992) Conversion of the Human 5-HT1D Beta Serotonin Receptor to the Rat 5-HT1B Ligand-Binding Phenotype by Thr355Asn Site Directed Mutagenesis. Biochem Pharmacol 44: pp 1917–1920.

Meyerhof W, Obermuller F, Fehr S and Richter D (1993) A Novel Rat Serotonin Receptor: Primary Structure, Pharmacology, and Expression Pattern in Distinct Brain Regions. DNA Cell Biol 12: pp 401–409.

Miller K, Weisberg E, Fletcher P W and Teitler M (1992a) Membrane-Bound and Solubilized Brain 5HT3 Receptors: Improved Radioligand Binding Assays Using Bovine Area Postrema or Rat Cortex and the Radioligands 3H-GR65630, 3H-BRL43694, and 3H-LY278584. Synapse 11: pp 58–66.

Miller K J, King A, Demchyshyn L, Niznik H and Teitler M (1992b) Agonist Activity of Sumatriptan and Metergoline at the Human 5-HT1D Beta Receptor: Further Evidence for a Role of the 5-HT1D Receptor in the Action of Sumatriptan. Eur J Pharmacol 227: pp 99–102.

Miller K J and Teitler M (1992) Quantitative Autoradiography of 5-CT-Sensitive (5-HT1D) and 5-CT-Insensitive (5-HT1E) Serotonin Receptors in Human Brain. Neurosci Lett 136: pp 223–226.

Miquel M C, Emerit M B, Bolanos F J, Schechter L E, Gozlan H and Hamon M (1990) Physicochemical Properties of Serotonin 5-HT3 Binding Sites Solubilized From Membranes of NG 108-15 Neuroblastoma-Glioma Cells. J Neurochem 55: pp 1526–1536.

Monsma Jr. F J, Shen Y, Ward R P, Hamblin M W and Sibley D R (1993) Cloning and Expression of a Novel Serotonin Receptor With High Affinity for Tricyclic Psychotropic Drugs. Mol Pharmacol 43: pp 320–327.

Neijt H C, Plomp J J and Vijverberg H P (1989) Kinetics of the Membrane Current Mediated by Serotonin 5-HT3 Receptors in Cultured Mouse Neuroblastoma Cells. J Physiol 411: pp 257–269.

Oksenberg D, Marsters S A, O'Dowd B F, Jin H, Havlik S, Peroutka S J and Ashkenazi A (1992) A Single Amino-Acid Difference Confers Major Pharmacological Variation Between Human and Rodent 5-HT1B Receptors. Nature 360: pp 161–163.

Olsen M A, Nawoschik S P, Schurman B R, Schmitt H L, Burno M, Smith D L and Schechter L E (1999) Identification of a Human 5-HT6 Receptor Variant Produced by Alternative Splicing. Brain Res Mol Brain Res 64: pp 255–263.

Pedigo N W, Yamamura H I and Nelson D L (1981) Discrimination of Multiple [3H]5-Hydroxytryptamine Binding Sites by the Neuroleptic Spiperone in Rat Brain. J Neurochem 36: pp 220–226.

Peroutka S J and McCarthy B G (1989) Sumatriptan (GR 43175) Interacts Selectively With 5-HT1B and 5-HT1D Binding Sites. Eur J Pharmacol 163: pp 133–136.

Plassat J L, Boschert U, Amlaiky N and Hen R (1992) The Mouse 5HT5 Receptor Reveals a Remarkable Heterogeneity Within the 5HT1D Receptor Family. EMBO J 11: pp 4779–4786.

Pritchett D B, Bach A W, Wozny M, Taleb O, Dal Toso R, Shih J C and Seeburg P H (1988) Structure and Functional Expression of Cloned Rat Serotonin 5HT-2 Receptor. EMBO J 7: pp 4135–4140.

Richardson B P, Engel G, Donatsch P and Stadler P A (1985) Identification of Serotonin M-Receptor Subtypes and Their Specific Blockade by a New Class of Drugs. Nature 316: pp 126–131.

Roth B L, Craigo S C, Choudhary M S, Uluer A, Monsma Jr. F J, Shen Y, Meltzer H Y and Sibley D R (1994) Binding of Typical and Atypical Antipsychotic Agents to 5-Hydroxytryptamine-6 and 5-Hydroxytryptamine-7 Receptors. J Pharmacol Exp Ther 268: pp 1403–1410.

Saudou F, Boschert U, Amlaiky N, Plassat J L and Hen R (1992) A Family of *Drosophila* Serotonin Receptors With Distinct Intracellular Signalling Properties and Expression Patterns. EMBO J 11: pp 7–17.

Shen Y, Monsma Jr. F J, Metcalf M A, Jose P A, Hamblin M W and Sibley D R (1993) Molecular Cloning and Expression of a 5-Hydroxytryptamine7 Serotonin Receptor Subtype. J Biol Chem 268: pp 18200–18204.

Smith C, Rahman T, Toohey N, Mazurkiewicz J, Herrick-Davis K and Teitler M (2006) Risperidone Irreversibly Binds to and Inactivates the H5-HT7 Serotonin Receptor. Mol Pharmacol 70: pp 1264–1270.

Sumikawa K, Houghton M, Emtage J S, Richards B M and Barnard E A (1981) Active Multi-Subunit ACh Receptor Assembled by Translation of Heterologous MRNA in *Xenopus* Oocytes. Nature 292: pp 862–864.

Suryanarayana S, Daunt D A, Von Zastrow M and Kobilka B K (1991) A Point Mutation in the Seventh Hydrophobic Domain of the Alpha 2 Adrenergic Receptor Increases Its Affinity for a Family of Beta Receptor Antagonists. J Biol Chem 266: pp 15488–15492.

Teitler M, Herrick-Davis K and Purohit A (2002) Constitutive Activity of G-Protein Coupled Receptors: Emphasis on Serotonin Receptors. Curr Top Med Chem 2: pp 529–538.

Titeler M, Herrick K, Lyon R A, McKenney J D and Glennon R A (1985) [3H]DOB: A Specific Agonist Radioligand for 5-HT2 Serotonin Receptors. Eur J Pharmacol 117: pp 145–146.

Titeler M, Lyon R A and Glennon R A (1988) Radioligand Binding Evidence Implicates the Brain 5-HT2 Receptor As a Site of Action for LSD and Phenylisopropylamine Hallucinogens. Psychopharmacology (Berl) 94: pp 213–216.

Vane J R (1957) A Sensitive Method for the Assay of 5-Hydroxytryptamine. Br J Pharmacol Chemother 12: pp 344–349.

Waeber C, Dietl M M, Hoyer D and Palacios J M (1989) 5.HT1 Receptors in the Vertebrate Brain. Regional Distribution Examined by Autoradiography. Naunyn Schmiedebergs Arch Pharmacol 340: pp 486–494.

Waeber C, Sebben M, Bockaert J and Dumuis A (1996) Regional Distribution and Ontogeny of 5-HT4 Binding Sites in Rat Brain. Behav Brain Res 73: pp 259–262.

Waeber C, Sebben M, Grossman C, Javoy-Agid F, Bockaert J and Dumuis A (1993) [3H]-GR113808 Labels 5-HT4 Receptors in the Human and Guinea-Pig Brain. Neuroreport 4: pp 1239–1242.

Waeber C, Sebben M, Nieoullon A, Bockaert J and Dumuis A (1994) Regional Distribution and Ontogeny of 5-HT4 Binding Sites in Rodent Brain. Neuropharmacology 33: pp 527–541.

Wainscott D B, Cohen M L, Schenck K W, Audia J E, Nissen J S, Baez M, Kursar J D, Lucaites V L and Nelson D L (1993) Pharmacological Characteristics of the Newly Cloned Rat 5-Hydroxytryptamine2F Receptor. Mol Pharmacol 43: pp 419–426.

Weinshank R L, Zgombick J M, Macchi M J, Branchek T A and Hartig P R (1992) Human Serotonin 1D Receptor Is Encoded by a Subfamily of Two Distinct Genes: 5-HT1D Alpha and 5-HT1D Beta. Proc Natl Acad Sci U S A 89: pp 3630–3634.

Weisberg E and Teitler M. (1992) Novel High-Affinity 3H-Serotonin Binding Sites in Rat and Bovine Brain Tissue. Drug Dev Res 26: pp 225–229.

Xiong W C and Nelson D L (1989) Characterization of a [3H]-5-Hydroxytryptamine Binding Site in Rabbit Caudate Nucleus That Differs From the 5-HT1A, 5-HT1B, 5-HT1C and 5-HT1D Subtypes. Life Sci 45: pp 1433–1442.

Yagaloff K A and Hartig P R (1985) 125I-Lysergic Acid Diethylamide Binds to a Novel Serotonergic Site on Rat Choroid Plexus Epithelial Cells. J Neurosci 5: pp 3178–3183.

Zgombick J M, Schechter L E, Macchi M, Hartig P R, Branchek T A and Weinshank R L (1992) Human Gene S31 Encodes the Pharmacologically Defined Serotonin 5-Hydroxytryptamine1E Receptor. Mol Pharmacol 42: pp 180–185.

Chapter 2

Distribution of D1-Like and D2 Receptors in the Monkey Brain: *Implications for Cognitive Function in Schizophrenia*

Ladislav Mrzljak[1], William E. Fieles[1], Amy M. Medd[1], Brian L. Largent[1] and Zafar U. Khan[2]

[1]Department of Neuroscience Biology, AstraZeneca Pharmaceuticals R&D, Wilmington, DE 19850, USA; [2]Department of Medicine, School of Medicine, Malaga, Spain

Abstract. Dopamine D1-like (D1 and D5) and D2 (D2S and D2L) receptors are widely distributed in neuronal circuits of the primate dorsolateral prefrontal cortex (dlPFC) which is critically implicated in cognitive function such as working memory. Working memory is dependent on normal dopaminergic function in the dlPFC and consistently impaired in schizophrenia. Therefore, D1 and D2 receptors are considered to be good drug targets for improvement of cognitive function and treatment of psychosis in schizophrenia. Current data show that D1 and D5 receptors have complementary localization and function in the brain and therefore a non-selective drug targeting both of these receptors may be optimal for improvement of cognitive function in schizophrenia. The differential distribution and function of short (D2S) and long (D2L) isoforms of the D2 receptor represent an opportunity to develop novel antipsychotics with improved efficacy and side effects.

2.1 Introduction

Cloning of five dopamine receptors (D1–D5) in the late 1980s and early 1990s (Gazi and Strange, 2002) enabled the development of the first subtype specific antibodies for analysis of dopamine receptor localization in the central nervous system (Levey et al., 1993; Sesack et al., 1994; Bergson et al., 1995; Mrzljak et al., 1996; Khan et al., 1998a, b, 2000; Ciliax et al., 2000). Application of subtype specific antibodies in immunohistochemistry at the light microscopic (LM) and electron microscopic (EM) level gave a great insight into the cellular and subcellular substrates associated with brain dopamine (DA) receptors. However, more than 10 years after the development of specific antibodies, a detailed picture about the distribution of DA receptors in brain neuronal circuits is still lacking, especially for the primate brain. This information would be of great importance to further understand DA receptors implicated in brain functions such as cognition, movement and motor behavior, goal directed behaviors, as well as in the neuroendocrine system in diseases like schizophrenia, mood disorders, addiction and Parkinson's disease.

Based on their structural, biochemical and pharmacological properties, DA receptors are classified into D1-like (D1 and D5) and D2-like (D2, D3 and D4) receptors. In this review we will focus on D1-like and D2 receptors in the nonhuman primate (macaque monkey) brain with special emphasis on the functional consequences of their distribution in the dorsolateral prefrontal cortex (dlPFC).

2.2 D1-Like Receptors (D1 and D5)

D1 (446 amino acids) and D5 (477 amino acids) receptor proteins inhibit adenylate cyclase activity and are encoded by two intronless genes (for review see Gingrich and Caron, 1993). D1-like receptors were chosen as promising drug targets for cognitive improvement in schizophrenia because it has been consistently shown that D1-like receptor agonists improve working memory function in the nonhuman primate models (Castner et al., 2004). Working memory is an important component of cognitive function, which is critically dependent on normal dopaminergic function in the dlPFC and is consistently impaired in schizophrenia (Goldman-Rakic et al., 2004). As shown by combined PET imaging/functional working memory testing in schizophrenic patients, D1-like receptor density upregulation in the dlPFC was accompanied by poor working memory performance (Abi-Dargham et al., 2002). It was suggested that the increase in D1 receptor density may be due to a decrease of DA levels in the dlPFC of schizophrenics and therefore hypothesized that an increase of dopamine function via application of D1-like agonists might normalize impaired working memory. Indeed, in the rhesus monkeys, D1-like antagonists impair while D1-like agonists improve working memory performance (Goldman-Rakic et al., 2004). D1-like agonists were able to normalize poor working memory performance in chronic haloperidol and aging rhesus monkey models (Castner et al., 2000; Castner and Goldman-Rakic, 2004). However, the agonists (ABT-431, A77636, SKF) and antagonists (SCH 23390 and SCH 39166) used in these models have equimolar potencies at both of the receptors, hence they are not subtype selective for either D1 or D5 receptors. Selective compounds for the functional analysis of D1 and D5 receptors using neurophysiological, neurochemical and behavioral assays are still under development. Complete knowledge about their subcellular localization in the brain of the macaque monkey may help pinpoint potential differential functional roles that these receptors play in the modulation of neuronal circuits involved in schizophrenia, including the d1PFC, basal ganglia and hippocampus. In the absence of D1 or D5 subtype selective pharmacological tools, the information about D1 or D5 receptor localization might also elucidate whether drugs selectively targeting either D1 or D5 receptors or both receptors (D1-like, D1 or D5) are necessary for cognition improvement in schizophrenia.

2.2.1 REGIONAL AND CELLULAR DISTRIBUTION

D1 and D5 receptors are widely distributed in the nonhuman primate central nervous system (Fig. 2.1, Table 2.1).

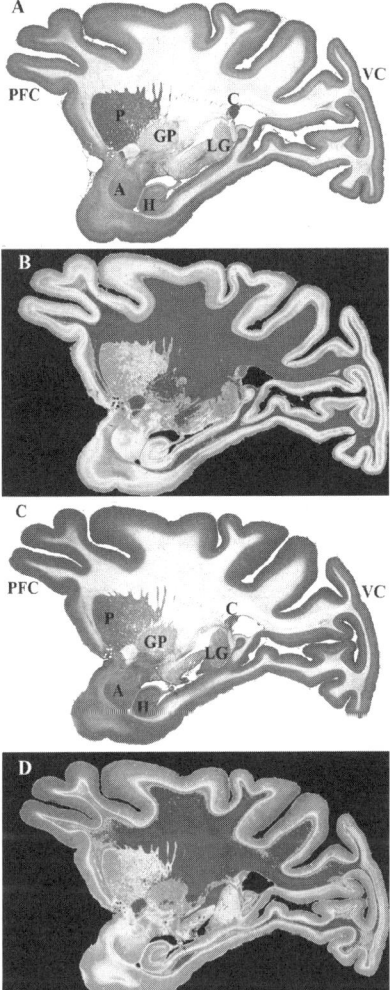

Figure 2.1 D1 (A–B) and D5 (B–C) receptor distribution in sagittal sections of macaque monkey brain as revealed by immunohistochemistry. On pseudo-colored images (B and D), red represents areas of highest level of protein expression and purple the lowest. Brain regions: PFC = prefrontal cortex, VC = visual cortex, P = putamen, C = head of caudate nucleus, GP = globus pallidus, A = amygdala, H = hippocampus and LG = lateral geniculate body. (See color insert.)

Both receptors are present in the cerebral cortex, however distribution of the D1 receptor has a more laminar pattern and is predominant in supragranular layers (II and III) and layer V, while distribution of the D5 receptor is more diffuse throughout the cerebral cortex (Fig. 2.1, Bergson et al., 1995; Ciliax et al., 2000; Khan et al., 2000). At the cellular level, D1 and D5 receptors were shown to be distributed both in pyramidal cells and interneurons of the dlPFC, where they often colocalize. D1 and D5 receptors are present in subpopulations of GABAergic neurons containing

containing calcium binding proteins such as parvalbumin, calbindin and calretinin (Fig. 2.2; Muly et al., 1998) In the striatum (caudate nucleus and putamen), D1 is the predominant D1-like receptor and is exclusively and abundantly expressed in medium medium spiny GABAergic neurons. D5 receptor is less abundant in striatum and is localized in large cholinergic interneurons, while medium sized neurons show less D5 immunoreactivity (Bergson et al., 1995; Ciliax et al., 2000). In contrast to the striatum, the globus pallidus shows moderate D5 but sparse D1 protein expression, as is the case in thalamic nuclei (Fig. 2.1; Bergson et al., 1995; Ciliax et al., 2000). The hippocampus and amygdala express higher levels of D5 than D1 receptor (Fig. 2.1, Bergson et al., 1995). Cholinergic neurons in the basal forebrain, similar to those in striatum, exclusively express the D5 receptor and have the highest amount of D5 receptor protein in the telencephalon (Table 2.1; Bergson et al., 1995; Ciliax et al., 2000). Selective distribution of the D5 receptor in cholinergic neurons of the basal forebrain suggests its role in modulation of acetylcholine release in the cortex and hippocampus as has been shown for this receptor in the mouse brain using D5 knockout animals (Laplante et al., 2004).

Table 2.1 D1 and D5 Receptor Expression in Macaque Monkey Brain: +++, strong expression; ++, moderate; +, low; −, weak or no expression.

Region	D1 receptor	D5 receptor
Striatum (GABA)	+++	++
N.accumbens	+++	++
G.Pallidus	-	+
Thalamus	-	+
Hippocampus	+	++
Amygdala	+	++
Basal forebrain (Ach)	-	+++
Cortex (layers III and V)	++	++

In the current literature, the D5 receptor is often described to be less abundant than the D1 receptor. This statement has to be taken with caution, because this comparison is not based on true D5 receptor protein quantification in nonhuman primate and human brain regions, but predominantly on an estimation of mRNA expression. For example, in the rat cortex, D1 receptor antibody immunoprecipitates 35% of the D1-like receptors labeled with [3H] SCH23390 (Khan et al., 2000). Together with the fact that dopamine is ten times more potent at the D5 than the D1 receptor, these data suggest that the role of the D5 receptor in the neuronal circuitry is not minor in comparison to the D1 receptor.

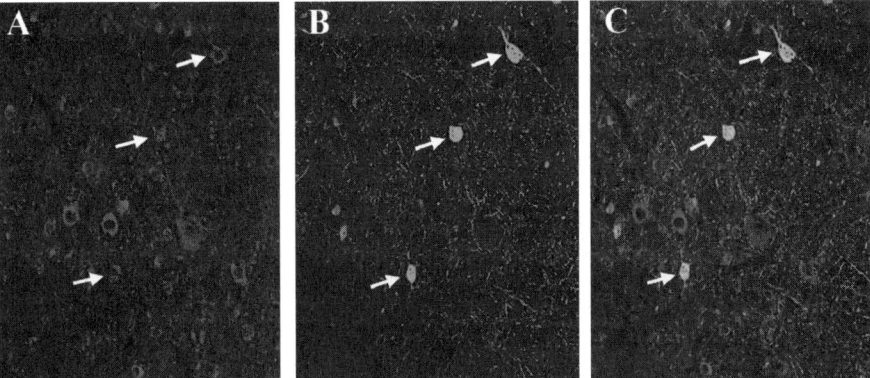

Figure 2.2 Localization of D5 receptor in parvalbumin positive interneurons (arrows) using immunofluorescence technique. (A) D5 containing neurons labeled with D5 antibody and secondary Cy3 linked antibody (red). (B) Same neurons (arrows) labeled with parvalbumin and secondary Fluor-Alexa 488 antibody (green). (C) Fluorophore overlay. (See color insert.)

2.2.2 SUBCELLULAR DISTRIBUTION

The first immunohistochemical results describing cellular and subcellular distribution of the D1 and D5 receptors in the primate dlPFC, hippocampus and basal ganglia were obtained using polyclonal antibodies developed against fusion proteins encompassing C-terminals portions of the D1 and D5 receptor proteins (Levey et al., 1993; Bergson et al., 1995; Ciliax et al., 2000). In these studies, strong subcellular compartmentalization of these receptors was observed: the D1 receptor was frequently observed in dendritic spines (Fig. 2.3A) while the D5 receptor was predominantly expressed in dendrites and rarely in dendritic spines (Bergson et al., 1995). Later studies in dlPFC of macaque monkeys and human cortex using shorter sequence antipeptide D5 antibodies (10 amino acids) showed more frequent distribution of the D5 receptor in dendritic spines (~20%, Fig. 2.3B; Khan et al., 2000; Mrzljak et al., 2006), suggesting that antipeptide antibodies may recognize more mature D5 protein epitopes transported to the distal neuronal processes (dendritic spines). Moreover, in the monkey dlPFC, D1 and D5 receptors are observed in the same dendritic spines (Bordelon et al., 2006). Still, the number of D1 positive spines greatly outnumber that of D5 positive spines (Mrzljak et al., 2006). D1 and D5 positive spines receive asymmetric inputs characteristic of excitatory, presumably glutamatergic synapses (Fig. 2.3A, B; Smiley et al., 1994; Bergson et al., 1995), and not dopaminergic input as shown by double labeling studies using D1 and tyrosine hydroxylase (TH) staining (Smiley et al., 1994). The source of excitatory glutamatergic input to D1 positive spines in the macaque monkey PFC is not known. Using preembedding immunogold techniques, the D1 receptor was shown to be either perisynaptic, at the edge of prominent postsynaptic densities of asymmetric synapses, or extrasynaptic, remote from the postsynaptic specialization at nonsynaptic plasma membrane (Paspalas and Goldman-Rakic, 2005). The association of D1 and D5 receptors with postsynaptic elements of asymmetric synapses suggested their role as receptors through which DA modulates incoming excitatory glutamatergic input to dendritic spines. Even without dopaminergic input at D1 and D5 positive spines, dopamine is believed to be supplied via "nonsynaptic" or "volume transmission" mediated via its diffusion through the neuropil from surrounding dopaminergic axons (for review see Agnati et al., 2000). Volume transmission may explain the subcellular localization of D1-like receptors because at the perisynaptic and extrasynaptic sites they may be more accessible to DA diffusion rather than at the synaptic specialization. On the other hand, synaptic localization of D1-like receptors may still be missed because the preembedding immunogold technique is not always convenient for localization of synaptic receptors (Nusser et al., 1996). Final judgment about their synaptic localization is pending postembedding immunogold studies.

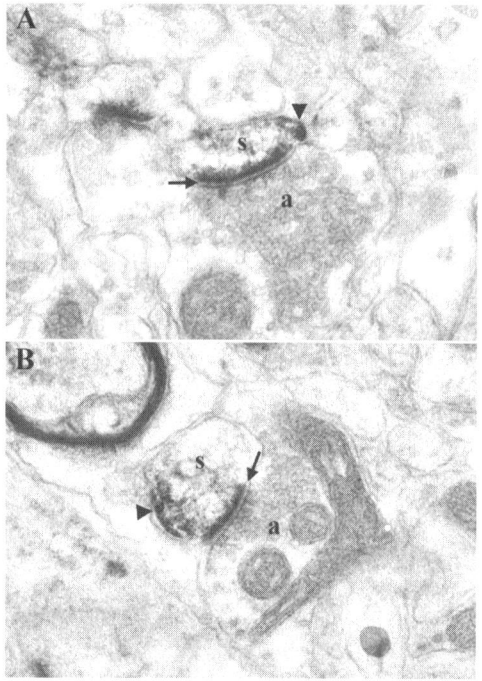

Figure 2.3 (A–B) Immunoperoxidase demonstration of D1 and D5 in the monkey prefrontal cortex area 9 at the electron microscopic level (EM). (A) D1 positive spine (s) receiving asymmetric input (arrow) from an unlabeled axon (a). Arrowhead points to diffuse immunoreaction product. (B) D5 positive spine (s) receiving asymmetric input (arrow) from an unlabeled axon (a). Arrowhead points to immunoreaction product.

The hypothesis that D1 receptors present in dendritic spines modulate glutamatergic neurotransmission was supported by the finding that the receptor physically and functionally interacts with N-methyl-D-aspartate (NMDA) receptors through coupling to NR1-1a and NR2A subunits (Lee et al., 2002). This interaction was not observed for D5 receptors (Lee et al., 2002; Pei et al., 2004). In dendritic spines, the NMDA receptor selectively upregulates D1 receptors (Scott et al., 2002) while dopamine D1 agonists in the rat prefrontal cortex potentiate NMDA currents (Chen et al., 2004). Since NMDA subunits are localized at the postsynaptic density of asymmetric synapses in dendritic spines (Petralia et al., 2000), and D1 receptors are found to be perisynaptic or extrasynaptic, physical interaction of NMDA subunits and D1 receptors could occur only perisynaptically, that is, at the edge of the postsynaptic specialization with a substantial lateral movement in the membrane. On the other hand, D1 receptors at the postsynaptic densities might not be detected by preembedding immunogold techniques.

In contrast to D1 receptors in spines of pyramidal neurons in the monkey PFC, which is presumably not a postsynaptic target of dopaminergic axons, D5 receptors, predominately localized in dendrites (Bergson et al., 1995; Table 2.2), may be either postsynaptic to dopaminergic axons, which frequently target dendritic shafts (Sesack et al., 1995), or act as a substrate for volume transmission. In addition, it has also been suggested that D5 receptors localized at perisomatic plasma membranes of PFC neurons, which form extrasynaptic microdomains with inositol 1,4,5-triphosphate-gated calcium stores of subsurface cisterns and mitochondria, are a dopaminoceptive substratum (Paspalas and Goldman-Rakic, 2004).

The significance of colocalization of D1 and D5 receptors in pyramidal and non-pyramidal neurons, as well as in subcellular compartments like dendritic spines (Bordelon et al., 2006), is not yet fully understood, although they might serve different modulatory influences on synaptic plasticity. For example, in mice lacking D1 receptors at corticostriatal projections, which target dendritic spines of striatal medium spiny neurons, D1 and D5 induce, long-term potentiation and long-term depression respectively (Centonze et al., 2003).

D1 and D5 receptors were traditionally considered to be postsynaptic receptors. However, the first ultrastructural immunohistochemical study in the monkey hippocampus and PFC has also shown their localization in axons forming symmetric and asymmetric synapses (Bergson et al., 1995). Recently, the D1 receptor was found in about 13% of PFC infragranular layer axons forming asymmetric excitatory synapses exclusively with dendritic spines (Table 2.2; Paspalas and Goldman-Rakic, 2005). It was suggested that these D1 heteroreceptors modulate recurrent excitation between deep layer pyramidal neurons, that is, persistent activity of neurons important for working memory. This hypothesis is supported by neurophysiological data that D1-like agonists presynaptically regulate recurrent excitation in the ferret PFC (Gao et al., 2001). However, D1 heteroreceptors in excitatory like axons targeting dendritic spines are not limited to deep layers of the monkey PFC but are also found in supragranular layers (layers II/III, Bergson et al., 1995), indicating that D1 hetero-receptors may be involved in wider control of excitatory neurotransmission in addition to recurrent inhibition. The first step in uncovering their role would be to determine the origin of axons bearing D1 receptors. Analysis of D5 receptor distribution in the monkey PFC layer III and V showed that this receptor is localized in a significantly larger portion of axons forming asymmetric, excitatory-like synapses (28–31%, Mrzljak et al., 2006; Fig. 2.4). In contrast, to the D1 receptor, synaptic targets of these axons were not only spines of pyramidal neurons (Fig. 2.4A) but also dendritic shafts with morphological characteristics of interneuronal (GABAergic) dendrites (Fig. 2.4B). This data suggest that in the nonhuman primate PFC, D5 receptors modulate excitatory input to both projection neurons as well as local circuit neurons in the supra- and infragranular layers. Similar to the D1 receptor, the origin of D5 positive excitatory axons in the monkey PFC is unknown. Among the possibilities are axons of association, commissural or thalamic projections to the PFC as well as axonal collaterals of pyramidal neurons in layers III and V.

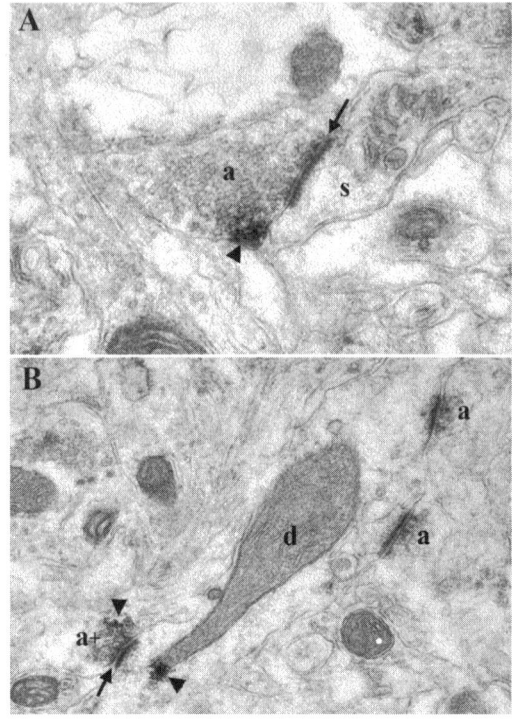

Figure 2.4 Presynaptic localization of D5 receptor: (A) D5 positive axon (a) forms an asymmetric synapse (arrow) onto immunonegative dendritic spine (s). Arrowhead points to extrasynaptic immunoreaction product. (B) D5 immunopositive dendrite (d, arrowhead) is receiving multiple asymmetric synapses from a D5 positive axon (a+, arrow; arrowhead points to immunoreaction product) and two D5 immunonegative axons(a) Multiple asymmetric (excitatory) synapses on dendrite (d) are morphological characteristic of a GABA interneuron dendrite.

Besides D1 and D5 positive excitatory-like axons, a smaller number of axons forming symmetric synapses on dendritic shafts were observed in the monkey hippocampus and PFC (Table 2.2; Bergson et al., 1995; Muly et al., 1998; Mrzljak et al., 2006). Such symmetric synapses have characteristics of inhibitory GABAergic synapses, suggesting an additional role of D1 and D5 as heteroreceptors on GABAergic axons.

Table 2.2 Subcellular Distribution of D1 and D5 Receptor in the Monkey dlPFC.

Neuropil structure	D1 receptor	D5 receptor
Spines with Glu input	+++	+ (~20%)
Dendritic shafts (Pyr + GABA)	+	+++ (~80%)
Axons forming Glu synapses	+/- (13%)	+ (~31%)
Axons forming GABA synap.	+/-	+/-

D1 receptor predominates in dendritic spines with excitatory glutamatergic (Glu) input while D5 receptor is most abundant in dendritic shafts of pyramidal (Pyr) and GABAergic interneurons (Mrzljak, et al., 2006). Thirteen percent of D1 positive axons forming glutamatergic synapses (Paspalas and Goldman-Rakic, 2004). +++, high frequency; +, moderate frequency; ±, low frequency.

2.2.3 D1 OR D5 AS DRUG TARGETS FOR COGNITION IMPROVEMENT IN SCHIZOPHRENIA

Cellular and subcellular distribution of D1 and D5 receptors is complex and suggests their complementary functions in the primate PFC. D5 is more abundant in the presynaptic domain, and is likely a heteroreceptor on glutamatergic axons modulating the excitation of pyramidal and GABAergic interneurons in the supra- and infragranular layers involved in working memory function. The D5 receptor also predominates in dendritic domains and basal forebrain neurons where it may be involved in acetylcholine release, which is known to have a procognitive effect. The D1 receptor predominantly modulates the excitatory glutamatergic neurotransmission in dendritic spines through interaction with NMDA receptors. In conclusion, a nonselective drug targeting both D1 and D5 receptor may be optimal for improvement of cognitive function such as working memory in schizophrenia.

2.3 D2 Receptors: D2S and D2L

The D2 receptor was the first cloned dopamine receptor (Bunzow et al., 1988). This receptor is involved in motor function, behavior, neuroendocrine and cardiovascular functions and is the major target of typical and atypical antipsychotic drugs (Gazi and Strange, 2002). The D2 receptor is linked to positive symptoms in schizophrenia and extrapyramidal side effects. Recent findings implicate the D2 receptor in modulation of specific components of working memory circuitry, which can be clearly distinguished from modulatory effects of the D1 receptor (Wang et al., 2004). For example, in the oculomotor delayed response task, the D2 receptor modulates phasic saccadic responses, but has no effect on the persistent mnemonic activity, which is regulated by the D1 receptor (Wang et al., 2004).

After cloning the D2 receptor, it was immediately found that this receptor exists in two isoforms generated by alternative splicing of the same gene: the D2 receptor short (D2S) and D2 receptor long (D2L) isoforms (Giros et al., 1989; Monsma et al., 1989). The longer form has 29 amino acids more than D2S in the third intracellular loop. Thus, the human D2S form has 414 and the D2L form has 443 amino acids.

2.3.1 CELLULAR AND SUBCELLULAR DISTRIBUTION

In the central nervous system, the D2 receptor has presynaptic autoreceptor and postsynaptic functions as demonstrated by neuroanatomical and functional studies (Gazi and Strange, 2002). Using the first subtype selective antibodies directed against the D2S and D2L receptors, it was shown that these isoforms are differentially distributed in the central nervous system of nonhuman primates. D2L is the predominant form in dopaminergic neurons in the midbrain, hypothalamus and axonal projections to the cerebral cortex, striatum and n.accumbens (Khan et al., 1998a, b). In contrast, D2L is expressed by neurons in the cerebral cortex, striatum and n.accumbens, which are targeted by dopaminergic axons (Khan et al., 1998a, b). At the ultrastructural level, D2S is localized presynaptically in dopaminergic fibers, while D2L is predominantly localized at postsynaptic sites such as dendritic spines. This study suggested, for the first time, that D2S is likely a dopamine autoreceptor, and D2L has a role as postsynaptic receptor.

2.3.2 FUNCTIONAL SIGNIFICANCE OF D2S AND D2L DISTRIBUTION

The neuroanatomical findings around D2S and D2L were supported by a series of neurophysiological, biochemical and pharmacological studies in mice lacking either the D2 receptor (D2L −/− and D2S −/−) or only the D2L receptor (D2L −/−). In D2L knockout mice, presynaptic function and normal DA release were preserved, implicating D2S in presynaptic autoreceptor function (Usiello et al., 2000). In contrast, the D2L form seemed to be responsible for haldol induced corticostriatal long-term potentiation (Centonze et al., 2004) and haldol induced catalepsy (Usiello et al., 2000) because both of these functions were missing in the mice lacking the D2 receptor (D2L −/− and D2S −/−) and in mice lacking D2L but expressing normal levels of the D2S isoform. These data, together with the finding that D2 long and D2S receptor can differentially contribute to the efficacy and side effects of typical and atypical antipsychotic drugs (Xu et al., 2002), open up possibilities for development of novel antipsychotic drugs with improved efficacy and side effects by using D2S and D2L pharmacophores.

2.4 Conclusions

D1-like and D2 (D2S and D2L) receptors have a widespread distribution in neuronal circuits of the dlPFC of nonhuman primates. However, to date the information about their precise linkage to the neuronal pathways, which are critical for the maintenance of normal working memory function in primates, such as corticocortical and thalamocortical pathways is still sparse. Such information and development of subtype specific pharmacological tools (compounds) targeting individual DA receptors is critical in order to further understand normal and impaired working memory function and develop therapeutic approaches for the treatment of schizophrenia.

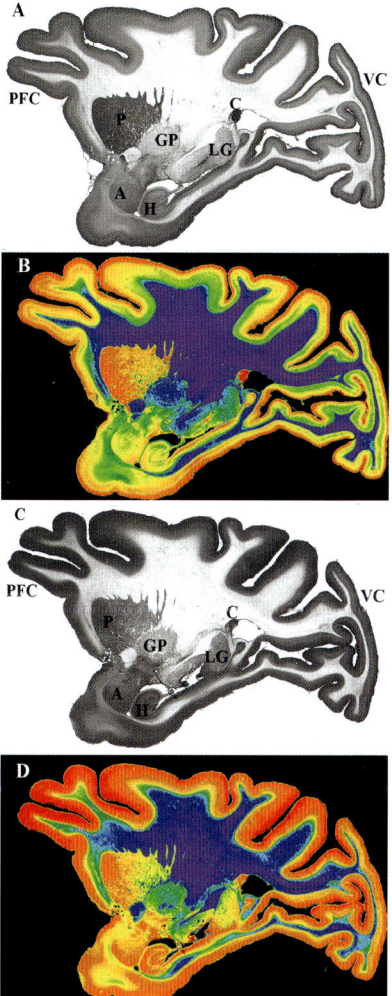

Figure 2.1 D1 (A–B) and D5 (B–C) receptor distribution in sagittal sections of macaque monkey brain as revealed by immunohistochemistry. On pseudo-colored images (B and D), red represents areas of highest level of protein expression and purple the lowest. Brain regions: PFC = prefrontal cortex, VC = visual cortex, P = putamen, C = head of caudate nucleus, GP = globus pallidus, A = amygdala, H = hippocampus and LG = lateral geniculate body.

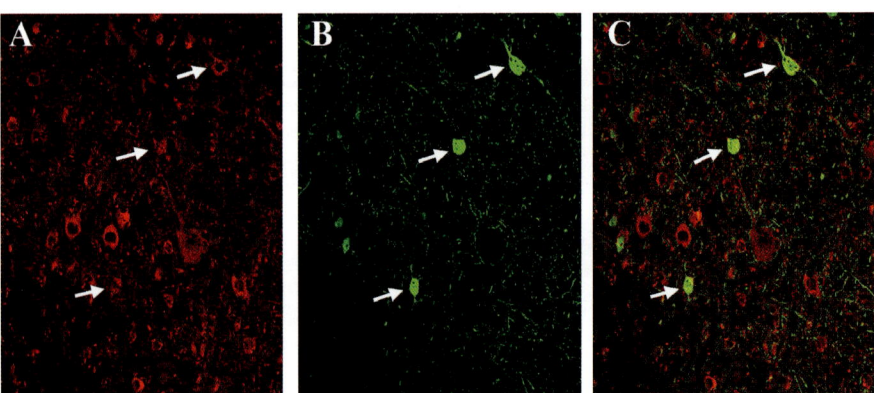

Figure 2.2 Localization of D5 receptor in parvalbumin positive interneurons (arrows) using immunofluorescence technique. (A) D5 containing neurons labeled with D5 antibody and secondary Cy3 linked antibody (red). (B) Same neurons (arrows) labeled with parvalbumin and secondary Fluor-Alexa 488 antibody (green). (C) Fluorophore overlay.

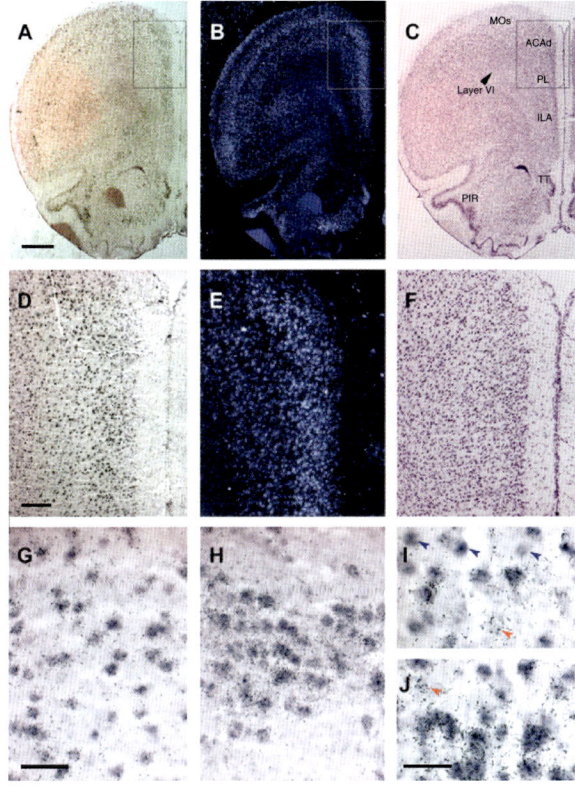

Figure 4.1 Expression of (A) 5-HT$_{1A}$ (dig-labeled oligonucleotides) and (B) 5-HT$_{2A}$ (dark field; ^{33}P-labeled oligonucleotides) receptors mRNA in the rat PFC. (C) An adjacent Nissl-stained section. Note the abundant presence of cells expressing both receptors in layers II–V, as well as in piriform cortex (PIR) and taenia tecta (TT). (D–F) High magnification of the cortical regions highlighted in (A–C). (D) and (E) show the presence of a large number of cells containing 5-HT$_{1A}$ and 5-HT$_{2A}$ receptor transcripts in cingulate (ACAd) and prelimbic (PL) cortex. Cells in deep layers (VI) express preferentially 5-HT$_{1A}$ receptor mRNA. (G–J) Coronal sections showing the expression of 5-HT$_{1A}$ and 5-HT$_{2A}$ receptor mRNAs in individual neurons. Occasional cell profiles containing only 5-HT$_{1A}$ (blue arrowheads) or 5-HT$_{2A}$ receptor mRNAs (red arrowheads) were seen in the infralimbic cortex (G and I), piriform cortex (H) and taenia tecta (J). Bar size are 1 mm in (A–C), 250 μm in (D–F), 50 μm in (G and H) and 30 μm in (I and J) [reproduced with permission from Amargós-Bosch et al. (2004)].

References

Abi-Dargham, A., Mawlawi, O., Lomberdo, I., Gil, R., Martinez, D., Huang, Y., Hwang, D.R., Keilp, J., Kochan, L., Van Heertum, R., Gorman, J.M. and Laruelle, M. (2002) Prefrontal dopamine D1 receptors and working memory in schizophrenia. J. Neurosci. 22, 3708–3719.

Agnati, L.F., Fuxe, K., Nicholson, C. and Sykova, E. (2000) Volume Transmission Revisited. Progress in Brain Research. Vol. 125, Elsevier, Amsterdam.

Bergson, C., Mrzljak, L., Smiley, J.F., Pappy, M., Levenson, R. and Goldman-Rakic, P.S. (1995) Regional, cellular and subcellular variations in the distribution of D1 and D5 dopamine receptors in primate brain. J. Neurosci. 15, 7821–7836.

Bordelon, J.R., Khan, Z.U. and Muly, E.C. (2006) D5 receptors are found with D1 in pyramidal cell spines of primate prefrontal cortex. Soc. Neurosci. Poster 332.18.

Bunzow, J.R., Van Tol, H.H.M., Grandy, D.K., Albert, P., Salon, J., Christie, M., Machida, C.A., Neve, K.A. and Civelli, O. (1988) Cloning and expression of a rat D2 dopamine receptor. Nature 336, 783–787.

Castner, S.A. and Goldman-Rakic, P.S. (2004) Enhancement of working memory in aged monkeys by a sensitizing regimen of dopamine D1 receptor stimulation. J. Neurosci. 24, 1446–1450.

Castner, S.A., Williams, G.V. and Goldman-Rakic, P.S. (2000) Reversal of antipsychotic induced working memory deficits by short-term dopamine D1 receptor stimulation. Science 287, 2020–2022.

Castner, S.A., Goldman-Rakic, P.S. and Williams, G.V. (2004) Animal models of working memory: insights for targeting cognitive dysfunction in schizophrenia. Psychopharmacology 174, 111–125.

Centonze, D., Grande, C., Saulle, E., Martin, A.B., Gubellini, P., Pavon, N., Pisani, A., Bernardi, G., Moratalla, R. and Calabresi, P. (2003) Distinct roles of D1 and D5 dopamine receptors in motor activity and striatal synaptic plasticity. J. Neurosci. 23, 8506–8512.

Centonze, D., Usiello, A., Costa, C., Picconi, B., Erbs, E., Bernardi, G., Borrelli, E. and Calabresi, P. (2004) Chronic haloperidol promotes corticostriatal long-term potentiation by targeting dopamine D2L receptors. J. Neurosci. 24, 8214–8222.

Chen, G., Greengard, P. and Yan, Z. (2004) Potentiation of NMDA receptor currents by dopamine D1 receptors in prefrontal cortex. Proc. Natl. Acad. Sci. U.S.A. 101, 2596–2600.

Ciliax, B.J., Nash, N., Heilman, C., Sunahara, R., Harney, A., Tiberi, M., Rye, D.B., Caron, M.G., Niznik, H.B. and Levey, A.L. (2000) Dopamine D5 receptor immunolocalization in rat and monkey brain. Synapse 37, 125–145.

Gao, W., Krimer, L.S. and Goldman-Rakic, P.S. (2001) Presynaptic regulation of recurrent excitation by D1 receptors in prefrontal circuits. Proc. Natl. Acad. Sci. U.S.A. 98, 295–300.

Gazi, L. and Strange, P.G. (2002) Dopamine receptors. In: M.N. Pangolos and C.H. Davies (Eds.), Understanding G Protein-coupled Receptors and Their Role in the CNS. Oxford University Press, Oxford, pp. 264–285.

Gingrich, J.A. and Caron, M.G. (1993) Recent advances in the molecular biology of dopamine receptors. Ann. Rev. Neurosci. 16, 299–321.

Giros, B., Sokoloff, P., Martres, M.-P., Riou, J.-F., Emorine, L.J. and Schwartz, J.-C. (1989) Alternative splicing directs the expression of two D2 receptor isoforms. Nature 342, 923–926.

Goldman-Rakic, P.S., Castner, S.A, Svensson, T.H., Siever, L.J. and Williams, G.V. (2004) Targeting the dopamine D1 receptor in schizophrenia: insights for cognitive dysfunction. Psychopharmacology 174, 3–16.

Khan, Z.U., Gutierrez, A., Martin, R., Penafiel, A., Rivera, A. and De La Calle, A. (1998a) Differential regional and cellular distribution of dopamine D2-like receptors. An immunocytochemical study of subtype specific antibodies in rat and human brain. J. Comp. Neurol. 402, 353–371.

Khan, Z.U., Mrzljak, L., Gutierrez, A., De La Calle, A. and Goldman-Rakic, P.S. (1998b) Prominence of the dopamine D2 short isoform in dopaminergic pathways. Proc. Natl. Acad. Sci. U.S.A. 95, 7731–7736.

Khan, Z.U., Gutierrez, A., Martin, R., Penafiel, A., Rivera, A. and De La Calle, A. (2000) Dopamine D5 receptors of rat and human brain. Neuroscience 100, 689–699.

Laplante, F., Sibley, D.R. and Quirion, R. (2004) Reduction in acetylcholine release in the hippocampus of dopamine D5 receptor-deficient mice. Neuropsychopharmacology 29, 1620–1627.

Lee, F.J.S, Xue, S., Pei, L., Vukusic, B., Chery, N., Wang, Y., Wang, Y.T., Niznik, H.B., Yu, A. and Liu, F. (2002) Dual regulation of NMDA receptor functions by direct protein–protein interactions with the dopamine D1 receptor. Cell 111, 219–230.

Levey, A.I., Hersch, S.M., Rye, D.B., Sunahara, R.K., Niznik, H.B., Kitt, C.A., Price, D.L., Maggio, R., Brann, M.R. and Ciliax, B.J. (1993) Localization of D1 and D2 dopamine receptors in brain with subtype-specific antibodies. Proc. Natl. Acad. Sci. U.S.A. 90, 8861–8865.

Monsma, F.J., McVittie, L.D., Gerfen, C.R., Mahan, L.C. and Sibley, D.R. (1989) Multiple dopamine D2 receptors produced by alternative RNA splicing. Nature 342, 926–929.

Mrzljak, L., Bergson, C., Pappy, M., Huff, R., Levenson, R. and Goldman-Rakic, P.S. (1996) Localization of dopamine D4 receptors in GABAergic neurons of the primate brain. Nature 381, 245–248.

Mrzljak, L., Fieles, W.E., Khan, Z.U. and Medd, A.M. (2006) Association of D5 dopamine receptors with pre- and postsynaptic elements of glutamatergic synapses. Soc. Neurosci. Poster. 332.22.

Muly III, C.E., Szigeti, K. and Goldman-Rakic, P.S. (1998) D1 receptor in interneurons of macaque prefrontal cortex: distribution and subcellular localization. J. Neurosci. 18, 10553–10565.

Nusser, Z., Sieghart, W., Stephenson, F.A. and Somogyi, P. (1996) The α-6 subunit of the GABAa receptor is concentrated in both inhibitory and excitatory synapses on cerebellar granule cells. J. Neurosci. 16, 103–114.

Paspalas, C.D. and Goldman-Rakic, P.S. (2004) Microdomains for dopamine volume neurotransmission in primate prefrontal cortex. J. Neurosci. 24, 5292–5300.

Paspalas, C.D. and Goldman-Rakic, P.S. (2005) Presynaptic D1 dopamine receptor in primate prefrontal cortex: target-specific expression in the glutamatergic synapse. J. Neurosci. 25, 1260–1267.

Pei, L., Lee, F.J.S., Moszczynska, A., Vukusic, B. and Liu, F. (2004) Regulation of dopamine D1 receptor function by physical interaction with the NMDA receptors. J. Neurosci. 24, 1149–1158.

Petralia, R.S., Rubio, M.E., Wang, Y.-X. and Wenthold, R.J. (2000) Regional and synaptic expression of ionotropic glutamate receptors. In: O.P. Ottersen and J. Storm-Mathisen (Eds.), Handbook of Chemical Neuroanatomy: Glutamate. Vol. 18, Elsevier, Amsterdam, pp. 145–182.

Scott, L., Sol Kruse, M., Forssberg, H., Brismar, H., Greengard, P. and Aperia, A. (2002) Selective up-regulation of dopamine D1 receptors in dendritic spines by NMDA receptor activation. Proc. Natl. Acad. Sci. U.S.A. 99, 1661–1664.

Sesack, S.R., Aoki, C. and Pickel, V.M. (1994) Ultrastructural localization of D2 receptor-like immunoreactivity in midbrain dopamine neurons and their striatal targets. J. Neurosci. 14, 88–106.

Sesack, S.R., Snyder, C.L. and Lewis, D.A. (1995) Axon terminals immunolabeled for dopamine or tyrosine hydroxylase synapse of GABA-immunoreactive dendrites in rat and monkey. J. Comp. Neurol. 363, 264–280.

Smiley, J.F., Levey, A.I., Ciliax, B.J. and Goldman-Rakic, P.S. (1994) D1 dopamine receptor immunoreactivity in human and monkey cerebral cortex: predominant and extrasynaptic localization in dendritic spines. Proc. Natl. Acad. Sci. U.S.A. 91, 5720–5724.

Usiello, A., Balk, J., Rouge-Pont, F., Picetti, R., Dierich, A., LeMeur, M., Piazza, P.V. and Borrelli, E. (2000) Distinct function of the two isoforms of dopamine D2 receptors. Nature 408, 199–203.

Wang, M., Vijayraghavan, S. and Goldman-Rakic, P.S. (2004) Selective D2 receptor actions on the functional circuitry of working memory. Science 303, 853–856.

Xu, R., Hranilovic, D., Fetsko, L.A., Bucan, M. and Wang, Y. (2002) Dopamine D2S and D2L receptors may differentially contribute to the actions of antipsychotic and psychotic agents in mice. Mol. Psychiatry 7, 1075–1082.

Chapter 3

Anatomical Characteristics of Norepinephrine Axons in the Prefrontal Cortex: *Unexpected Findings That May Indicate Low Activity State in Naïve Animals*

Lee Ann H. Miner and Susan R. Sesack
Departments of Neuroscience and Psychiatry, University of Pittsburgh, Pittsburgh, PA 15260, USA.

3.1 Importance of the PFC NE Innervation for Normal Function and Clinical Conditions

The catecholamine norepinephrine (NE) critically regulates information processing within the central nervous system. Along with dopamine (DA) and serotonin (5-HT; 5-hydroxytryptamine), the NE system forms an essential component of the modulatory brainstem innervation that ascends directly to the cerebral cortex without first being relayed through the thalamus. Within the prefrontal cortex (PFC) in particular, NE is known to modulate the essential cognitive and affective functions of this region, with animal studies demonstrating that normal NE innervation to the PFC is necessary for working memory, attention, and arousal (Robbins, 1984; Aston-Jones et al., 1999; Berridge, 2001; Berridge and Waterhouse, 2003; Arnsten and Li, 2005; Lapiz and Morilak, 2006).

Alterations in central NE systems have been observed in patients with depression, posttraumatic stress disorder (PTSD), and attention deficit hyperactivity disorder (ADHD) (Southwick et al., 1993; Pliszka et al., 1996; Klimek et al., 1997; Solanto, 1998; Biederman and Spencer, 1999; Southwick et al., 1999; Zhu et al., 1999) (*see also the Chapters 19, 20, and 21 in this volume*). These conditions afflict 7–10% of the population, including 3–5% of children, and strain society through lost productivity and poor quality of life. In addition, many more Americans suffer from the effects of chronic levels of stress in their daily lives. Many clinical studies suggest that dysfunction within the PFC may contribute to the symptoms of mental disorders (Drevets et al., 1992; Ernst et al., 1994; Mostofsky et al., 2002; Rauch et al., 2003; Shin et al., 2001; Soares and Mann, 1997; Solanto, 1998; Zametkin et al., 1990), and several studies propose more specifically that alterations of NE transmission in the PFC directly contribute to the pathophysiology of affective disorders and ADHD

(Meana et al., 1992; Aston-Jones et al., 1994; Charney et al., 1995; Bremner et al., 1996; Pliszka et al., 1996; Callado et al., 1998; Solanto, 1998; Russell et al., 2000).

The essential role of the NE system in mental health is further supported by findings that drugs blocking the reuptake of NE through the NE transporter (NET) effectively treat mood disorders and ADHD (Nelson, 1999; Frazer, 2000; Kent, 2000; Moller, 2000; Bymaster et al., 2002; Spencer et al., 2002; Michelson et al., 2003). Moreover, the therapeutic properties of these drugs may be mediated in part within the PFC itself (Tanda et al., 1994; Frazer, 2000; Bymaster et al., 2002). Reuptake through the NET is a key regulator of NE transmission, in that it controls the temporal dynamics of extracellular NE, the spatial range of NE diffusion, and the synthetic load of NE terminals. Although alterations in NE transmission within the PFC have been associated with both the etiology and treatment of mental disorders, the precise bases for these illnesses and the modes of action of the therapeutic drugs used to treat them are poorly understood. This necessitates a greater understanding of the cellular and molecular control of NE axon terminals within this region.

Our laboratory has initiated a series of studies examining the ultrastructural and phenotypical characteristics of NE axons in the PFC. The initial investigations revealed rather unusual features of these fibers that deviate from traditional expectations of how key proteins are distributed within monoamine axons. This chapter provides a brief historical background on the anatomy of the cortical NE innervation and a review of the atypical characteristics that our research has revealed. We then offer a working hypothesis that seeks to explain these features in a manner amenable to experimental testing. Finally, we review some of the data we have collected to test the model and outline future experiments for challenging the hypothesis.

3.2 General Anatomy of the NE Innervation to the Cerebral Cortex

3.2.1 ORIGIN FROM THE LOCUS COERULEUS (LC)

The general anatomical characteristics of the NE innervation to the cerebral cortex have been well established (Descarries et al., 1984; Fallon and Loughlin, 1987; Berridge and Waterhouse, 2003) and will be summarized only briefly here. Early studies utilizing histochemistry in combination with lesions, retrograde tracers, or anterograde autoradiographic tracing (Ungerstedt, 1971; Pickel et al., 1974; Levitt and Moore, 1978; Morrison et al., 1979; Morrison et al., 1981; Fallon and Loughlin, 1982; Loughlin et al., 1982; Jones and Yang, 1985) have reliably established that the cerebral cortex receives a moderately dense NE innervation from the LC. Evidence for a cortical NE innervation from other brainstem NE cell groups is lacking. Within the LC, cells projecting to the cortex are distributed across the dorsal compact division (Mason and Figiber, 1979; Fallon and Loughlin, 1982; Waterhouse et al., 1983; Loughlin et al., 1986a, b), and the majority are situated within the ipsilateral hemisphere (Ader et al., 1980; Room et al., 1981; Waterhouse et al., 1983; Simpson et al., 1997). Cells projecting to disparate cortical areas are somewhat interspersed with each other (Loughlin et al., 1982). However, a rough anterior to posterior topography exists, with the more anterior LC neurons projecting to anterior cortical areas and vice versa (Waterhouse et al., 1983; Grzanna and Fritschy, 1991; Simpson et al., 1997). In addition, more lateral cortical areas tend to receive input from more ventral LC cells (Waterhouse et al., 1983). Such findings suggest that, in opposition to traditional views of NE homogeneity, subdivisions of LC may influence the cortex differentially (Waterhouse et al., 1983). It should be noted that only the frontal cortex receives inputs from both the dorsal and ventral LC divisions (Waterhouse et al., 1983).

While an individual cell within the LC may not innervate the entire cerebral cortex, they do send collateral branches to areas of the cortex that are functionally and cytoarchitectonically diverse (Fallon and Loughlin, 1982; Loughlin et al., 1982). Some NE axons also branch into both cortical hemispheres (Ader et al., 1980; Room et al., 1981). In general, axons of individual LC cells collateralize extensively within the anterior to posterior dimension of cortex and much less in the medial to lateral domain (Loughlin et al., 1982). LC cells projecting to cortex collateralize not only to other cortical areas but also to widely distributed subcortical regions (Nagai et al., 1981; Room et al., 1981; Steindler, 1981; Simpson et al., 1997). In one important study, LC neurons in the rat were reported to innervate multiple brain regions that comprise a functional circuit (Simpson et al., 1997). Specifically, individual LC cells were shown to innervate the contralateral trigeminal nucleus, the ipsilateral ventrobasal nucleus of the thalamus, and the ipsilateral barrel fields in the primary somatosensory cortex (Simpson et al., 1997). In this way, NE projections can provide a coordinated regulation of cell excitability in all areas receiving sensory information from the periphery. Within the barrel cortex, one study reported that the NE innervation to the barrels was greater than to the surround (Lidov et al., 1978), although a more recent study using more sensitive techniques did not find this to be the case (Simpson et al., 2006).

3.2.2 PATTERNS OF TERMINATION

Ascending cortical projections arising from the LC course through the dorsal bundle (Ungerstedt, 1971). The majority enters the cortex through the frontal pole and travels longitudinally in layer VI to more caudal areas of cortex, all the while giving off arborizations that ascend into the more superficial layers (Morrison et al., 1979; Morrison et al., 1981). In general, the laminar patterns of NE innervation to diverse areas of cerebral cortex are rather similar (Audet et al., 1988). Indeed, common patterns of arborization have been observed within the rat cortex (Levitt and Moore, 1978; Morrison et al., 1978), with layer I containing a dense plexus of fibers, and layers II and III having less dense, radially oriented axons. Layers IV and V have moderately dense, short, twisted fibers, and layer VI has long NE axons coursing in a rostro-caudal direction. The primate cortex exhibits greater regional heterogeneity than the rat (Morrison et al., 1982; Lindvall and Björklund, 1984; Morrison et al., 1984), with the density of fibers being lower in the frontal association cortices (Morrison et al., 1984).

3.2.3 LIGHT MICROSCOPIC DESCRIPTIONS OF AXONS LABELED FOR TH AND DBH

The most reliable method for identifying the innervation patterns and morphological features of NE axons in the cerebral cortex has been immunocytochemical detection of synthetic enzymes or of NE itself. The synthetic pathway for DA and NE (Fig. 3.1) involves the hydroxylation of tyrosine by tyrosine hydroxylase (TH) into 3,4 dihydroxyphenylalanine (DOPA) (Cooper et al., 1991), which is then converted into DA by the relatively ubiquitous enzyme aromatic L-amino acid decarboxylase (AADC) (Christenson et al., 1972). In NE terminals, following transport of DA into vesicles by the vesicular monoamine transporter (Henry et al., 1998; Hoffman et al., 1998), the enzyme dopamine-β-hydroxylase (DβH) converts DA into NE (Fig. 3.1) (Cooper et al., 1991). Hence, DA neurons should contain only TH, whereas NE cells should express both TH and DβH. Virtually all LC neurons contain both synthetic enzymes and their associated mRNA (Asan, 1993; Hoffman et al., 1998; Lorang et al., 1994; Pickel et al., 1975a, b; Swanson and Hartman, 1975; Tillet and Kitahama, 1998; Verhofstad et al., 1979), and only one class of small cells has been found that does not contain NE (Loughlin et al., 1986b). These most likely correspond to the GABA neurons that are more often observed in the pericoerulear area (Aston-Jones et al., 2004).

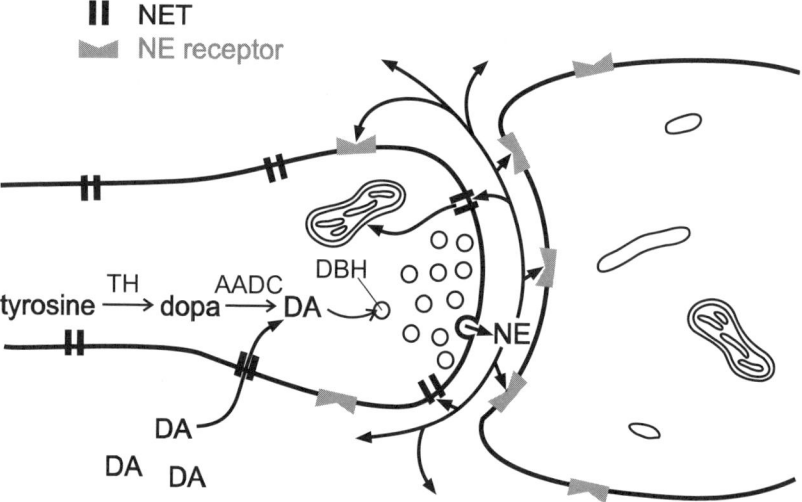

Figure 3.1 Schematic drawing illustrating the traditional view of NE synthesis, release, and reuptake in terminal axons. The synthetic pathway for NE synthesis dictates that the enzymes TH, AADC, and DβH should all be present. The latter enzyme is localized to synaptic vesicles and converts DA to NE within these organelles. NE released by exocytosis acts on pre- and postsynaptic receptors and can diffuse away from the synaptic cleft. In theory, NET localized to the plasma membrane captures NE and restricts its diffusion. NET also clears extracellular DA, which may serve as a precursor for NE synthesis.

Despite the presence of both TH and DβH in LC soma, a large number of studies over three decades have indicated that TH is not always present in axons that express other markers of the NE phenotype. Indeed, a pronounced mismatch between axons immunoreactive for TH versus DβH has been noted throughout several divisions of the human and nonhuman primate cerebral cortex (Lewis et al., 1987; Gaspar et al., 1989). Estimates from human tissue indicate that only 10–50% of DβH-immunoreactive (-ir) fibers are colocalized with TH; in the human PFC (area 9), the extent of this colocalization reaches only 10% (Gaspar et al., 1989). Moreover, electrolytic lesions of the LC virtually eliminate DβH-ir but not TH-ir axons from several regions of the monkey cortex (Lewis et al., 1987), indicating that the majority of TH-labeled fibers are DA axons. Although the exact basis for this relative absence of detectable TH in many cortical NE projections is not known, numerous light microscopic studies in rats and monkeys have consistently reported this finding (Pickel et al., 1975c; Lewis et al., 1987; Lewis et al., 1988; Noack and Lewis, 1989), and similar observations have been made in the amygdala (Asan, 1993). Many of these studies used methods for signal amplification and were conducted at the light microscopic level where antibody penetration can be maximized. Moreover, these findings are consistent with biochemical and immunocytochemical assays indicating that cortical NE axons contain less TH than do cortical DA axons (Pickel et al., 1975c; Hökfelt et al., 1977; Emson and Koob, 1978; Schmidt and Bhatnagar, 1979; Asan, 1993).

It is important to note that immunoreactivity for TH has occasionally been shown to label a substantial proportion of NE axons in some cortical areas that typically do not receive abundant DA inputs, for example, the primary visual cortex and hippocampus (Milner and Bacon, 1989; Noack and Lewis, 1989; Aoki, 1992). However, others have noted low levels of detectable TH in DβII-ir fibers even in areas that are sparsely innervated by DA axons (Lewis et al., 1987; Asan, 1993). It is of course possible that most cortical NE axons contain levels of TH that are undetectable by immunocytochemistry but are nevertheless sufficient for synthesizing NE. However, the same immunocytochemical techniques can readily detect DβH within NE axons (Lewis and Morrison, 1989; Aoki et al., 1998; Miner et al., 2003), TH within PFC DA fibers (Sesack et al., 1995), and the coexpression of TH and phenylethanolamine N-methyltransferase (PNMT) within adrenergic axons (Asan, 1993). Finally, studies showing a strong influence of gonadal hormone levels on the density of TH-ir but not DβH-ir axons in the rat cortex (Kritzer, 2000, 2003) argue for these enzymes being largely confined to separate processes as identified by light microscopic techniques. Consequently, immunocytochemical studies of the PFC often use antibodies directed against TH to selectively label DA axons. This is particularly useful, given that the dopamine transporter (DAT) is not a good marker for DA axons in this region (Ciliax et al., 1995; Sesack et al., 1998) and that antibodies against DA require fixative conditions that are incompatible with most other antigens (Séguéla et al., 1988).

3.3 Postsynaptic Actions of NE

3.3.1 LOCALIZATION OF ADRENOCEPTORS

In situ hybridization studies report that mRNA for alpha (α) and beta (β) adrenergic receptors is present within the cerebral cortex (Nicholas et al., 1993a, b). Autoradiographic studies also support the presence of α1, α2, and β adrenoceptors (Wamsley et al., 1981; Wamsley, 1984; Goldman-Rakic et al., 1990; Berridge and Waterhouse, 2003). Ligands for α2 receptors have revealed a relatively low concentration of binding over all cortical layers with a slightly greater density in superficial than deep layers (Young and Kuhar, 1980; Wamsley et al., 1981). NE receptors of the β subtype are concentrated in the superficial layers of parietal cortex and in lamina I of cingulate cortex. These are mainly β1 receptors (Palacios and Kuhar, 1980; Palacios and Kuhar, 1982; Palacios and Wamsley, 1983, 1984). In the rat somatosensory cortex, adrenergic receptor subtypes have been shown to be preferentially localized to cortical barrels (Vos et al., 1990). In addition to receptor binding studies, functional pharmacological experiments also indicate the presence of NE receptors of the α1 and α2 subtypes within the PFC (see below).

The ultrastructural immunocytochemical work of Aoki and colleagues has localized NE receptors within the rat visual cortex and monkey PFC (Aoki, 1992; Aoki et al., 1994; Venkatesan et al., 1996; Aoki et al., 1998). These experiments indicate that the majority of NE receptors in the cortex are of the α2A and β subtypes. Receptors of the β-adrenergic type are commonly found within astrocytes, providing a hypothesized means for NE to regulate glutamate transmission by modulating glial uptake (Aoki, 1992). Receptors of the α2A subclass are distributed to perikarya of both pyramidal and nonpyramidal cells. These receptors are also found along dendritic shafts, dendritic spines, intervaricose portions of axons, and glial processes. The morphological features of immunolabeled axons are characteristic of both NE and non-NE fibers, suggesting both autoreceptor and heteroceptor functions of these receptors (Aoki et al., 1998). Although discrete subcellular labeling methods were not employed in these studies, the data were interpreted as consistent with spinous α2A receptors mediating synaptic actions of NE at spiny pyramidal neurons. Within other immunoreactive structures, α2A receptors are found mainly at extrasynaptic sites and so are more likely to be activated via volume transmission (Aoki et al., 1998). The same observations were made for the low levels of α2B and α2C receptors that could be detected by electron microscopy.

3.3.2 PHYSIOLOGICAL FINDINGS

The physiological effects of NE on postsynaptic cells are still the subject of much study (Van Dengen, 1981; Foote et al., 1983; Foote, 1985; Berridge and Waterhouse, 2003) (*see also the Chapters 7, 14, 13, and 17 in this volume*). A complete synopsis of this work is beyond the scope of our chapter and will be reviewed only briefly here. The earliest investigations into the basic actions of NE reported either strong inhibitory effects or a mix of inhibitory and some excitatory responses evoked by NE in the rat and monkey cortex (Morruzzi and Hart, 1955; Krnjevic and Phillips,

1963a, b; Johnson et al., 1969; Bunney and Aghajanian, 1976; Reader et al., 1979; Sawaguchi and Matsumura, 1985). Some of the discrepancies in this literature may have reflected the choice of anesthetic (Johnson et al., 1969; Bunney and Aghajanian, 1976) and the actual concentration of NE at the synapse (Armstrong-James and Fox, 1983). The complexity of postsynaptic responses to NE is further illustrated by studies addressing how NE modulates cortical responses evoked by other inputs. These investigations have been conducted in primary sensory regions and revealed that NE typically enhances the "signal to noise" ratio by boosting the excitatory or inhibitory responses evoked by visual, auditory, or somatosensory stimuli while reducing or leaving unaffected the spontaneous firing rate of these cells (Foote et al., 1975; Waterhouse and Woodward, 1980; Waterhouse et al., 1980, 1981, 1982; Kasamatu and Heggelund, 1982; Waterhouse et al., 1988, 1990; McLean and Waterhouse, 1994; Waterhouse et al., 1998). By means of this "gating" function, NE enhances the efficacy of sensory transmission (Waterhouse et al., 1998; Berridge and Waterhouse, 2003), and both α and β receptors appear to contribute to these effects (Waterhouse et al., 1981, 1982). A similar modulatory action of NE was reported in the rat PFC following electrical stimulation of the LC (Mantz et al., 1988). Although NE typically enhances evoked responses in cortical cells in vivo, experiments conducted in vitro indicate that NE reduces excitatory postsynaptic potentials or currents at NMDA and non-NMDA receptors (Law-Tho et al., 1993; Liu et al., 2006). These effects appear to be mediated via both α1 and α2 receptor mechanisms through separate signaling pathways (Liu et al., 2006).

   In awake monkeys performing delayed-response tasks, evidence suggests that NE acting on α2 receptors facilitates the sustained firing of PFC neurons during the delay period. More specifically, application of α2 antagonists, but not blockers of α1 or β-adrenergic receptors, attenuates delay activity and alters the spatial tuning of oculomotor delayed-response performance (Sawaguchi, 1998). Reciprocal studies have also reported the facilitating actions of α2 receptor agonists on Go-No Go activity in PFC neurons (Li and Kubota, 1998). These electrophysiological studies agree well with behavioral performance measures, in which α2 receptors have been repeatedly shown to improve measures of PFC function (Arnsten and Goldman-Rakic, 1985; Li and Mei, 1994; Arnsten et al., 1996; Tanila et al., 1996; Avery et al., 2000; Ma et al., 2003; Wang et al., 2004). Additional pharmacological studies have also documented the disruption of working memory functions by α1 receptors in the PFC (Arnsten et al., 1998; Mao et al., 1999; Arnsten and Li, 2005). The latter actions of NE have been hypothesized to interfere with PFC operation during periods of stress when central NE systems are activated (Birnbaum et al., 1999; Arnsten, 2000; Arnsten and Li, 2005). However, it should be noted that studies in rodents suggest that α1 receptors may also work to improve cognitive function within the PFC (Lapiz and Morilak, 2006).

3.4 Ultrastructural Features of Cortical NE Axons

3.4.1 GENERAL MORPHOLOGY OF NE AXONS AND SYNAPSES

Initial ultrastructural investigations of cortical NE axons were conducted using autoradiographic visualization of NE uptake or immunocytochemical localization of DβH. Many of these studies reported that the majority of NE varicosities in the cortex (80–95%) were not observed to form conventional synaptic specializations (Descarries et al., 1977; Beaudet and Descarries, 1978, 1984; Aoki et al., 1998). These findings suggested that some cortical LC varicosities might release transmitter outside of classical junctions (Descarries et al., 1977; Beaudet and Descarries, 1978, 1984). Such extrasynaptic release events implied that NE was being transmitted over a broad area and gave rise to the notion of this transmitter system as relatively diffuse and nonspecific. However, other investigators using similar labeling methods and serial reconstruction reported that a substantial proportion of cortical NE axons did form conventional synapses (Olschowka et al., 1981; Molliver et al., 1982; Parnavelas et al., 1985; Papadopoulos and Parnavelas, 1991). Subsequently, antibodies directed against NE itself became available and were used to specifically address the synaptic incidence of cortical NE axon varicosities. Some laboratories continued to report a high synaptic incidence for NE axon terminals in the visual and frontoparietal cortices (Papadopoulos et al., 1989). However, Descarries and colleagues employed extensive serial reconstruction of numerous profiles in the frontal, parietal, and occipital cortices and concluded that only 17–26% of NE varicosities show evidence of synapses with classical morphological features (Séguéla et al., 1990). To some extent, these discrepancies may reflect the strictness of criteria used to define synaptic junctions. However, the same approach indicated that as many as 90% of DA axons in the medial pregenual PFC form evident synapses (Séguéla et al., 1988). Hence, the exact extent of synapse formation varies by transmitter and region (Séguéla et al., 1988; Séguéla et al., 1989; Séguéla et al., 1990), suggesting that cortical monoamine fibers, including NE axons, are designed for both specific synaptic transmission and extrasynaptic release of a more diffuse nature (Descarries and Mechawar, 2000).

3.4.2 NEW OBSERVATIONS REVEALED BY IMMUNOLABELING FOR NET IN THE PFC

Our laboratory wished to reexamine the ultrastructural characteristics of NE axons in the PFC once antibodies against the NET became available. NET represents a selective anatomical marker for central NE systems, given that NET mRNA is expressed only by NE neurons and by no other cell type. Moreover, the fact that all LC neurons express NET mRNA and protein (Lorang et al., 1994; Hoffman et al., 1998; Savchenko et al., 2003) indicates that immunoreactivity for NET can be used to identify NE axons in the PFC, which receives input solely from the LC. The sensitivity and reliability of this marker for labeling axons with the morphological features of NE profiles was established by our initial studies (Schroeter et al., 2000; Miner et al., 2003). We used immunogold-silver labeling for NET, as this method provides greater subcellular localization than immunoperoxidase (Sesack et al., 2006). Moreover, immunolabeling for NET allowed us to perform dual labeling studies to specifically address the reported absence of TH from many cortical NE axons (described above). For this purpose, we used the immunoperoxidase method to detect TH, as this approach involves signal amplification at high sensitivity (Sesack et al., 2006). Localization of NET is preferable to DβH in combination with TH, as DβH antibodies must penetrate both plasma and vesicular membranes to reach the antigen and are therefore less sensitive for ultrastructural studies where only minimal detergent can be used to enhance penetration (Aoki et al., 1998).

Our initial ultrastructural characterization of the rat PFC (Schroeter et al., 2000; Miner et al., 2003) revealed that cortical axons labeled for NET exhibited innervation patterns, morphological features, and synaptic incidence that matched previous descriptions of fibers singly labeled for NE or DβH (Beaudet and Descarries, 1978, 1984; Loughlin et al., 1982; Séguéla et al., 1990). However, we were surprised to observe that the majority of NE varicosities (85–90%) expressed NET predominantly within the cytoplasm (approximately 75% of gold particles) and minimally on the plasma membrane (Fig. 3.2A). Moreover, approximately half of NET-ir profiles exhibited no plasmalemmal gold particles (i.e., 100% cytoplasmic; Fig. 3.2B,C), although full serial reconstruction might have revealed some NET on the plasma membrane in each case. These findings stand in marked contrast to the predominantly plasmalemmal distribution of SERT within many (though not all) 5-HT axons in the PFC (Fig. 3.2D) (Miner et al., 2000) and of DAT in the striatum (Fig. 3.2E) (Nirenberg et al., 1996; Hersch et al., 1997). Nevertheless, the finding that NET was predominantly intracellular agrees with similar observations made in peripheral NE neurons, in cultured NE cells, and in heterologous expression systems (Melikian et al., 1994; Melikian et al., 1996; Apparsundaram et al., 1998a; Schroeter et al., 2000; Ren et al., 2001; Sung et al., 2003).

The same majority population of NET+ varicosities in the rat PFC lacked any evidence of immunoperoxidase reaction product for TH (Fig. 3.2A–C) (Miner et al., 2003). The other less abundant population of profiles immunolabeled for NET (10–15%) did display detectable levels of TH (Fig. 3.2F), and it is remarkable how well this proportion matched previous light microscopic estimates of the presence of TH in NE axons in the human PFC (Gaspar et al., 1989) In the rat PFC, axons dually labeled for NET and TH consistently showed NET predominantly localized to the plasma membrane (Fig. 3.2F), exactly the reverse of what was observed for NET profiles lacking TH. These observations suggest some mechanistic link between the presence of TH and the insertion of NET into the plasma membrane, although the specific means by which this occurs are not known.

Although our initial studies did not involve serial reconstruction, our estimates from single sections indicated that NET+ axons formed synapses only 25% of the time. Also in good agreement with prior electron microscopic studies, these synapses were mainly of the symmetric type (Fig. 3.2B), a morphological feature typically associated with inhibitory or modulatory function, and the principal target was dendritic shafts or spines (Fig. 3.2B). Remarkably, these synaptic profiles were entirely composed of the majority and not the minority population of NET-ir axons. Hence, no axons dually immunoreactive for NET and TH were observed to form synaptic junctions in naïve animals (Miner et al., 2003).

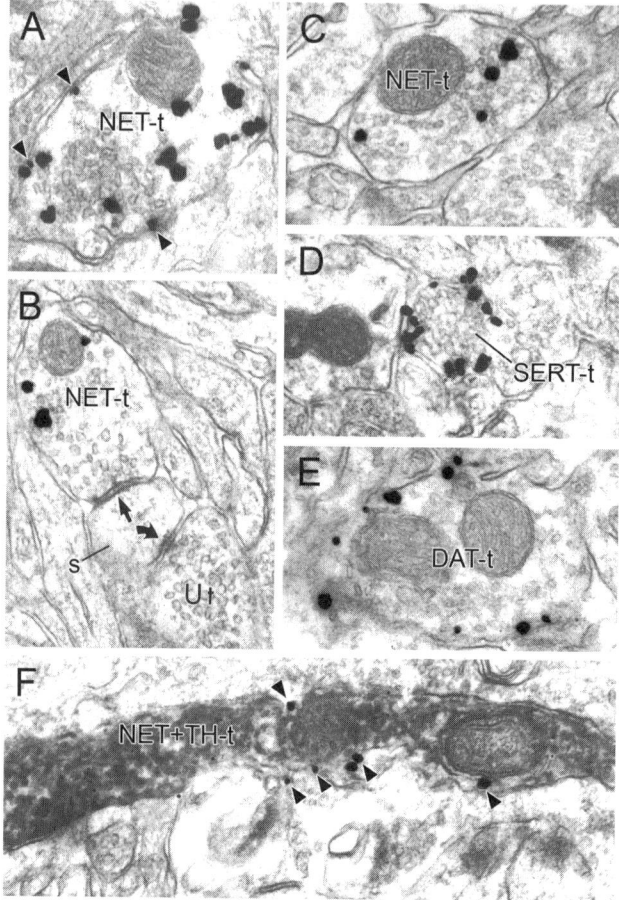

Figure 3.2 Electron microscopic images of monoamine transporter localization in the PFC (A–D, F) or dorsal striatum (E) of the rat. A–C: Immunogold–silver labeling for NET is localized within axon terminals (NET-t) and mainly distributed to the cytoplasm. Occasionally, gold–silver particles for NET are found on the plasma membrane (*arrowheads in A*), although some profiles exhibit only cytoplasmic NET (B,C). One NET+ profile forms a symmetric synapse (*straight arrow in B*) onto a spine (s) that receives an additional asymmetric synapse (*curved arrow*) from an unlabeled terminal (Ut). For panels A–C, note the absence of immunoperoxidase reaction product for TH in the NET+ axons. D,E: Immunogold–silver labeling for SERT in most cortical 5-HT axons (SERT-t in D) and for DAT in striatal DA fibers (DAT-t in E) is mainly found along the plasma membrane. F: In the PFC of naïve animals, only 10–15% of NET+ axons also exhibit immunoperoxidase labeling for TH (NET+TH-t). In this case, gold–silver particles for NET are found mainly on the plasma membrane (*arrowheads*).

These initial observations of NET and TH distribution in the PFC do not match traditional models for how NE is thought to be synthesized, released, and recaptured

(Figs. 3.1 and 3.3), although the relative absence of TH does agree with the light microscopic findings described earlier. The absence of detectable TH is not due to the presence of a distinct isoform of the enzyme in PFC NE axons, as rodents express only a single isoform (Brown et al., 1987; Ichikawa et al., 1991). In theory, phosphorylation state might affect antibody recognition. However, the monoclonal antibody used was not raised against the N-terminal regulatory domain where phosphorylation takes place (Campbell et al., 1986; Haycock 1990). Nevertheless, there have been reports of differences in the extent to which various antibodies can detect TH in cortical NE fibers (Noack and Lewis, 1989), suggesting that molecular variations in the TH enzyme may yet be detected between DA and NE axons.

It is of course possible that TH is present in NE axons at levels below the sensitivity of detection, particularly in ultrastructural studies that permit only minimal detergent to aid antibody penetration. Nevertheless, despite the difficulties associated with electron microscopic detection of DβH, we were able to develop a method that successfully localized DβH to NET+ axons in the PFC (Miner et al., 2003). These findings agree with the results of light microscopic colocalization studies (Savchenko et al., 2003). Hence, we believe that the absence of detectable TH in the majority of PFC NE axons indicates either an absence or a markedly low function of this enzyme in these fibers, making it important to understand why this is the case and how these terminals can synthesize NE. It is possible that the absence of appreciable TH in most PFC NE axons reflects a low activity state associated with minimal NE synthesis and release. Alternatively, or in addition, the ability of NET to efficiently transport DA (Horn, 1973; Raiteri et al., 1977; Gu et al., 1996) provides one potential mechanism by which NE axons may acquire synthetic precursor.

3.5 Working Model

Our results and the findings from in vitro expression systems (described below) suggest the following hypothesis for explaining the unusual characteristics of NE axons in the PFC. Specifically, we believe that the expression of detectable TH and the plasmalemmal distribution of NET in the PFC are determined at least in part by the level of activity in the cortical NE system and by the availability of extracellular DA (Fig. 3.3). This model implies that activity in the PFC NE system varies along a continuum that may be artificially low in laboratory rats, which undergo few of the challenges that would be expected to elicit the full capacity of the LC system in the wild. Moreover, the model suggests that functional interactions between the cortical NE and DA systems may have evolved to a symbiotic point, in which DA axons are relieved of the energy requirement for reuptake in exchange for relinquishing precursor DA to NE axon terminals. Components of this hypothesis can be tested by creating experimental situations that increase the demand for NE transmission and/or alter the availability of extracellular DA in the cortex.

3.5.1 ACTIVITY STATE IN THE NE SYSTEM

The majority of NE terminals in the rat PFC appear to be in a low activity state relative to their maximal capacity, in that they cannot synthesize NE precursors on their own (apparent absence of TH), nor can they maximize the creation of NE from captured DA (low plasmalemmal NET). The minority population of NE profiles seems to function in a high activity state, because their greater expression of TH and plasmalemmal NET allow them to synthesize NE and recycle both released NE and DA. Perhaps these characteristics and the proportion of each axon type are sufficient to meet the demands of the PFC NE system in naïve laboratory rats. If this is the case, then alterations in the degree of activity in NE neurons would be predicted to alter the subcellular distribution of NET and the expression of TH.

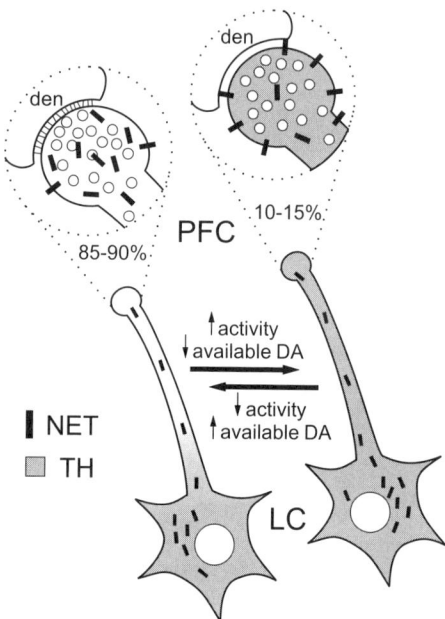

Figure 3.3 Schematic drawing illustrating the working hypothesis constructed to explain the relative absence of detectable TH in the majority of NE axons in the PFC and the predominant localization of NET to the cytoplasm. Our model predicts that enhanced neuronal activity in this system, coupled with or independent of reduced availability of extracellular DA, will increase the plasmalemmal insertion of NET and recruit TH from cell bodies in the LC. Within the PFC, NE axons are shown either in synaptic contact (*left*; intercleft filaments indicate a synaptic junction) or in nonsynaptic apposition (*right*) to dendrites (*den*).

Increasing evidence from in vitro studies indicates that cellular activity, extrinsic signals, and intracellular pathways can regulate the surface distribution of NET (Blakely and Bauman, 2000). For instance, plasmalemmal NET is augmented in response to cellular depolarization (Savchenko et al., 2003). In addition, NET colocalizes with the t-SNARE syntaxin 1A, a protein that supports the plasmalemmal trafficking of NET (Sung et al., 2003), suggesting that NET surface expression is associated with vesicular release. Finally, the surface distribution of NET is diminished in response to activation of protein kinase C or phorbol ester treatment (Apparsundaram et al., 1998a, b; Jayanthi et al., 2004). These studies imply that NET participates in regulated trafficking events coordinated with, or directly linked to, the activity of NE cells and vesicle exocytosis (Blakely and Bauman, 2000). However, as noted above, experiments are needed to test whether acute activation of the cortical LC projection increases the plasmalemmal expression of NET in the intact PFC. We do not expect such acute treatments to enhance TH expression, which will require time for de novo synthesis of enzyme and transport to nerve terminals.

3.5.2 EFFECTS OF ACUTE AND CHRONIC STRESS ON PFC NE AXONS

Stress presents a natural means for increasing the activity of NE neurons in intact animals. In this regard, it is well established that exposure to acute stress can increase both the firing activity of LC neurons (Korf et al., 1973; Abercrombie and Jacobs, 1987; Abercrombie et al., 1988) and extracellular NE levels in several brain regions, including the PFC (Abercrombie et al., 1988, 1989; Rossetti et al., 1989). However, humans and animals often experience repeated or prolonged stressors, and it has been suggested that chronic unavoidable stress may be useful as an animal model of depressed behavior (Gambarana et al., 2001; Ossowska et al., 2001). Exposure to chronic stress alters an organism's response to subsequent stressors (Anisman and Zacharko, 1990), and the same is true for the activity of LC neurons (Stanford, 1995; Zigmond et al., 1995). For example, the evoked responsivity of LC spike firing is enhanced following chronic exposure to cold or repeated exposure to foot or tail shock (Simson and Weiss, 1988; Curtis et al., 1995; Conti and Foote, 1996; Mana and Grace, 1997; Jedema et al., 2001; Jedema and Grace, 2003). Repeated foot shock, restraint, or continuous cold exposure also reportedly increases the basal firing of LC neurons, although this effect seems not as pronounced as the enhancement of evoked activity (Simson and Weiss, 1988; Pavcovich and Ramirez, 1991; Mana and Grace 1997).

The enzymatic activity of TH is also increased in the LC following chronic exposure to cold, social stress, social isolation, restraint, or foot shock (Zigmond et al., 1974; Stone et al., 1978; Angulo et al., 1991; Nisenbaum et al., 1991; Watanabe et al., 1995; Rusnak et al., 1998). Accordingly, the levels of TH mRNA and/or protein are also increased in the LC in response to various chronic stressors (Mamalaki et al., 1992; Watanabe et al., 1995; Wang et al., 1998). Despite these observations, no increase in basal TH activity in terminal regions such as the hippocampus has been observed following chronic stress (Nisenbaum et al., 1991). However, the ability of acute novel stressors to evoke increased TH activity in hippocampal NE terminals is enhanced following chronic stress exposure (Nisenbaum et al., 1991).

Whether chronic stress increases basal or evoked TH activity in the PFC has never been examined in biochemical studies. Moreover, because of high TH levels in DA axons, enzyme activity assays of the PFC are not interpretable with regard to changes that might occur exclusively within NE axons. Hence, our approach of examining dual ultrastructural labeling of NET and TH has the advantage of determining whether detectable TH levels change specifically within individual NE axons. We therefore tested the hypothesis that TH levels would increase within PFC

NE axons in animals exposed to two weeks of cold stress. We chose cold stress because of its continuous, inescapable nature and its demonstrated impact on the activity of LC neurons (Jedema et al., 2001; Jedema and Grace, 2003). We found that the percentage of axons immunoreactive for NET that also expressed TH increased from 13% in control rats to 32% in chronically stressed animals (Miner et al., 2006). This finding demonstrates the ability of our ultrastructural immunolabeling method to detect changes in TH expression. Furthermore, the results indicate that PFC NE axons have a greater synthetic capacity following chronic activation of this system, consistent with our working hypothesis.

Given the excitability changes observed in LC neurons and the enhanced TH expression in PFC NE terminals, it is expected that chronic stress would also increase extracellular levels of NE within the PFC, as reported for some other cortical areas (Adell et al., 1988). However, chronic exposure to cold stress does not change basal extracellular levels of NE in the PFC, despite the fact that efflux in response to an acute novel stressor is enhanced (Gresch et al., 1994; Finlay et al., 1995, 1997; Jedema et al., 1999). This apparent discrepancy may be explained in part by a greater proportion of NET associated with the plasma membrane, which would act to maintain basal extracellular levels of NE. In our chronically cold-stressed rats, we found that the proportion of NET distributed to the plasma membrane was indeed increased from 29% plasmalemmal gold labeling in control rats to 51% in stressed animals (Miner et al., 2006). Most importantly, in the PFC of chronically stressed rats we found examples of NE axons that contained TH and exhibited NET in a perisynaptic location, that is, near the edges of synapses. Such images match traditional views of how cortical NE axons operate (see Fig. 3.1) but are only seen in stressed and not control animals. Hence, stress seems to produce an increased responsiveness of the PFC NE system by inducing a greater synthetic capacity and a more dynamic regulation of NE transmission. These data are consistent with our working hypothesis and help to resolve the seemingly conflicting responses to chronic stress exhibited by markers of activity in somatic versus terminal regions.

3.5.3 EFFECTS OF SELECTIVE NE UPTAKE INHIBITORS

Drugs that selectively block the reuptake of NE through NET (selective NE reuptake inhibitors, SNRIs) are effective antidepressants (Nelson, 1999; Frazer, 2000; Kent, 2000; Moller, 2000). However, their therapeutic action takes 2–3 weeks to manifest. Therefore, direct blockade of NET does not fully account for their clinical efficacy, and compensatory changes in response to the drug are crucial to their beneficial effects (Frazer and Benmansour, 2002). Unfortunately, the precise nature of the changes occurring following chronic SNRI treatment has not been fully elucidated.

In general, the effects of long-term treatment with SNRIs on NE cell activity, TH expression, and NET localization are opposite to those of chronic stress and consistent with a decreased activity of LC neurons. For instance, chronic (at least 1 week) treatment with SNRIs consistently: (i) decreases the spontaneous firing rate and sensory-evoked burst firing of LC neurons (Svensson and Udin, 1978; Huang et al., 1980; McMillen et al., 1980; Grant and Weiss, 2001; Szabo and Blier, 2001), (ii) decreases TH immunoreactivity in the LC (Komori et al., 1992), and (iii) reduces autoradiographic binding to NET in a number of brain areas, including the PFC and LC (Bauer and Tejani-Butt, 1992; Hebert et al., 2001; Frazer and Benmansour, 2002;

Benmansour et al., 2004). The reduction in NET binding does not appear to be due to a decrease in NET mRNA, which is actually elevated (Szot et al. 1993; Shores et al., 1994; Zhu et al., 2002; Benmansour et al., 2004).

Chronic administration of SNRIs also decreases the uptake of NE (Benmansour et al., 2004) and produces a concomitant elevation of basal extracellular levels that exceeds what is evoked by acute treatment (Nutt et al., 1997; Sacchetti et al., 1999; Invernizzi et al., 2001; Page and Lucki, 2002). This suggests that adaptive changes in the PFC occurring in response to chronic SNRI treatment are acting to sustain higher levels of extracellular NE. One mechanism that might account for the seeming discordance between reduced somal and enhanced terminal measures of NE transmission is a decrease in the plasmalemmal localization of NET on PFC NE axons. Our laboratory is initiating experiments to specifically test this hypothesis at the ultrastructural level. Moreover, chronic antidepressants attenuate many of the cellular changes induced by chronic stress, and such actions may contribute to their therapeutic efficacy (Adell et al., 1989; Nestler et al., 1990; Melia et al., 1992a, b; Papp et al., 1996). Hence, we also hypothesize that chronic antidepressant treatment will attenuate the morphological alterations in NET localization and TH expression that are evoked by chronic stress (Miner et al., 2006).

3.5.4 AVAILABILITY OF EXTRACELLULAR DA AS A SUBSTRATE FOR NE SYNTHESIS

The most recent class of drugs designed to alleviate the symptoms of ADHD selectively block the NET (Bymaster et al., 2002; Spencer et al. 2002), and their therapeutic effects are hypothesized to involve the LC NE system and its projections to the PFC (Solanto, 1998; Berridge, 2001; Bymaster et al., 2002). In addition, it is likely that the therapeutic efficacy of these compounds involves an interaction between the NE and DA systems specifically within the PFC (Tanda et al., 1994; Tanda et al., 1997; Bymaster et al., 2002). In addition to transporting NE, NET can efficiently transport DA (Horn, 1973; Raiteri et al., 1977; Gu et al., 1996), making it possible for NE terminals to capture DA from the extracellular space and use it to synthesize NE. In this case, NE terminals would not require high levels of TH under normal conditions.

These findings suggest that the availability of extracellular DA in the PFC provides another mechanism through which the subcellular distribution of NET can be modulated. This hypothesis is supported by the relative lack of DAT in the PFC (Ciliax et al., 1995; Sesack et al., 1998), the ability of extracellular DA to diffuse over long distances in this region (Garris et al., 1993; Garris and Wightman, 1994; Cass and Gerhardt, 1995), and the demonstrated capacity of NET to clear PFC DA (Jordan et al., 1994; Tanda et al., 1994; Gresch et al., 1995; Tanda et al., 1997; Yamamoto and Novotney, 1998; Bymaster et al., 2002) (*see also the Chapter 12 in this volume*). This hypothesized scenario is further supported by the fact that NET binding is normal in DβH knockout mice, despite the absence of NE (Weinshenker et al., 2002); this might be expected if NET were operating to transport DA in these animals. Interestingly, NET binding in terminal regions is decreased in TH knockout mice, which lack both NE and DA (Weinshenker et al., 2002). Together, these studies imply that the functional regulation of NET depends on the combined availability of extracellular DA and NE.

An obvious limitation of this part of our working model is that reliance on extracellular DA would seem to demand a reasonable level of plasmalemmal NET in order to acquire sufficient amounts of precursor DA. Our observation that NET is predominantly localized within the cytoplasm (Miner et al., 2003) makes it unclear how these axons would meet a sudden increased demand for NE release. Nevertheless, in vitro studies indicate that NET can undergo relatively quick insertion into the plasma membrane (Apparsundaram et al., 1998a; Savchenko et al., 2003), suggesting that heightened demand for NE might indeed be met by scavenging the available pool of extracellular DA via this mechanism. Under chronic stress conditions, the fact that we observed more axons with plasmalemmal NET but no detectable TH (Miner et al., 2006) is consistent with this idea.

Furthermore, we have speculated that the availability of extracellular DA is one reason why the magnitude of increased plasmalemmal NET and TH expression has not been greater following chronic stress (Miner et al., 2006). Given the putative role of DA in the regulation of NET distribution, it would be meaningful to investigate whether eliminating the extracellular pool of DA within the PFC by lesioning the DA cell bodies in the VTA would increase plasmalemmal NET localization and/or TH expression either in control or in chronically stressed rats.

3.5.5 ALTERNATIVE POSSIBILITIES AND GENERAL FUTURE STUDIES

An important question for future investigation is the extent to which observations regarding plasmalemmal NET and detectable TH are specific to the PFC. LC neurons collateralize extensively to multiple cortical and subcortical targets (Steindler, 1981; Loughlin et al., 1982), making it likely that the findings in the PFC are predictive of other brain regions. Indeed, neurochemical studies in the hippocampus are consistent with a minimal contribution of NET to baseline NE levels (Abercrombie et al., 1988). Nevertheless, there is some degree of topography within the LC projection system (Loughlin et al., 1986a; Simpson et al., 1997), and local factors such as the levels of available DA may further influence the extent of plasmalemmal NET and TH expression. Whether the low content of axonal TH and predominant intracellular distribution of NET generalizes to the ascending projections of non-LC neurons is also an important question. That this is likely is suggested by our finding of similar ultrastructural features of NE axons in the VTA (Liprando et al., 2004), a subcortical structure that receives most of its NE innervation from the A1 and A2 cell groups (Mejias-Aponte et al., 2004).

It is likely that variability in individual responses contributes to the ultrastructural localization of NET and coexpression of TH in PFC NE axons, as supported by our findings with chronic cold stress exposure (Miner et al., 2006). This is to be expected, given that preexisting individual differences such as genetic predisposition are likely to influence one's vulnerability to stress and subsequent development of depressive or anxiety disorders (Kendler et al., 1995). We predict that rats with different genetic susceptibility to stress (Guitart et al., 1993; Pare, 1994; Tejani-Butt et al., 1994; Rex et al., 1996; Lopez-Rubalcava and Lucki, 2000; Pardon et al., 2002) will show differential changes in ultrastructural characteristics of NE axons in the PFC. It is also possible that the magnitude of effects of chronic stress (or chronic SNRI treatment) on NET localization and TH expression might exhibit correlations with the behavioral responsivity of individual rats within a strain. For example, rats exhibiting low levels of novelty-induced locomotion (Rosario and Abercrombie, 1999) are more behaviorally responsive to chronic stress exposure (Cordero et al., 2003). While basal extracellular levels of NE in terminal regions do not differ between low and high responders to novelty, high responders show an enhanced NE efflux evoked by an acute stressor (Rosario and Abercrombie, 1999). Hence, it would be interesting to determine whether novelty responsivity predicts which individual rats are more susceptible to anatomical measures of chronic stress effects on PFC NE axons.

One of the most difficult issues raised by our studies is the question of whether chronic stress-induced changes in NET and TH expression in the PFC represent adaptive or maladaptive alterations. Two weeks of continuous cold exposure is a substantial stressor, although rats maintain normal body temperature and continue to gain weight, albeit at a slower rate than control animals (Moore et al., 2001; Miner et al., 2006). It is important to compare the results of these studies with other chronic stress paradigms that may be less physiologically demanding, such as predator exposure or intermittent unpredictable stress. To further explore whether ultrastructural changes in cortical NE axons reflect normal or pathological processes, it will also be useful to examine the effects of enriched environments that challenge animals without stressing them. We would also expect activity in this system to vary across the sleep–wake cycle (Feenstra et al., 2000; Berridge and Waterhouse, 2003). The fact that we consistently examined rats during their light cycle may have contributed to the observation of ultrastructural features consistent with low activity. Hence, it would be interesting to examine cortical NE axons in rats housed under a reversed light–dark condition.

3.6 Summary

In summary, NE axons in the PFC of normal rats exhibit ultrastructural characteristics that would not be predicted from classical views of monoamine neurotransmission. The low levels of NET inserted into the plasma membrane and the absence of detectable TH in most cortical NE axons suggest that these fibers are able to synthesize and release transmitter only under low activity conditions when extracellular DA is readily available. We have constructed a working hypothesis predicting that increased activity in the cortical LC system, with or without concomitant reductions in available DA, will drive intracellular NET to the plasma membrane and recruit TH from the cell bodies. Our findings that chronic stress increases plasmalemmal NET and detectable TH are consistent with this model. However, additional studies such as those outlined here are needed to test the impact of acute LC activation, chronic treatment with SNRIs, and removal of cortical DA on these ultrastructural measures. Moreover, important goals of future research will be to determine whether these functional anatomical changes are consistent across NE target regions, are predicted by susceptibility factors, and are reflective of adaptive or maladaptive plasticity.

References

Abercrombie, E.D. and Jacobs, B.L. (1987) Single-unit response of noradrenergic neurons in the locus coeruleus of freely-moving cats. I. Acutely presented stressful and nonstressful stimuli. J. Neurosci. 7, 2837–2843.

Abercrombie, E.D., Keller, R.W. and Zigmond, M.J. (1988) Characterization of hippocampal norepinephrine release as measured by microdialysis perfusion: pharmacological and behavioral studies. Neuroscience 27, 897–904.

Abercrombie, E.D., Keefe, K.A., DiFrischia, D.S. and Zigmond, M.J. (1989) Differential effects of stress on in vivo dopamine release in striatum, nucleus accumbens, and medial frontal cortex. J. Neurochem. 52, 1655–1658.

Adell, A., Garcia-Marquez, C., Armario, A. and Gelpi, E. (1988) Chronic stress increases serotonin and noradrenaline in the rat brain and sensitizes their responses to further acute stress. J. Neurochem. 50, 1678–1681.

Adell, A., Garcia-Marquez, C., Armario, A. and Gelpi, E. (1989) Chronic administration of clomipramine prevents the increase in serotonin and noradrenalin induced by chronic stress. Psychopharmacology 99, 22–26.

Ader, J.-P., Room, P., Postema, F. and Korf, J. (1980) Bilateral diverging axon collaterals and contralateral projections from rat locus coeruleus neurons. J. Neural Transm. 49, 207–218.

Angulo, J.A., Printz, D., Ledoux, M. and McEwen, B.S. (1991) Isolation stress increases tyrosine hydroxylase mRNA in the locus coeruleus and midbrain and decreases proenkephalin mRNA in the striatum and nucleus accumbens. Molec. Brain Res. 11, 301–308.

Anisman, H. and Zacharko, R.M. (1990) Multiple neurochemical and behavioral consequences of stressors: implications for depression. Pharmacol Therapeutics 46, 119–136.

Aoki, C. (1992) Beta-adrenergic receptors: astrocytic localization in the adult visual cortex and their relation to catecholamine axon terminals as revealed by electron microscopic immunocytochemistry. J. Neurosci. 12, 781–792.

Aoki, C., Go, C.-G., Venkatesan, C. and Kurose, H. (1994) Perikaryal and synaptic localization of alpha2A-adrenergic receptor-like immunoreactivity. Brain Res. 650, 181–204.

Aoki, C., Venkatesan, C., Go, C.-G., Forman, R. and Kurose, H. (1998) Cellular and subcellular sites for noradrenergic action in the monkey dorsolateral prefrontal cortex as revealed by the immunocytochemical localization of noradrenergic receptors and axons. Cereb. Cortex 8, 269–277.

Apparsundaram, S., Schroeter, S., Giovanetti, E. and Blakely, R.D. (1998a) Acute regulation of norepinephrine transport. II. PKC-modulated surface expression of human norepinephrine transporter proteins. J. Pharm. Exp. Ther. 287, 744–751.

Apparsundaram, S., Galli, A., DeFelice, L.J., Hartzell, H.C. and Blakely, R.D. (1998b) Acute regulation of norepinephrine transport. I. Protein kinase C-linked muscarinic receptors influence transport capacity and transporter density in SK-N-SH cells. J. Pharm. Exp. Ther. 287, 733–743.

Armstrong-James, M. and Fox, K. (1983) Effects of iontophoresed noradrenaline on the spontaneous activity of neurons in rat primary somatosensory cortex. J. Physiol. 355, 427–447.

Arnsten, A. and Goldman-Rakic, P. (1985) a2-Adrenergic mechanisms in prefrontal cortex associated with cognitive decline in aged non-human primates. Science 230, 1273–1276.

Arnsten, A., Steere, J. and Hunt, R. (1996) The contribution of alpha2-noradrenergic mechanisms to prefrontal cortical cognitive function. Arch. Gen. Psychiatry 53, 448–455.

Arnsten, A.F. (2000) Stress impairs prefrontal cortical function in rats and monkeys: Role of dopamine D1 and norepinephrine a-1 receptor mechanisms. Prog. Brain Res. 126, 183–192.

Arnsten, A.F. and Li, B.M. (2005) Neurobiology of executive functions: Catecholamine influences on prefrontal cortical functions. Biol. Psychiatry 57, 1377–1384.

Arnsten, A.F., Steere, J.C., Jentsch, D.J. and Li, B.M. (1998) Noradrenergic influences on prefrontal cortical cognitive function: opposing actions at postjunctional alpha-1 versus alpha-2-adrenergic receptors. Adv. Pharmacol. 42, 764–767.

Asan, E. (1993) Comparative single and double immunolabelling with antisera against cate-cholamine biosynthetic enzymes: criteria for the identification of dopaminergic, noradren-ergic and adrenergic structures in selected rat brain areas. Histochemistry 99, 427–442.

Aston-Jones, G., Fajkowski, J. and Cohen, J. (1999) Role of the locus coeruleus in attention and behavioral flexibility. Biol. Psychiatry 46, 1309–1320.

Aston-Jones, G., Zhu, Y. and Card, J.P. (2004) Numerous GABAergic afferents to locus ceruleus in the pericerulear dendritic zone: possible interneuronal pool. J. Neurosci. 24, 2313–2321.

Aston-Jones, G., Valentino, R.J., VanBockstaele, E.J. and Meyerson, A.T. (1994) Locus coeruleus, stress, and PTSD: neurobiological and clinical parallels. In: M.M. Murbur (ed.), Catecholamine Function in PTSD: Emerging Concepts. American Psychiatric Press, Washington, DC, pp. 17–62.

Audet, M.A., Doucet, G., Oleskevich, S. and Descarries, L. (1988) Quantified regional and laminar distribution of the noradrenaline innervation in the anterior half of the adult rat cerebral cortex. J. Comp. Neurol. 274, 307–318.

Avery, R.A., Franowicz, J.S., Studholme, C., van Dyck, C.H. and Arnsten, A.F. (2000) The alpha-2A-adrenoceptor agonist, guanfacine, increases regional cerebral blood flow in dor-solateral prefrontal cortex of monkeys performing a spatial working memory task. Neuro-psychopharmacology 23, 240–249.

Bauer, M.E. and Tejani-Butt, S.M. (1992) Effects of repeated administration of desipramine or electroconvulsive shock on norepinephrine uptake sites measured by [3H] nisoxetine autoradiography. Brain Res. 582, 208–214.

Beaudet, A. and Descarries, L. (1978) The monoamine innervation of rat cerebral cortex: synaptic and nonsynaptic axon terminals. Neuroscience 3, 851–860.

Beaudet, A. and Descarries, L. (1984) Fine structure of monoamine axon terminals in cerebral cortex. In: L. Descarries, T.R. Reader and H.H. Jasper (eds.), Monoamine Innervation of the Cerebral Cortex. Liss, New York, pp. 77–93.

Benmansour, S., Altamirano, A.V., Jones, D.J., Sanchez, T.A., Gould, G.G., Pardon, M.-C., Morilak, D.A. and Frazer, A. (2004) Regulation of the norepinephrine transporter by chronic administration of antidepressants. Biol. Psychiatry 55, 313–316.

Berridge, C.W. (2001) Arousal and attention-related actions of the locus coeruleus-noradrenergic system: potential target in the therapeutic actions of amphetamine-like stimulants. In: M.V. Solanto, A.F.T. Arnsten and F.X. Castellanos (eds.), Stimulant Drugs and ADHD. Oxford University, New York, pp. 158–184.

Berridge, C.W. and Waterhouse, B.D. (2003) The locus coeruleus-noradrenergic system: modulation of behavioral state and state-dependent cognitive processes. Brain Res. Rev. 42, 33–84.

Biederman, J. and Spencer, T. (1999) Attention-deficit/hyperactivity disorder (ADHD) as a noradrenergic disorder. Biol. Psychiatry 46, 1234–1242.

Birnbaum, S.G., Gobeske, K.T., Auerbach, J., Taylor, J.R. and Arnsten, A.F. (1999) A role for norepinephrine in stress-induced cognitive deficits: a-1-adrenoceptor mediation in the pre-frontal cortex. Biol. Psychiatry 46, 1266–1274.

Blakely, R.D. and Bauman, A.L. (2000) Biogenic amine transporters: regulation in flux. Curr. Opin. Neurobiol. 10, 328–336.

Bremner, J.D., Krystal, K.H., Soutwick, S.M. and Charney, D.S. (1996) Noradrenergic mechanisms in stress and anxiety. II. Clinical studies. Synapse 23, 39–51.

Brown, E.R., Coker, G.T. and O'Malley, K.L. (1987) Organization and evolution of the rat tyrosine hydroxylase gene. Biochemistry 26, 5208–5212.

Bunney, B.S. and Aghajanian, G.K. (1976) Dopamine and norepinephrine innervated cells in the rat prefrontal cortex: pharmacological differentiation using microiontophoretic techniques. Life Sci. 19, 1783–1792.

Bymaster, F.P., Katner, J.S., Nelson, D.L., Hemrick-Luecke, S.K., Threlkeld, P.G., Heiligenstein, J.H., Morin, S.M., Gehlert, D.R. and Perry, K.W. (2002) Atomoxetine increases extracellular levels of norepinephrine and dopamine in the prefrontal cortex of the rat: a potential mechanism for efficacy in attention deficit/hyperactivity disorder. Neuropsychopharmacology 27, 699–711.

Callado, L.F., Meana, J.J. and Grijalba, B. (1998) Selective increase of alpha2-adrenoceptor agonist binding sites in brains of depressed suicide victims. J. Neurochem. 70, 1114–1123.

Campbell, D.G., Hardie, D.G. and Vulliet, P.R. (1986) Identification of the four phosphorylation sites in the N-terminal region of tyrosine hydroxylase. J. Biol. Chem. 261, 10489–10492.

Cass, W.A. and Gerhardt, G.A. (1995) In vivo assessment of dopamine uptake in rat medial prefrontal cortex: comparison with dorsal striatum and nucleus accumbens. J. Neurochem. 65, 201–207.

Charney, D.S., Bremner, J.D. and Redmond, D.E. (1995) Noradrenergic substrates for anxiety and fear. In: F.E. Bloom and D.J. Kupfer (eds.), Psychopharmacology: The Fourth Generation of Progress. Raven Press, New York, pp. 387–395.

Christenson, J.G., Dairman, W. and Udenfriend, S. (1972) On the identity of DOPA decarboxylase and 5-hydroxytryptophan decarboxylase. Proc. Natl. Acad. Sci. 69, 343–347.

Ciliax, B., Heilman, C., Demchyshyn, L., Pristupa, Z., Ince, E., Hersch, S., Niznik, H. and Levey, A. (1995) The dopamine transporter: immunocytochemical characterization and localization in brain. J. Neurosci. 15, 1714–1723.

Conti, L.H. and Foote, S.L. (1996) Reciprocal cross-desensitization of locus coeruleus electrophysiological responsivity to corticotrophin-releasing factor and stress. Brain Res. 722, 19–29.

Cooper, J.R., Bloom, F.E. and Roth, R.H. (1991) The Biochemical Basis of Neuropharmacology, 6th Edition. New York: Oxford University Press.

Cordero, M.I., Kruyt, N.D. and Sandi, C. (2003) Modulation of contextual fear conditioning by chronic stress in rats is related to individual differences in behavioral reactivity to novelty. Brain Res. 970, 242–245.

Curtis, A.L., Pavcovich, L.A., Grigoriadis, D.E. and Valentino, R.J. (1995) Previous stress alters corticotropin-releasing factor neurotransmission in the locus coeruleus. Neuroscience 65, 541–550.

Descarries, L. and Mechawar, N. (2000) Ultrastructural evidence for diffuse transmission by monoamine and acetylcholine neurons of the central nervous system. Prog. Brain Res. 125, 27–47.

Descarries, L., Watkins, K. and Lapierre, Y. (1977) Noradrenergic axon terminals in the cerebral cortex of rat. III. Topometric ultrastructural analysis. Brain Res. 133, 197–222.

Descarries, L., Reader, T.A. and Jasper, H.H., eds (1984) Monoamine Innervation of Cerebral Cortex. Alan R. Liss, New York.

Drevets, W.C., Videen, T.O., Price, J.L., Preskorn, S.H., Carmichael, S.T. and Raichle, M.E. (1992) A functional anatomical study of unipolar depression. J. Neurosci. 12, 3628–3641.

Emson, P.C. and Koob, G.F. (1978) The origin and distribution of dopamine-containing afferents to the rat prefrontal cortex. Brain Res. 142, 249–267.

Ernst, M., Liebenauer, L., King, A., Fitzgerald, G.A., Cohen, R.A. and Zametkin, A.J. (1994) Reduced brain metabolism in hyperactive girls. J. Am. Acad. Child Adolesc. Psychiatry 33, 858–868.

Fallon, J.H. and Loughlin, S.E. (1982) Monoamine innervation of the forebrain: collateralization. Brain Res. Bull. 9, 295–307.

Fallon, J.H. and Loughlin, S.E. (1987) Monoamine innervation of cerebral cortex and a theory of the role of monoamines in cerebral cortex and basal ganglia. In: E.G. Jones and A. Peters (eds.), Cerebral Cortex. Plenum, New York, pp. 41–127.

Feenstra, M.G.P., Botterblom, M.H.A. and Matenbroek, S. (2000) Dopamine and noradrenaline efflux in the prefrontal cortex in light and dark period: effects of novelty and handling and comparisons to the nucleus accumbens. Neuroscience 100, 741–748.

Finlay, J.M., Zigmond, M.J. and Abercrombie, E.D. (1995) Increased dopamine and norepinephrine release in medial prefrontal cortex induced by acute and chronic stress: effects of diazepam. Neuroscience 64, 619–628.

Finlay, J.M., Jedema, H.P., Rabinovic, A.D., Mana, M.J., Zigmond, M.J. and Sved, A.F. (1997) Impact of corticotropin-releasing hormone on extracellular norepinephrine in prefrontal cortex after chronic cold stress. J. Neurochem. 69, 144–150.

Foote, S.L. (1985) Anatomy and physiology of brain monoamine systems. Psychiatry 3, 1–14.

Foote, S.L., Bloom, F.E. and Oliver, A.P. (1975) Effects of putative neurotransmitters on neuronal activity in monkey auditory cortex. Brain Res. 86, 229–242.

Foote, S.L., Bloom, F.E. and Aston-Jones, G. (1983) Nucleus locus coeruleus: new evidence of anatomical and physiological specificity. Physiol. Rev. 63, 844–914.

Frazer, A. (2000) Norepinephrine involvement in antidepressant action. J. Clin. Psychiat. 61 (Suppl 10), 25–30.

Frazer, A. and Benmansour, S. (2002) Delayed pharmacological effects of antidepressants. Mol. Psychiatry 7, S23–S28.

Gambarana, C., Scheggi, S., Tagliamonte, A., Tolu, P. and DeMontis, M.G. (2001) Animal models for the study of antidepressant activity. Br. Res. Protocols 7, 11–20.

Garris, P.A. and Wightman, R.M. (1994) Different kinetics govern dopaminergic transmission in the amygdala, prefrontal cortex, and striatum: an in vivo voltametric study. J. Neurosci. 14, 442–450.

Garris, P.A., Collins, L.B., Jones, S.R. and Wightman, R.M. (1993) Evoked extracellular dopamine in vivo in the medial prefrontal cortex. J. Neurochem. 61, 637–647.

Gaspar, P., Berger, B., Febvret, A., Vigny, A. and Henry, J.P. (1989) Catecholamine innervation of the human cerebral cortex as revealed by comparative immunohistochemistry of tyrosine hydroxylase and dopamine-beta-hydroxylase. J. Comp. Neurol. 279, 249–271.

Goldman-Rakic, P.S., Lidow, M.S. and Gallager, D.W. (1990) Overlap of dopaminergic, adrenergic, and serotoninergic receptors and complimentarity of their subtypes in the primate prefrontal cortex. J. Neurosci. 10, 2125–2138.

Grant, M.M. and Weiss, J.M. (2001) Effects of chronic antidepressant drug administration and electroconvulsive shock on locus coeruleus electrophysiological activity. Biol. Psychiatry 49, 117–129.

Gresch, P.J., Sved, A.F., Zigmond, M.J. and Finlay, J.M. (1994) Stress-induced sensitization of dopamine and norepinephrine efflux in medial prefrontal cortex of the rat. J. Neurochem. 63, 575–583.

Gresch, P.J., Sved, A.F., Zigmond, M.J. and Finlay, J.M. (1995) Local influences of endogenous norepinephrine on extracellular dopamine in rat medial prefrontal cortex. J. Neurochem. 65, 111–116.

Grzanna, R. and Fritschy, J.-M. (1991) Efferent projections of different subpopulations of central noradrenergic neurons. In: C.D. Barnes and O. Pompeiano (eds.), Progress in Brain Research. Elsevier, New York, pp. 89–101.

Gu, H.H., Wall, S. and Rudnick, G. (1996) Ion coupling stoichiometry for the norepinephrine transporter in membrane vesicles from stably transfected cells. J. Biol. Chem. 271, 6911–6919.

Guitart, X., Kogan, J.H., Berhow, M., Terwilliger, R.Z., Aghajanian, G.K. and Nestler, E.J. (1993) Lewis and Fischer rat strains display differences in biochemical, electrophysiological and behavioral parameters: studies in the nucleus accumbens and locus coeruleus of drug naive and morphine-treated animals. Brain Res. 611, 7–17.

Haycock, J.W. (1990) Involvement of serine-31 in phosphorylation of tyrosine hydroxylase in PC12 cells. J. Biol. Chem. 265, 11682–11691.

Hebert, C., Habimana, A., Elie, R. and Reader, T.A. (2001) Effects of chronic antidepressant treatments on 5-HT and NA transporters in rat brain: an autoradiographic study. Neurochem. Int. 38, 63–74.

Henry, J.P., Sagne, C., Bedet, C. and Gasnier, B. (1998) The vesicular monoamine transporter: from chromaffin granule to brain. Neurochem. Int. 32, 227–246.

Hersch, S.M., Yi, H., Heilman, C.J., Edwards, R.H. and Levey, A.I. (1997) Subcellular localization and molecular topology of the dopamine transporter in the striatum and substantia nigra. J. Comp. Neurol. 388, 211–227.

Hoffman, B.J., Hansson, S.R., Mezey, E. and Palkovits, M. (1998) Localization and dynamic regulation of biogenic amine transporters in the mammalian central nervous system. Front. Neuroendocrinol. 19, 187–231.

Hökfelt, T., Johansson, O., Fuxe, K., Goldstein, M. and Park, D. (1977) Immunohistochemical studies on the localization and distribution of monoamine neuron systems in the rat brain. II. Tyrosine hydroxylase in the telencephalon. Med. Biol. 55, 21–40.

Horn, A.S. (1973) Structure-activity relations for the inhibition of catecholamine uptake into synaptosomes from noradrenaline and dopaminergic neurons in rat brain homogenates. Br. J. Pharmacol. 47, 332–338.

Huang, Y.H., Maas, J.W. and Hu, G.H. (1980) The time course of noradrenergic pre- and postsynaptic activity during chronic desipramine treatment. Eur. J. Pharmacol. 68, 41–47.

Ichikawa, S., Sasaoka, T. and Nagatsu, T. (1991) Primary structure of mouse tyrosine hydroxylase deduced from its cDNA. Biochem. Biophys. Res. Commun. 176, 1610–1616.

Invernizzi, R.W., Parini, S., Sacchetti, G., Fracasso, C., Caccia, S., Annoni, K. and Samanin, R. (2001) Chronic treatment with reboxetine by osmotic pumps facilitates its effect on extracellular noradrenaline and may desensitize alpha2-adrenoceptors in the prefrontal cortex. Br. J. Pharmacol. 132, 183–188.

Jayanthi, L.D., Samuvel, D.J. and Ramamoorthy, S. (2004) Regulated internalization and phosphorylation of the native norepinephrine transporter in response to phorbol esters: evidence for localization in lipid rafts and lipid raft-mediated internalization. J. Biol. Chem. 279, 19315–19326.

Jedema, H., Sved, A., Zigmond, M. and Finlay, J. (1999) Sensitization of norepinephrine release in medial prefrontal cortex: effect of different chronic stress protocols. Brain Res. 830, 211–217.

Jedema, H.P. and Grace, A.A. (2003) Chronic exposure to cold stress alters electrophysiological properties of locus coeruleus neurons recorded in vitro. Neuropsychopharmacology 28, 63–72.

Jedema, H.P., Finlay, J.M., Sved, A.F. and Grace, A.A. (2001) Chronic cold exposure potentiates CRH-evoked increases in electrophysiologic activity of locus coeruleus neurons. Biol. Psychiatry 49, 351–359.

Johnson, E.S., Roberts, M.H.T. and Straughan, D.W. (1969) The responses of cortical neurones to monoamines under differing anaesthetic conditions. J. Physiol. 203, 261–280.

Jones, B.E. and Yang, T.Z. (1985) The efferent projections from the reticular formation and the locus coeruleus studied by anterograde and retrograde axonal transport in the rat. J. Comp. Neurol. 242, 56–92.

Jordan, S., Kramer, G.L., Zukar, P.K., Moeller, M. and Petty, F. (1994) In vivo biogenic amine efflux in medial prefrontal cortex with imipramine, fluoxetine, and fluvoxamine. Synapse 18, 294–297.

Kasamatu, T. and Heggelund, P. (1982) Single cell responses in cat visual cortex to visual stimulation during iontophoresis of noradrenaline. Exp. Brain Res. 45, 317–327.

Kendler, K.S., Kessler, R.C., Walters, E.E., MacLean, C., Neale, M.C., Heath, A.C. and Eaves, L.J. (1995) Stressful life events, genetic liability, and onset of an episode of major depression in women. Am. J. Psychiatry 152, 833–842.

Kent, J.M. (2000) SNaRIs, NaSSAs, and NaRIs: new agents for the treatment of depression. Lancet 355, 911–918.

Klimek, V., Stockmeier, C., Overholser, J., Meltzer, H.Y., Kalka, S., Dilley, G. and Ordway, G.A. (1997) Reduced levels of norepinephrine transporters in the locus coeruleus in major depression. J. Neurosci. 17, 8451–8458.

Komori, K., Kunimi, Y., Yamaoka, K., Ito, T., Kasahara, Y. and Nagatsu, I. (1992) Semiquantitative analysis of immunoreactivities of tyrosine hydroxylase and aromatic L-amino acid decarboxylase in the locus coeruleus of desipramine-treated rats. Neurosci. Lett. 147.

Korf, J., Aghaganian, G.K. and Roth, R.H. (1973) Increased turnover of norepinephrine in the rat cerebral cortex during stress: role of locus coeruleus. Neuropsychopharmacology 12, 933–938.

Kritzer, M.F. (2000) Effects of acute and chronic gonadectomy on the catecholamine innervation of the cerebral cortex in adult male rats: insensitivity of axons immunoreactive for dopamine-β-hydroxylase to gonadal steroids, and differential sensitivity of axons immunoreactive for tyrosine hydroxylase to ovarian and testicular hormones. J. Comp. Neurol. 427, 617–633.

Kritzer, M.F. (2003) Long term gonadectomy affects the density of tyrosine hydroxylase but not dopamine beta hydroxylase, choline acetyltransferase or serotonin immunoreactive axons in the medial prefrontal cortices of adult male rats. Cereb. Cortex 13, 282–296.

Krnjevic, L. and Phillips, J.W. (1963a) Iontophoretic studies of neurons in the mammalian cerebral cortex. J. Physiol. 165.

Krnjevic, L. and Phillips, J.W. (1963b) Actions of certain amines on cerebral cortical neurons. Brit J Pharmacol Chemother 20, 471–490.

Lapiz, M.D.S. and Morilak, D.A. (2006) Noradrenergic modulation of cognitive function in rat medial prefrontal cortex as measured by attentional set shifting capability. Neuroscience 137, 1039–1049.

Law-Tho, D., Crepel, F. and Hirsch, J.C. (1993) Noradrenaline decreases transmission of NMDA- and non-NMDA-receptor mediated monosynaptic EPSPs in rat prefrontal neurons in vitro. Eur. J. Neurosci. 5, 1494–1500.

Levitt, P. and Moore, R.Y. (1978) Noradrenaline neuron innervation of the neocortex in the rat. Brain Res. 139, 219–231.

Lewis, D.A. and Morrison, J.H. (1989) Noradrenergic innervation of monkey prefrontal cortex: A dopamine-b-hydroxylase immunohistochemical study. J. Comp. Neurol. 282, 317–330.

Lewis, D.A., Foote, S.L., Goldstein, M. and Morrison, J.H. (1988) The dopaminergic innervation of monkey prefrontal cortex: a tyrosine hydroxylase immunocytochemical study. Brain Res. 449, 225–243.

Lewis, D.A., Campbell, M.J., Foote, S.L., Goldstein, M. and Morrison, J.H. (1987) The distribution of tyrosine hydroxylase-immunoreactive fibers in primate neocortex is widespread but regionally specific. J. Neurosci. 7, 279–290.

Li, B.M. and Mei, Z.T. (1994) Delayed-response deficit induced by local injection of the alpha 2 adrenergic antagonist yohimbine into the dorsolateral prefrontal cortex in young adult monkeys. Behav. Neural Biol. 62, 134–139.

Li, B.M. and Kubota, K. (1998) Alpha-2 adrenergic modulation of prefrontal cortical neuronal activity related to a visual discrimination task with GO and NO-GO performances in monkeys. Neurosci. Res. 31, 83–95.

Lidov, H.G.W., Rice, F.L. and Molliver, M.E. (1978) The organization of the catecholamine innervation of somatosensory cortex: the barrel field of the mouse. Brain Res. 153, 577–584.

Lindvall, O. and Björklund, A. (1984) General organization of cortical monoamine systems. In: L. Descarries, T. Reader and H. Jasper (eds.), Monoamine Innervation of the Cerebral Cortex. Liss, New York, pp. 9–40.

Liprando, L.A., Miner, L.A.H., Blakely, R.D., Lewis, D.A. and Sesack, S.R. (2004) Ultra-structural interactions between terminals expressing the norepinephrine transporter and dopamine neurons in the rat and monkey ventral tegmental area. Synapse 52, 233–244.

Liu, W., Yuen, E.Y., Allen, P.B., Feng, J., Greengard, P. and Yan, Z. (2006) Adrenergic modulation of NMDA receptors in prefrontal cortex is differentially regulated by RGS proteins and spinophilin. Proc. Natl. Acad. Sci. 103, 18338–18343.

Lopez-Rubalcava, C. and Lucki, I. (2000) Strain differences in the behavioral effects of anti-depressant drugs in the forced swim test. Neuropsychopharmacology 22, 191–199.

Lorang, D., Amara, S.G. and Simerly, R.B. (1994) Cell-type-specific expression of cate-cholamine transporters in the rat brain. J. Neurosci. 14, 4903–4914.

Loughlin, S.E., Foote, S.L. and Fallon, J.H. (1982) Locus coeruleus projections to the cortex: topography, morphology and collaterization. Brain Res. Bull. 9, 287–294.

Loughlin, S.E., Foote, S.L. and Bloom, F.E. (1986a) Efferent projections of nucleus locus coeruleus: topographic organization of cells of origin demonstrated by three dimensional reconstruction. Neuroscience 18, 291–306.

Loughlin, S.E., Foote, S.L. and Grzanna, R. (1986b) Efferent projections of nucleus locus coeruleus: morphologic subpopulations have different efferent targets. Neuroscience 18, 307–319.

Ma, C.L., Qi, X.L., Peng, J.Y. and Li, B.M. (2003) Selective deficit in no-go performance induced by blockade of prefrontal cortical alpha 2-adrenoceptors in monkeys. Neuroreport 14, 1013–1016.

Mamalaki, E., Kvetnansky, R., Brady, L.S., Gold, P.W. and Herkenham, M. (1992) Repeated immobilization stress alters tyrosine hydroxylase, corticotropin-releasing hormone and corticosteroid receptor messenger ribonucleic acid levels in rat brain. J. Neuroendocrinol. 4, 689–699.

Mana, M.J. and Grace, A.A. (1997) Chronic cold stress alters basal and evoked electrophysio-logical activity of rat locus coeruleus neurons. Neuroscience 81, 1055–1064.

Mantz, J., Millla, C., Glowinski, J. and Thierry, A.M. (1988) Differential effects of ascending neurons containing dopamine and noradrenaline in control of spontaneous activity and of evoked responses in the rat prefrontal cortex. Neuroscience 27, 517–526.

Mao, Z.M., Arnsten, A.F. and Li, B.M. (1999) Local infusion of an alpha-1 adrenergic agonist into the prefrontal cortex impairs spatial working memory performance in monkeys. Biol. Psychiatry 46, 1259–1265.

Mason, S.T. and Figiber, H.C. (1979) Regional topography within noradrenergic locus coeruleus as revealed by retrograde transport of horseradish peroxidase. J. Comp. Neurol. 187, 703–724.

McLean, J. and Waterhouse, B.D. (1994) Noradrenergic modulation of cat area 17 neuronal responses to moving visual stimuli. Brain Res. 667, 83–97.

McMillen, B.A., Warnack, W., German, D.C. and Shore, P.A. (1980) Effects of chronic desip-ramine treatment on rat brain noradrenergic responses to alpha-adrenergic drugs. Eur. J. Pharmacol. 61, 239–246.

Meana, J.J., Baruren, F. and Garcia-Sevilla, J.A. (1992) Alpha2-adrenoceptors in the brain of suicide victims: increased receptor density associated with major depression. Biol. Psy-chiatry 31, 471–490.

Mejias-Aponte, C.A., Zhu, Y. and Aston-Jones, G. (2004) Noradrenergic innervation of mid-brain dopamine neurons: prominent inputs from A1 and A2 cell groups. Soc. Neurosci. Abstr. 465.4.

Melia, K.R., Nestler, E.J. and Duman, R.S. (1992a) Chronic imipramine treatment normalizes levels of tyrosine hydroxylase in the locus coeruleus of chronically stress rats. Psy-chopharmacology 108, 23–26.

Melia, K.R., Rasmussen, K., Terwilliger, R.Z., Haycock, J.W., Nestler, E.J. and Duman, R.S. (1992b) Coordinate regulation of the cyclic AMP system with firing rate and expression of tyrosine hydroxylase in the rat locus coeruleus: effects of chronic stress and drug treatment. J. Neurochem. 58, 494–502.

Melikian, H.E., Ramamoorthy, S., Tate, C.G. and Blakely, R.D. (1996) Inability to N glycosylate the human norepinephrine transporter reduces protein stability, surface trafficking, and transport activity but not ligand recognition. Mol. Pharm. 50, 266–276.

Melikian, H.E., McDonald, J.K., Gu, H., Rudnick, G., Moore, K.R. and Blakely, R.D. (1994) Human norepinephrine transporter. Biosynthetic studies using a site-directed polyclonal antibody. J. Biol. Chem. 269, 12290–12297.

Michelson, D., Adler, L., Spencer, T., Reimherr, F.W., West, S.A., Allen, A.J., Kelsey, D., Wernicke, J., Dietrich, A. and Milton, D. (2003) Atomoxetine in adults with ADHD: two randomized, placebo-controlled studies. Biol. Psychiatry 53, 112–120.

Milner, T.A. and Bacon, C.E. (1989) Ultrastructural localization of tyrosine hydroxylase-like immunoreactivity in the rat hippocampal formation. J. Comp. Neurol. 281, 479–495.

Miner, L.A.H., Schroeter, S., Blakely, R.D. and Sesack, S.R. (2000) Ultrastructural localization of the serotonin transporter in superficial and deep layers of the rat prelimbic prefrontal cortex and its spatial relationship to dopamine terminals. J. Comp. Neurol. 427, 220–234.

Miner, L.A.H., Schroeter, S., Blakely, R.D. and Sesack, S.R. (2003) Ultrastructural localization of the norepinephrine transporter in superficial and deep layers of the rat prelimbic prefrontal cortex and its spatial relationship to probable dopamine terminals. J. Comp. Neurol. 466, 478–494.

Miner, L.A.H., Jedema, H.P., Moore, F.W., Blakely, R.D., Grace, A.A. and Sesack, S.R. (2006) Chronic stress increases the plasmalemmal distribution of the norepinephrine transporter and coexpression of tyrosine hydroxylase in norepinephrine axons in the prefrontal cortex. J. Neurosci. 26, 1571–1578.

Moller, H.-J. (2000) Are all antidepressants the same? J. Clin. Psychiat. 61 (Suppl 10), 24–27.

Molliver, M.E., Grzanna, R., Lidov, H.G.W., Morrison, J.H. and Olschowka, J.A. (1982) Monoamine systems in the cerebral cortex In: V, Chan-Palay and S.L. Palay (eds.), Cytochemical Methods in Neuroanatomy. Liss, New York, pp. 255–277.

Moore, H., Rose, H.J. and Grace, A.A. (2001) Chronic cold stress reduces the spontaneous activity of ventral tegmental dopamine neurons. Neuropsychopharmacology 24, 410–419.

Morrison, J.H., Molliver, M.E. and Grzanna, R. (1979) Noradrenergic innervation of cerebral cortex: widespread effects of locus cortical lesions. Science 202, 313–316.

Morrison, J.H., Foote, S.L. and Bloom, F.E. (1984) Regional, laminar, developmental and functional characteristics of noradrenaline and serotonin innervation patterns in monkey cortex. In: L. Descarries, T.A. Reader and H.H. Jasper (eds.), Monoamine Innervation of Cerebral Cortex. Liss, New York, pp. 61–75.

Morrison, J.H., Grzanna, R., Molliver, M.E. and Coyle, J.T. (1978) The distribution and orientation of noradrenergic fibers in the neocortex of the rat: an immunofluorescence study. J. Comp. Neurol. 181, 17–40.

Morrison, J.H., Molliver, M.E., Grzanna, R. and Coyle, J.T. (1981) The intracortical trajectory of the coeruleo-cortical projection in the rat: a tangentially organized afferent. Neuroscience 6, 139–158.

Morrison, J.H., Foote, S.L., O'Conner, D. and Bloom, F.E. (1982) Laminar, tangential and regional organization of the noradrenergic innervation of monkey cortex: dopamine b-hydroxylase immunohistochemistry. Brain Res. Bull. 9, 309–319.

Morruzzi, A.S. and Hart, E.R. (1955) Evoked cortical responses under the influence of hallucinogens and related drugs. Electroencephalgr. Clin. Neurophysiol. 1.

Mostofsky, S.H., Cooper, K.L., Kates, W.R., Denckla, M.B. and Kaufmann, W.E. (2002) Smaller prefrontal and premotor volumes in boys with attention-deficit/hyperactivity disorder. Biol. Psychiatry 52, 785–794.

Nagai, T.K., Satoh, K., Imamoto, K. and Maeda, T. (1981) Divergent projections of catecholamine neurons of the locus coeruleus as revealed by fluorescent retrograde double
labeling technique. Neurosci. Lett. 23, 117–123.

Nelson, J.C. (1999) A review of the efficacy of serotonergic and noradrenergic reuptake inhibitors for treatment of major depression. Biol. Psychiatry 46, 1301–1308.

Nestler, E.J., McMahon, A., Sabban, E.L., Tallman, J.F. and Duman, R.S. (1990) Chronic
antidepressant administration decreases the expression of tyrosine hydroxylase in the rat
locus coeruleus. Proc. Natl. Acad. Sci. 87, 7522–7526.

Nicholas, A.P., Pieribone, V.A. and Hokfelt, T. (1993a) Cellular localization of messenger
RNA for beta-1 and beta-2 adrenergic receptors in rat brain: an in situ hybridization study.
Neuroscience 56, 1023–1039.

Nicholas, A.P., Pieribone, V. and Holkfelt, T. (1993b) Distribution of mRNA for alpha2-
adrenergic receptor subtypes in rat brain: an in situ hybridization study. J. Comp. Neurol.
328, 575–594.

Nirenberg, M.J., Vaughan, R.A., Uhl, G.R., Kuhar, M.J. and Pickel, V.M. (1996) The dopamine transporter is localized to dendritic and axonal plasma membranes of nigrostriatal
dopaminergic neurons. J. Neurosci. 16, 436–447.

Nisenbaum, L.K., Zigmond, M.J., Sved, A.F. and Abercrombie, E.D. (1991) Prior exposure to
chronic stress results in enhanced synthesis and release of hippocampal norepinephrine in
response to a novel stressor. J. Neurosci. 11, 1478–1484.

Noack, H.J. and Lewis, D.A. (1989) Antibodies directed against tyrosine hydroxylase differentially recognize noradrenergic axons in monkey neocortex. Brain Res. 500, 313–324.

Nutt, D.J., Lalies, M.D., Lione, L.A. and Hudson, A.L. (1997) Noradrenergic mechanisms in
the prefrontal cortex. J. Psychopharmacol. 11, 163–168.

Olschowka, J.A., Molliver, M.E., Grzanna, R., Rice, F.L. and Coyle, J.T. (1981) Ultrastructural demonstration of noradrenergic synapses in the rat central nervous system by dopamine-b-hydroxylase immunocytochemistry. J. Histochem. Cytochem. 29, 271–280.

Ossowska, G., Nowak, G., Kata, R., Klenk-Majewska, B., Canilczuk, Z. and Zebrowska-
Lupina, I. (2001) Brain monoamine receptors in a chronic unpredictable stress model in
rats. J. Neural Transm. 108, 311–319.

Page, M.E. and Lucki, I. (2002) Effects of acute and chronic reboxetine treatment on stress-
induced monoamine efflux in the rat frontal cortex. Neuropsychopharmacology 27, 237–
247.

Palacios, J.M. and Kuhar, M.J. (1980) Beta-adrenergic-receptor localization by light microscopic autoradiography. Science 208, 1378–1380.

Palacios, J.M. and Kuhar, M.J. (1982) Beta-adrenergic receptor localization by light microscopic autoradiography. Neurochem. Int. 4, 473–490.

Palacios, J.M. and Wamsley, J.K. (1983) Microscopic localization of adrenoreceptors. In: G.
Kinos (ed.), Adrenoreceptors and Catecholamine Action, Part B. Wiley, New York, pp.
295–313.

Palacios, J.M. and Wamsley, J.K. (1984) Catecholamine receptors. In: A. Bjorklund, T. Hokfelt and M.J. Kuhar (eds.), Handbook of Chemical Neuroanatomy, Vol. 3.: Classical
Transmitters and Transmitter Receptors in the CNS, Part II. Elsevier, New York, pp. 325–
351.

Papadopoulos, G., Parnavelas, J. and Bujis, R. (1989) Light and electron microscopic immunocytochemical analysis of the noradrenaline innervation of the rat visual cortex. J.
Neurocytol. 18, 1–10.

Papadopoulos, G.C. and Parnavelas, J.G. (1991) Monoamine systems in the cerebral cortex:
evidence for anatomical specificity. Prog. Neurobiol. 36, 195–200.

Papp, M., Moryl, E. and Wilner, P. (1996) Pharmacological validation of the chronic mild
stress model of depression. Eur. J. Pharmacol. 296, 129–136.

Pardon, M.-C., Gould, G.G., Garcia, A., Phillips, L., Cook, M.C., Miller, S.A., Mason, P.A.
and Morilak, D.A. (2002) Stress reactivity of the brain noradrenergic system in three rat

strains differing in their neuroendocrine and behavioral responses to stress: implications for susceptibility to stress-related neuropsychiatric disorders. Neuroscience 115, 229–242.

Pare, W.P. (1994) Open field, learned helplessness, defensive burying and forced-swim test in WKY rats. Physiol. Behav. 55, 433–439.

Parnavelas, J., Moises, H. and Speciale, S. (1985) The monoaminergic innervation of the rat visual cortex. Proc. Royal Soc. Lond. 223, 319–329.

Pavcovich, L.A. and Ramirez, O.A. (1991) Time course of uncontrollable stress in locus coeruleus neuronal activity. Brain Res. Bull. 26, 17–21.

Pickel, V.M., Segal, M. and Bloom, F.E. (1974) A radioautographic study of the efferent pathways of the nucleus locus coeruleus. J. Comp. Neurol. 155, 15–42.

Pickel, V.M., Joh, T.H. and Reis, D.J. (1975a) Ultrastructural localization of tyrosine hydroxylase in noradrenergic neurons of brain. Proc. Natl. Acad. Sci. 72, 659–663.

Pickel, V.M., Joh, T.H. and Reis, D.J. (1975b) Immunohistochemical localization of tyrosine hydroxylase in brain by light and electron microscopy. Brain Res. 85, 295–300.

Pickel, V.M., Joh, T.H., Field, P.M., Becker, C.G. and Reis, D.J. (1975c) Cellular localization of tyrosine hydroxylase by immunocytochemistry. J. Histochem. Cytochem. 23, 1–12.

Pliszka, S.R., McCracken, J.T. and Maas, J.W. (1996) Catecholamines in attention-deficit hyperactivity disorder: current prospectives. J. Am. Acad. Child Adolesc. Psychiatry 35, 264–272.

Raiteri, M., Del Carmine, R., Bertollini, A. and Levi, G. (1977) Effect of sympathomimetic amines on the synaptosomal transport of noradrenaline, dopamine and 5-hydroxytryptamine. Eur. J. Pharmacol. 41, 133–143.

Rauch, S.L., Shin, L.M., Segal, E., Pitman, R.K., Carson, M.A., McMullin, K., Whalen, P.J. and Makris, N. (2003) Selectively reduced regional cortical volumes in post-traumatic stress disorder. Neuroreport 14, 913–916.

Reader, T., Ferron, A., Descarries, L. and Jasper, H. (1979) Modulatory role for biogenic amines in the cerebral cortex. Microiontophoretic studies. Brain Res. 160, 217–229.

Ren, Z.G., Porzgen, P., Zhang, J.M., Chen, X.R., Amara, S.G., Blakely, R.D. and Sieber-Blum, M. (2001) Autocrine regulation of norepinephrine transporter expression. Mol. Cell. Neurosci. 17, 539–550.

Rex, A., Sondern, U., Voight, J.P., Franck, S. and Fink, H. (1996) Strain differences in fear-motivated behavior of rats. Pharmacol. Biochem. Behav. 54, 107–111.

Robbins, T.W. (1984) Cortical noradrenaline, attention and arousal. Psychol. Med. 14, 13–21.

Room, P., Postema, F. and Korf, J. (1981) Divergent axon collaterals of rat locus coeruleus neurons: demonstration by a fluorescent double labeling technique. Brain Res. 221.

Rosario, L.A. and Abercrombie, E.D. (1999) Individual differences in behavioral reactivity: correlation with stress-induced norepinephrine efflux in the hippocampus of Sprague-Dawley rats. Brain Res. Bull. 48, 595–602.

Rossetti, Z.L., Pani, L., Portas, C. and Gessa, G. (1989) Brain dialysis provides evidence for D2-dopamine receptors modulating noradrenaline release in the rat frontal cortex. Eur. J. Pharmacol. 163, 393–395.

Rusnak, M., Zorad, S., Buckendahl, P., Sabban, E.L. and Kvetnansky, R. (1998) Tyrosine hydroxylase mRNA levels in locus coeruleus of rats during adaption to long-term immobilization stress exposure. Mol. Chem. Neuropath. 33, 249–258.

Russell, V., Allie, S. and Wiggins, T. (2000) Increased noradrenergic activity in prefrontal cortex slices of an animal model for attention-deficit hyperactivity disorder - the spontaneously hypertensive rat. Behav. Brain Res. 117, 69–74.

Sacchetti, G., Bernini, M., Bianchetti, A., Parini, S., Invernizzi, R.M. and Samanin, R. (1999) Studies on the acute and chronic effects of reboxetine on extracellular noradrenaline and other monoamines in the rat brain. Br. J. Pharmacol. 128, 1332–1338.

Savchenko, V., Sung, U. and Blakely, R.D. (2003) Cell surface trafficking of the antidepressant-sensitive norepinephrine transporter revealed with an ectodomain antibody. Mol. Cell. Neurosci. 24, 1131–1150.

Sawaguchi, T. (1998) Attenuation of delay-period activity of monkey prefrontal neurons by an alpha2-adrenergic antagonist during an oculomotor delayed-response task. J. Neurophysiol. 80, 2200–2205.

Sawaguchi, T. and Matsumura, M. (1985) Laminar distribution of neurons sensitive to acetylcholine, noradrenaline and dopamine in the dorsolateral prefrontal cortex of the monkey. Neurosci. Res. 2, 255–273.

Schmidt, R.H. and Bhatnagar, R.K. (1979) Assessment of the effects of neonatal subcutaneous 6-hydroxydopamine on noradrenergic and dopaminergic innervation of the cerebral cortex. Brain Res. 166, 309–319.

Schroeter, S., Apparsundaram, S., Wiley, R.G., Miner, L.A.H., Sesack, S.R. and Blakely, R.D. (2000) Immunolocalization of the cocaine- and antidepressant-sensitive 1-norepinephrine transporter. J. Comp. Neurol. 420, 211–232.

Séguéla, P., Watkins, K.C. and Descarries, L. (1988) Ultrastructural features of dopamine axon terminals in the anteromedial and the suprarhinal cortex of adult rat. Brain Res. 442, 11–22.

Séguéla, P., Watkins, K.C. and Descarries, L. (1989) Ultrastructural relationships of serotonin axon terminals in the cerebral cortex of the adult rat. J. Comp. Neurol. 289, 129–142.

Séguéla, P., Watkins, K.C., Geffard, M. and Descarries, L. (1990) Noradrenaline axon terminals in adult rat neocortex: an immunocytochemical analysis in serial thin sections. Neuroscience 35, 249–264.

Sesack, S.R., Snyder, C.L. and Lewis, D.A. (1995) Axon terminals immunolabeled for dopamine or tyrosine hydroxylase synapse on GABA-immunoreactive dendrites in rat and monkey cortex. J. Comp. Neurol. 363, 264–280.

Sesack, S.R., Miner, L.A.H. and Omelchenko, N. (2006) Pre-embedding immunoelectron microscopy: applications for studies of the nervous system. In: L. Zaborszky, F.G. Wouterlood and J.L. Lanciego (eds.), Neuroanatomical Tract-Tracing 3: Molecules, Neurons, Systems. Springer, New York, pp. 6–71.

Sesack, S.R., Hawrylak, V.A., Matus, C., Guido, M.A. and Levey, A.I. (1998) Dopamine axon varicosities in the prelimbic division of the rat prefrontal cortex exhibit sparse immunoreactivity for the dopamine transporter. J. Neurosci. 18, 2697–2708.

Shin, L.M., Whalen, P.J., Pitman, R.K., Bush, G., Macklin, M.l., Lasko, N.B., Orr, S.P., McInerney, S.C. and Rauch, S.L. (2001) An fMRI study of anterior cingulate function in posttraumatic stress disorder. Biol. Psychiatry 50, 932–942.

Shores, M.M., Szot, P. and Veith, R.C. (1994) Desipramine-induced increase in norepinephrine transporter mRNA is not mediated via alpha2 receptors. Molec. Brain Res. 27, 337–341.

Simpson, K.L., Waterhouse, B.D. and Lin, R.C. (2006) Characterization of neurochemically specific projections from the locus coeruleus with respect to somatosensory-related barrels. Anat. Rec. Part A 288A, 166–173.

Simpson, K.L., Altman, D.W., Wang, L., Kirifides, M.L., Lin, R.C. and Waterhouse, B.D. (1997) Lateralization and functional organization of the locus coeruleus projection to the trigeminal somatosensory pathway in rat. J. Comp. Neurol. 385, 135–147.

Simson, P.E. and Weiss, J.M. (1988) Altered activity of the locus coeruleus in an animal model of depression. Neuropsychopharmacology 1, 287–294.

Soares, J.C. and Mann, J.J. (1997) The functional neuroanatomy of mood disorders. J. Psychiatric Res. 31, 393–432.

Solanto, M.V. (1998) Neuropsychopharmacological mechanisms of stimulant drug action in attention-deficit hyperactivity disorder: a review and integration. Behav. Brain Res. 94, 127–152.

Southwick, S.M., Bremner, J.D., Rasmusson, A., Morgan, D.A., Arnsten, A.F.T. and Charney, D.S. (1999) Role of norepinephrine in the pathophysiology and treatment of posttraumatic stress disorder. Biol. Psychiatry 46, 1192–1204.

Southwick, S.M., Krystal, J.H., Morgan, C.A., Johnson, D.R., Nagy, L.M., Nicolaou, A., Heninger, G.R. and Charney, D. (1993) Abnormal noradrenergic function in posttraumatic stress disorder. Arch. Gen. Psychiatry 50, 266–274.

Spencer, T., Biederman, J., Coffey, B., Geller, D., Crawford, M., Bearman, S.K., Tarazi, B. and Faraone, S.V. (2002) A double-blind comparison of desipramine and placebo in children and adolescents with chronic Tic disorder and comorbid attention-deficit/hyperactivity disorder. Arch. Gen. Psychiatry 59, 649–656.

Stanford, S.C. (1995) Central noradrenergic neurons and stress. Pharmacol Therapeutics 68, 297–242.

Steindler, D.A. (1981) Locus coeruleus neurons have axons that branch to the forebrain and cerebellum. Brain Res. 223, 367–373.

Stone, E.A., Freedman, L.S. and Morgano, L.E. (1978) Brain and adrenal tyrosine hydroxylase activity after chronic footshock stress. Pharmacol. Biochem. Behav. 9, 551–553.

Sung, U., Apparsundaram, S., Galli, A., Kahlig, K.M., Savchenko, V., Schroeter, S., Quick, M.W. and Blakely, R.D. (2003) A regulated interaction of syntaxin 1A with the antidepressant-sensitive norepinephrine transporter establishes catecholamine clearance capacity. J. Neurosci. 23, 1697–1709.

Svensson, T.H. and Udin, T. (1978) Feedback inhibition of brain noradrenaline neurons by tricyclic antidepressants: alpha-receptor mediation. Science 202, 1089–1091.

Swanson, L.W. and Hartman, B.K. (1975) The central adrenergic system. An immunofluorescence study of the location of cell bodies and their efferent connections in the rat utilizing dopamine-B-hydroxylase as a marker. J. Comp. Neurol. 163, 467–487.

Szabo, S.T. and Blier, P. (2001) Effect of the selective noradrenergic reuptake inhibitor reboxetine on the firing activity of noradrenaline and serotonin neurons. Eur. J. Neurosci. 13, 2077–2087.

Szot, P., Ashliegh, E.A., Kohen, R., Petrie, E., Dorsa, D.M. and Veith, R. (1993) Norepinephrine transporter mRNA is elevated in the locus coeruleus following short- and long-term desipramine treatment. Brain Res. 618, 308–312.

Tanda, G., Pontier, F.E., Frau, R. and DiChiara, G. (1997) Contribution of blockade of the noradrenaline carrier to the increase of extracellular dopamine in the rat prefrontal cortex by amphetamine and cocaine. Eur. J. Neurosci. 9, 2077–2085.

Tanda, G.L., Carboni, E., Frau, R. and DiChiara, G. (1994) Increase of extracellular dopamine in the prefrontal cortex: portrait of drugs with antidepressant potential? Psychopharmacology 115, 285–288.

Tanila, H., Rama, P. and Carlson, S. (1996) The effects of prefrontal intracortical microinjections of an alpha-2 agonist, alpha-2 antagonist and lidocaine on the delayed alternation performance of aged rats. Brain Res. Bull. 40, 117–119.

Tejani-Butt, S.M., Pare, W.P. and Yang, J. (1994) Effect of repeated novel stressors on depressive behavior and brain norepinephrine receptor system in Sprague-Dawley and Wistar Kyoto (WKY) rats. Brain Res. 649, 27–35.

Tillet, Y. and Kitahama, K. (1998) Distribution of central catecholaminergic neurons: a comparison between ungulates, humans and other species. Histol. Histopathol. 13, 1163–1177.

Ungerstedt, U. (1971) Stereotaxic mapping of the monoamine pathways in the rat brain. Acta Physiol. Scand. 367 [Suppl.], 1–48.

Van Dengen, P. (1981) The central norepinephrine transmission and the locus coeruleus: a review of the data. Prog. Neurobiol. 16, 117–143.

Venkatesan, C., Song, X.Z., Go, C.G., Kurose, H. and Aoki, C. (1996) Cellular and subcellular distribution of a 2A -adrenergic receptors in the visual cortex of neonatal and adult rats. J. Comp. Neurol. 365, 79–95.

Verhofstad, A.A., Hokfelt, T., Goldstein, M., Steinbusch, H.W. and Joosten, H.W. (1979) Appearance of tyrosine hydroxylase, aromatic amino-acid decarboxylase, dopamine beta-hydroxylase and phenylethanolamine N-methyltransferase during the ontogenesis of the adrenal medulla: an immunohistochemical study in the rat. Cell Tiss. Res. 200, 1–13.

Vos, P., Kaufmann, D., Hand, P.J. and Wolfe, B.B. (1990) Beta 2-adrenergic receptors are colocalized and coregulated with "whisker barrels" in rat somatosensory cortex. Proc. Natl. Acad. Sci. 87, 5114–5118.

Wamsley, J.K. (1984) Autoradiographic localization of cortical biogenic amine receptors. In: L. Descarries, T.A. Reader and H.H. Jasper (eds.), Monoamine Innervation of Cerebral Cortex. Liss, New York, pp. 153–174.

Wamsley, J.K., Palacios, J.M., Young, W.S. and Kuhar, M.J. (1981) Autoradiographic determination of neurotransmitter receptor distributions in the cerebral and cerebellar cortices. J. Histochem. Cytochem. 29, 125–135.

Wang, M., Tang, Z.X. and Li, B.M. (2004) Enhanced visuomotor associative learning following stimulation of alpha 2A-adrenoceptors in the ventral prefrontal cortex in monkeys. Brain Res. 1024, 176–182.

Wang, P., Kitayama, I. and Nomura, J. (1998) Tyrosine hydroxylase gene expression in the locus coeruleus of depression-model rats and rats exposed to short- and long-term forced walking stress. Life Sci. 62, 2083–8092.

Watanabe, Y., McKittrick, C.R., Blanchard, D.C., Blanchard, R.J., McEwen, B.S. and Sakai, R.R. (1995) Effects of chronic social stress on tyrosine hydroxylase mRNA and protein levels. Molec. Brain Res. 32, 176–180.

Waterhouse, B., Lin, C., Burne, R. and Woodward, D. (1983) The distribution of neocortical projection neurons in the locus coeruleus. J. Comp. Neurol. 217, 418–431.

Waterhouse, B.D. and Woodward, D.J. (1980) Interaction of norepinephrine with cerebrocortical activity evoked by stimulation of somatosensory afferent pathways in the rat. Exp. Neurol. 67, 11–34.

Waterhouse, B.D., Moises, H.C. and Woodward, D.J. (1980) Noradrenergic modulation of somatosensory cortical neuronal responses to iontophoretically applied putative neurotransmitters. Exp. Neurol. 69, 30–49.

Waterhouse, B.D., Moises, H.C. and Woodward, D.J. (1981) Alpha receptor mediated facilitation of somatosensory cortical neuronal responses to excitatory synaptic inputs and iontophoretically applied acetylcholine. Neuropharmacology 20, 907–920.

Waterhouse, B.D., Moises, H.C. and Woodward, D.J. (1982) Norepinephrine enhancement of inhibitory synaptic mechanisms in cerebellum and cerebral cortex: mediation by beta adrenergic receptors. J. Pharm. Exp. Ther. 221, 495–506.

Waterhouse, B.D., Moises, H.C. and Woodward, D.J. (1998) Phasic activation of the locus coeruleus enhances responses of primary sensory cortical neurons to peripheral receptive field stimulation. Brain Res. 790, 33–44.

Waterhouse, B.D., Azizi, S.A., Burne, R.A. and Woodward, D.J. (1988) New evidence for a gating action of norepinephrine in central neuronal circuits of mammalian brain. Brain Res. Bull. 21, 425–432.

Waterhouse, B.D., Azizi, S.A., Burne, R.A. and Woodward, D.J. (1990) Modulation of rat cortical area 17 neuronal responses to moving visual stimuli during norepinephrine and serotonin microiontophoresis. Brain Res. 514, 276–292.

Weinshenker, D., White, S.S., Javors, M.A., Palmiter, R.D. and Szot, P. (2002) Regulation of the norepinephrine transporter abundance by catecholamines and desipramine in vivo. Brain Res. 946, 239–246.

Yamamoto, B.K. and Novotney, S. (1998) Regulation of extracellular dopamine by the norepinephrine transporter. J. Neurochem. 71, 274–280.

Young, W.S. and Kuhar, M.J. (1980) Noradrenergic a1 and a2 receptors: light microscopic autoradiographic localization. Proc. Natl. Acad. Sci. 77, 1696–1700.

Zametkin, A.J., Nordahl, T.E., Gross, M., King, A.C., Semple, W.E., Rumsey, J., Hamburger, S. and Cohen, R.M. (1990) Cerebral glucose metabolism in adults with hyperactivity of childhood onset. N. Engl. J. Med. 323, 1361–1366.

Zhu, M.-Y., Kim, C.-H., Hwang, D.-Y., Baldessarini, R. and Kim, K.-S. (2002) Effects of desipramine treatment on norepinephrine transporter gene expression in the cultured SK-N-BE(2)M17 cells and rat brain tissue. J. Neurochem. 82, 146–153.

Zhu, M.-Y., Klimek, V., Dilley, G.E., Haycock, J.W., Stockmeier, C., Overholser, J.C., Meltzer, H.Y. and Ordway, G.A. (1999) Elevated levels of tyrosine hydroxylase in the locus coeruleus in major depression. Biol. Psychiatry 46, 1275–1286.

Zigmond, M.J., Finlay, J.M. and Sved, A.F. (1995) Neurochemical studies of central noradrenergic responses to acute and chronic stress. In: M.J. Friedman, D.S. Charney and A.Y. Deutsch (eds.), Neurobiological and Clinical Consequences of Stress: From Normal Adaptations to PTSD. Lippencott-Raven, Philadelphia, pp. 45–60.

Zigmond, R.E., Schon, F. and Iversen, L.L. (1974) Increased tyrosine hydroxylase activity in the locus coeruleus of rat brain after reserpine treatment and cold stress. Brain Res. 70, 547–552.

Chapter 4

Serotonin Modulation of Cortical Activity

Pau Celada and Francesc Artigas
Department of Neurochemistry and Neuropharmacology, Institut d' Investigacions
  Biomèdiques de Barcelona (CSIC), IDIBAPS, 08036 Barcelona, Spain

4.1 Introduction

Serotonin (5-hydroxytryptamine, 5-HT) is one of the phylogenetically older molecules used in cellular communications. It is present in the central nervous system
(CNS) of vertebrate and invertebrate animals, and plays the role of neurotransmitter/
neuromodulator. It also operates as a developmental signal in the CNS and regulates a variety of physiological functions in the periphery, such as intestinal motility,
platelet aggregation, and vasoconstriction.

  The central 5-HT system is involved in a large number of functions resulting
from its widespread innervation of the entire neuraxis. The axons of serotonergic
neurons of the midbrain raphe nuclei (dorsal and median; DR and MnR, respectively) reach almost every brain region. Action potentials traveling along these axons
release 5-HT, which in turn impacts on several pre- and postsynaptic 5-HT receptors,
coupled to different signal transduction mechanisms. Currently, 14 different 5-HT
receptor subtypes have been identified. They are from seven different families: 5-HT_1
(5-HT_{1A}, 5-HT_{1B}, 5-HT_{1D}, 5-HT_{1E}, 5-HT_{1F}), 5-HT_2 (5-HT_{2A}, 5-HT_{2B}, 5-HT_{2C}), 5-HT_3,
5-HT_4, 5-HT_5 (5-HT_{5A}, 5-HT_{5B}), $5\text{-HT}_{6,}$ and 5-HT_7. With the exception of the 5-HT_3
receptor, all 5-HT receptors belong to the superfamily of G-protein-coupled recep-
tors. The 5-HT_3 receptor is a pentameric ligand-gated ion channel composed of several subunits (up to five different ones have been identified).

  Given such a widespread innervation of the brain and the richness of signals
evoked by 5-HT, it is not surprising that the 5-HT system is the target of many drugs
used to treat brain diseases. For instance, most antidepressant drugs increase the
extracellular level of 5-HT by blocking the 5-HT transporter, leading to an enhancement of the basal 5-HT tone to pre- and postsynaptic 5-HT receptors. This effect is
believed to mediate the therapeutic action of currently available antidepressant drugs.

On the other hand, drugs of abuse such as cocaine, amphetamine, or MDMA (ecstasy) target monoaminergic transporters, including the 5-HT one. Furthermore hallucinogens like LSD, DOI, DOB, or DOM are $5-HT_2$ receptor agonists whereas atypical antipsychotic drugs act preferential as antagonists of these receptors.

Among the various 5-HT receptors, the $5-HT_1$ family has probably received the major attention because of its high density expression in several limbic ($5-HT_{1A}$) and motor ($5-HT_{1B}$) brain regions. They are located at both pre- and postsynaptic levels and regulate the overall ($5-HT_{1A}$) as well as the local ($5-HT_{1B}$) activity of the 5-HT system. $5-HT_{1B}$ receptors are also involved in the regulation of different transmitters release including dopamine, glutamate, GABA, and acetylcholine. Moreover, $5-HT_{1A}$ receptors are highly expressed in the prefrontal cortex (PFC) suggesting that 5-HT is involved in the control of mood and emotions as well as cognitive processes. In this section, we will summarize and discuss recent findings regarding the role of 5-HT in the modulation of cortical activity. We will also cover some basic aspects about the anatomy, physiology, neurochemistry, and neuropharmacology of the 5-HT system relevant to the scope of the present chapter (extensively reviewed by Jacobs and Azmitia, 1992; Barnes and Sharp, 1999; Adell et al., 2002).

4.2 The Cortical 5-HT System: Receptor Localization

There is growing evidence indicating that the 5-HT pathways originating from the dorsal and median raphe nuclei (DR and MnR, respectively) are critically involved in cortical functions. 5-HT appears to be critical during the development of the somatosensory cortex and formation of the barrel cortex. In the adult brain, the axons of 5-HT neurons innervate a large number of cortical areas, including the entorhinal and cingulate cortices, which contain a moderate-high density of 5-HT receptors. However, the frontal lobe is the richest cortical area of 5-HT terminals and 5-HT receptors.

Unlike dopamine, whose function has been extensively studied over the last decade, the role of 5-HT in the PFC remains largely unknown. Indeed, the widespread localization of 5-HT receptors (particularly of the $5-HT_{1A}$, $5-HT_{2A}$, and $5-HT_{2C}$ subtypes) and the high density of 5-HT axons (greater than in any other cortical area) in this cortical region suggest an important role of 5-HT in cognitive and emotional processes dependent on PFC functions. The scarce information available to date supports this view. In fact, selective depletion of 5-HT in the orbitofrontal subdivision of the PFC impairs learning flexibility in nonhuman primates (Clarke et al., 2004). As previously observed for D1 receptors (Williams and Goldman-Rakic, 1995), the blockade of $5-HT_{2A}$ receptors in the dorsolateral PFC also avoids the increase in neuronal activity during working memory performance (Williams et al., 2002). More importantly, a recent study indicates a strong association between allelic variants of $5-HT_{2A}$ receptor with memory capacity in humans (De Quervain et al., 2003). Hallucinogens like LSD or DOI are $5-HT_{2A}$ receptor agonists suggesting a role of 5-HT in the processing of external (sensory) and internal information through the activation of $5-HT_{2A}$ receptors. On the other hand, $5-HT_{1A}$ agonists display anxiolytic/antidepressant effects in animal models (De Vry, 1995) whereas $5-HT_{1A}$ receptor antagonists reverse drug-induced cognitive deficits (Harder and Ridley, 2000; Mello e Souza et al., 2001; Misane and Ögren, 2003).

Understanding the precise anatomical and cellular distribution of 5-HT receptors is one of the key basic information relevant for the interpretation of physiological and behavioral data concerning the role of 5-HT in the cortex. Early studies using receptor autoradiography and in situ hybridization enabled to identify the presence of various 5-HT receptors in cortical areas, notably the 5-HT_{1A}, 5-HT_{2A}, and 5-HT_{2C} subtypes (Pazos and Palacios, 1985; Pazos et al., 1985; Pompeiano et al., 1992, 1994). Further studies had identified the presence of other receptor subtypes, yet in lower density as compared to 5-HT_1 and 5-HT_2 families.

5-HT_{1A} receptors are particularly enriched in the medial PFC (mPFC), entorhinal cortex and to a lesser extent, cingulate, and retrosplenial cortices. They are also densely expressed in the hippocampus, septum, and the raphe nuclei. In the latter location the receptor is almost exclusively expressed by 5-HT neurons and functions as autoreceptors. 5-HT_{1A} receptors are localized in the plasma membrane of perikarya and dendrites of 5-HT neurons (Riad et al., 2000).

PET scan studies using a radiolabeled selective antagonist ($[^{11}\text{C}]$-WAY-100635) have shown a totally similar distribution in human brain, with an enrichment of the signal in the temporal and frontal lobes, cingulate cortex and the MnR (Martinez et al., 2001). Interestingly, and similar to what has been observed in rodents, there is a marked rostro-caudal gradient of cortical of 5-HT_{1A} receptor levels, with the largest expression in the PFC. Likewise, nonhuman and human primates brains show a large abundance of 5-HT_{2A} receptors, with an enrichment in the frontal regions (Pompeiano et al., 1994; Burnet et al., 1995; López-Giménez Vilaró et al., 1998; Hall et al., 2000; Amargós-Bosch et al., 2004). Lower levels of 5-HT_{2A} were found in the ventro-caudal part of CA3, medial mammillary nucleus, striatum (dorsal and ventral) and several brainstem nuclei (Pompeiano et al., 1994; Burnet et al., 1995; López-Giménez et al., 1998). As for 5-HT_{1A} receptors, there is a good agreement between the autoradiographic and in situ hybridization signals, which indicates that the receptor is expressed mainly in the somatodendritic region. Similar anatomical distributions have been reported in human brain using the selective antagonist ligand M100907 in vivo (PET scan) and in vitro (autoradiography) (Hall et al., 2000).

5-HT_{1A} and 5-HT_{2A} receptors are present in a high proportion of cells in some cortical regions. Double in situ hybridization studies have shown that around 50% of the pyramidal neurons (labeled with vGluT1 mRNA) and 20–30% of the GABAergic interneurons (labeled with GAD65/67 mRNA) express 5-HT_{1A} and/or 5-HT_{2A} receptor mRNAs in various areas of the PFC (Santana et al., 2004) (Table 4.1).

Figure 4.1 shows the localization of 5-HT_{1A} and 5-HT_{2A} receptors in the PFC of the rat. Interestingly, 5-HT_{1A} and 5-HT_{2A} receptor transcripts are heavily coexpressed in rat and mouse PFC. Approximately 80% of the cells expressing 5-HT_{1A} receptor mRNA also express the 5-HT_{2A} receptor mRNA in all PFC areas examined, except in layer VI and the lower part of the infralimbic area, where the density of cells expressing 5-HT_{2A} receptors is lower (Amargós-Bosch et al., 2004).

The abundant coexpression raises questions about the physiological role of the simultaneous occurrence of inhibitory (5-HT_{1A}) and excitatory (5-HT_{2A}) receptors responding to 5-HT in the same cortical neurons. Various hypothesis have been examined (Amargós-Bosch et al., 2004; Puig et al., 2005), but perhaps one of the most convincing explanations is the putative localization of both receptors in different cellular compartments. Immunohistochemical studies by several groups consistently show a predominant location of 5-HT_{2A} receptors in the apical dendrites (and to a lower extent, cell bodies) of cortical pyramidal neurons (Jakab and Goldman-Rakic, 1998, 2000; Jansson et al., 2001; Martín-Ruiz et al., 2001). However, the subcellular location of 5-HT$_{1A}$ receptors in cortical neurons remains controversial due to the use of different antibodies.

Table 4.1 Proportion of Pyramidal and Local GABAergic Neurons that Express mRNAs Encoding 5-HT$_{1A}$ and 5-HT$_{2A}$ Receptors. Data are means of three rats and represent the percentage of counted cells expressing the mRNAs of each 5-HT receptor in pyramidal (vGluT1 mRNA-positive) and GABAergic (GAD mRNA-positive) cells. Cortical areas designated according to Paxinos and Watson (1998) and Swanson (1998). MOs, secondary motor area; ACAd, dorsal anterior cingulate area; PrL, prelimbic area; ILA, infralimbic area; PIR, piriform cortex; TT, tenia tecta. Layer VIa denotes deep areas of the sensorimotor cortex at prefrontal level (Swanson, 1998). [a]The data of the ILA correspond to its more ventral part, which shows a remarkable low level of 5-HT$_{2A}$ receptor, whereas cell counts from its dorsal part are more similar to those of PrL. $*P < 0.05$ versus PrL, TT and PIR; $**P < 0.05$ versus the rest of areas, except layer VIa ($P = 0.9$); $***P < 0.05$ versus the rest of areas; $^{+}P < 0.05$ versus the rest of areas except ILA; $^{++}P < 0.05$ versus ACAd and PrL (Tukey test post-ANOVA). Data from Santana et al. (2004).

	Pyramidal neurons		GABAergic neurons	
	5-HT$_{1A}$ mRNA	5-HT$_{2A}$ mRNA	5-HT$_{1A}$ mRNA	5-HT$_{2A}$ mRNA
MOs	54 ± 4	60 ± 2	28 ± 6	28 ± 10
ACAd	54 ± 3	66 ± 5	22 ± 4	32 ± 2
PrL	61 ± 2	51 ± 3	20 ± 1	34 ± 1
ILA[a]	40 ± 4*	12 ± 1**	22 ± 4	22 ± 3
TT	63 ± 6	81 ± 3***	24 ± 1	24 ± 2
PIR	60 ± 2	50 ± 3	21 ± 6	24 ± 2
Layer VI	54 ± 3	26 ± 3^{+}	23 ± 4	11 ± 3^{++}

A homogenous labeling of cell bodies and dendrites was initially reported (Kia et al., 1996). However, recent immunohistochemical studies performed in rodent and primates (human and nonhuman primates) brain tissues (Azmitia et al., 1996) show an exclusive labeling of 5-HT$_{1A}$ the axon hillock of pyramidal neurons (De Felipe et al., 2001; Czyrak et al., 2003; Cruz et al., 2004). This location suggests that 5-HT would be able to establish axo-axonic contacts with pyramidal neurons similar to that by chandelier interneurons, which in turn would impact the generation of action potentials. Therefore, 5-HT axons reaching apical dendrites would modulate glutamatergic inputs onto pyramidal cells (Aghajanian and Marek, 1997; Puig et al., 2003) whereas those reaching to the axonal hillocks would control the probability of generation of nerve impulses through activation of 5-HT$_{1A}$ receptors. 5-HT$_{1B}$ and 5-HT$_{1D}$ receptors show a widespread brain distribution, with a relative low abundance in the cortex.

5-HT$_{1B}$ receptors are highly expressed in the basal ganglia and hippocampal formation, particularly the subiculum (Pazos and Palacios, 1985; Offord et al., 1988). They are negatively coupled to adenylate cyclase and activation of 5-HT$_{1B}$ by selective

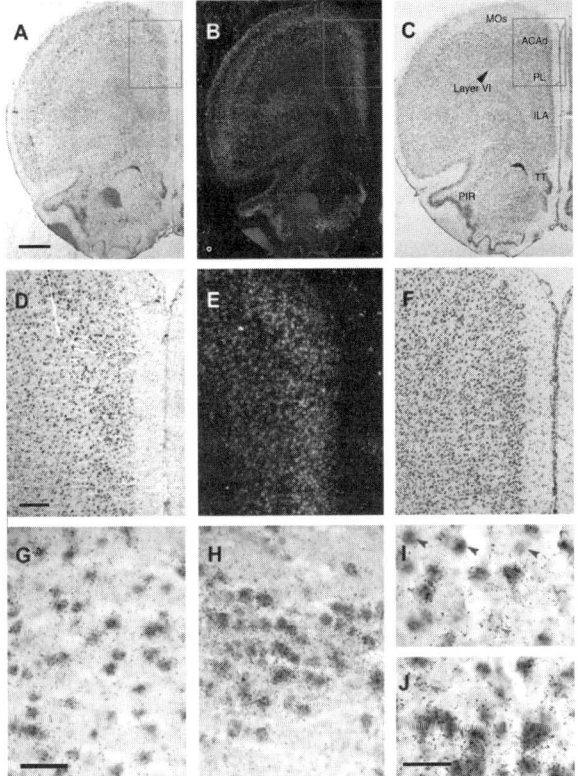

Figure 4.1 Expression of (A) 5-HT$_{1A}$ (dig-labeled oligonucleotides) and (B) 5-HT$_{2A}$ (dark field; ^{33}P-labeled oligonucleotides) receptors mRNA in the rat PFC. (C) An adjacent Nissl-stained section. Note the abundant presence of cells expressing both receptors in layers II–V, as well as in piriform cortex (PIR) and taenia tecta (TT). (D–F) High magnification of the cortical regions highlighted in (A–C). (D) and (E) show the presence of a large number of cells containing 5-HT$_{1A}$ and 5-HT$_{2A}$ receptor transcripts in cingulate (ACAd) and prelimbic (PL) cortex. Cells in deep layers (VI) express preferentially 5-HT$_{1A}$ receptor mRNA. (G–J) Coronal sections showing the expression of 5-HT$_{1A}$ and 5-HT$_{2A}$ receptor mRNAs in individual neurons. Occasional cell profiles containing only 5-HT$_{1A}$ (blue arrowheads) or 5-HT$_{2A}$ receptor mRNAs (red arrowheads) were seen in the infralimbic cortex (G and I), piriform cortex (H) and taenia tecta (J). Bar size are 1 mm in (A–C), 250 μm in (D–F), 50 μm in (G and H) and 30 μm in (I and J) [reproduced with permission from Amargós-Bosch et al. (2004)]. (See color insert.)

agonists decreases the forskolin-stimulated adenylate cyclase levels (for review see Sari, 2004).

The comparison of autoradiographic, in situ hybridization and immunohisto-chemical studies data revealed that 5-HT$_{1B}$ receptors are located at both pre- (i.e., on

5-HT axons) and postsynaptically to 5-HT neurons (Riad et al., 2000). Presynaptic 5-HT$_{1B}$ autoreceptors, however, represent a small proportion of the entire population of 5-HT$_{1B}$ receptors in the brain because the lesion of 5-HT neurons does not result in significant reduction of 5-HT$_{1B}$ receptor density (Compan et al., 1998). Despite the low density expression of cortical 5-HT$_{1B}$ receptors, electrophysiological studies have identified 5-HT$_{1B}$ receptor-mediated actions in the cingulate cortex of the rat (Tanaka and North, 1993; see below). The other two members of the 5-HT$_1$ family (5-HT$_{1E}$ and 5-HT$_{1F}$ receptors) are also present in the cortex, particularly in the entorhinal cortex. Unfortunately, the low abundance and the lack of selective pharmacological tools have hampered the study of 5-HT$_{1E}$ and 5-HT$_{1F}$ receptors actions in cortical neurons.

5-HT$_{2B}$ receptors are expressed in a very low density in the brain. Conversely, 5-HT$_{2C}$ receptors (formerly named 5-HT$_{1C}$ receptors) are highly expressed in the choroid plexus (where they were initially identified) and several cortical and subcortical regions in the rodent brain including the PFC, the striatal complex (ventral – nucleus accumbens – and dorsal striatum), hippocampus, amygdala, and the substantia nigra. 5-HT$_{2C}$ receptors are also expressed in the human cortex, yet their abundance relative to other brain areas appears to be lower than in rat brain (Clemett et al., 2000; Pandey et al., 2006). At the cellular level, cortical 5-HT$_{2C}$ receptors were found mainly in pyramidal neurons (Clemett et al., 2000), even though a recent study also indicates that around 50% of the 5-HT$_{2C}$ receptor immunoreactivity is present in GABAergic interneurons (Liu et al., submitted).

5-HT$_3$ receptors are moderately abundant in the neocortex and other telecephalic regions, such as the olfactory cortex, the hippocampus, and the amygdala. Interestingly, most cortical 5-HT$_3$ receptor mRNA is located in GABAergic interneurons, as assessed by in situ hybridization (Morales and Bloom, 1997; Puig et al., 2004). These are calbindin- and calretinin- (but not parvalbumin-) containing GABAergic neurons and are located in superficial cortical layers (I–III) (Morales and Bloom, 1997; Puig et al., 2004) (Fig. 4.2).

5-HT$_4$ receptors are abundant in the olfactory tubercle, some structures of the basal ganglia (caudate putamen, ventral striatum), medial habenula, hippocampal formation, and amygdala. In the neocortex, only low levels of 5-HT$_4$ receptor and its encoding mRNA were observed as assessed by autoradiography and in situ hybridization, respectively (Waeber et al., 1994; Vilaró et al., 1996, 2005). However, more recent studies using double in situ hybridization suggest that a moderate proportion (~15%) of pyramidal neurons in the PFC express the 5-HT$_4$ receptor transcript as revealed by double in situ hybridization (Vilaró et al., unpublished observations).

Among the different 5-HT receptors, the distribution of 5-HT$_5$ receptors in the brain is perhaps the one that remains unclear and often contradictory. Both 5-HT$_{5A}$ and 5-HT$_{5B}$ receptor subtypes were found in rodents whereas only the 5-HT$_{5A}$ was identified in the human brain. A relative high level of 5-HT$_{5A}$ mRNA was observed in the hippocampus, the medial habenula, and the raphe nucleus. Their presence in MnR 5-HT neurons raises the possibility that 5-HT$_{5A}$ receptor may influence the activity of 5-HT neurons, and thus the levels of 5-HT in the target structures. The 5-HT$_{5B}$ receptor mRNA, on the other hand, was found throughout the entire rat brain, with higher levels in the hippocampus, hypothalamus, pons as well as the neocortex (Erlander et al., 1993).

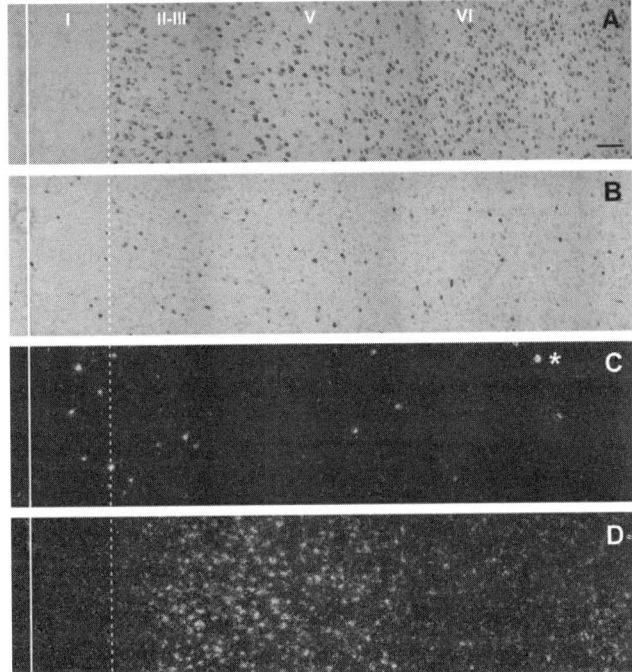

Figure 4.2 Expression of vGluT1 (A), GAD (B), 5-HT₃ (C), and 5 HT₂ₐ (D) mRNAs through layers I–VI the rat prelimbic PFC. The continuous vertical line denotes the location of the midline whereas the dotted line indicates the border between layers I and II. Pyramidal neurons are visualized by the vGluT1 mRNA only in layers II–VI, whereas the GAD mRNA-positive cells are present in all layers. Note the segregation of cells expressing 5-HT₃ (C) and 5-HT₂ₐ receptors (D). 5-HT₃ receptor mRNA was found only in few cells in layers I–III, particularly between layers I and II. However, they represent 40% of GABAergic neurons in layer I. On the other hand, cells in these locations, particularly in layer I, do not express 5-HT₂ₐ receptors. The asterisk denotes an artifact of the emulsion, seen in the dark field. Scale bar = 150 μm [reproduced with permission from Puig et al. (2004)].

The richest brain areas in 5-HT₆ receptor mRNA are the dorsal and ventral striatum as well as adjacent areas including the nucleus accumbens, olfactory tubercles and the islands of Calleja. High levels of 5-HT₆ receptor mRNA was also found in the hypothalamus and the hippocampus, whereas a moderate expression level was observed in the cerebral cortex, the substantia nigra, and the spinal cord (Gerard et al., 1996). A similar receptor protein distribution was observed in immunohisto-chemical studies, even though the PFC shows an immunochemical labeling density greater than that expected from the mRNA data (Gerard et al., 1997).

The expression of 5-HT₇ mRNA as well as the receptor binding site brain seems to be more restricted to the thalamus and the hippocampus in the rodent, even though a moderate level of expression was also found in the septum, the hypothalamus, the centromedial amygdala, the periaquaductal gray, and superior colliculus (Gustafson et al., 1996). A similar distribution of 5-HT$_7$ receptor was recently reported in the human brain (Martin-Cora and Pazos, 2004). Interestingly, the 5-HT$_7$ receptors are also localized in the raphe nuclei from both rodent and human brains. This finding has raised interests of targeting 5-HT$_7$ receptors as a potential new mechanism to control brain's 5-HT levels by regulating the neuronal activity of the ascending 5-HT systems.

In summary, 5-HT released from axons innervating the cerebral cortex can modulate the activity of cortical neurons through several distinct receptors. However, with few exceptions (see above), little is known about the cellular phenotype of the neurons expressing 5-HT receptors, their precise distribution in cortical layers and the proportion of neurons of each type (e.g., pyramidal, stellate, or GABAergic neurons) expressing the receptor subtypes. A better understanding of the actual cellular distribution of 5-HT receptors are necessary and critical steps to identify the mechanisms as well as the functional impact of 5-HT on local cortical circuitries involved in cognitive functions.

4.3 Role of 5-HT Receptors on Cortical Activity

4.3.1 5-HT$_{1A}$ RECEPTORS

The 5-HT$_{1A}$ receptor has been characterized biochemically and electrofisiologically as being coupled to the Gi/o family of heterotrimeric G-proteins. The Gi/o proteins coupled to 5-HT$_{1A}$ receptors are composed of pertussis toxin sensitive αi/αo subunits. This coupling mechanism has been demonstrated in vivo and in vitro in 5-HT neurons from the dorsal raphe (Innis and Aghajanian, 1987). In hippocampal pyramidal neurons, a similar G-protein also couples 5-HT$_{1A}$ and GABA$_B$ receptors to potassium channels (Andrade et al., 1986).

5-HT$_{1A}$ receptors are coupled to potassium and – to a lesser extent – calcium channels. Electrophysiological studies performed in brain slices containing the dorsal raphe had indicated that the 5-HT-mediated inhibition is mediated by an enhancement of the inward rectifying potassium conductance (Aghajanian and Lakoski, 1984; Williams et al., 1988). Exogenous application of 5-HT and 5-HT$_{1A}$ agonists also elicit membrane potential hyperpolarization and decrease the membrane input resistance in dorsal raphe 5-HT neurons in vitro (Aghajanian and Lakoski, 1984; Sprouse and Aghajanian, 1987), leading to an overall reduction in the probability of action potential firing. Similar effects to 5-HT$_{1A}$ receptor activation were observed in pyramidal neurons from the hippocampus (Andrade and Nicoll, 1987).

Early in vivo electrophysiological recordings of cortical neurons showed that iontophoretic application of 5-HT can result in either excitation or inhibition of neuronal firing depending on the cell types (Krnjevic and Phillis, 1963; Roberts and Straughan, 1967), even though the major effect of 5-HT is inhibition (Krnjevic and Phillis, 1963; Reader et al., 1979; Ashby et al., 1994; Zhang et al., 1994). Depending on the receptor subtypes involved, application of 5-HT in the neocortex can result in neuronal membrane potential hyperpolarization and deplarizaiton (fig 4.3) (Davies et al., 1987; Tanaka and North, 1993). A similar biphasic effect was observed in the entorhinal cortex (Ma et al., 2007). Both the inhibitory and excitatory action of 5-HT are often observed in the same cell (Davies et al., 1987; Tanaka and North, 1993; Ma et al., 2007), probably due to activation of 5-HT_{1A} and 5-HT_2 receptors, respectively (Araneda and Andrade, 1991), as result of a high degree of 5-HT_{1A} and 5-HT_{2A} receptors coexpression (Fig. 4.1) (Amargós-Bosch et al., 2004). A similar response to 5-HT_{1A} and 5-HT_{2A} receptors stimulation was also found in neocortical neurons recorded from human's brain slices (Newberry et al., 1999).

Systemic administration of selective 5-HT_{1A} receptor agonists suppresses the firing activity of 5-HT neurons in the DR and MnR (Blier and de Montigny, 1987; Hajós et al., 1995; Casanovas et al., 2000) and hippocampus (Tada et al., 1999). However, these agents appear to have a more complex effect on cortical (PFC) pyramidal neurons, with either a biphasic effect (increase in firing activity followed by decrease at higher doses) or affect different neuronal populations in a distinct manner (most neurons are excited by 5-HT_{1A} agonists) (Borsini et al., 1995; Hajós et al., 1999; Diaz-Mataix et al., 2006).

4.3.2 $5\text{-HT}_{1B/1D}$ RECEPTORS

Anatomical and pharmacological evidence indicate that 5-HT_{1B} receptors are localized in the axons of different cerebral pathways and it exerts an inhibitory action on neurotransmitter release (Sari, 2004). Furthermore, electrophysiological evidence of the role of 5-HT_{1B} receptors in neuronal function is based on the assessment of inhibitory actions on evoked synaptic potentials or currents in target neurons whereas neurochemical studies have examined direct effects on neurotransmitter release.

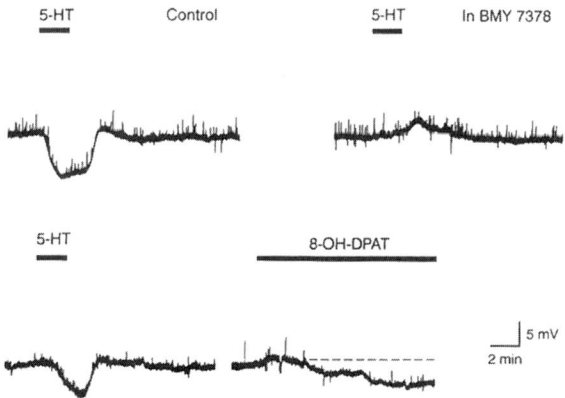

Figure 4.3 Hyperpolarizing effects of 5-HT on medial prefrontal cortex. Bath application of 5-HT (30 μM) elicited a hyperpolarizing response which was blocked by BMY 7378 (3 μM). The lower trace shows that the selective 5-HT_{1A} agonist 8-OH-DPAT (30 μM) induced a membrane hyperpolarization comparable to that elicited by 5-HT (30 μM) [reproduced with permission from Araneda and Andrade (1991)].

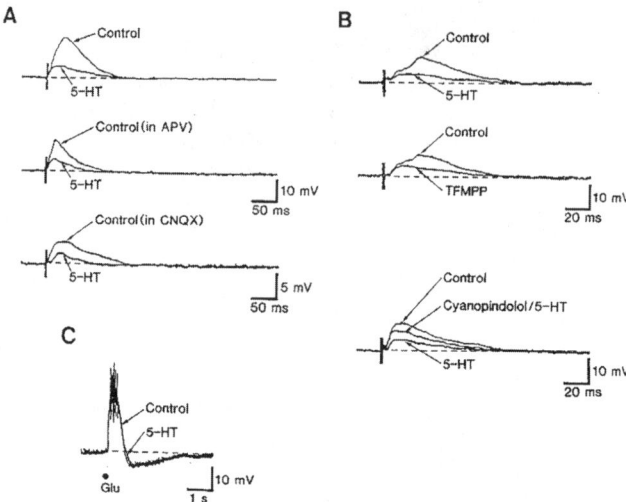

Figure 4.4 5-HT$_{1B}$ receptors inhibit excitatory synaptic transmission in cingulate cortex. (A) White matter stimulation elicited excitatory postsynaptic potentials (EPSP) dependent on NMDA and non-NMDA currents in cingulate cortex pyramidal neurons. 5-HT partially depressed EPSPs amplitude. (B) This effect was mimicked by the nonselective 5-HT$_{1B}$ agonist TFMPP and was antagonized by the 5-HT$_{1B}$ antagonist cyanopindolol. (C) The depressing action of 5-HT appears to be mediated by a presynaptic mechanism since the depolarizing effect of glutamate was unaffected by 5-HT application [reproduced with permission from Tanaka and North (1993)].

The control of glutamate release by 5-HT$_{1B}$ receptors has been described in different brain areas. In slices of cingulate cortex, 5-HT, acting on 5-HT$_{1B}$ receptors, reduced the amplitude of NMDA and non-NMDA components of synaptic potentials recorded intracellularly in layer V pyramidal neurons (Tanaka and North, 1993) (Fig. 4.4). It has been also reported that 5-HT$_{1B}$ receptors mediate the 5-HT suppression of evoked fast excitatory postsynaptic current (evEPSC) in layer V pyramidal neurons elicited nearby electrical stimulation of cortical afferents (Lambe and Aghajanian, 2004).

4.3.3 5-HT$_{2A}$ RECEPTORS

5-HT$_{2A}$ receptors are coupled to phospholipase through Gq proteins. Their activation entrains the production of IP3, diacylgligerol and the mobilization of intracellular Ca^{2+}. Furthermore, it is widely recognized that 5-HT$_{2A}$ is the main receptor through which 5-HT has excitatory actions.

There is a substantial overlap between the localization of 5-HT axon terminals and 5-HT$_2$ receptors in the rat brain (Blue et al., 1988). 5-HT$_{2A}$ receptors are mainly localized to pyramidal neurons and GABAergic interneurons in mPFC (Willins et al., 1997; Jakab and Goldman-Rakic, 1998; Santana et al., 2004). 5-HT$_{2A}$ receptors are highly expressed in large and medium-size parvalbumin- and calbindin-containing interneurons involved in the feed-forward inhibition of pyramidal neurons (Jakab and Goldman-Rakic, 2000). In the rat frontoparietal cortex, 5-HT axons are parallel to the apical dendrites of pyramidal neurons expressing 5-HT$_{2A}$ receptors (Jansson et al., 2001). Additionally, a lower proportion of 5-HT$_{2A}$ receptors were found at the presynaptic sites (Jakab and Goldman-Rakic, 1998; Miner et al., 2003).

Activation of 5-HT$_{2A}$ receptors exerts complex effects on the activity of PFC neurons. Microiontophoretic application of DOI suppressed firing activity of putative pyramidal neurons in anesthetized rats but enhanced the excitatory effect of glutamate at low ejection currents (Ashby et al., 1989a, 1990). In vitro recordings of in PFC slices have revealed that 5-HT$_{2A}$ receptor activation increases spontaneous EPSCs and depolarizes the membrane potential of identified pyramidal neurons (Fig. 4.5A and B) (Araneda and Andrade, 1991; Tanaka and North, 1993; Aghajanian and Marek, 1997, 1999a, b; Zhou and Hablitz, 1999). In addition, 5-HT can elicit 5-HT$_{2A}$ mediated IPSCs through activation of GABA synaptic inputs (Fig. 4.5C and D) (Zhou and Hablitz, 1999), an effect that could be accounted by its effect through activation of 5-HT$_{2A}$ receptors in GABAergic interneurons (Santana et al., 2004).

5-HT increases pyramidal cell excitability seems to be mediated by the inhibition of the afterhyperpolarizating current (IAHP), typically observed after a burst of spikes in response to 5-HT$_2$ receptors activation (Araneda and Andrade, 1991; Andrade, 1998). Accordingly, recent studies conducted in layer V pyramidal neurons of the PFC have identified 5-HT$_{2A}$ as the primary receptor involved in the inhibitory effect of 5-HT on the slow afterhyperpolarizing current (IsAHP) (Villalobos et al., 2005). As AHP is involved in determining neuronal excitability, such an inhibition could contribute to regulate the firing pattern of cortical neurons. Interestingly, the 5-HT$_{2A/2C}$ receptor agonist DOI markedly affects the firing of PFC pyramidal neurons in vivo (Puig et al., 2003). Systemic DOI administration increased, decreased, or left unaffected pyramidal neurons in PFC by a 5-HT$_{2A}$ receptor-dependent mechanism. As observed for 5-HT (Zhou and Hablitz, 1999), inhibitory actions of DOI seems to be dependent on GABA$_A$ receptor tone.

There seems to be a tight link between 5-HT$_{2A}$ receptors and glutamatergic transmission. Hence, the excitatory effects of DOI appear to involve interaction with glutamatergic transmission because DOI increases the excitatory action of glutamate on prefrontal neurons (Ashby et al., 1989a, 1990). Likewise, the 5-HT$_{2A}$ receptor-mediated EPSCs induced by 5-HT were occluded by AMPA and mGluR II receptors activation (Aghajanian and Marek, 1997, 1999a, b). Moreover, the modulation of prefrontal NMDA transmission by 5-HT and DOB appears to involve pre- and post-synaptic 5-HT$_{2A}$ receptors (Arvanov et al., 1999). The observation that the selective mGluRII agonist LY-379268 reversed the excitatory effect of DOI on pyramidal neurons in vivo is consistent with these in vitro observations.

It has been suggested that 5-HT activates 5-HT$_{2A}$ receptors putatively located on thalamocortical terminals in mPFC to release glutamate and evoke EPSCs in pyramidal cells (Aghajanian and Marek, 1997). This interpretation was based on a number of observations, including the fact that this effect was antagonized by AMPA receptor antagonists and mGluR receptor agonists (Aghajanian and Marek, 1997, 1999a, b). Similarly, the increase in mPFC pyramidal cell firing evoked by systemic DOI administration was dependent on glutamate inputs (Puig et al., 2003). However, the lesion of the thalamus (including the dorsomedial and centromedial nuclei which project to mPFC) (Berendse and Groenewegen, 1991; Fuster, 1997) did not abolish the excitatory effects of DOI on mPFC pyramidal neurons (Puig et al., 2003). Likewise, electron microscopy analyses have failed to identify 5-HT$_{2A}$ receptors in excitatory axonal terminals in the PFC (most are located postsynaptically; Miner et al., 2003). Overall, these data have raised doubts about the presynaptic mechanisms responsible for the excitatory actions of 5-HT$_{2A}$ receptors and suggest that postsynaptic 5-HT$_{2A}$ receptors may underlie somehow the facilitatory effect of 5-HT on cortical excitatory synaptic transmission.

5-HT$_{2A}$ receptors seem to exert a more marked depolarizing action in early stages of postnatal development, since this effect diminishes with age. Both 5-HT$_{2A}$ and 5-HT$_7$ receptors appears to underlie the depolarizing effect of 5-HT (see below) (Zhang, 2003; Beique et al., 2004).

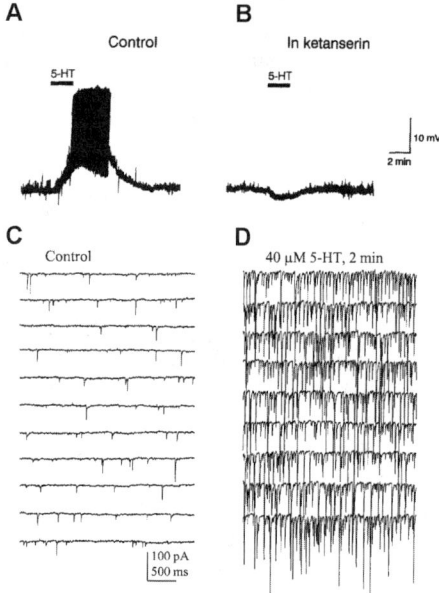

Figure 4.5 (A) Bath application of 5-HT evoked a depolarization and induced firing in a pyramidal neuron of the PFC. (B) Effect of ketanserin (5-HT$_{2A/2C}$ receptor antagonist) on the 5-HT-induced membrane depolarization. Notice that in the presence of ketanserin 5-HT elicited a small hyperpolarization. (C and D) Spontaneous IPSCs recorded in a PFC pyramidal neuron in presence of AP-V and CNQX to block glutamatergic inputs. (D) IPSCs evoked by the bath application of 5-HT in the same recording conditions. Note the large increase in the frequence and amplitude of IPSCs induced by 5-HT [Reproduced with permission from Araneda and Andrade (1991) (A and B) and Zhou and Hablitz (1999) (C and D)].

There is almost no evidence of the role of 5-HT$_{2C}$ receptors in neocortex. In the piriform cortex, it was reported that 5-HT could activate pyramidal neurons via 5-HT$_{2C}$ receptors and GABAergic neurons via 5-HT$_{2A}$ receptors (Sheldon and Aghajanian, 1991). However, the depolarizing action of 5-HT in layer V pyramidal neurons of the mPFC does not seem to depend on 5-HT$_{2C}$ receptor activation since it was not blocked by the selective antagonist SB 242084 (Beique et al., 2004).

4.3.4 5-HT$_3$ RECEPTORS

5-HT can also mediate rapid excitatory postsynaptic responses through activation of 5-HT$_3$ receptors, a ligand-gated ion channel. These receptors may underlie some of the cortical effects of 5-HT since several 5-HT$_3$ receptor antagonists display procognitive actions (Staubli and Xu, 1995). These agents have been also reported to display anxiolytic and antipsychotic activity in animal models (Higgins and Kilpatrick, 1999), and to improve the therapeutic action of antipsychotic drugs in schizophrenia patients (Sirota et al., 2000). Likewise, the atypical antipsychotic drug clozapine is an antagonist of 5-HT$_3$ receptors (Watling et al., 1989; Edwards et al., 1991).

Early microiontophoretic studies showed that activation of 5-HT$_3$ receptors suppress pyramidal cells activity in rat PFC through a direct postsynaptic mechanism (Ashby et al., 1989b, 1991, 1992). However, recent in vitro studies indicate that 5-HT$_3$ receptors may also increase pyramidal neurons IPSCs by facilitating local GABA release from inhibitory interneurons (Zhou and Hablitz, 1999; Férézou et al., 2002; Xiang and Prince, 2003). The latter observation is consistent with the presence of 5-HT$_3$ receptors in GABAergic interneurons from several regions of the rat telencephalon including the PFC (Morales and Bloom, 1997; Puig et al., 2004). Likewise, 5-HT$_3$ receptors are expressed by a subpopulation of calbindin- and calretinin-positive cortical interneurons in the macaque (Jakab and Goldman-Rakic, 2000). Furthermore, the excitatory effect of 5-HT on mPFC GABAergic interneurons firing involves activation of 5-HT$_3$ receptors (see below, Puig et al., 2004).

4.3.5 Other 5-HT RECEPTORS

The effect of 5-HT$_4$–5-HT$_7$ receptors on cortical neuronal activity remains unclear. For instance, it has been suggested that 5-HT$_7$ receptor may play a role during early postnatal development of cortical circuits. Whole cell patch clamp recordings conducted in PFC brain slices have showed a shift in the effect of 5-HT on membrane potential across different stages of postnatal development concurrent with changes in the expression and function of 5-HT$_{1A}$, 5-HT$_{2A}$, and 5-HT$_7$ receptors. Hence, 5-HT in early postnatal days elicits a marked depolarizing effect on pyramidal cells (dependent on 5-HT$_{2A}$ and 5-HT$_7$ receptors), which progressively shifts to hyperpolarization (mediated by 5-HT$_{1A}$ receptors). This change appears to be due to the loss of 5-HT$_7$ receptors and increased function of 5-HT$_{1A}$ receptors (Beique et al., 2004).

4.3.6 IN VIVO ACTIONS OF ENDOGENOUS SEROTONIN ON CORTICAL NEURONS

Despite the wealth of in vitro evidence showing the effects of 5-HT on cortical neuronal activity, the endogenous effect of 5-HT and its receptors in the regulation of cortical inhibitory and excitatory responses in vivo is not fully identified. For instance, electrical stimulation of the medial forebrain bundle inhibits hippocampal pyramidal neurons, an effect that could be reversed by 5-HT$_{1A}$ receptor blockade (Chaput and de Montigny, 1988).

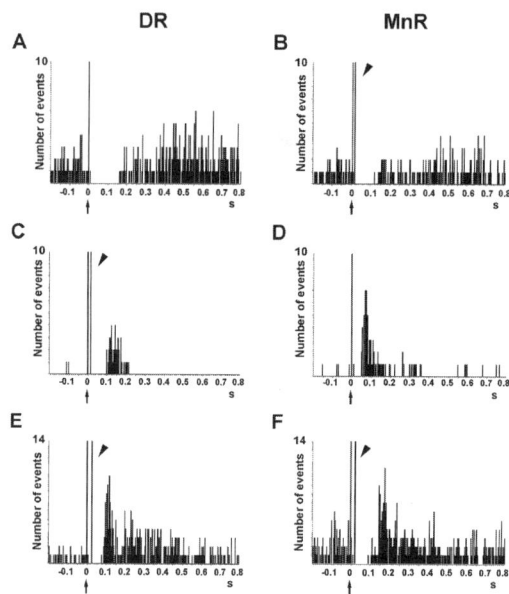

Figure 4.6 Representative examples of the responses evoked in pyramidal neurons of the anterior cingulate and prelimbic areas of the mPFC by the electrical stimulation of the DR and MnR. (A and B) Short-latency long duration inhibitory responses. (C and D) Pure excitatory responses often recorded in pyramidal neurons with a low baseline firing rate. These orthodromic excitations have a similar duration after DR or MnR stimulation but the latency is significantly lower after MnR stimulation, as in the example seen in the figure. (E and F) Examples of orthodromic excitatory response preceded by short-latency inhibitions (biphasic responses) recorded in PFC pyramidal neurons. Note the presence of antidromic spikes in the units recorded (B, C, E, and F), resulting from the presence of a dense connectivity between the PFC and the DR/MnR. The units in (A) and (D) were antidromically activated at currents higher than those used to evoke a pyramidal response. Each peristimulus time histogram consists of 110 triggers (2 min). Bin size = 4 ms. The arrow in the abcissa denotes the stimulus artifact (time 0). The arrowheads indicate antidromic spikes. Reproduced with permission from Puig et al. 2005.

Similarly, electrical stimulation of the DR/MnR nuclei (~1 spike/sec) elicited a powerful inhibitory response in the majority of PFC neurons recorded in vivo, an effect that could be blocked by the selective 5-HT$_{1A}$ antagonist WAY-100635 (Hajós et al., 2003; Amargós-Bosch et al., 2004; Puig et al., 2005). DR/MnR stimulation evoked inhibitory responses in two thirds of the cases in pyramidal neurons of the mPFC, identified by antidromic stimulation from midbrain. The rest of responses were orthodromic excitations, either pure (13%) or preceded by short-latency inhibitions (20%) (Puig et al., 2005). On the other hand, the DR/MnR-evoked excitatory responses in the PFC were blocked by the selective 5-HT$_{2A}$ receptor antagonist

M100907 (Amargós-Bosch et al., 2004; Puig et al., 2005). Despite that ~80% of PFC neurons coexpress 5-HT$_{1A}$ and 5-HT$_{2A}$ receptors (Amargós-Bosch et al., 2004), the in vivo electrophysiological data is in good agreement with the predominant inhibitory effects of 5-HT observed in cortical neurons recorded in vitro (see above). The putative differential location of 5-HT$_{1A}$ and 5-HT$_{2A}$ receptors in different subcellular compartments of pyramidal neurons may perhaps account for the greater proportion of inhibitory responses. Should 5-HT$_{1A}$ receptor be localized on axonal hillocks, endogenous 5-HT would have a profound suppressing effect on action potential generation. Alternatively, a direct effect of 5-HT$_{1A}$ receptor on ionic conductances ($\uparrow$ K$^+$ currents) may favor the inhibitory action over the excitation mediated by 5-HT$_{2A}$ receptor, which results from an indirect and long lasting changes of similar ionic conductances into the opposite direction ($\downarrow$ K$^+$ currents, $\uparrow$ Ca^{2+} currents).

In vivo responses elicited in mPFC pyramidal neurons by raphe stimulation also involve activation of local GABAergic interneurons by 5-HT since they were blocked by the selective 5-HT$_{1A}$ antagonist WAY-100635 as well as by the GABA$_A$ antagonist picrotoxinin (Puig et al., 2005). This GABAergic mediated effect may indeed result from the antidromic activation of local (mPFC) GABA neurons by axon collaterals of pyramidal neurons projecting to midbrain. Likewise, 5-HT may also activate PFC GABAergic interneurons through 5-HT$_{2A}$ receptors and 5-HT$_3$, which in turn would lead to decrease pyramidal cell firing, as observed in vitro (Zhou and Hablitz, 1999). A third possibility seems also likely to explain the inhibition given the very short latency of inhibitory responses (9 ms on average) induced by DR/MnR stimulation. This latency is shorter than the time required for action potentials to travel along 5-HT axons from midbrain to PFC (orthodromic potentials; ~25 ms) or pyramidal neurons axons from PFC to midbrain (antidromic potentials; ~15 ms). Overall, these observations would be consistent with the existence of a monosynaptic GABAergic projection neuron from the MnR to the mPFC, as suggested by some anatomical evidence (Jankowski and Sesack, 2002). This pathway would be analogous to the ascending GABAergic pathway from the ventral tegmental area to the mPFC (Carr and Sesack, 2000), and thus a common pattern of control of PFC function by monoaminergic nuclei.

There is also evidence that 5-HT$_3$ receptors can activate GABA interneurons in the rat PFC in vivo. Physiological stimulation of the raphe nuclei increases firing of local GABAergic neurons in superficial layers (I–III) of the prelimbic and cingulate areas. These responses are clearly distinguishable from the 5-HT$_{2A}$ receptor-mediated excitation because the 5-HT mediate effects have shorter onset latency and duration and they are blocked by the 5-HT$_3$ receptor antagonists ondansetron and tropisetron (Puig et al., 2004).

In summary, several pre- and postsynaptic mechanisms may account for the predominantly PFC inhibition elicited by raphe stimulation, which are not mutually exclusive as discussed above. To the best of our knowledge, in vivo responses to other 5-HT receptors such as 5-HT$_{2C}$, 5-HT$_4$, 5-HT$_6$, or 5-HT$_7$ have not been reported yet so far, perhaps because of their lower levels of expression in the cortex.

4.4 Conclusions

The assessment of in vivo and in vitro actions of 5-HT on cortical neurons has revealed a complex pattern of effects. 5-HT can hyperpolarize pyramidal neurons through the activation of 5-HT$_{1A}$ receptors resulting from the opening of G-protein-coupled inward rectifying K$^+$ channels. This effect would decrease the firing activity of pyramidal neurons. Concurrently, 5-HT can depolarize the same neurons through 5-HT$_{2A}$ receptors and increase their excitability. Both 5-HT$_{1A}$ and 5-HT$_{2A}$ receptors appear to be the main players for the actions of 5-HT in the cerebral cortex.

In situ hybridization studies have also revealed the concurrent presence of 5-HT$_{1A}$ and 5-HT$_{2A}$ receptor mRNAs in a large proportion (~80%) of neurons in the PFC. Although several hypotheses have been postulated, it is yet unclear what factors determine the excitatory or inhibitory response to 5-HT in a given proportion of pyramidal neurons. Similarly, the role of these two receptors in the modulation of GABAergic interneuronal activity remains to be determined. Anatomical studies clearly indicate that a significant proportion of interneurons located in layers II–VI express both 5-HT$_{1A}$ and/or 5-HT$_{2A}$ receptors.

On the contrary, there is a reasonable knowledge on the role of 5-HT$_3$ receptors in the control of GABAergic interneurons function. It seems like there is a segregation of 5-HT$_{1A}$/5-HT$_{2A}$ receptors on one side, expressed mostly in parvalbumin- and calbindin-containing interneurons, whereas the 5-HT$_3$ receptors are expressed in calretinin- (and to a lesser extent) calbindin-containing interneurons. Moreover, interneurons expressing 5-HT$_3$ receptors are localized mostly through layers I–III, which suggests a role in the regulation of inputs reaching the tufts and upper segments of the apical dendrites of pyramidal neurons.

Another 5-HT receptor for which a role (yet still poorly characterized) has been attributed is the 5-HT$_{1B}$ receptor, whose activation by 5-HT can modulate presynaptically GABAergic and glutamatergic inputs onto pyramidal neurons. Figure 4.7 summarize the synaptic interactions within the PFC – raphe circuit highlighting the most important receptors responsible for the 5-HT actions in PFC neurons as well as and their presumed subcellular localization.

Unfortunately, there is a poor knowledge of the actions of 5-HT on other receptor subtypes, some of which are expressed in significant amounts in the neocortex. Evidence indicates that 5-HT$_{2C}$ receptors have an important impact in the piriform cortex, but in the neocortex. The neuronal depolarization induced by 5-HT$_7$ receptor activation disappears few weeks after birth, which suggests a role in development but not in adulthood. Further studies are indeed required to clarify the complex role of 5-HT in cortical functions. Current and new knowledge in this area will help to understand the involvement of 5-HT system in cortical functions, particularly in the PFC, a brain region highly enriched of 5-HT elements and critically involved in cognition and emotional control.

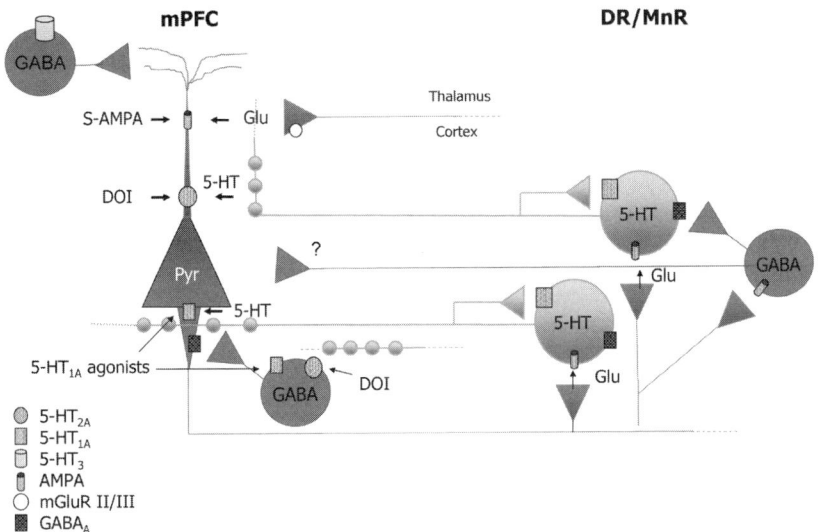

Figure 4.7 Diagram illustrating the relationship between the mPFC and the DR: role of 5-HT$_{1A}$ and 5-HT$_{2A}$ receptors. Pyramidal neurons in the mPFC project densely to the DR/MnR and modulate the activity of serotonergic neurons via direct and indirect mechanisms (Celada et al., 2001). In turn, endogenous 5-HT modulates pyramidal cell activity through the activation of various receptors subtypes, of which 5-HT$_{1A}$ and 5-HT$_{2A}$ receptors play a major role. The latter receptors are particularly enriched in apical dendrites of pyramidal neurons where they can facilitate excitatory inputs. In a lesser extend, 5-HT$_{2A}$ receptors is expressed by large GABA interneurons. Pyramidal 5-HT$_{1A}$ receptors may be localized in the axon hillock, together with GABA$_A$ receptors activated by chandelier axons (Azmitia et al., 1996; De Felipe et al., 2001; Cruz et al., 2004) or in the somatodendritic compartment (Riad et al., 2000). It is possible that 5-HT axons reaching the cortex at different levels may exert distinct effects on pyramidal neurons, depending on the subcellular interactions between certain 5-HT neurons or neuronal clusters within the DR/MnR and 5-HT$_{1A}$- or 5-HT$_{2A}$-receptor-rich compartments (De Felipe et al., 2001; Jansson et al., 2001). Also, 5-HT axons reaching to the upper layers may activate 5-HT$_3$ receptors in GABAergic interneurons, and thus regulate the inputs onto the tufts and most distal segments of apical dendrites of deep-layer pyramidal neurons. 5-HT$_{1B}$ receptors are present on serotonergic axons (not shown) and in axons of other neuronal types (e.g., glutamatergic) where they regulate neurotransmitter release and modulate synaptic activity. The scheme shows also the putative GABAergic projections from DR/MnR to the mPFC suggested by electrophysiological and anatomical studies.

Acknowledgments

Work supported by grants SAF 2004-05525 and FIS (PI 060264). Support from the Spanish Ministry of Health, Instituto de Salud Carlos III, Red de Enfermedades Mentales (REM-TAP Network) is also acknowledged.

References

Adell, A., Celada, P., Abellán, M.T. and Artigas, F. (2002) Origin and functional role of the extracellular serotonin in the midbrain raphe nuclei. Brain Res. Rev. 39, 154–180.

Aghajanian, G.K. and Lakoski, J.M. (1984) Hyperpolarization of serotonergic neurons by serotonin and LSD: studies in brain slices showing increased K^+ conductance. Brain Res. 305, 181–185.

Aghajanian, G.K. and Marek, G.J. (1997) Serotonin induces excitatory postsynaptic potentials in apical dendrites of neocortical pyramidal cells. Neuropharmacology 36, 589–599.

Aghajanian, G.K. and Marek, G.J. (1999a) Serotonin-glutamate interactions: a new target for antipsychotic drugs. Neuropsychopharmacology 21, S122–S133.

Aghajanian, G.K. and Marek, G.J. (1999b) Serotonin, via 5-HT_{2A} receptors, increases EPSCs in layer V pyramidal cells of prefrontal cortex by an asynchronous mode of glutamate release. Brain Res. 825, 161–171.

Amargós-Bosch, M., Bortolozzi, A., Puig, M.V., Serrats, J., Adell, A., Celada, P., Toth, M., Mengod, G. and Artigas, F. (2004) Co-expression and *in vivo* interaction of serotonin1a and serotonin2a receptors in pyramidal neurons of prefrontal cortex. Cereb. Cortex 14, 281–299.

Andrade, R. (1998) Regulation of membrane excitability in the central nervous system by serotonin receptor subtypes. Ann. N.Y. Acad. Sci. 861, 190–203.

Andrade, R. and Nicoll, R.A. (1987) Pharmacologically distinct actions of serotonin on single pyramidal neurons of the rat hippocampus recorded *in vitro*. J. Physiol. 394, 99–124.

Andrade, R., Malenka, R.C. and Nicoll, R.A. (1986) A G-protein couples serotonin and GABAB receptors to the same channel in hippocampus. Science 234, 1261–1265.

Araneda, R. and Andrade, R. (1991) 5-Hydroxytryptamine-2 and 5-hydroxytryptamine-1A receptors mediate opposing responses on membrane excitability in rat association cortex. Neuroscience 40, 399–412.

Arvanov, V.L., Liang, X., Magro, P., Roberts, R. and Wang, R.Y. (1999) A pre- and postsynaptic modulatory action of 5-HT and the 5-HT2A/ 2C receptor agonist DOB on NMDA-evoked responses in the rat medial prefrontal cortex. Eur. J. Neurosci. 11, 2917–2934.

Ashby Jr., C.R., Edwards, E., Harkins, K. and Wang, R.Y. (1989a) Effects of (±)-DOI on medial prefrontal cortical cells: a microiontophoretic study. Brain Res. 498, 393–396.

Ashby Jr., C.R., Edwards, E., Harkins, K. and Wang, R.Y. (1989b) Characterization of 5-hydroxytryptamine3 receptors in the medial prefrontal cortex: a microiontophoretic study. Eur. J. Pharmacol. 173, 193–196.

Ashby, Jr., C.R., Jiang, L.H., Kasser, R.J. and Wang, R.Y. (1990) Electrophysiological characterization of 5-hydroxytryptamine-2 receptors in the rat medial prefrontal cortex. J. Pharmacol. Exp. Ther. 252, 171–178.

Ashby Jr., C.R., Minabe, Y., Edwards, E. and Wang, R.Y. (1991) 5-HT3-like receptors in the rat medial prefrontal cortex: an electrophysiological study. Brain Res. 550, 181–191.

Ashby, Jr., C.R., Edwards, E. and Wang, R.Y. (1992) Action of serotonin in the medial prefrontal cortex: mediation by serotonin3-like receptors. Synapse 10, 7–15.

Ashby, Jr., C.R., Edwards, E. and Wang, R.Y. (1994) Electrophysiological evidence for a functional interaction between 5-HT(1A) and 5-HT(2A) receptors in the rat medial prefrontal cortex: an iontophoretic study. Synapse 17, 173–181.

Azmitia, E.C., Gannon, P.J., Kheck, N.M. and Whitaker-Azmitia, P.M. (1996) Cellular localization of the 5-HT_{1A} receptor in primate brain neurons and glial cells. Neuropsychopharmacology 14, 35–46.

Barnes, N.M. and Sharp, T. (1999) A review of central 5-HT receptors and their function. Neuropharmacology 38, 1083–1152.

Berendse, H.W. and Groenewegen, H.J. (1991) Restricted cortical termination fields of the midline and intralaminar thalamic nuclei in the rat. Neuroscience 42, 73–102.

Beique, J.C., Campbell, B., Perring, P., Hamblin, M.W., Walker, P., Mladenovic, L. and Andrade, R. (2004) Serotonergic regulation of membrane potential in developing rat prefrontal cortex: coordinated expression of 5-hydroxytryptamine (5-HT)1A, 5-HT2A, and 5-HT7 receptors. J. Neurosci. 24, 4807–4817.

Blier, P. and de Montigny, C. (1987) Modification of 5-HT neuron properties by sustained administration of the 5-HT1A agonist gepirone: electrophysiological studies in the rat brain. Synapse 1, 470–480.

Blue, M.E., Yagaloff, K.A., Mamounas, L.A., Hartig, P.R. and Molliver, M.E. (1988) Correspondence between 5-HT$_2$ receptors and serotonergic axons in rat neocortex. Brain Res. 453, 315–328.

Borsini, F., Giraldo, E., Monferini, E., Antonini, G., Parenti, M., Bietti, G. and Donetti, A. (1995) BIMT 17, a 5-HT$_{2A}$ receptor antagonist and 5-HT$_{1A}$ receptor full agonist in rat cerebral cortex. Naunyn-Schmiedeberg's Arch. Pharmacol. 352, 276–282.

Burnet, P.W., Eastwood, S.L., Lacey, K. and Harrison, P.J. (1995) The distribution of 5-HT$_{1A}$ and 5-HT$_{2A}$ receptor mRNA in human brain. Brain Res. 676, 157–168.

Carr, D.B. and Sesack, S.R. (2000) Projections from the rat prefrontal cortex to the ventral tegmental area: target specificity in the synaptic associations with mesoaccumbens and mesocortical neurons. J. Neurosci. 20, 3864–3873.

Casanovas, J.M., Berton, O., Celada, P. and Artigas, F. (2000) In vivo actions of the selective 5-HT$_{1A}$ receptor agonist BAY x 3702 on serotonergic cell firing and release. Naunyn-Schmiedebergs Arch. Pharmacol. 362, 248–254.

Celada, P., Puig, M.V., Casanovas, J.M., Guillazo, G. and Artigas, F. (2001) Control of dorsal raphe serotonergic neurons by the medial prefrontal cortex: involvement of serotonin-1A, GABA(A), and glutamate receptors. J. Neurosci. 21, 9917–9929.

Chaput, Y. and de Montigny, C. (1988) Effects of the 5-hydroxytryptamine receptor antagonist, BMY 7378, on 5-hydroxytryptamine neurotransmission: electrophysiological studies in the rat central nervous system. J. Pharmacol. Exp. Ther. 246, 359–370.

Clarke, H.F., Dalley, J.W., Crofts, H.S., Robbins, T.W. and Roberts, A.C. (2004) Cognitive inflexibility after prefrontal serotonin depletion. Science 304, 878–880.

Clemett, D.A., Punhani, T., Duxon, M.S., Blackburn, T.P. and Fone, K.C. (2000) Immunohistochemical localisation of the 5-HT$_{2C}$ receptor protein in the rat CNS. Neuropharmacology 39, 123–132.

Compan, V., Segu, L., Buhot, M.C. and Daszuta, A. (1998) Selective increases in serotonin 5-HT1B/1D and 5-HT2A/2C binding sites in adult rat basal ganglia following lesions of serotonergic neurons. Brain Res. 793, 103–111.

Cruz, D.A., Eggan, S.M., Azmitia, E.C. and Lewis, D.A. (2004) Serotonin1A receptors at the axon initial segment of prefrontal pyramidal neurons in schizophrenia. Am. J. Psychiatry 161, 739–742.

Czyrak, A., Czepiel, K., Mackowiak, M., Chocyk, A. and Wedzony, K. (2003) Serotonin 5-HT1A receptors might control the output of cortical glutamatergic neurons in rat cingulate cortex. Brain Res. 989, 42–51.

Davies, M.F., Deisz, R.A., Prince, D.A. and Peroutka, S.J. (1987) Two distinct effects of 5-hydroxytryptamine on single cortical neurons. Brain Res. 423, 347–352.

De Felipe, J., Arellano, J.I., Gomez, A., Azmitia, E.C. and Muñoz, A. (2001) Pyramidal cell axons show a local specialization for GABA and 5-HT inputs in monkey and human cerebral cortex. J. Comp. Neurol. 433, 148–155.

De Quervain, D.J., Henke, K., Aerni, A., Coluccia, D., Wollmer, M.A., Hock, C., Nitsch, R.M. and Papassotiropoulos, A. (2003) A functional genetic variation of the 5-HT$_{2A}$ receptor affects human memory. Nat. Neurosci. 6, 1141–1142.

De Vry, J. (1995) 5-HT$_{1A}$ receptor agonists: recent developments and controversial issues. Psychopharmacology 121, 1–26.

Diaz-Mataix, L., Artigas, F. and Celada, P. (2006) Activation of pyramidal cells in rat medial prefrontal cortex projecting to ventral tegmental area by a 5-HT$_{1A}$ receptor agonist. Eur. Neuropsychopharmacol. 16, 288–296.

Edwards, E., Ashby, C.R. and Wang, R.Y. (1991) The effect of typical and atypical antipsychotic drugs on the stimulation of phosphoinositide hydrolysis produced by the 5-HT$_3$ receptor agonist 2-methyl-serotonin. Brain Res. 545, 276–278.

Erlander, M.G., Lovenberg, T.W., Baron, B.M., de Lecea, L., Danielson, P.E., Racke, M., Slone, A.L., Siegel, B.W., Foye, P.E., Cannon, K., Burns, J.E. and Sutcliffe J.G. (1993) Two members of a distinct subfamily of 5-hydroxytryptamine receptors differentially expressed in rat brain. Proc. Natl. Acad. Sci. U.S.A. 90, 3452–3456.

Férézou, I., Cauli, B., Hill, E.L., Rossier, J., Hamel, E. and Lambolez, B. (2002) 5-HT$_3$ receptors mediate serotonergic fast synaptic excitation of neocortical vasoactive intestinal peptide/cholecystokinin interneurons. J. Neurosci. 22, 7389–7397.

Fuster, J.M. (1997) *The prefrontal cortex. Anatomy, physiology and neuropsychology of the frontal lobe.* Philadelphia: Lippincott-Raven.

Gerard, C., el Mestikawy, S., Lebrand, C., Adrien, J., Ruat, M., Traiffort, E., Hamon, M. and Martres, M.P. (1996) Quantitative RT-PCR distribution of serotonin 5-HT$_6$ receptor mRNA in the central nervous system of control or 5,7 dihydroxytryptamine-treated rats. Synapse 23, 164–173.

Gerard, C., Martres, M.P., Lefevre, K., Miquel, M.C., Verge, D., Lanfumey, L., Doucet, E., Hamon, M. and el Mestikawy, S. (1997) Immuno-localization of serotonin 5-HT$_6$ receptor-like material in the rat central nervous system. Brain Res. 746, 207–219.

Gustafson, E.L., Durkin, M.M., Bard, J.A., Zgombick, J. and Branchek, T.A. (1996) A receptor autoradiographic and in situ hybridization analysis of the distribution of the 5-ht7 receptor in rat brain. Br. J. Pharmacol. 117, 657–666.

Hajós, M., Gartside, S.E. and Sharp, T. (1995) Inhibition of median and dorsal raphe neurones following administration of the selective serotonin reuptake inhibitor paroxetine. Naunyn-Schmied. Arch. Pharmacol. 351, 624–629.

Hajós, M., Hajos-Korcsok, E. and Sharp, T. (1999) Role of the medial prefrontal cortex in 5-HT$_{1A}$ receptor-induced inhibition of 5-HT neuronal activity in the rat. Br. J. Pharmacol. 126, 1741–1750.

Hajós, M., Gartside, S.E., Varga, V. and Sharp, T. (2003) *In vivo* inhibition of neuronal activity in the rat ventromedial prefrontal cortex by midbrain-raphe nuclei: role of 5-HT$_{1A}$ receptors. Neuropharmacology 45, 72–81.

Hall, H., Farde, L., Halldin, C., Lundkvist, C., Sedvall, G. (2000) Autoradiographic localization of 5-HT(2A) receptors in the human brain using [3H]M100907 and [11C]M100907. Synapse 38, 421–431.

Harder, J.A. and Ridley, R.M. (2000) The 5-HT$_{1A}$ antagonist WAY 100 635 alleviates cognitive impairments induced by dizocilpine (MK-801) in monkeys. Neuropharmacology 39, 547–552.

Higgins, G.A. and Kilpatrick, G.J. (1999) 5-HT(3) receptor antagonists. Expert Opin. Investig. Drugs 8, 2183–2188.

Innis, R.B. and Aghajanian, G.K. (1987) Pertussis toxin blocks 5-HT$_{1A}$ and GABA$_B$ receptor-mediated inhibition of serotonergic neurons. Eur. J. Pharmacol. 143, 195–204.

Jacobs, B.L. and Azmitia, E.C. (1992) Structure and function of the brain serotonin system. Physiol. Rev. 72, 165–229.

Jakab, R.L. and Goldman-Rakic, P.S. (1998) 5-Hydroxytryptamine(2A) serotonin receptors in the primate cerebral cortex: possible site of action of hallucinogenic and antipsychotic drugs in pyramidal cell apical dendrites. Proc. Natl. Acad. Sci. U.S.A. 95, 735–740.

Jakab, R.L. and Goldman-Rakic, P.S. (2000) Segregation of serotonin 5-HT$_{2A}$ and 5-HT$_3$ receptors in inhibitory circuits of the primate cerebral cortex. J. Comp. Neurol. 417, 337–348.

Jankowski, M.P. and Sesack, S.R. (2002) Electron microscopic analysis of the GABA projection from the dorsal raphe nucleus to the prefrontal cortex in the rat. Soc. Neurosci. Abs. 587.8.

Jansson, A., Tinner, B., Bancila, M., Verge, D., Steinbusch, H.W., Agnati, L.F. and Fuxe, K. (2001) Relationships of 5-hydroxytryptamine immunoreactive terminal-like varicosities to 5-hydroxytryptamine-2A receptor-immunoreactive neuronal processes in the rat forebrain. J. Chem. Neuroanat. 22, 185–203.

Krnjevic, K. and Phillis, J.W. (1963) Iontophoretic studies of neurones in the mammalian cerebral cortex. J. Physiol. 165, 274–304.

Kia, H.K., Brisorgueil, M.J., Hamon, M., Calas, A. and Vergé, D. (1996) Ultrastructural localization of 5-hydroxytryptamine(1A) receptors in the rat brain. J. Neurosci. Res. 46, 697–708.

Lambe, E.K. and Aghajanian, G.K. (2004) *Serotonin (5-HT) supresses electrophysiological effects by hallucinogens in rat prefrontal cortex.* Program No. 394.3. Abstract Viewer/ Intinerary Planner. Washington, DC: Society for Neuroscience, 2004.

Liu, S., Bubar, M.J., Lanfranco, M.F., Hillman, G.R. and Cunningham, K.A. Serotonin (2C) receptor localization in GABA neurons of the rat medial prefrontal cortex: implications for understanding the neurobiology of addiction. Neuroscience. 2007 Apr 27; [Epub, ahead of print]

López-Giménez, J.F., Vilaró, M.T., Palacios, J.M. and Mengod, G. (1998) [3H] MDL100,907 labels 5-HT2A serotonin receptors selectively in primate brain. Neuropharmacology 37, 1147–1158.

Ma, L., Shalinsky, M.H., Alonso, A. and Dickson, C.T. (2007) Effects of serotonin on the intrinsic membrane properties of layer II medial entorhinal cortex neurons. Hippocampus 17, 114–129.

Martín-Ruiz, R., Puig, M.V., Celada, P., Shapiro, D.A., Roth, B.L., Mengod, G. and Artigas, F. (2001) Control of serotonergic function in medial prefrontal cortex by serotonin-2A receptors through a glutamate-dependent mechanism. J. Neurosci. 21, 9856–9866.

Martin Cora, F.J., Pazos, A. (2004) Autoradiographic distribution of 5-HT₇ receptors in the human brain using [3H]mesulergine: comparison to other mammalian species. Br. J. Pharmacol. 141, 92–104.

Martinez, D., Hwang, D.R., Mawlawi, O., Slifstein, M., Kent, J., Simpson, N., Parsey, R.V., Hashimoto, T., Huang, Y.Y., Shinn, A., VanHeertum, R., Abidargham, A., Caltabiano, S., Malizia, A., Cowley, H., Mann, J.J. and Laruelle, M. (2001) Differential occupancy of somatodendritic and postsynaptic 5HT(1A) receptors by pindolol: a dose-occupancy study with [C-11]WAY 100635 and positron emission tomography in humans. Neuropsychopharmacology 24, 209–229.

Mello e Souza, T., Rodrigues, C., Souza, M.M., Vinade, E., Coitinho, A., Choi, H. and Izquierdo, I. (2001) Involvement of the serotonergic type 1A (5-HT1A) receptor in the agranular insular cortex in the consolidation of memory for inhibitory avoidance in rats. Behav. Pharmacol. 12, 349–353.

Miner, L.A.H., Backstrom, J.R., Sanders-Bush, E. and Sesack, S.R. (2003) Ultrastructural localization of serotonin-2A receptors in the middle layers of the rat prelimbic prefrontal cortex. Neuroscience 116, 107–117.

Misane, I. and Ögren, S.O. (2003) Selective 5-HT₁A antagonists WAY 100635 and NAD-299 attenuate the impairment of passive avoidance caused by scopolamine in the rat. Neuropsychopharmacology 28, 253–264.

Morales, M. and Bloom, F.E. (1997) The 5-HT₃ receptor is present in different subpopulations of GABAergic neurons in the rat telencephalon. J. Neurosci. 17, 3157–3167.

Newberry, N.R., Footitt, D.R., Papanastassiou, V. and Reynolds, D.J. (1999) Actions of 5-HT on human neocortical neurones *in vitro*. Brain Res. 833, 93–100.

Offord, S.J., Ordway, G.A. and Frazer, A. (1988) Application of (125I)iodocyanopindolol to measure 5-hydroxytryptamine1B receptors in the brain of the rat. J. Pharmacol. Exp. Ther. 244, 144–153.

Pandey, G.N., Dwivedi, Y., Ren, X., Rizavi, H.S., Faludi, G., Sarosi, A. and Palkovits, M. (2006) Regional distribution and relative abundance of serotonin(2c) receptors in human brain: effect of suicide. Neurochem. Res. 31, 167–176.

Paxinos, G. and Watson, C. (1998) *The rat brain in stereotaxic coordinates*. 4th edn. Sydney: Academic Press.

Pazos, A. and Palacios, J.M. (1985) Quantitative autoradiographic mapping of serotonin receptors in the rat brain. I. Serotonin-1 receptors. Brain Res. 346, 205–230.

Pazos, A., Cortés, R. and Palacios, J.M. (1985) Quantitative autoradiographic mapping of serotonin receptors in the rat brain. II. Serotonin-2 receptors. Brain Res. 346, 231–249.

Pompeiano, M., Palacios, J.M. and Mengod, G. (1992) Distribution and cellular localization of mRNA coding for 5-HT$_{1A}$ receptor in the rat brain: correlation with receptor binding. J. Neurosci. 12, 440–453.

Pompeiano, M., Palacios, J.M. and Mengod, G. (1994) Distribution of the serotonin 5-HT$_2$ receptor family mRNAs: comparison between 5-HT$_{2A}$ and 5-HT$_{2C}$ receptors. Mol. Brain. Res. 23, 163–178.

Puig, M.V., Celada, P., Díaz-Mataix, L. and Artigas. F. (2003) *In vivo* modulation of the activity of pyramidal neurons in the rat medial prefrontal cortex by 5-HT$_{2A}$ receptors. Relationship to thalamocortical afferents. Cereb. Cortex 13, 1870–1882.

Puig, M.V., Santana, N., Celada, P., Mengod, G. and Artigas, F. (2004) *In vivo* excitation of GABA interneurons in the medial prefrontal cortex through 5-HT$_3$ receptors. Cereb. Cortex 14, 1365–1375.

Puig, M.V., Artigas, F. and Celada, P. (2005) Modulation of the activity of pyramidal neurons in rat prefrontal cortex by raphe stimulation *in vivo*: involvement of serotonin and GABA. Cereb. Cortex 15, 1–14.

Reader, T.A., Ferron, A., Descarries, L. and Jasper, H.H. (1979) Modulatory role for biogenic amines in the cerebral cortex. Microiontophoretic studies. Brain Res. 160, 217–229.

Riad, M., Garcia, S., Watkins, K.C., Jodoin, N., Doucet, E., Langlois, X., El Mestikawy, S., Hamon, M. and Descarries, L. (2000) Somatodendritic localization of 5-HT$_{1A}$ and preterminal axonal localization of 5-HT$_{1B}$ serotonin receptors in adult rat brain. J. Comp. Neurol. 417, 181–194.

Roberts, M.H. and Straughan, D.W. (1967) Excitation and depression of cortical neurones by 5-hydroxytryptamine. J. Physiol. 193, 269–294.

Santana, N., Bortolozzi, A., Serrats, J., Mengod, G. and Artigas, F. (2004) Expression of serotonin1A and serotonin2A receptors in pyramidal and GABAergic neurons of the rat prefrontal cortex. Cereb. Cortex 14, 1100–1109.

Sari, Y. (2004) Serotonin(1B) receptors: from protein to physiological function and behavior. Neurosci. Biobehav. Rev. 28, 565–582.

Sheldon, P.W. and Aghajanian, G.K. (1991) Excitatory responses to serotonin (5-HT) in neurons of the rat piriform cortex: evidence for mediation by 5-HT1C receptors in pyramidal cells and 5-HT2 receptors in interneurons. Synapse 9, 208–218.

Sprouse, J.S. and Aghajanian, G.K. (1987) Electrophysiological responses of serotonergic dorsal raphe neurons to 5-HT$_{1A}$ and 5-HT$_{1B}$ agonists. Synapse 1, 3–9.

Staubli, U. and Xu, F.B. (1995) Effects of 5-HT$_3$ receptor antagonism on hippocampal theta rhythm, memory, and LTP induction in the freely moving rat. J. Neurosci. 15, 2445–2452.

Sirota, P., Mosheva, T., Shabtay, H., Giladi, N. and Korczyn, A.D. (2000) Use of the selective serotonin 3 receptor antagonist ondansetron in the treatment of neuroleptic-induced tardive dyskinesia. Am. J. Psychiatry 157, 287–289.

Swanson, L.W. (1998) Brain Maps: Structure of the Rat Brain. Elsevier. Amsterdam.

Tada, K., Kasamo, K., Ueda, N., Suzuki, T., Kojima, T. and Ishikawa, K. (1999) Anxiolytic 5-hydroxytryptamine1A agonists suppress firing activity of dorsal hippocampus CA1 pyramidal neurons through a postsynaptic mechanism: single-unit study in unanesthetized, unrestrained rats. J. Pharmacol. Exp. Ther. 288, 843–848.

Tanaka, E. and North, R.A. (1993) Actions of 5 hydroxytryptamine on neurons of the rat cingulate cortex. J. Neurophysiol. 69, 1749–1757.

Vilaró, M.T., Cortes, R., Gerald, C., Branchek, T.A., Palacios, J.M. and Mengod, G. (1996) Localization of 5-HT$_4$ receptor mRNA in rat brain by *in situ* hybridization histochemistry. Brain Res. Mol. Brain Res. 43, 356–360.

Vilaró, M.T., Cortés, R. and Mengod, G. (2005) Serotonin 5-HT$_4$ receptors and their mPRNAs in rat and guinea pig brain: distribution and effect of neurotoxic lesions. J. Comp. Neurol. 484, 418–439.

Villalobos, C., Beique, J.C., Gingrich, J.A. and Andrade, R. (2005) Serotonergic regulation of calcium-activated potassium currents in rodent prefrontal cortex. Eur. J. Neurosci. 22, 1120–1126.

Waeber, C., Sebben, M., Nieoullon, A., Bockaert, J. and Dumuis, A. (1994) Regional distribution and ontogeny of 5-HT$_4$ binding sites in rodent brain. Neuropharmacology 33, 527–541.

Watling, K.J., Beer, M.S. and Stanton, J.A. (1989) Effects of clozapine and other neuroleptics on binding of [3H]-Q ICS 205–930 to central 5-HT$_3$ recognition sites. Br. J. Pharmacol. 98(Suppl.), 813P.

Williams, G.V. and Goldman-Rakic P.S. (1995) Modulation of memory fields by dopamine D1 receptors in prefrontal cortex. Nature 376, 572–575.

Williams, J.T., Colmers, W.F. and Pan, Z.Z. (1988) Voltage- and ligand-activated inwardly rectifying currents in dorsal raphe neurons *in vivo*. J. Neurosci. 8, 3499–3506.

Williams, G.V., Rao, S.G. and Goldman-Rakic, P.S. (2002) The physiological role of 5-HT$_{2A}$ receptors in working memory. J. Neurosci. 22, 2843–2854.

Willins, D.L., Deutch, A.Y. and Roth, B.L. (1997) Serotonin 5-HT$_{2A}$ receptors are expressed on pyramidal cells and interneurons in the rat cortex. Synapse 27, 79–82.

Xiang, Z. and Prince, D.A. (2003) Heterogeneous actions of serotonin on interneurons in rat visual cortex. J. Neurophysiol. 89, 1278–1287.

Zhang, Z.W. (2003) Serotonin induces tonic firing in layer V pyramidal neurons of rat prefrontal cortex during postnatal development. Neuroscience 23(8), 3373–3384.

Zhang, J.Y., Ashby, C.R. and Wang, R.Y. (1994) Effect of pertussis toxin on the response of rat medial prefrontal cortex cells to the iontophoresis of serotonin receptor agonists. J. Neural. Transm-Gen. Sect. 95, 165–172.

Zhou, F.M. and Hablitz, J.J. (1999) Activation of serotonin receptors modulates synaptic transmission in rat cerebral cortex. J. Neurophysiol. 82, 2989–2999.

Chapter 5

Serotonergic Regulation of NMDA Receptor Trafficking and Function in Prefrontal Cortex

Eunice Y. Yuen and Zhen Yan
Department of Physiology and Biophysics, School of Medicine and Biomedical Sciences,
   State University of New York at Buffalo, Buffalo, NY 14214, USA

5.1 Introduction

Serotonin is synthesized in brain stem raphe nuclei and has widespread projections in the brain. One of the major targets is prefrontal cortex (PFC), which is located anterior of the frontal lobe. PFC transmits information to and from multiple brain regions including cortex, thalamus, hypothalamus, brain stem, basal ganglia, and limbic areas. PFC contains two primary neuronal populations: glutamatergic pyramidal projection neurons (~80%) and GABAergic interneurons (~20%). Working memory, a form of on-line holding and mental manipulation of information, has been found to require the sustained activity of PFC neurons (Goldman-Rakic, 1995). Functional study in primate shows that serotonin enhances firing of some PFC neurons during the delay period of a delayed-response task, a behavioral test for working memory (Williams et al., 2002).

Serotonin exerts its action via activation of seven distinct subtype receptors (Andrade, 1998). The 5-HT$_1$-class receptors (5-HT$_{1A,1B,1D,1E,1F}$) couples to Gi/o proteins to inhibit adenylate cyclase/PKA. The 5-HT$_2$-class receptor (5-HT$_{2A,2B,2C}$) couples to Gq proteins to activate the turnover of phospholipids. The 5-HT$_3$ receptor is a ligand-gated ion channel. The 5-HT$_4$-class receptors (5-HT$_{4,5,6}$) couples to Gs proteins to activate adenylate cyclase/PKA. The function of 5-HT$_5$-class receptor is unknown. Multiple 5-HT receptor subtypes are expressed in PFC pyramidal neurons with 5-HT$_{1A}$ and 5-HT$_{2A}$ receptors being the most prevalent and abundant (Feng et al., 2001).

Recent evidence has suggested the significance of 5-HT$_{1A}$ receptors in regulating cognition and emotion. For instance, treatment of 5-HT$_{1A}$ antagonist alleviates the cognitive deficits induced by NMDAR hypofunction (Harder and Ridley, 2000). Levels of 5-HT$_{1A}$ receptors are elevated in schizophrenia patients (Sumiyoshi et al., 1996). 5-HT$_{1A}$ receptors knockout mice exhibit anxiety-like behavior (Ramboz et al.,

1998), which is reversed by overexpression of 5-HT$_{1A}$ receptors in the forebrain (Gross et al., 2002). Thus, 5-HT$_{1A}$ receptors are proposed as one of the drug targets in treating mental disorders like schizophrenia and depression (Schreiber and De Vry, 1993; Gross and Hen, 2004).

Because 5-HT$_{1A}$ receptor plays a key role in cognitive diseases, it is important to understand the targets of 5-HT$_{1A}$ signaling. Postsynaptic 5-HT$_{1A}$ receptors are concentrated in the dendritic spine and shaft of cortical neurons (Kia et al., 1996), where glutamate receptors are enriched. Thus, it is likely that 5-HT$_{1A}$ receptors may control PFC excitability by regulating glutamate receptors. NMDA receptor is one of the glutamate subtype receptors, which mediates numerous neuronal events ranging from neurodevelopment to synaptic plasticity to excitotoxicity (Dingledine et al., 1999). Administration of noncompetitive NMDAR antagonists or knock-down of NMDARs produces schizophrenia-like psychosis in human and animal models (Javitt and Zukin, 1991; Jentsch and Roth, 1999). Thus, dysfunction of NMDAR is implicated in the pathophysiology of many mental disorders. Summarized here are some findings demonstrating that serotonin, via activation of distinct receptor-mediated signaling, regulates NMDAR trafficking and functions in PFC pyramidal neurons.

5.2 Regulation of NMDAR by 5-HT$_{1A}$ Signaling

To study the impact of 5-HT$_{1A}$ signaling on NMDARs, we examined the effect of 5-HT$_{1A}$ agonist on NMDAR currents in PFC pyramidal neurons.

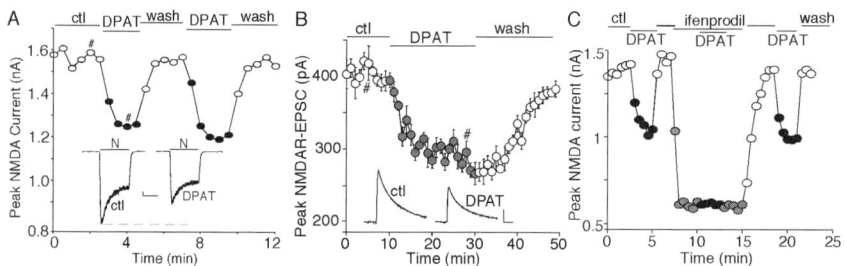

Figure 5.1 Activation of 5-HT$_{1A}$ receptors reduces NMDAR-mediated ionic and synaptic currents in PFC pyramidal neurons. (A) Plot of peak NMDAR currents showing the effect of 5-HT$_{1A}$ agonist (8-OH-DPAT, 40 μM) in isolated PFC neurons. *Inset*: Representative traces at the indicated time points (#). *Scale bars*: 100 pA, 1 s. (B) Plot of peak NMDAR-EPSCs showing the effect of 8-OH-DPAT (20 μM) in PFC slices. *Inset*: Representative traces (average of three trials) at the indicated time points (#). *Scale bars*: 100 pA, 100 ms. (C) Plot of peak NMDAR currents as a function of time and 8-OH-DPAT application in the absence and presence of the selective NR2B inhibitor ifenprodil (3 μM). N, NMDA.

As shown in Fig. 5.1A, application of the 5-HT$_{1A}$ receptor agonist 8-OH-DPAT reversibly reduced the amplitude of NMDA (100 μM)-evoked currents in acutely isolated PFC neurons. This modulation was mimicked by the endogenous transmitter

5-HT, and was blocked by selective 5-HT$_{1A}$ antagonists, such as NAN-190 and WAY-100635 (Yuen et al., 2005), suggesting that it is mediated by 5-HT$_{1A}$ receptors. Moreover, 8-OH-DPAT decreased NMDAR-EPSC amplitude in PFC slices (Fig. 5.1B), suggesting that synaptic NMDARs are also regulated by 5-HT$_{1A}$ receptors.

In mature cortical neurons, NMDARs are composed of NR1/NR2A or NR1/NR2B, which differ in subcellular localization and channel properties (Tovar and West-brook, 1999; Cull-Candy et al., 2001). As shown in Fig. 5.1C, the effect of 5-HT$_{1A}$ on NMDAR currents was abrogated by ifenprodil, an NR2B antagonist, suggesting that 5-HT$_{1A}$ receptors primarily target NR2B-containing NMDARs.

In addition to the regulation of NMDAR channel properties by protein kinase/phosphatase, increasing evidence has proposed that NMDARs are highly mobile and its function can be controlled by receptor trafficking (Wenthold et al., 2003). After NR1 and NR2 subunits assemble into a functional complex, NMDA receptors exit endoplasmic reticulum (ER) and are rapidly transported on microtubule tracks in dendritic shaft (Washbourne et al., 2002). As shown in Fig. 5.2A, application of the microtubule depolymerizer nocodazole mimicked and occluded the reducing effect of 5-HT$_{1A}$ on NMDAR currents. Moreover, dialysis of the microtubule stabilizer Taxol abolished the 5-HT$_{1A}$ effect on NMDAR currents (Yuen et al., 2005). These data suggest that 5-HT$_{1A}$ may regulate NMDAR currents by affecting microtubule-based transport of NMDARs.

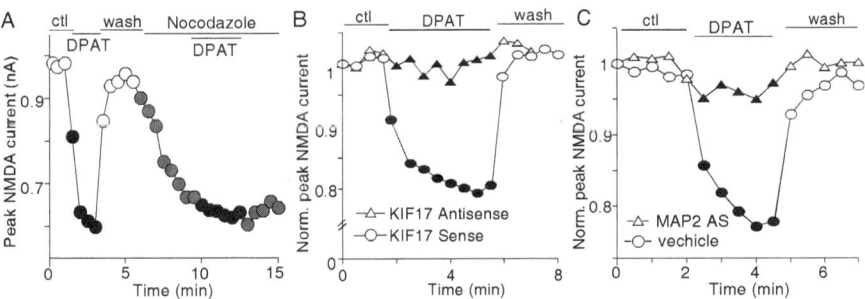

Figure 5.2 5-HT$_{1A}$ regulates NMDARs through a mechanism involving the transport of NMDARs on microtubules by the kinesin motor protein KIF17. (A) Plot of peak NMDAR currents showing the effect of 8-OH-DPAT in the absence and presence of the microtubule depolymerizer nocodazole (30 μM). (B, C) Plots of normalized peak NMDA currents showing the effect of 5-HT$_{1A}$ in cultured PFC neurons transfected with KIF17 antisense versus sense oligonucleotides (B) or MAP2 antisense oligonucleotides versus vehicle (C).

Microtubules are assembled from α- and β-tubulins. They are highly dynamic and undergo rapid, GTP-dependent transition between growing and shrinking states (Cleveland, 1982). Kinesin, a family of motor proteins, mediates anterograde transport on microtubules from the cell body to axons and dendrites. To date, 45 members of kinesin have been identified (Hirokawa and Takemura, 2005). Among them, KIF17 is a neuron-specific kinesin that transports NR2B-containing NMDA receptors via a PDZ protein mLin complex on microtubules (Setou et al., 2000; Guillaud et al., 2003). Animals with overexpression of KIF17 show enhanced spatial and working memory (Wong et al., 2002). These findings suggest that the microtubule-based transport of NMDA receptors plays an important role in neuronal functions which may contribute to cognitive behaviors. As shown in Fig. 5.2B, transfection of KIF17 antisense to knockdown KIF17 expression in cultured PFC neurons blocked the 5-HT$_{1A}$ inhibition of NMDAR currents, suggesting that the 5HT$_{1A}$ modulation of NMDAR currents involves the transport of NR2B-containing NMDA receptors along microtubules in dendrites by motor proteins KIF17.

One way to regulate microtubule stability is through the interaction with micro-tubule-binding protein. MAP2, one member of MAP family proteins that is dendrite specific (Bernhardt and Matus, 1984; Caceres et al., 1984), binds and stabilizes microtubules dependent on its phosphorylation state (Brugg and Matus, 1991; Sanchez et al., 2000). As shown in Fig. 5.2C, knocking down MAP2 with antisense oligonucleotides in PFC cultures blocked the inhibitory effect of 5-HT$_{1A}$ on NMDAR currents, suggesting the involvement of MAP2 in the 5-HT$_{1A}$ modulation.

How does 5-HT$_{1A}$ signaling affect the microtubule-based transport of NMDA receptors? One possibility is that activation of 5-HT$_{1A}$ receptor changes microtubule stability and thus interfere with the transport of NMDAR on microtubule. Biochemical results show that 5-HT$_{1A}$ agonist enhanced the level of free tubulin, which is an indi-cator of depolymerized microtubules (Fig. 5.3A; lane 1–3). This enhancement was blocked by 5-HT$_{1A}$ receptor antagonist or microtubule stabilizer (Fig. 5.3A; lane 4–7).

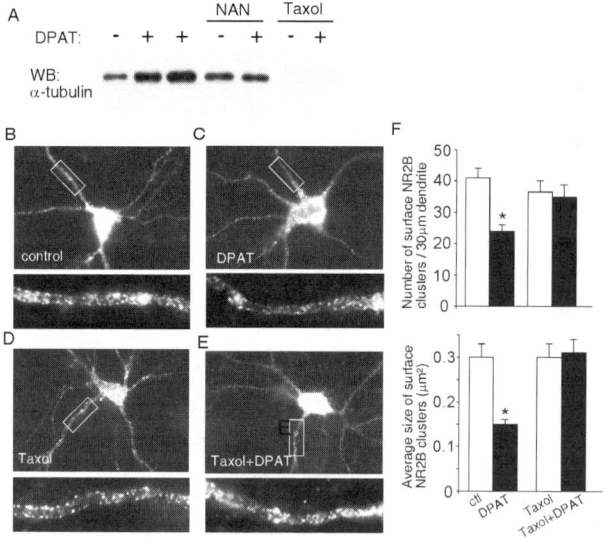

Figure 5.3 5-HT$_{1A}$ receptors induce microtubule depolymerization and reduce surface NR2B clusters. (A) Immunoblot of free tubulin in lysates of PFC cultures treated without or with 8-OH-DPAT (40 μM) in the absence or presence of 5-HT$_{1A}$ antagonist NAN-190 (40 μM) or microtubule stabilizer Taxol (10 μM). (B-E) Immunostaining of surface NR2B subunits in transfected PFC neurons treated without or with 8-OH-DPAT (40 μM) in the absence or presence of Taxol (10 μM). (F) Quantitation of surface NR2B clusters with various treatment.

If 5-HT$_{1A}$ disrupts the microtubule-mediated dendritic delivery of NR2B-containing NMDARs, the surface level of NR2B subunit should be reduced by 5-HT$_{1A}$ agonist. To test this, immunostaining was performed in PFC neurons transfected with GFP-tagged NR2B subunits at the N-terminus. Surface NR2B is detected by anti-GFP primary antibody followed by rhodamine-conjugated secondary antibody in nonpermeabilized conditions. As shown in Fig. 5.3B–E, the 5-HT$_{1A}$ agonist 8-OH-DPAT significantly reduced the density and size of surface NR2B clusters, which was prevented by the microtubule stabilizer Taxol. These results suggest that 5-HT$_{1A}$ suppresses NMDAR functions by disrupting the MT-mediated NMDAR transport and thus reducing the surface functional NMDA receptors (Fig. 5.3F).

Since the 5-HT$_{1A}$ modulation of NMDAR currents depends on the regulation of MAP2 and microtubule stability, we would like to know what molecules link 5-HT$_{1A}$ receptors to the MAP2–microtubule complex. Two key signaling molecules that regulate MAP2 phosphorylation are ERK and CaMKII (Sanchez et al., 2000). To examine the involvement of ERK in 5-HT$_{1A}$ modulation of NMDARs, a dominant negative mutant form of MEK1 (Mansour et al., 1994) was transfected in PFC cultures. As shown in Fig. 5.4A, the effect of 8-OH-DPAT on NMDAR currents was significantly abolished in neurons transfected with dnMEK1, comparing to neurons transfected with wild-type MEK1.

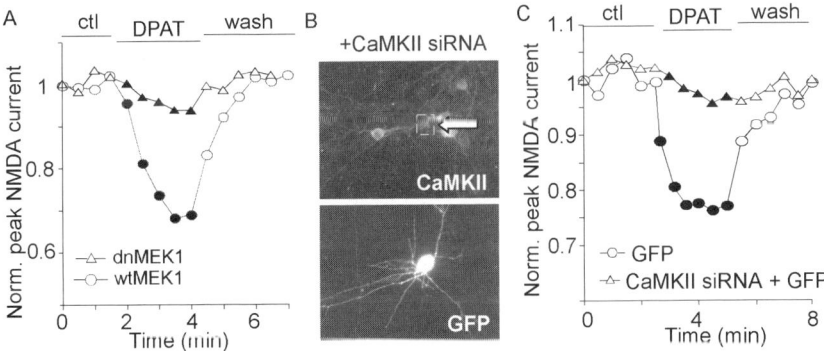

Figure 5.4 Inhibition of ERK or CaMKII activity prevents 5-HT$_{1A}$ reduction of NMDAR currents. (A) Plot of peak NMDAR currents with 8-OH-DPAT application in cells transfected with wild-type (*wt*) or dominant negative (*dn*) MEK1. (B) Immunostaining with α-CaMKII in cultured PFC neurons cotransfected with CaMKII siRNA and GFP. *Arrow* indicates GFP+ cells. (C) Plot of peak NMDAR currents showing the effect of 8-OH-DPAT in a GFP+ neuron transfected without or with CaMKII siRNA.

To examine the involvement of CaMKII in 5-HT$_{1A}$ signaling, CaMKII expression was suppressed by transfecting an siRNA directed against α-CaMKII in cultured PFC neurons. As shown in Fig. 5.4B, the expression of CaMKII was abolished in GFP-positive neurons transfected with CaMKII siRNA. Moreover, the 8-OH-DPAT reduction of NMDAR current was blocked in CaMKII siRNA-transfected neurons

(Fig. 5.4C). These data suggest that both ERK and CaMKII play important roles in 5-HT$_{1A}$ modulation of NMDAR trafficking.

5.3 Regulation of NMDAR by 5-HT$_{2A}$ Signaling

In addition to 5-HT$_{1A}$ receptors, another serotonin receptor subtype that is enriched in PFC pyramidal neurons is 5-HT$_{2A}$ (Feng et al., 2001). 5-HT$_{2A}$ receptor has long been implicated in mood regulation because classics hallucinogens (LSD) are ago- nists (Kroeze and Roth, 1998) and atypical antipsychotics are antagonists of 5-HT$_{2A}$ receptors (Meltzer, 1999). It is intriguing that 5-HT$_{1A}$ and 5-HT$_{2A}$ receptors often have opposing actions on common substrates. For example, activation of 5-HT$_{1A}$- receptors results in neuronal inhibition by increasing potassium currents (Araneda and Andrade, 1991) and decreasing calcium currents (Penington and Kelly, 1990). In contrast, 5-HT$_{2A}$ receptor stimulation leads to neuronal excitation by suppressing potassium currents (Andrade, 1998) and enhancing presynaptic glutamate release (Aghajanian and Marek, 1999). Furthermore, clinical studies have shown that simul- taneous blockade of 5-HT$_{2A}$ receptors and application of antidepressants, serotonin- specific reuptake inhibitors (SSRIs), has greater therapeutic effect for depressed patients than SSRI action alone (Marek et al., 2003).

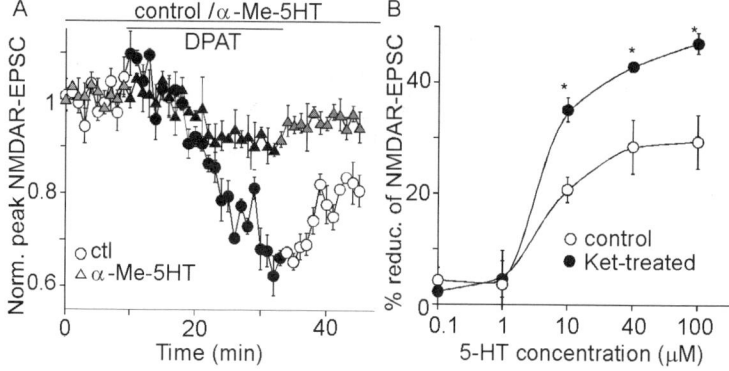

Figure 5.5 5-HT$_{2A}$ receptors oppose the 5-HT$_{1A}$-induced reduction of NMDAR-mediated synaptic currents in PFC pyramidal neurons. (A) Plot of normalized peak NMDAR-EPSCs with application of 8-OH-DPAT in the absence or presence of 5-HT$_{2A}$ receptor agonist α-Me- 5-HT. (B) Dose-response curves summarizing the percentage reduction of NMDAR-EPSCs by different concentrations of 5-HT without or with treatment of 5-HT$_{2A}$ antagonist ketanserin (Ket). *p < 0.01, ANOVA.

Next, we examined whether 5-HT$_{1A}$ and 5-HT$_{2A}$ have opposite effects on NMDAR channels. As shown in Fig. 5.5A, application of 8-OH-DPAT reduces the amplitude of NMDAR-EPSC in PFC slices. However, this effect was significantly attenuated by the 5-HT$_{2A/C}$ agonist α-Me-5-HT. Then, what would happen when both

5-HT$_{1A}$ and 5-HT$_{2A}$ receptors are coactivated by 5-HT? If two receptor signals oppose each other, blocking activation of one receptor would unmask the effect of the other. In agreement with this, blocking 5-HT$_{2A}$ receptors with the antagonist ketanserin significantly augmented the reducing effect of 5-HT on NMDA-EPSCs (Fig. 5.5B). These findings suggest that 5-HT$_{2A}$ receptors counteract the 5-HT$_{1A}$ action on NMDAR-mediated synaptic transmission.

Because 5-HT$_{1A}$ disrupts NMDAR transport by destabilizing microtubules, we examined whether microtubule is the convergent target of 5-HT$_{1A}$ and 5-HT$_{2A}$ signaling. As shown in Fig. 5.6A, application of 8-OH-DPAT increased the level of free (depolymerized) tubulin, which was attenuated by α-Me-5-HT, indicating that 5-HT$_{2A}$ opposes the 5-HT$_{1A}$ action on microtubule stability.

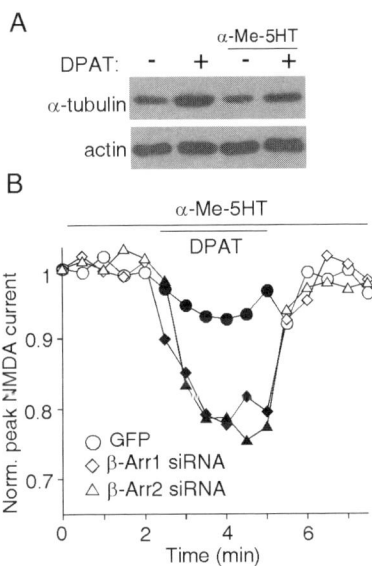

Figure 5.6 The counteracting effect of 5-HT$_{2A}$ on 5-HT$_{1A}$ regulation of NMDARs depends on microtubule stability and β-arrestin. (A) Immunoblot of free tubulin in lysates of cultured PFC neurons treated with 8-OH-DPAT (40 μM) in the absence or presence of α-Me-5-HT (20 μM). (B) Plot of normalized peak NMDAR currents showing the effect of α-Me-5-HT (20 μM) on DPAT (20 μM) regulation of NMDAR response in GFP+ neurons transfected with GFP alone or GFP plus β-arrestin1/2 siRNA.

Since activation of 5-HT$_{1A}$ receptor destabilizes microtubule by suppressing ERK activity, we tested whether ERK is the common substrate of 5-HT$_{1A}$ and 5-HT$_{2A}$ receptors. We found that activation of 5-HT$_{2A}$ receptor induced phosphorylation of ERK (Thr202/Tyr 204) (data not shown). Because some G protein-coupled receptors have been found to activate ERK via the scaffold protein β-arrestin (Lefkowitz and Shenoy, 2005), we examined the involvement of β-arrestin in 5-HT$_{2A}$ activation of ERK. We found that in neurons transfected with β-arrestin1/2 siRNA, 5-HT$_{2A}$ failed to activate ERK and failed to oppose the 5-HT_{1A} effect on NMDAR currents (Fig. 5.6B). These results suggest that 5-HT_{2A} opposes the 5-HT_{1A} regulation of NMDAR through β-arrestin/ERK pathway.

Based on the experimental results, we propose a model demonstrating the role of serotonin in regulating NMDAR trafficking and function (Fig. 5.7). 5-HT_{1A} signaling, through inhibition of PKA, suppresses CaMKII and ERK activity. In contrast, 5-HT_{2A} signaling, via activation of β-arrestin pathway, enhances ERK activity. As a result, these signals converge to regulate MAP2 phosphorylation and microtubule stability, which consequently control the microtubule/KIF17-mediated dendritic transport of NMDARs and the NMDAR currents.

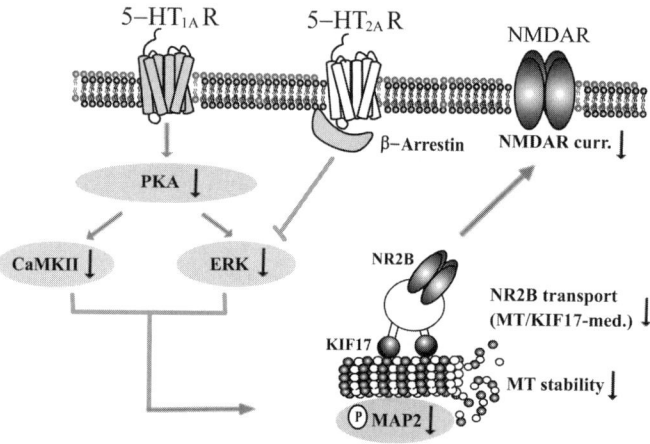

Figure 5.7 Schematic diagram illustrating the serotonergic regulation of NMDAR trafficking in PFC pyramidal neurons. MT: microtubule.

5.4 Discussion

Despite intensive research, the cellular and molecular mechanism by which serotonin regulates PFC functions remains unclear. The present findings show that 5-HT_{1A} signaling suppresses NMDA receptor functions in PFC pyramidal neurons. This regulation has functional implications as follows. First, loss of 5-HT_{1A} inhibition of NMDA receptor functions may lead to the hyperactivation of glutamatergic neurons, thereby increasing glutamate release, which may underlie the anxiety-like behavior observed in 5-HT_{1A} knock-out mice (Heisler et al., 1998; Ramboz et al., 1998). Second, the reduction of NMDA-mediated transmission may also play a key role in the 5-HT_{1A} inhibition of long-term potentiation (Sakai and Tanaka, 1993; Edagawa et al., 1998), a synaptic model of learning that involves CaMKII and ERK (Malenka and Nicoll, 1999). Third, increasing evidence indicates that neurons require microtubule-based transport systems to deliver vital cellular cargoes to support their functions, and defects in neuronal transport are implicated in several neurological disorders, such as Alzheimer's disease (Goldstein, 2003; Baas and Qiang, 2005). Thus, the 5-HT$_{1A}$ regulation of microtubule-mediated transport of NMDA receptors may contribute to the pathophysiology of mental disorders.

One of the unique features of serotonin is the ability of a single transmitter to generate diverse functions by coupling to distinct subtype receptors. Our data suggests that coactivation of 5-HT$_{1A}$ and 5-HT$_{2A}$ receptor oppositely modulates the NMDAR functions in PFC pyramidal neurons. This involves the 5-HT$_{2A}$-induced ERK activation through β-arrestins, opposing the 5-HT$_{1A}$ effect on microtubule-based transport of NMDARs. This counteractive effect enables serotonin to achieve a precise and dynamic regulation of glutamate signaling within PFC network. Since the excitability of PFC neurons is greatly influenced by serotonin, the present results provide important insights into the serotonergic regulation of synaptic transmission and plasticity in PFC, which may shed light on finding new approaches to treat neurological disorders.

Acknowledgments

This work was supported by National Institutes of Health Grants MH63128, NS48911, and AG21923, National Science Foundation Grant IBN-0117026, and a National Alliance for Research on Schizophrenia and Depression Independent Investigator Award (Z.Y.) and Young Investigator Award (E.Y.).

References

Aghajanian GK, Marek GJ (1999) Serotonin, via 5-HT2A receptors, increases EPSCs in layer V pyramidal cells of prefrontal cortex by an asynchronous mode of glutamate release. Brain Res 825:161–171.

Andrade R (1998) Regulation of membrane excitability in the central nervous system by serotonin receptor subtypes. Ann N Y Acad Sci 861:190–203.

Araneda R, Andrade R (1991) 5-Hydroxytryptamine2 and 5-hydroxytryptamine 1A receptors mediate opposing responses on membrane excitability in rat association cortex. Neuroscience 40:399–412.

Baas PW, Qiang L (2005) Neuronal microtubules: when the MAP is the roadblock. Trends Cell Biol 15:183–187.

Bernhardt R, Matus A (1984) Light and electron microscopic studies of the distribution of microtubule-associated protein 2 in rat brain: a difference between dendritic and axonal cytoskeletons. J Comp Neurol 226:203–221.

Brugg B, Matus A (1991) Phosphorylation determines the binding of microtubule-associated protein 2 (MAP2) to microtubules in living cells. J Cell Biol 114:735–743.

Caceres A, Banker G, Steward O, Binder L, Payne M (1984) MAP2 is localized to the dendrites of hippocampal neurons which develop in culture. Brain Res 315:314–318.

Cleveland DW (1982) Treadmilling of tubulin and actin. Cell 28:689–691.

Cull-Candy S, Brickley S, Farrant M (2001) NMDA receptor subunits: diversity, development and disease. Curr Opin Neurobiol 11:327–335.

Dingledine R, Borges K, Bowie D, Traynelis SF (1999) The glutamate receptor ion channels. Pharmacol Rev 51:7–61.

Edagawa Y, Saito H, Abe K (1998) 5-HT1A receptor-mediated inhibition of long-term potentiation in rat visual cortex. Eur J Pharmacol 349:221–224.

Feng J, Cai X, Zhao J, Yan Z (2001) Serotonin receptors modulate GABA(A) receptor channels through activation of anchored protein kinase C in prefrontal cortical neurons. J Neurosci 21:6502–6511.

Goldman-Rakic PS (1995) Cellular basis of working memory. Neuron 14:477–485.

Goldstein LS (2003) Do disorders of movement cause movement disorders and dementia? Neuron 40:415–425.

Gross C, Hen R (2004) The developmental origins of anxiety. Nat Rev Neurosci 5:545–552.

Gross C, Zhuang X, Stark K, Ramboz S, Oosting R, Kirby L, Santarelli L, Beck S, Hen R (2002) Serotonin1A receptor acts during development to establish normal anxiety-like behaviour in the adult. Nature 416:396–400.

Guillaud L, Setou M, Hirokawa N (2003) KIF17 dynamics and regulation of NR2B trafficking in hippocampal neurons. J Neurosci 23:131–140.

Harder JA, Ridley RM (2000) The 5-HT1A antagonist, WAY 100 635, alleviates cognitive impairments induced by dizocilpine (MK-801) in monkeys. Neuropharmacology 39:547–552.

Heisler LK, Chu HM, Brennan TJ, Danao JA, Bajwa P, Parsons LH, Tecott LH (1998) Elevated anxiety and antidepressant-like responses in serotonin 5-HT1A receptor mutant mice. Proc Natl Acad Sci USA 95:15049–15054.

Hirokawa N, Takemura R (2005) Molecular motors and mechanisms of directional transport in neurons. Nat Rev Neurosci 6:201–214.

Javitt DC, Zukin SR (1991) Recent advances in the phencyclidine model of schizophrenia. Am J Psychiatry 148:1301–1308.

Jentsch JD, Roth RH (1999) The neuropsychopharmacology of phencyclidine: from NMDA receptor hypofunction to the dopamine hypothesis of schizophrenia. Neuropsychopharmacology 20:201–225.

Kia HK, Miquel MC, Brisorgueil MJ, Daval G, Riad M, El Mestikawy S, Hamon M, Verge D (1996) Immunocytochemical localization of serotonin1A receptors in the rat central nervous system. J Comp Neurol 365:289–305.

Kroeze WK, Roth BL (1998) The molecular biology of serotonin receptors: therapeutic implications for the interface of mood and psychosis. Biol Psychiatry 44:1128–1142.

Lefkowitz RJ, Shenoy SK (2005) Transduction of receptor signals by beta-arrestins. Science 308:512–517.

Malenka RC, Nicoll RA (1999) Long-term potentiation--a decade of progress? Science 285:1870–1874.

Mansour SJ, Matten WT, Hermann AS, Candia JM, Rong S, Fukasawa K, Vande Woude GF, Ahn NG (1994) Transformation of mammalian cells by constitutively active MAP kinase kinase. Science 265:966–970.

Marek GJ, Carpenter LL, McDougle CJ, Price LH (2003) Synergistic action of 5-HT2A antagonists and selective serotonin reuptake inhibitors in neuropsychiatric disorders. Neuropsychopharmacology 28:402–412.

Meltzer HY (1999) The role of serotonin in antipsychotic drug action. Neuropsychopharmacology 21:106S–115S.

Penington NJ, Kelly JS (1990) Serotonin receptor activation reduces calcium current in an acutely dissociated adult central neuron. Neuron 4:751–758.

Ramboz S, Oosting R, Amara DA, Kung HF, Blier P, Mendelsohn M, Mann JJ, Brunner D, Hen R (1998) Serotonin receptor 1A knockout: an animal model of anxiety-related disorder. Proc Natl Acad Sci USA 95:14476–14481.

Sakai N, Tanaka C (1993) Inhibitory modulation of long-term potentiation via the 5-HT1A receptor in slices of the rat hippocampal dentate gyrus. Brain Res 613:326–330.

Sanchez C, Diaz-Nido J, Avila J (2000) Phosphorylation of microtubule-associated protein 2 (MAP2) and its relevance for the regulation of the neuronal cytoskeleton function. Prog Neurobiol 61:133–168.

Schreiber R, De Vry J (1993) 5-HT1A receptor ligands in animal models of anxiety, impulsivity and depression: multiple mechanisms of action? Prog Neuropsychopharmacol Biol Psychiatry 17:87–104.

Setou M, Nakagawa T, Seog DH, Hirokawa N (2000) Kinesin superfamily motor protein KIF17 and mLin-10 in NMDA receptor-containing vesicle transport. Science 288:1796–1802.

Sumiyoshi T, Stockmeier CA, Overholser JC, Dilley GE, Meltzer HY (1996) Serotonin1A receptors are increased in postmortem prefrontal cortex in schizophrenia. Brain Res 708:209–214.

Tovar KR, Westbrook GL (1999) The incorporation of NMDA receptors with a distinct subunit composition at nascent hippocampal synapses in vitro. J Neurosci 19:4180–4188.

Washbourne P, Bennett JE, McAllister AK (2002) Rapid recruitment of NMDA receptor transport packets to nascent synapses. Nat Neurosci 5:751–759.

Wenthold RJ, Prybylowski K, Standley S, Sans N, Petralia RS (2003) Trafficking of NMDA receptors. Annu Rev Pharmacol Toxicol 43:335–358.

Williams GV, Rao SG, Goldman-Rakic PS (2002) The physiological role of 5-HT2A receptors in working memory. J Neurosci 22:2843–2854.

Wong RW, Setou M, Teng J, Takei Y, Hirokawa N (2002) Overexpression of motor protein KIF17 enhances spatial and working memory in transgenic mice. Proc Natl Acad Sci USA 99:14500–14505.

Yuen EY, Jiang Q, Chen P, Gu Z, Feng J, Yan Z (2005) Serotonin 5-HT1A receptors regulate NMDA receptor channels through a microtubule-dependent mechanism. J Neurosci 25:5488–5501.

Chapter 6

Cross-Modulation Between GABA$_B$ and 5-HT Receptors: *A Link Between Anxiety and Depression?*

Juan C. Pineda and José L. Góngora-Alfaro
Departamento de Neurociencias, centro de Investigaciones Regionales "Dr. Hideyo Noguchi",
  Universidad Autonoma de Yucatan, Merida, Yucatan 97000, Mexico

Abstract. Pyramidal neurons in the sensory-motor cortex express multiple types of metabotropic receptors. Simultaneous application of serotonin (5-HT) and GABA$_B$ agonists produces a reduction of the neurotransmitter release probability throughout the activation of the GABA$_B$ and 5-HT$_{1A}$ receptors. Since some of these receptors may be coexpressed in a set of neurons, we examined the consequence of the simultaneous activation of GABA$_B$ and 5-HT receptors and investigated their influence on neurotransmitter release probability in the sensorimotor cortex of the rat using extracellular stimulation with a paired pulse protocol. We found that the effect of 5-HT is occluded when the GABA$_B$ receptor is previously activated with baclofen. The effect is mimicked by the 5-HT$_2$ agonist DOI and prevented by the 5-HT$_2$ antagonist ritanserin. Since prefrontal cortical 5-HT terminals may contact "en passant" fibers and release the 5-HT by volume transmission, and because 5-HT$_{1A}$ and 5-HT$_{2A/C}$ receptors are coexpressed in pyramidal neurons, they may reciprocally modify each other's metabolic pathways, leading to the production of nonlinear interactions in this cortical area. The potential implications of the cross-talk between 5-HT and GABA$_B$ receptors are discussed in terms of the consequence of the use of selective serotonin reuptake inhibitors in the treatment of the depression.

6.1 Introduction

G protein-coupled receptors (GPCRs), ion channels, and other proteins embedded in the mosaic fluid of the plasma membrane can move laterally to interact with neighbor proteins present in the same area, forming homo- or heteromultimers whose functions differ from the parent molecules. Protein–protein modulation at the membrane level is a form of "cross-talk" between signaling pathways, which may also involve downstream molecules such as second messengers, kinases, and other intracellular effectors. In neurons, intramembrane protein cross-talk can regulate the function of neurotransmitter receptors and ion channels leading to changes in neuronal excitability, firing pattern, neurotransmitter release probability, and other important neural functions.

The interest in studying the functional significance of intramembrane protein cross-talk in neurons has steadily grown during the past decade, as indicated by the increasing number of reports in the literature (for a review see Cordeaux and Hill, 2002). This is not surprising since advancements in the understanding of the cross-talk interactions between different G-proteins has the potential to lead to the development of more selective and specific drugs for neuropsychiatric diseases, such as anxiety disorders, depressive illness, schizophrenia, and others.

Cross-talk, observed among the G-protein coupled receptors for the biogenic amines noradrenaline (NA), dopamine (DA), and serotonin (5-HT), might be favored not only by the existence of a large variety of receptor subtypes, but also by axonal convergence and by "en passant" synaptic contacts allowing simultaneous activation of different receptors.

6.2 The Impact of the Diversity of 5-HT Receptors

Among such biogenic amines, the 13 G protein-coupled 5-HT receptor subtypes appear exquisitely likely candidates for the cross-talk process. One interesting feature of the high diversity of 5-HT receptors is the fact that many neurons coexpress two or more receptor subtypes which usually activate different signaling pathways that exert mutually opposing effects. What could be the functional meaning of the coexistence of multiple 5-HT receptor subtypes in the same cell? The answer to this question is being tackled by means of diverse experimental approaches, comprising both cellular and behavioral studies. A few examples will be reviewed with the purpose of illustrating the possible complexity of the cross-talk between different 5-HT receptor subtypes.

The 1C11 cell line has the peculiarity of acquiring a complete serotonergic phenotype within 4 days. Starting on the second day of differentiation, these cells express functional $5\text{-HT}_{1B/1D}$ and 5-HT_{2B} receptors, whereas 5-HT_{2A} receptors are expressed until day 4 (Tournois et al., 1998). In 2-day old 1C11 cells, activation of $5\text{-HT}_{1B/1D}$ receptors inhibits the forskolin-stimulated production of cAMP. This effect is antagonized by concomitant stimulation of 5-HT_{2B} receptors, which are coupled to the phospholipase A_2-mediated release of arachidonic acid (AA). The inhibitory effect of 5-HT_{2B} receptors on $5\text{-HT}_{1B/1D}$ signaling is blocked by indomethacin, indicating that the cross-talk between these receptors is mediated by a cyclooxygenase-dependent AA metabolite (Tournois et al., 1998), whose likely target is the enzyme adenylyl cyclase (Evans et al., 2001). By day 4, however, activation of the newly expressed 5-HT_{2A} receptors in 1C11 cells antagonizes the negative regulation exerted by 5-HT_{2B} on the $5\text{-HT}_{1B/1D}$ function (Tournois et al., 1998).

Although the studies with stable cell lines have provided valuable information about the pathways involved in the cross-talk between G protein-coupled receptors, studies with adult animals have given more clues about the physiological meaning of this process. In rodents, for example, up to 60% of pyramidal neurons in the prefrontal cortex express 5-HT_{1A} and 5-HT_{2A} receptor mRNAs. Of particular relevance is the fact that both mRNAs show a high degree (~80%) of colocalization (Amargós-Bosch et al., 2004). Similarly, the 5-HT_{1A} and 5-HT_{2A} receptor proteins are coexpressed by more than 95% of the oxytocin and corticotropin releasing factor neurons of the hypothalamic paraventricular nucleus (PVN) (Zhang et al., 2004). In these neuronal populations, the activation of one of these receptor subtypes usually inhibits the effect elicited by activation of the other receptor. Thus, electrical stimulation of the serotonergic raphe-cortical pathway evokes 5-HT$_{1A}$-mediated inhibitions and 5-HT$_{2A}$-mediated excitations in medial prefrontal pyramidal neurons (Amargós-Bosch et al., 2004). The dual action of endogenous 5-HT on these neurons was unveiled in an elegant experiment where the dominant inhibitory response evoked in a pyramidal neuron by dorsal raphe stimulation could be switched to a predominant excitatory effect after systemic administration of a 5-HT$_{1A}$ antagonist. In turn, the excitatory response was reversed by administration of a 5-HT$_{2A}$ antagonist (Amargós-Bosch et al., 2004). In the case of PVN neurons, systemic administration of the 5-HT$_{2A/2C}$ receptor agonist (-)-1-(2,5-dimethoxy-4-iodophenyl) 2-aminopropane (DOI) produced a delayed and long-lasting inhibition of the increase in the plasma levels of oxytocin and adrenocorticotropic hormone (ACTH) induced by subsequent administration of the 5-HT$_{1A}$ agonist (+)8-OH-DPAT, applied 2.5 h later (Zhang et al., 2001, 2004). Similarly, a single systemic injection of the 5-HT$_{1A}$ receptor agonist significantly reduced the oxytocin and ACTH increases mediated by 5-HT$_{2A}$ receptor activation (Carrasco et al., 2007).

The cross-talk process is not limited to the 5-HT receptor subtypes, whose signaling pathways can also converge with the intracellular messages evoked by the simultaneous activation of nonserotonergic GPCRs coexisting in the same cell. Thus, the long-lasting increase in glutamatergic synaptic transmission elicited in slices of the basolateral amygdala by activation of β-adrenergic receptors (βARs) could be prevented by the concurrent activation of the 5-HT$_{1A}$ receptors, which by itself reduced the amplitude of the evoked synaptic potentials (Wang et al., 1999). Experiments in cortical synaptosomes revealed that this cross-talk was consequence of a 5-HT$_{1A}$-mediated inhibition of the synaptosomal Ca^{2+} influx induced by βARs stimulation, confirming that both receptors colocalize in the same nerve endings (Wang et al., 2002). It was concluded that the interaction between the two receptor signaling pathways occurs at a locus downstream from cAMP production, possibly at the level of voltage-dependent Ca^{2+} influx.

We recently presented evidence that the cross-talk between three different G-protein coupled receptors can modulate in a complex way the short-term synaptic plasticity in excitatory synapses of the layers 2/3 of the somatosensory cortex (Torres-Escalante et al., 2004). In these experiments, activation of presynaptic 5-HT$_{1A}$ receptors induced paired pulse (PP) facilitation, which reflects inhibition of glutamatergic transmission. The same effect was elicited by activation of GABA$_B$ receptors, albeit with higher efficiency. Surprisingly the large PP facilitation induced by GABA$_B$ activation was reversed by application of the same concentration of 5-HT. This particular effect was now mediated by 5-HT$_2$ receptors, and not by 5-HT$_{1A}$ receptors like PP facilitation. What could be the functional significance of the dual opposite action of 5-HT on cortical glutamatergic synapses in the presence or in the absence of GABA$_B$ receptors activation? Any answer to this unresolved question could be relevant for the understanding of the pathophysiology of mood disorders in which 5-HT$_{1A}$, 5-HT$_{2A}$, and GABA$_B$ receptors have been implied, in particular, it could help to understand the mechanisms that underlie the antidepressive effects of 5-HT$_{2A}$ and GABA$_B$ antagonists (Marek et al., 2003; Slattery et al., 2005).

Since the cross talk between presynaptic receptors can induce short- and long-term changes in neurotransmitter release probability, we will briefly discuss some aspects of synaptic plasticity before analyzing in further detail some possible mechanisms by which the 5-HT_{1A}, 5-HT_{2A}, and $GABA_B$ receptors interact to modulate cortical glutamatergic transmission.

6.3 Synaptic Plasticity Regulations by 5-HT Receptors

Synaptic plasticity can be caused by a change in the probability of neurotransmitter release from a synaptic terminal. Experimentally, synaptic plasticity can be measured by recording changes in the amplitude of the postsynaptic potential or in the neurotransmitter release probability, evoked by a variety of stimulation protocols or physiological conditions, and then classified according to the positive or negative direction of the modulation (facilitation vs. depression). A second classification of synaptic plasticity is based on its temporal course: short-term (milliseconds to seconds) versus long-term (from hours to years, Thomson, 2000).

It is generally accepted that the short-term plasticity is due to presynaptic modulation of the neurotransmitter release probability. The short-term synaptic plasticity has been considered by some authors as the physiological substrate of the working memory (Hempel et al., 2000). It has also been proposed that short-term plasticity participates in the stabilization and reconfiguration of motor circuits as well as in the initiation, maintenance, and modulation of motor programs (Colino et al., 2002).

Many classes of chemical transmitters are able to produce (presynaptic) modulation of the neurotransmitter release probability. Some of them may be simultaneously present in a single terminal. Murakoshi and coworkers suggested that different neuromodulators may induce plastic changes in the same single synaptic terminal of the neocortex (Murakoshi et al., 2001). Using the paired pulse protocol, they showed that the synaptic potentials evoked by a minimal stimulation, presumably affecting one or very few synaptic terminals, were modulated presynaptically by the application of different neuromodulators such as 5-HT, the $GABA_B$ agonist baclofen, and muscarine (Hempel et al., 2000; Murakoshi et al., 2001).

Given that all the modulators tested by Murakoshi and coworkers exert their effects through metabotropic receptors activating intracellular signaling pathways that may converge at some point, a question was raised on what would be the influence of simultaneous activation of 5-HT and $GABA_B$ receptors on the synaptic potentials. We addressed this question by measuring extracellular synaptic potential evoked by afferent stimulation, recorded in layer 2–3 from the somatosensory cortex (Torres-Escalante et al., 2004).

As shown in Fig. 6.1, both 5-HT and the $GABA_B$ agonist baclofen increased the S_2/S_1 ratio obtained by using the paired pulse protocol. Note that the facilitation produced by baclofen was larger than that produced by the indolamine. The effect of 5-HT was mediated through the 5-HT_{1A} receptors since it was blocked by the 5-HT_{1A} antagonist NAN-190, and mimicked by the selective $5\text{-HT}_{1A/7}$ agonist 8-OH-DPAT (see Torres-Escalante et al., 2004, Fig. 2).

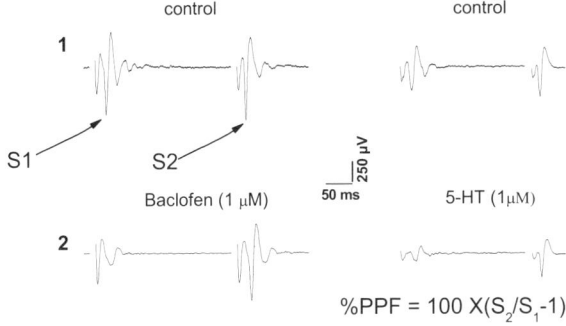

Figure 6.1 Baclofen and 5-HT reduce neurotransmitter release probability at the somatosensory cortex. Traces of layer 5-evoked field potential recorded from layers 2–3 using a paired pulse protocol showing before (1) and after (2) application of baclofen or 5-HT. Note that both neuromodulators increase the paired pulse facilitation (PPF) as calculated by the equation at the bottom, where S_2 and S_1 are the amplitudes between the peaks of the synaptic potentials and the basal line. An increment in the S_2/S_1 ratio is generally accepted as an index of reduction in the neurotransmitter release probability.

In order to explore the possible cross-talk between the 5-HT$_{1A}$ and the GABA$_B$ receptors, we applied simultaneously 5-HT and baclofen. In these experiments, the effect of a relatively low concentration of 5-HT (1 μM) alone was tested first. The indolamine increased the S_2/S_1 ratio about 100%, and this effect was reversible after washing. Next, baclofen (1 μM) was applied. Compared with the effect of 5-HT, the GABA$_B$ agonist showed a greater efficacy, increasing the S_2/S_1 ratio about 500%. Interestingly, when the maximal effect of baclofen was obtained and 5-HT was applied again sequentially, rather than producing facilitation, the amine reverted the action of the GABA$_B$ agonist, reducing the S_2/S_1 ratio back to the level previously produced by 5-HT alone (Torres-Escalante et al., 2004). Thus coactivation of the 5-HT and GABA$_B$ receptors reverted the strong influence of baclofen, reducing the S_2/S_1 ratio to the effect induced previously by 5-HT alone. One possible explanation for this effect is that GABA$_B$ activation occluded the effect of 5-HT on 5-HT$_{1A}$ receptors, unmasking the excitatory influence of the 5-HT$_{2A/C}$ receptors that may be active in the basal condition.

To test this hypothesis, we repeated the same experiment in the presence of the 5-HT$_2$ antagonist ritanserin. In this experimental condition, 5-HT was unable to prevent the PP facilitation induced by GABA$_B$ stimulation (Torres-Escalante et al., 2004). As shown in Fig. 6.2, the 5-HT$_2$ agonist DOI mimics the effect 5-HT when applied in presence of baclofen (Pineda and Góngora-Alfaro, unpublished observation). Based on these observations, we raised the hypothesis that during basal conditions 5-HT is activating 5-HT$_{1A}$ receptors and their associated Gi/o proteins. The same effect could be produced by activation of GABA$_B$ receptors, except that the coupling efficiency of 5-HT and GABA$_B$ receptors to their respective Gi/o proteins are different as indicated by the larger PP facilitation effect of baclofen as compared to 5-HT.

There is evidence that the $^{2\ 3}$ subunits released upon activation of the Gi/o protein can augment Gq-mediated responses, such as phospholipase C (PLC) activation

(Cordeaux et al., 2002). The activation of PLC promotes the synthesis of the second messenger inositol triphosphate (IP_3) and diacylglycerol (DAG) (Barnes and Sharp, 1999). In turn, IP_3 induces the release of Ca^{2+} from intracellular stores, and this ion acts in synergy with DAG to activate protein kinase C (PKC). There is evidence that activation of PKC reduces the hyperpolarization and the outward currents induced by stimulation of $GABA_B$ receptors expressed in Xenopus oocytes (Taniyama et al., 1991). Since the $5\text{-}HT_2$ receptors mediate their effects through Gq proteins, and can stimulate PLC (Barnes et al., 1999), we postulated that the activation of presynaptic $GABA_B$ receptors coupled to Gi/o proteins caused a more efficient PPF that occluded the effect of 5-HT on $5\text{-}HT_{1A}$ receptors, and simultaneously released the βγ dimers in a membrane domain close to PLC that sensitized the enzyme to $5\text{-}HT_2$ activation. Consequently, the reversal of baclofen-induced PPF by 5-HT would be explained by an increase of PLC activity that stimulates PKC, and desensitizes the $GABA_B$ receptors.

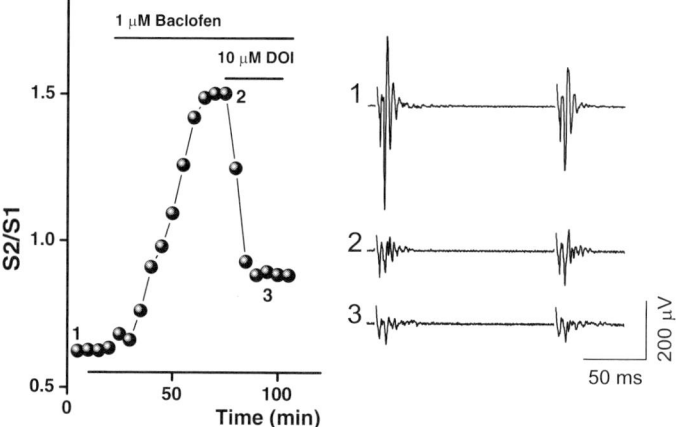

Figure 6.2 The $5\text{-}HT_2$ agonist DOI mimics the effect of 5-HT on the facilitation induced by baclofen. *Right:* temporal course of the PPF obtained from the sensorimotor cortex. *Left:* average evoked response obtained from 30 recordings at 0.1 Hz of frequency. The numbers indicate the time points when the records were obtained. Note that the strong amplitude reduction produced by baclofen is not reverted by DOI.

Since the above evidence was obtained using extracellular recordings (i.e., field potentials), it is difficult to determine whether the observed interaction between $GABA_B$ and 5-HT receptors occurs at the single neuronal level or from interactions at several units expressing different 5-HT receptor subtypes. However, evidences suggest that cross-talking between $5\text{-}HT_1$ and $5\text{-}HT_2$ receptors may occur in single pyramidal neurons. In fact, in situ hybridization studies have shown that both 5-HT receptor subtypes are coexpressed in a high proportion of pyramidal neurons from the prefrontal cortex (Amargós-Bosch et al., 2004). Intracellular recordings in pyramidal neurons of the sensory motor cortex (Foehring et al., 2002) and the amygdala (Sokal et al., 2005; Xiang et al., 2005) have revealed that 5-HT can induce 5-HT$_{1A}$-mediated hyperpolarizing responses and 5-HT$_{2A}$-mediated increases in firing frequency. Perhaps the most convincing data showing that 5-HT can exert opposing effects in the same cortical cell is provided by experiments where stimulation of the serotonergic raphe-cortical pathway evoked hyperpolarizing and depolarizing responses that could be blocked by selective 5-HT$_{1A}$ and 5-HT$_{2A}$ antagonists, respectively (Amargós-Bosch et al., 2004).

The GABAergic system is almost ubiquitous in the brain, and exerts its influence throughout two types of receptors – GABA$_A$ and GABA$_B$ receptors. The GABA$_B$ receptor activates a Gi/o-protein similar to the 5-HT$_{1A}$ receptor. These metabotropic GABA$_B$ receptors predominantly function as heterodimers of GABA$_{B(1)}$ and GABA$_{B(2)}$ subunits, but GABA$_{B(1)}$ can also form functional receptors in the absence of GABA$_{B(2)}$ (Mombereau et al., 2005). Mice lacking the GABA$_{B(1)}$ subunit have altered behavioral responses in tests for anxiety and depression. Interestingly, recent works have proposed the GABA system as target for novel antidepressants and mood stabilizers. Slattery and coworkers have suggested that GABA$_B$ receptor antagonists can mediate antidepressant-like behaviors, an effect that could be prevented by 5-HT depletion (Pilc and Nowak, 2005; Slattery et al., 2005). However, the mechanisms by which GABA$_B$ receptors may interact with the 5-HT system to produce the antidepressive influence remain unclear.

Selective 5-HT reuptake inhibitors (SSRI) such as fluoxetine and citalopram are among the antidepressive drugs most frequently used today. Recently it has been suggested that the therapeutic effect of these drugs is mediated by the 5-HT$_{1A}$ and 5-HT$_{2A/C}$ receptors present in the prefrontal cortex. Since these receptors are coexpressed in single pyramidal neurons and 5-HT is released by en passant axonal terminals producing volume transmission, there is the possibility that both receptors are activated simultaneously (Ridet and Privat, 2000). The result of this simultaneous activation, as it may happen after the application of the SSRIs, would depend on the balance between the inhibitory and excitatory effects of the pathways associated to these receptors. Indeed, it has been shown that agonists from the 5-HT$_{1A}$ receptors (Celada et al., 2004) and antagonists from the 5-HT$_2$ receptors produce antidepressive influences (Marek et al., 2003).

6.4 Implications

The possible interaction between 5-HT$_{1A}$ and 5-HT$_2$ receptors may underlie the therapeutic influence of 5-HT on depressive states. The cross-talk between the two associated pathways described in this chapter may explain some peculiar features of the action of the SSRI drugs, due to the simultaneous activation of the two types of receptors. If 5-HT receptor activation occurs when the GABA$_B$ receptors are already hyperactive, like it may occur during anxiety states frequently appearing during depressive episodes (Hirschfeld, 2001), a process of cross-talk analogous to the one observed in our experimental condition may occur, precluding the correct functioning of 5-HT receptors. Perhaps the cross-talking phenomena observed between GABA$_B$ and 5-HT receptors underlie the delayed therapeutic effects of SSRIs antidepressive drugs. There is the possibility that the action of SSRIs only reach its full effectiveness in a phase in which the anxiety process associated with clinical depression is reduced, and consequently the activation of $GABA_B$ receptors are also reduced (Celada et al., 2004; Pilc et al., 2005). This explanation would account for the antidepressive influence observed for the $GABA_B$ antagonists and the observed 5-HT dependence of this effect (Nakagawa et al., 1996; Slattery et al., 2005).

Acknowledgments

We would like to thank Jorge Ortigoza, Marco Atzori, and Kuei Y. Tseng for editorial reviews.

6.6 References

Amargós-Bosch, M., Bortolozzi, A., Puig, M. V., Serrats, J., Adell, A., Celada, P., Toth, M., Mengod, G. and Artigas, F. (2004) Co-expression and in vivo interaction of serotonin1A and serotonin2A receptors in pyramidal neurons of prefrontal cortex. Cereb Cortex. 14, 281-299.

Barnes, N. M. and Sharp, T. (1999) A review of central 5-HT receptors and their function. Neuropharmacology 38, 1083-1152.

Carrasco, G. A., Van de Kar, L. D., Jia, C., Xu, H., Chen, Z., Chadda, R., Garcia, F., Muma, N. A. and Battaglia, G. (2007) Single exposure to a serotonin 1A receptor agonist, (+)8-hydroxy-2-(di-n-propylamino)-tetralin, produces a prolonged heterologous desensitization of serotonin 2A receptors in neuroendocrine neurons in vivo. J Pharmacol Exp Ther. 320, 1078-1086.

Celada, P., Puig, M., Amargos-Bosch, M., Adell, A. and Artigas, F. (2004) The therapeutic role of 5-HT$_{1A}$ and 5-HT$_{2A}$ receptors in depression. J Psychiatry Neurosci. 29, 252-265.

Colino, A., Munoz, J. and Vara, H. (2002) [Short term synaptic plasticity]. Rev Neurol. 34, 593-599.

Cordeaux, Y. and Hill, S. J. (2002) Mechanisms of cross-talk between G-protein-coupled receptors. Neurosignals. 11, 45-57.

Evans, K. L., Cropper, J. D., Berg, K. A. and Clarke, W. P. (2001) Mechanisms of regulation of agonist efficacy at the 5-HT$_{1A}$ receptor by phospholipid-derived signaling components. J Pharmacol Exp Ther. 297, 1025-1035.

Foehring, R. C., van Brederode, J. F., Kinney, G. A. and Spain, W. J. (2002) Serotonergic modulation of supragranular neurons in rat sensorimotor cortex. J Neurosci. 22, 8238-8250.

Hempel, C. M., Hartman, K. H., Wang, X. J., Turrigiano, G. G. and Nelson, S. B. (2000) Multiple forms of short-term plasticity at excitatory synapses in rat medial prefrontal cortex. J Neurophysiol. 83, 3031-3041.

Hirschfeld, R. M. (2001) The comorbidity of major depression and anxiety disorders: Recognition and management in primary care. Prim Care Companion J Clin Psychiatry. 3, 244-254.

Marek, G. J., Carpenter, L. L., McDougle, C. J. and Price, L. H. (2003) Synergistic action of 5-HT$_{2A}$ antagonists and selective serotonin reuptake inhibitors in neuropsychiatric disorders. Neuropsychopharmacology. 28, 402-412.

Mombereau, C., Kaupmann, K., Gassmann, M., Bettler, B., van der Putten, H. and Cryan, J. F. (2005) Altered anxiety and depression-related behaviour in mice lacking GABA$_{B(2)}$ receptor subunits. Neuroreport. 16, 307-310.

Murakoshi, T., Song, S. Y., Konishi, S. and Tanabe, T. (2001) Multiple G-protein-coupled receptors mediate presynaptic inhibition at single excitatory synapses in the rat visual cortex. Neurosci Lett. 309, 117-120.

Nakagawa, Y., Ishima, T., Ishibashi, Y., Yoshii, T. and Takashima, T. (1996) Involvement of GABA$_B$ receptor systems in action of antidepressants: baclofen but not bicuculline attenuates the effects of antidepressants on the forced swim test in rats. Brain Res. 709, 215-220.

Pilc, A. and Nowak, G. (2005) GABAergic hypotheses of anxiety and depression: focus on GABA-B receptors. Drugs Today (Barc). 41, 755-766.

Ridet, I. and Privat, A. (2000) Volume transmission. Trends Neurosci. 23, 58-59.

Slattery, D. A., Desrayaud, S. and Cryan, J. F. (2005) GABA$_B$ receptor antagonist-mediated antidepressant-like behavior is serotonin-dependent. J Pharmacol Exp Ther. 312, 290-296.

Sokal, D. M., Giarola, A. S. and Large, C. H. (2005) Effects of GABA$_B$, 5-HT$_{1A}$, and 5-HT$_2$ receptor stimulation on activation and inhibition of the rat lateral amygdala following medial geniculate nucleus stimulation in vivo. Brain Res. 1031, 141-150.

Taniyama, K., Takeda, K., Ando, H., Kuno, T. and Tanaka, C. (1991) Expression of the GABA$_B$ receptor in Xenopus oocytes and inhibition of the response by activation of protein kinase C. FEBS Lett. 278, 222-224.

Thomson, A. M. (2000) Facilitation, augmentation and potentiation at central synapses. Trends Neurosci. 23, 305-312.

Torres-Escalante, J. L., Barral, J. A., Ibarra-Villa, M. D., Perez-Burgos, A., Gongora-Alfaro, J. L. and Pineda, J. C. (2004) 5-HT$_{1A}$, 5-HT$_2$, and GABA$_B$ receptors interact to modulate neurotransmitter release probability in layer 2/3 somatosensory rat cortex as evaluated by the paired pulse protocol. J Neurosci Res. 78, 268-278.

Tournois, C., Mutel, V., Manivet, P., Launay, J. M. and Kellermann, O. (1998) Cross-talk between 5-hydroxytryptamine receptors in a serotonergic cell line. Involvement of arachidonic acid metabolism. J Biol Chem. 273, 17498-17503.

Wang, S. J., Coutinho, V. and Sihra, T. S. (2002) Presynaptic cross-talk of beta-adrenoreceptor and 5-hydroxytryptamine receptor signalling in the modulation of glutamate release from cerebrocortical nerve terminals. Br J Pharmacol. 137, 1371-1379.

Wang, S. J., Cheng, I. I. and Gean, P. W. (1999) Cross-modulation of synaptic plasticity by beta-adrenergic and 5-HT$_{1A}$ receptors in the rat basolateral amygdala. J Neurosci. 19, 570-577.

Xiang, Z., Wang, L. and Kitai, S. T. (2005) Modulation of spontaneous firing in rat subthalamic neurons by 5-HT receptor subtypes. J Neurophysiol. 93, 1145-1157.

Zhang, Y., D'Souza, D., Raap, D. K., Garcia, F., Battaglia, G., Muma, N. A. and Van de Kar, L. D. (2001) Characterization of the functional heterologous desensitization of hypothalamic 5-HT$_{1A}$ receptors after 5-HT$_{2A}$ receptor activation. J Neurosci. 21, 7919-7927.

Zhang, Y., Gray, T. S., D'Souza, D. N., Carrasco, G. A., Damjanoska, K. J., Dudas, B., Garcia, F., Zainelli, G. M., Sullivan Hanley, N. R., Battaglia, G., Muma, N. A. and Van de Kar, L. D. (2004) Desensitization of 5-HT$_{1A}$ receptors by 5-HT$_{2A}$ receptors in neuroendocrine neurons in vivo. J Pharmacol Exp Ther. 310, 59-66.

Chapter 7

Serotonin Involvement in Plasticity of the Visual Cortex

Qiang Gu
Department of Neurobiology and Anatomy, Wake Forest University School of Medicine,
  Medical Center Boulevard, Winston-Salem, NC 27157, USA

7.1 Introduction

Serotonin (5-hydroxytryptamine, 5-HT) is widely distributed in the nervous system, and has been implicated in many aspects of behavioral and physiological regulation, including the control of blood pressure, body temperature, sleep, pain perception, sensory processing, anxiety, impulsivity, aggression, depression, sex and feeding behaviors, learning, and memory (Curzon, 1988; Wilkinson and Dourish, 1991; Lucki, 1992; Westenberg et al., 1996). A number of studies over the past decades provided evidence that serotonin also is involved in structural and functional remodeling of cortical circuits. Neurons in cortical areas that process sensory information such as vision, audition, and somatic sensation can modify their response properties following prolonged alterations in input activity, especially during early postnatal life. For instance, visual experience plays a pivotal role in shaping visual cortex structure and function during postnatal development. In addition, it has been also shown that nonvisual inputs to the visual cortex are important regulators for visual cortex plasticity (Gu, 2002, 2003). Neurotransmission of serotonin in the visual cortex is considered one of the nonvisual inputs that may serve as the neurochemical basis of attention, arousal, and motivation. In the following sections, serotonin-induced neuronal responses in the visual cortex will be briefly described. The contribution of serotonin to ocular dominance plasticity in the visual cortex and possible underlying mechanisms will then be discussed.

7.2 Serotonergic Actions in the Visual Cortex

One function of serotonergic transmission in the visual cortex appears to modify neuronal responsiveness to visual input. Visual cortex excitability was modulated by electrical stimulation of the raphe nucleus (Moyanova and Dimov, 1986; Gasanov et al., 1989), the origin of serotonergic projection to the cortex (Lidov et al., 1980). Similar to the electrical stimulation of raphe nucleus, which caused release of serotonin from its axon terminals, direct application of 5-HT to visual cortical neurons in vivo showed either facilitation or suppression of neuronal activity (Krnjevic and Phillis, 1963; Reader, 1978; Waterhouse et al., 1990). Because many 5-HT receptor subtypes are present in the neocortex, these diverse 5-HT effects are presumably dependent on postsynaptic composition of 5-HT subtype receptors as well as interactions of the serotonergic system and other transmitter systems. Among the various 5-HT subtype receptors, activation of $5-HT_1$ receptors results in increased conductance of potassium ions across the cell membrane and a subsequent decrease in neuronal excitability (McCormick et al., 1993). In contrast, $5-HT_2$ and $5-HT_4$ receptors increase cell excitability and firing rate by decreasing resting conductance of potassium ions and hence depolarizing the cell membrane (Andrade and Chaput, 1991; Panicker et al., 1991; Bockaert et al., 1992). The function of $5-HT_3$ receptors is to mediate fast synaptic transmission (Roerig et al., 1997; Ferezou et al., 2002). Activation of the $5-HT_3$ receptor decreases the amplitude and lateral extent of excitation in the visual cortex of the ferret (Roerig and Katz, 1997), probably resulting from an increase in GABAergic synaptic activity (Xiang and Prince, 2003). The expression of $5-HT_3$ receptors in subpopulations of GABAergic interneurons in the cortex provided anatomical evidence that serotonergic terminals can directly interact with GABAergic neurons (Tecott et al., 1993; Morales and Bloom, 1997; Paspalas and Papadopoulos, 2001; Ferezou et al., 2002). The serotonergic actions on visual information processing may be related behaviorally to arousal and attention.

7.3 Serotonergic Contributions to Ocular Dominance Plasticity

The primary visual cortex is the earliest stage of the visual pathway at which inputs of both eyes are mixed together. Visual cortical neurons have response specifications that are more complex than retinal ganglion cells and relay neurons in the lateral geniculate nucleus (LGN), including selective responses based on binocularity, orientation and direction of the visual stimulus (Hubel and Wiesel, 1962, 1963). Wiesel and Hubel were the first to demonstrate that the proper development of the visual cortex is dependent on visual experience. In a series of experiments (Wiesel and Hubel, 1963a, b, 1965), they showed that depriving kittens of vision in one eye by suturing eyelids early in their lives had three major consequences. First, the animals displayed little or no visual capacities through the deprived eye after the eyelids were re-opened. Second, cells in the LGN layer connected to the deprived eye were smaller than those cells receiving input from the normal eye. Third, cortical cells lost functional connections with the deprived eye, and the vast majority of cortical cells encountered were activated only by the stimulation of the nondeprived eye. These findings have been widely replicated, and similar effects of monocular deprivation have been observed in other species, such as monkey (Baker et al., 1974; Crawford et al., 1975; Hubel et al., 1977), rat (Maffei et al., 1992; Fagiolini et al., 1994), mouse (Gordon and Stryker, 1996), and ferret (Issa et al., 1999). In monocularly deprived animals, the strong input from the normal eye competes with the weak one from the deprived eye at the cortical level, so that input connections to the normal eye are retained and input connections to the deprived eye are decayed. Because the responsiveness of cortical neurons to visual stimulation can be easily manipulated by visual experience, ocular dominance plasticity has become an in vivo model for studying experience-dependent modifications of neural circuits (Hubel and Wiesel, 1970; Movshon and van Sluyters, 1981; Sherman and Spear, 1982; Frégnac and Imbert, 1984; Rauschecker, 1991; Daw, 1994, 1995; Singer, 1995; Katz and Shatz, 1996).

A contribution of 5-HT to ocular dominance plasticity was implicated when serotonin axons in kitten primary visual cortex were destroyed by chronic infusion of 5,7-dihydroxytryptamine (5,7-DHT) (Gu and Singer, 1995). In normal kittens, most visual cortical neurons respond to visual stimulation of either eye. Intracortical infusion of 5,7-DHT disrupted ocular dominance plasticity in the visual cortex. Despite monocular eyelid suture, neurons in the visual cortex remained binocular, while most neurons in the control (saline-treated) hemispheres displayed a normal shift of ocular dominance toward the nondeprived eye (Fig. 7.1).

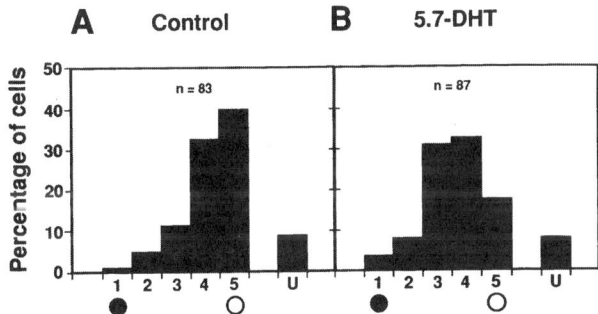

Figure 7.1 Ocular dominance (OD) histograms compiled from the primary visual cortex in three kittens after an intracortical infusion of saline (A) or of 5,7-DHT (B) coincident with 7 days of monocular deprivation. "n" refers to the number of recorded cells. Neurons assigned to OD categories 1 and 5 were monocular and activated by only the right eye ($\bullet$) or the left ($\circ$) eye, respectively. Neurons in OD category 3 were activated equally by stimulation of either eye. Cells in OD categories 2 and 4 were binocular, but their responses were clearly dominated by either the right or the left eye, respectively. The category labeled U contains those cells that could not be activated with the visual stimuli used. The OD distributions in the control and the 5,7-DHT-treated hemispheres differed significantly ($p < 0.005$, χ^2-test) (Gu and Singer, 1995).

In a different set of experiments, two broad serotonergic receptor antagonists, ketanserin and methysergide (Bradley et al., 1986), were infused either alone or together into the visual cortex of kittens. This combined intracortical infusion reduced ocular dominance plasticity and significantly blocked the expected shift in ocular dominance, but each of them infused alone had no effect (Gu and Singer, 1995). In addition, effects of intracortical infusion of mesulergine, a specific serotonin 5-HT$_{2c}$ receptor antagonist, on ocular dominance plasticity were examined in kittens. Similarly, mesulergine reduced ocular dominance shifts in visual cortex of monocularly deprived kittens (Fig. 7.2). These results suggest that serotonin contributes to ocular dominance plasticity, and that the 5-HT$_{2c}$ receptor subtype plays a key role in experience-dependent synaptic modifications in kitten visual cortex.

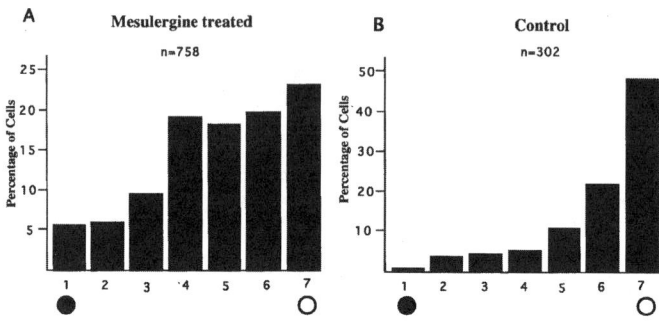

Figure 7.2 Ocular dominance (OD) histograms compiled from 1,060 cells recorded in the primary visual cortex of 19 kittens that had received an intracortical infusion of mesulergine (A) or vehicle (B) coincident with 7 days of monocular deprivation. "*n*" refers to the number of recorded cells. Cells of group 1 were activated by the deprived eye (●); for cells of group 2, there was marked dominance of the deprived eye; for group 3, slight dominance. For cells in group 4, there was no obvious difference between the two eyes. In group 5, the nondeprived eye dominated slightly and in group 6, markedly. Cells in group 7 were activated only by the non-deprived eye (○). The difference between the OD distributions in the control and mesulergine-treated hemispheres was statistically significant ($p < 0.001$, χ^2-test) (Wang et al., 1997).

7.4 Serotonergic Effects on Plasticity in Visual Cortex Slices

In parallel to in vivo electrophysiological examinations of cortical plasticity, visual cortex slices have been utilized in studies of long-term potentiation (LTP) and long-term depression (LTD), which are measures of long-lasting changes in synaptic efficacy (Artola and Singer, 1987; Komatsu et al., 1988; Perkins and Teyler, 1988; Kirkwood and Bear, 1994; Kojic et al., 1997). While in vitro studies of synaptic plasticity have demonstrated certain similarities to in vivo studies, such as age-dependent decline of plasticity in the visual cortex, they also have distinct characteristics. For instance, the outcome of slice preparations depends on more variables, such as the stimulation-recording path and the composition of the slice incubation medium. A number of studies were carried out using visual cortex slices to determine a potential role of serotonin in visual cortex plasticity. The results were more diverse than those of in vivo studies.

In slices from the visual cortex of 40- to 80-day-old kittens, application of serotonin in the incubation medium markedly facilitated the induction of both LTP and LTD in layer 4, with underlying white matter stimulations (Kojic et al., 1997). The effect of serotonin was completely blocked by mesulergine (Kojic et al., 1997,

2001). These in vitro results are consistent with in vivo results, and suggest that serotonin promotes visual cortex plasticity and that the 5-HT$_{2c}$ receptor is a major contributor for mediating serotonergic facilitation. It has been shown that 5-HT$_{2c}$ receptors are not evenly distributed in kitten visual cortex; they are transiently expressed in alternate 5-HT$_{2c}$ receptor-rich and 5-HT$_{2c}$ receptor-poor zones that have a similar dimension as that of ocular dominance columns (Dyck and Cynader, 1993a). Further correlation analyses in kitten visual cortex revealed that serotonin promoted LTP in 5-HT$_{2c}$ receptor-rich zones, whereas serotonin promoted LTD in 5-HT$_{2c}$ receptor-poor zones (Kojic et al., 2000). These results imply a molecular mechanism, by which serotonin may facilitate activity-dependent segregation of ocular dominance columns through 5-HT$_{2c}$ receptors (Fig. 7.3).

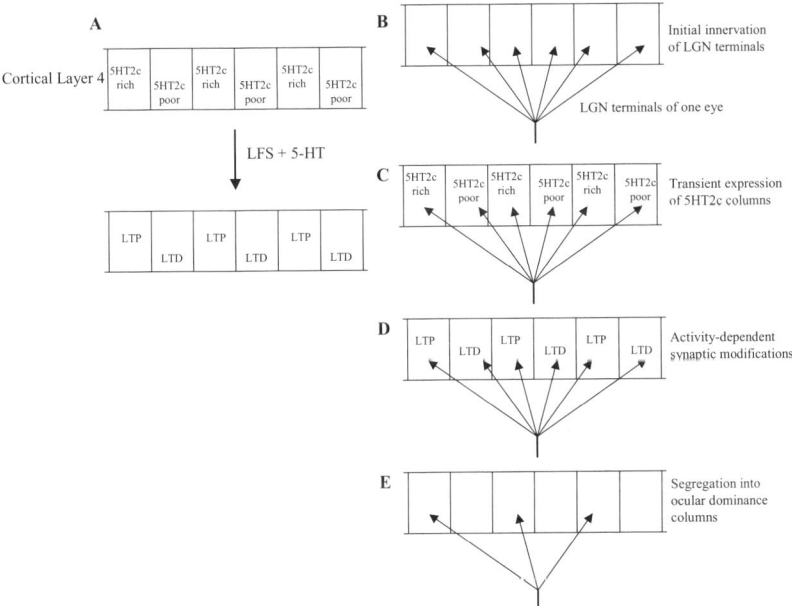

Figure 7.3 Possible mechanism by which a transient expression pattern of 5-HT$_{2c}$ receptors could leads to segregation of LGN terminals in the visual cortex. (A) Low-frequency stimulation (LFS) in the presence of 5-HT-induced LTP within 5-HT$_{2c}$ receptor-rich columns, whereas LTD was induced within 5-HT$_{2c}$ receptor-poor columns (Kojic et al., 2000). LGN terminals initially innervate the entire cortical layer 4 (B). However, because of differentially expressed 5-HT$_{2c}$ receptors within layer 4 during visual cortex development (C), the same input could induce either LTP or LTD depending on postsynaptic 5-HT$_{2c}$ receptor levels (D). LGN terminals with LTP will gain functional strength and stay, while LGN terminals with LTD will loss functional connectivity and withdraw, which lead ultimately to segregation of LGN terminals into ocular dominance columns (E).

During visual cortex development, LGN terminals representing the two eyes are initially mixed and evenly distributed in cortical layer 4. They later become separated into discrete columns that represent either the left or the right eye. Transient expression of 5-HT_{2c} receptors in "rich" and "poor" columns during visual cortex development could provide spatial and temporal instructions that allow axon terminals in the 5-HT_{2c} receptor-rich zone to stay (via LTP) and axon terminals in the 5-HT_{2c} receptor-poor zone to withdraw (via LTD). This model, based on the expression of particular transmitter receptors, may help to explain the segregation of input axon terminals and the formation of ocular dominance columns in the visual cortex. However, several issues need to be resolved before this model can be fully established: (i) which eye dominance columns do 5-HT_{2c} receptor-rich columns represent, the left or right eye?; (ii) What would be the complementary transmitter/receptor systems or neurochemical molecules that contribute to the other eye's dominance columns?; (iii) Is the correlation of these neurochemical markers and eye dominance columns genetically determined, or is it dependent on other factors during early visual cortex development? A number of neurochemical markers have already been found to display columnar distribution patterns in the developing visual cortex (Schoen et al., 1990; Dyck and Cynader, 1993a, b; Trepel et al., 1998; Murphy et al., 2001). The challenging tasks for future investigations would be to identify molecular cues that contribute to the formation of ocular dominance columns and to correlate the molecular cues with the respective eye input terminals.

Studies of serotonergic effects on synaptic plasticity using visual cortex slices derived from rats generated somewhat different results. Recording in layers 2 and 3 with stimulation in layer 4 in 3- to 5-week-old rat visual cortex indicated that either serotonin (Edagawa et al., 1998a), 8-hydroxy-2-(N,N-dipropylamino)tetralin (8-OH-DPAT) (Edagawa et al., 1998b) (a 5-HT_{1A} receptor agonist), or 1-(2,5-dimethyl-4-iodophenyl)-2-aminopropane (DOI) (Edagawa et al., 2000) (a 5-HT_2 receptor agonist) prevented the induction of LTP, whereas pindolol (a 5-HT_1 receptor antagonist) or ritanserin (a $5\text{-HT}_2/5\text{-HT}_7$ receptor antagonist) could abolish the effect of serotonin or 8-OH-DPAT or DOI (Edagawa et al., 1998a, b, 2000). Bath application of methysergide-induced LTP in cortical layers 2 and 3 following white matter stimulation in 5-week-old rat visual cortex, which in the absence of methysergide showed no LTP after high-frequency stimulation in the white matter (Edagawa et al., 2001). In addition, depletion of serotonin by 5,7-DHT (Edagawa et al., 2001) or by para-chloroamphetamine (Kim et al., 2006) increased LTP induction in visual cortex slices of 5-week-old rats. These results suggest that serotonin has an inhibitory effect on LTP in rat visual cortex. However, when recorded in layer 5 with stimulation in layer 4 in 15- to 25-day-old rat visual cortex, bath application of ketanserin, a 5-HT_2 receptor antagonist, reduced the incidence of LTP (Komatsu, 1996), suggesting that activation of 5-HT_2 receptors is required to initiate LTP in this pathway.

Together, the available evidence suggests that serotonin facilitates LTP and LTD in kitten visual cortex slices (Kojic et al., 1997, 2000, 2001), while in rat visual cortex slices serotonin may inhibit (Edagawa et al., 1998a, b, 2000, 2001; Kim et al., 2006) or facilitate LTP (Komatsu, 1996) depending on the specific protocol used. It remains to be determined whether serotonin has any effect on LTD in rat visual cortex slices, since the plasticity of ocular dominance is correlated with homosynaptic LTD rather than LTP in the visual cortex (Rittenhouse et al., 1999; Heynen et al., 2003). The apparent discrepancies concerning 5-HT effects on LTP between cat and rat visual cortex slices were attributed to a species difference as well as different synaptic pathways tested (Edagawa et al., 2001). This is conceivable, since serotonergic innervation in the visual cortex showed different developmental profiles between cats and rats. For example, the level of serotonin in 4-week-old kitten visual cortex, which is at its highest level of cortical plasticity, is already at the peak (Jonsson and Kasamatsu, 1983), while in rat visual cortex it continuously increases until adulthood (Edagawa et al., 2001). In addition, postnatal expression of serotonin subtype receptors in the visual cortex of rats and cats showed different laminar distribution patterns and development profiles (Marcinkiewicz et al., 1984; Pazos and Palacios, 1985; Gozlan et al., 1990; Dyck and Cynader, 1993a). Since different stimulation and recording protocols have been applied in electrophysiological studies of cats' and rats' visual cortex slices, it would be more appropriate to compare results when a unified protocol is employed and the same synaptic pathway (e.g., stimulation in the white matter and recording in layer 4) is examined.

7.5 Mechanisms of Serotonin in the Plasticity of Ocular Dominance

Serotonin affects membrane potentials at the cellular level, either directly through ion-gated channel or indirectly through intracellular second messengers, and consequently to modulate membrane excitability. Serotonin also interacts with other transmitter systems, such as glutametergic and GABAergic transmission. A mechanism associated with NMDA receptor-gated synaptic modifications may be considered to explain the facilitatory action of serotonin in ocular dominance plasticity. In vitro experiments have shown that serotonin enhances depolarizing responses to excitatory amino acids in the neocortex of cats (Nedergaard et al., 1987) and rats (Reynolds et al., 1988). This response could be achieved by reducing potassium ion conductance (Andrade and Chaput, 1991; Panicker et al., 1991; Bockaert et al., 1992), leading to a slow membrane depolarization that in turn enhances the flux of calcium ions from extra- to intracellular compartments through NMDA receptors.

Another possibility is that synergistic interactions at the level of second messengers enhance the intracellular signals induced by the excitatory visual input. For instance, activation of 5-HT$_{2c}$ receptors stimulate phospholipid turnover (Hoyer and Martin, 1997), which in turn could contribute to enhance intracellular effects via inositol triphosphate through calcium release from intracellular stores and protein kinase C activation via diacyl glycerol. Through the augmentation of intracellular responses following NMDA receptor activation, input activity can then induce activity-dependent modifications of synaptic connections (Choi et al., 2005). Thus, activation of serotonergic receptors could increase the probability that retinal input drives cortical neurons above the threshold that needs to be reached for the weakening of the deprived visual afferents (Frégnac et al., 1988).

Evidences also indicate that 5-HT axons can directly connect with GABAergic interneurons (Paspalas and Papadopoulos, 2001) and influence GABAergic inhibition in the visual cortex (Roerig et al., 1997; Xiang et al., 1998; Edagawa et al., 2000). Therefore, another possible mechanism of 5-HT in cortical plasticity is to regulate neuronal inhibition, an important factor for determining the threshold for activity-dependent synaptic modifications (Hensch et al., 1998; Fagiolini and Hensch, 2000).

7.6 Conclusions

The available evidence suggests that the role of 5-HT in the cerebral cortex is to modulate the excitability of cortical neurons, in order to gate information processes, to enhance the signal-to-noise ratio, and to determine the threshold for activity-dependent synaptic modifications. The involvement of 5-HT in plasticity of the visual cortex has been investigated in vivo and in vitro; 5-HT appears to facilitate the plasticity of ocular dominance in the developing visual cortex. Among 5-HT receptor subtypes, the 5-HT$_{2c}$ receptor emerges to be a predominant player in ocular dominance plasticity. A prerequisite for cortical plasticity is the activation of NMDA receptors in the cortex (Singer et al., 1988; Collingridge and Singer, 1990; Bear, 1996). However, activation of NMDA receptor-dependent processes appears to be only a necessary, but not sufficient, condition to induce activity-dependent modifications. For example, monocular light stimulation alone cannot induce an ocular dominance shift in the visual cortex of anaesthetized and paralyzed animals. Therefore, the occurrence of adaptive changes in the visual cortex depends not only on the visual input, but also on the functional state of the visual cortex. The latter is mainly controlled by nonvisual input systems, which can enhance visual input responses to overcome the threshold for activity-dependent physiological and anatomical changes. The nonvisual inputs could be the neuromodulatory inputs from subcortical regions, since these transmitters can influence the state of arousal, attention, and motivation. There is also evidence that serotonin could directly act on GABAergic neurons in the visual cortex. Thus, the contribution of 5-HT to ocular dominance plasticity could be mediated through interactions with both NMDA- and GABAergic receptor-gated processes.

Acknowledgments

Supported by NIH EY014892 (Q.G.). The author thanks Karen Klein for her editorial contributions to this manuscript.

References

Andrade, R. and Chaput, Y. (1991) The electrophysiology of serotonin receptor subtypes. In: S.J. Peroutka (Ed.), *Serotonin Receptor Subtypes: Basic and Clinical Aspects*. Wiley-Liss, New York, pp. 103–124.

Artola, A. and Singer, W. (1987) Long-term potentiation and NMDA receptors in rat visual cortex. Nature 330, 649–652.

Baker, F.H., Grigg, P. and Von Noorden, G.K. (1974) Effects of visual deprivation and strabismus on the response of neurones in the visual cortex of the monkey, including studies on the striate and prestriate cortex in the normal animal. J. Brain Res. 66, 185–208.

Bear, M.F. (1996) Progress in understanding NMDA-receptor-dependent synaptic plasticity in the visual cortex. J. Physiol. (Paris) 90, 223–227.

Bockaert, J., Fozard, J.R., Dumuis, A. and Clarke, D.E. (1992) The 5-HT4 receptor: a place in the sun. Trends Pharmacol. Sci. 13, 141–145.

Bradley, P.B., Engel, G., Feniuk, W., Fozard, J.R., Humphrey, P.P.A., Middlemiss, D.N., Mylecharane, E.J., Richardson, B.P. and Saxena, P.R. (1986) Proposals for the classification and nomenclature of functional receptors for 5-hydroxytryptamine. Neuropharmacology 25, 563–576.

Choi, S.Y., Chang, J., Jiang, B., Seol, G.H., Min, S.S., Han, J.S., Shin, H.S., Gallagher, M. and Kirkwood, A. (2005) Multiple receptors coupled to phospholipase C gate long-term depression in visual cortex. J. Neurosci. 25, 11433–11443.

Collingridge, G.L. and Singer, W. (1990) Excitatory amino acid receptors and synaptic plasticity. Trends Pharmacol. Sci. 11, 290–296.

Crawford, M.L.J., Blake, R., Cool, S.J. and von Noorden, G.K. (1975) Physiological consequences of unilateral and bilateral eye closure in macaque monkeys: some further observations. Brain Res. 84, 150–154.

Curzon, G. (1988) Serotonergic mechanisms of depression. Clin. Neuropharmacol. 11 (Suppl. 2), S11–S20.

Daw, N.W. (1994) Mechanisms of plasticity in the visual cortex. The Friedenwald Lecture. Invest. Ophthalmol. Vis. Sci. 35, 4168–4179.

Daw, N.W. (1995) *Visual Development*. Plenum Publishing, New York.

Dyck, R.H. and Cynader, M.S. (1993a) Autoradiographic localization of serotonin receptor subtypes in cat visual cortex: transient regional, laminar, and columnar distributions during postnatal development. J. Neurosci. 13, 4316–4338.

Dyck, R.H. and Cynader, M.S. (1993b) An interdigitated columnar mosaic of cytochrome oxidase, zinc, and neurotransmitter-related molecules in cat and monkey visual cortex. Proc. Natl. Acad. Sci. U.S.A. 90, 9066–9069.

Edagawa, Y., Saito, H. and Abe, K. (1998a) Serotonin inhibits the induction of long-term potentiation in rat primary visual cortex. Prog. Neuropsychopharmacol. Biol. Psychiatry 22, 983–997.

Edagawa, Y., Saito, H. and Abe, K. (1998b) 5-HT1A receptor-mediated inhibition of long-term potentiation in rat visual cortex. Eur. J. Pharmacol. 349, 221–224.

Edagawa, Y., Saito, H. and Abe, K. (2000) The serotonin 5-HT2 receptor-phospholipase C system inhibits the induction of long-term potentiation in the rat visual cortex. Eur. J. Neurosci. 12, 1391–1396.

Edagawa, Y., Saito, H. and Abe, K. (2001) Endogenous serotonin contributes to a developmental decrease in long-term potentiation in the rat visual cortex. J. Neurosci. 21, 1532–1537.

Fagiolini, M. and Hensch, T.K. (2000) Inhibitory threshold for critical-period activation in primary visual cortex. Nature 404, 183–186.

Fagiolini, M., Pizzorusso, T., Berardi, N., Domenici, L. and Maffei, L. (1994) Functional postnatal development of the rat primary visual cortex and the role of visual experience: dark rearing and monocular deprivation. Vis. Res. 34, 709–720.

Ferezou, I., Cauli, B., Hill, E.L., Rossier, J., Hamel, E. and Lambolez, B. (2002) 5-HT3 receptors mediate serotonergic fast synaptic excitation of neocortical vasoactive intestinal peptide/cholecystokinin interneurons. J. Neurosci. 22, 7389–7397.

Frégnac, Y. and Imbert, M. (1984) Development of neuronal selectivity in primary visual cortex of cat. Physiol. Rev. 64, 325–434.

Frégnac, Y., Shulz, D., Thorpe S. and Bienenstock, E. (1988) A cellular analogue of visual cortical plasticity. Nature 333, 367–370.

Gasanov, G.G., Mamedov, Z.G. and Samedova, N.F. (1989) Changes in reactivity of neurons of the visual cortex under influence of the posterolateral hypothalamus and the nuclei of the midbrain raphe. Neurosci. Behav. Physiol. 19, 169–175.

Gordon, J.A. and Stryker, M.P. (1996) Experience-dependent plasticity of binocular responses in the primary visual cortex of the mouse. J. Neurosci. 16, 3274–3286.

Gozlan, H., Daval, G., Verge, D., Spampinato, U., Fattaccini, C.M., Gallissot, M.C., el Mestikawy, S. and Hamon, M. (1990) Aging associated changes in serotoninergic and dopaminergic pre- and postsynaptic neurochemical markers in the rat brain. Neurobiol. Aging 11, 437–449.

Gu, Q. (2002) Neuromodulatory transmitter systems in the cortex and their role in cortical plasticity. Neuroscience 111, 815–835.

Gu, Q. (2003) Contribution of acetylcholine to visual cortex plasticity. Neurobiol. Learn. Mem. 80, 291–301.

Gu, Q. and Singer, W. (1995) Involvement of serotonin in neuronal plasticity of visual cortex. Eur. J. Neurosci. 7, 1146–1153.

Hensch, T.K., Fagiolini, M., Mataga, N., Stryker, M.P., Baekkeskov, S. and Kash, S.F. (1998) Local GABA circuit control of experience-dependent plasticity in developing visual cortex. Science 282, 1504–1508.

Heynen, A.J., Yoon, B.J., Liu, C.H., Chung, H.J., Huganir, R.L. and Bear, M.F. (2003) Molecular mechanism for loss of visual cortical responsiveness following brief monocular deprivation. Nat. Neurosci. 6, 854–862.

Hoyer, D. and Martin, G. (1997) 5-HT receptor classification and nomenclature: towards a harmonization with the human genome. Neuropharmacology 36, 419–428.

Hubel, D.H. and Wiesel, T.N. (1962) Receptive fields, binocular interaction and functional architecture in the cat's visual cortex. J. Physiol. (Lond.) 160, 106–154.

Hubel, D.H. and Wiesel, T.N. (1963) Receptive fields of cells in striate cortex of very young, visually inexperienced kittens. J. Neurophysiol. 26, 994–1002.

Hubel, D.H. and Wiesel, T.N. (1970) The period of susceptibility to the physiological effects of unilateral eye closure in kittens. J. Physiol. (Lond.) 206, 419–436.

Hubel, D.H., Wiesel, T.N. and LeVay, S. (1977) Plasticity of ocular dominance columns in monkey striate cortex. Philos. Trans. R. Soc. London Ser. B 278, 377–409.

Issa, N.P., Trachtenberg, J.T., Chapman, B., Zahs, K.R. and Stryker, M.P. (1999) The critical period for ocular dominance plasticity in the ferret's visual cortex. J. Neurosci. 19, 6965–6978.

Jonsson, G. and Kasamatsu, T. (1983) Maturation of monoamine neurotransmitters and receptors in cat occipital cortex during postnatal critical period. Exp. Brain Res. 50, 449–458.

Katz, L.C. and Shatz, C.J. (1996) Synaptic activity and the construction of cortical circuits. Science 274, 1133–1138.

Kim, H.S., Jang, H.J., Cho, K.H., Hahn, S.J., Kim, M.J., Yoon, S.H., Jo, Y.H., Kim, M.S. and Rhie, D.J. (2006) Serotonin inhibits the induction of NMDA receptor-dependent long-term potentiation in the rat primary visual cortex. Brain Res. 1103, 49–55.

Kirkwood, A. and Bear, M.F. (1994) Homosynaptic long-term depression in the visual cortex. J. Neurosci. 14, 3404–3412.

Kojic, L., Gu, Q., Douglas, R.M. and Cynader, M.S. (1997) Serotonin facilitates synaptic plasticity in kitten visual cortex: an in vitro study. Dev. Brain Res. 101, 299–304.

Kojic, L., Dyck, R., Gu, Q., Douglas, R.M., Matsubara, J. and Cynader, M.S. (2000) Columnar distribution of serotonin-dependent plasticity within kitten striate cortex. Proc. Natl. Acad. Sci. U.S.A. 97, 1841–1844.

Kojic, L., Gu, Q., Douglas, R.M. and Cynader, M.S. (2001) Laminar distribution of cholinergic- and serotonergic-dependent plasticity within kitten visual cortex. Dev. Brain Res. 126, 157–162.

Komatsu, Y. (1996) GABAB receptors, monoamine receptors, and postsynaptic inositol trisphosphate-induced Ca^{2+} release are involved in the induction of long-term potentiation at visual cortical inhibitory synapses. J. Neurosci. 16, 6342–6352.

Komatsu, Y., Fujii, K., Maeda, J., Sakaguchi, H. and Toyama, K. (1988) Long-term potentiation of synaptic transmission in kitten visual cortex. J. Neurophysiol. 59, 124–141.

Krnjevic, K. and Phillis, J.W. (1963) Iontophoretic studies of neurons in the mammalian cerebral cortex. J. Physiol. (Lond.) 165, 274–304.

Lidov, H.G.W., Grzanna, R. and Molliver, M.E. (1980) The serotonin innervation of the cerebral cortex in the rat – an immunohistochemical analysis. Neuroscience 5, 207–227.

Lucki, I. (1992) 5-HT1 receptors and behavior. Neurosci. Biobehav. Rev. 16, 83–93.

Maffei, L., Berardi, N., Domenici, L., Parisi, V. and Pizzorusso, T. (1992) Nerve growth factor (NGF) prevents the shift in ocular dominance distribution of visual cortical neurons in monocularly deprived rats. J. Neurosci. 12, 4651–4662.

Marcinkiewicz, M., Verge, D., Gozlan, H., Pichat, L. and Hamon, M. (1984) Autoradiographic evidence for the heterogeneity of 5-HT1 sites in the rat brain. Brain Res. 291, 159–163.

McCormick, D.A., Wang, Z. and Huguenard, J. (1993) Neurotransmitter control of neocortical neuronal activity and excitability. Cereb. Cortex 3, 387–398.

Morales, M. and Bloom, F.E. (1997) The 5-HT3 receptor is present in different subpopulations of GABAergic neurons in the rat telencephalon. J. Neurosci. 17, 3157–3167.

Movshon, J.A. and van Sluyters, R.C. (1981) Visual neural development. Ann. Rev. Psychol. 32, 477–522.

Moyanova, S. and Dimov, S. (1986) Modulation of visual excitability cycles in some brain structures by high-frequency stimulation of raphe dorsal nucleus in cats. Acta Physiol. Pharmacol. Bulg. 12, 17–25.

Murphy, K.M., Duffy, K.R., Jones, D.G. and Mitchell, D.E. (2001) Development of cytochrome oxidase blobs in visual cortex of normal and visually deprived cats. Cereb. Cortex 11, 122–135.

Nedergaard, S. Engberg, I. and Flatman, J.A. (1987) The modulation of excitatory amino acid responses by serotonin in the cat neocortex. Cell. Mol. Neurobiol. 7, 367–379.

Panicker, M.M., Parker, I. and Miledi, R. (1991) Receptors of the serotonin 1C subtype expressed from cloned DNA mediate the closing of K+ membrane channels encoded by brain mRNA. Proc. Natl. Acad. Sci. U.S.A. 88, 2560–2562.

Perkins IV, A.T. and Teyler, T.J. (1988) A critical period for long-term potentiation in the developing rat visual cortex. Brain Res. 439, 222–229.

Paspalas, C.D. and Papadopoulos, G.C. (2001) Serotoninergic afferents preferentially innervate distinct subclasses of peptidergic interneurons in the rat visual cortex. Brain Res. 891, 158–167.

Pazos, A. and Palacios, J.M. (1985) Quantitative autoradiographic mapping of serotonin receptors in the rat brain. I. Serotonin-1 receptors. Brain Res. 346, 205–230.

Rauschecker, J.P. (1991) Mechanisms of visual plasticity: Hebb synapses, NMDA receptors, and beyond. Physiol. Rev. 71, 587–615.

Reader, T.A. (1978) The effects of dopamine, noradrenaline and serotonin in the visual cortex of the cat. Experimentia 34, 1586–1588.

Reynolds, J.N., Baskys, A. and Carlen, P. (1988) The effects of serotonin on N-methyl-D-aspartate and synaptically evoked depolarizations in rat neocortex. Brain Res. 456, 286–292.

Rittenhouse, C.D., Shouval, H.Z., Paradiso, M.A. and Bear, M.F. (1999) Monocular deprivation induces homosynaptic long-term depression in visual cortex. Nature 397, 347–350.

Roerig, B. and Katz, L.C. (1997) Modulation of intrinsic circuits by serotonin 5-HT3 receptors in developing ferret visual cortex. J. Neurosci. 17, 8324–8338.

Roerig, B., Nelson, D.A. and Katz, L.C. (1997) Fast synaptic signaling by nicotinic acetylcholine and serotonin 5-HT3 receptors in developing visual cortex. J. Neurosci. 17, 8353–8362.

Sherman, S.M. and Spear, P.D. (1982) Organization of visual pathways in normal and visually deprived cats. Physiol. Rev. 62, 738–855.

Schoen, S.W., Leutenecker, B., Kreutzberg, G.W. and Singer, W. (1990) Ocular dominance plasticity and developmental changes of 5′-nucleotidase distributions in the kitten visual cortex. J. Comp. Neurol. 296, 379–392.

Singer, W. (1995) Development and plasticity of cortical processing architectures. Science 270, 758–764.

Singer, W., Artola, A., Greuel, J. and Gu, Q. (1988) Gating of NMDA-receptor mediated neuronal plasticity. In: E.A. Cavalheiro, J. Lehman and L. Turski (Eds.), *Frontiers in Excitatory Amino Acid Research*. Alan R. Liss, New York, pp. 443–450.

Tecott, L.H., Maricq, A.V. and Julius, D. (1993) Nervous system distribution of the serotonin 5-HT3 receptor mRNA. Proc. Natl. Acad. Sci. U.S.A. 90, 1430–1434.

Trepel, C., Duffy, K.R., Pegado, V.D. and Murphy, K.M. (1998) Patchy distribution of NMDAR1 subunit immunoreactivity in developing visual cortex. J. Neurosci. 18, 3404–3415.

Wang, Y.-C., Gu, Q. and Cynader, M.S. (1997) Blockade of serotonin-2c receptors by mesulergine reduces ocular dominance plasticity in kitten visual cortex. Exp. Brain Res. 114, 321–328.

Waterhouse, B.D., Azizi, S.A., Burne, R.A. and Woodward, D.J. (1990) Modulation of rat cortical area 17 neuronal responses to moving visual stimuli during norepinephrine and serotonin microiontophoresis. Brain Res. 514, 276–292.

Westenberg, H.G., Murphy, D.L. and Den Boer, J.A. (1996) *Advances in the Neurobiology of Anxiety Disorders*. Wiley, New York.

Wiesel, T.N. and Hubel, D.H. (1963a) Effects of visual deprivation on morphology and physiology of cells in the cat's lateral geniculate body. J. Neurophysiol. 26, 978–993.

Wiesel, T.N. and Hubel, D.H. (1963b) Single-cell responses in striate cortex of kittens deprived of vision in one eye. J. Neurophysiol. 26, 1003–1017.

Wiesel, T.N. and Hubel, D.H. (1965) Comparison of the effects of unilateral and bilateral eye closure on cortical unit responses in kittens. J. Neurophysiol. 28, 1029–1040.

Wilkinson, L.O. and Dourish, C.T. (1991) Serotonin and animal behavior. In: S.J. Peroutka (Ed.), *Serotonin Receptor Subtypes: Basic and Clinical Aspects*. Wiley-Liss, New York, pp. 147–210.

Xiang, Z. and Prince, D.A. (2003) Heterogeneous actions of serotonin on interneurons in rat visual cortex. J. Neurophysiol. 89, 1278–1287.

Xiang, Z., Huguenard, J.R. and Prince, D.A. (1998) GABAA receptor-mediated currents in interneurons and pyramidal cells of rat visual cortex. J. Physiol. (Lond.) 506, 715–730.

Chapter 8

Cellular Mechanisms of Working Memory and its Modulation by Dopamine in the Prefrontal Cortex of Primates and Rats

Guillermo Gonzalez-Burgos[1], Sven Kröner[2] and Jeremy K. Seamans[3]

[1]Department of Psychiatry, University of Pittsburgh, PA 15213, USA; [2]Department of Physiology & Neuroscience, Medical University of South Carolina, Charleston, SC, USA; [3]Brain Research Centre, University of British Columbia, Vancouver, BC, Canada

8.1 Introduction

It is often assumed that the massive expansion of the prefrontal cortex (PFC) in humans and nonhuman primates was a pivotal determinant of evolutionary success. Whereas dorsolateral prefrontal cortical areas in macaque monkeys and humans share multiple characteristics with regard to cytoarchitecture, hodology (Petrides and Pandya, 1999), and DA innervation (Lewis and Sesack, 1997), the organization of the frontal cortex and its dopaminergic innervation are vastly different in rats and primates (Berger et al., 1991; Preuss, 1995; Williams and Goldman-Rakic, 1998; Uylings et al., 2003). This raises the question of whether certain properties of the primate PFC are not present, or only represented in a rudimentary fashion in other species. Seminal findings from brain lesion and behavioral electrophysiology studies determined that different subregions of the PFC engage in distinct cognitive functions. Among these findings is the observation that cells in the dorsolateral PFC regions exhibit persistent activity in response to external stimuli, after the stimuli have ceased. This activity may reflect the ability to temporarily hold items in working memory (WM), a cognitive function required for the temporal organization of behavior (Fuster, 1997). Persistent activity is believed to be maintained through recurrent excitation in assemblies of PFC neurons. In the primate PFC, such cell assemblies may be organized into interconnected stripe-like clusters, similar to the columns in primary sensory cortex. By means of persistent firing, the PFC is thought to transiently store stimulus-related information that is task-relevant, for example, "what" an object is, or "where" it is located. Thus, primate PFC neurons are said to have "memory fields," analogous to the receptive fields in sensory cortices (Goldman-Rakic, 1995; Rao et al., 1997). Importantly, delay-period activity and the properties of the memory fields in primate PFC are strongly modulated by dopamine (DA) (Williams and Goldman-Rakic, 1995; Sawaguchi, 2001). Indeed, it is commonly assumed that disruption of WM function by local DA depletion (Brozoski et al., 1979) or by local DA receptor blockade in the PFC (Sawaguchi and Goldman-Rakic, 1991) results from interference with persistent firing. Whether the ability to exhibit persistent activity associated with WM is a property conserved across species or unique to primate PFC neurons and networks is not well understood. Persistent activity is found in various regions of the primate brain, including parietal and temporal cortices, as well as the thalamus (Fuster and Alexander, 1971; Miller et al., 1996; Chafee and Goldman-Rakic, 1998). However, persistent activity of primate PFC neurons has been distinguished from that in other brain regions because in PFC it is sustained even in the presence of distractors (Constantinidis and Steinmetz, 1996; Miller et al., 1996). Moreover, behavioral errors correlate with failure of PFC neurons to sustain activity during the delay (Funahashi et al., 1989) and sustained activity follows changes in delay duration (Fuster and Alexander, 1971). The presence of delay-related activity in the rat PFC which was originally demonstrated in 1990 (Batuev et al., 1990), but recently has received renewed attention, raises a number of questions. Are there clear features that distinguish single-cell properties in primate versus rat PFC? Is persistent activity in rat PFC similar to that found in primate dorsolateral PFC, even though an anatomical homologue of this region may not exist in rats? What is the cellular architecture that produces delay-related activity in the rat PFC, given the lack of evidence for columns, stripes, or other types of functional cell clusters in rat PFC (Fisk and Wyss, 1999; Gabbott et al., 2003)? In primate PFC neurons, persistent activity is influenced by task-specific rules, by the type of information processed and by the prospective versus retrospective nature of the information. Does delay activity in rat PFC exhibit the same richness of modulation? Moreover, is the neuromodulation of delay activity by DA similar across species? Does DA exert the same biophysical effects on PFC neurons across species?

In this chapter we first present a simplified and abbreviated view of the complex organization of the PFC in rodents and primates with an emphasis on those properties that might represent the substrate of putative WM functions. We will then compare the primate dorsolateral and rat medial PFC modulation by dopaminergic afferents. Finally, we give an overview of the properties of PFC neurons in primate and rat PFC during WM tasks.

8.2 Anatomical Architecture of the PFC

Among various mammalian species the PFC is commonly defined as the cortical area with strong reciprocal connections with the mediodorsal thalamic nucleus (MD) (Preuss, 1995; Uylings et al., 2003). However, the MD also projects to other areas of the frontal cortex, and in addition the PFC receives substantial innervation from various other thalamic nuclei (Giguere and Goldman-Rakic, 1988; Uylings et al., 2003). In primates, the MD projects to Brodmann areas 8, 9, 10, 45, and 46, which comprise the dorsolateral PFC, and also to the orbitofrontal/ventromedial PFC areas 11, 12, and 13. In rats, MD projections to the medial frontal cortex terminate in the prelimbic (PL) and infralimbic (IL) cortex, as well as the ventral anterior cingulate areas. In addition, the orbitofrontal cortex and the laterally located agranular insular cortex receive strong projections from the MD (Preuss, 1995; Uylings et al., 2003).

The PFC has unique widespread and reciprocal connections with the sensory, motor and association neocortical systems, and with a number of subcortical structures that provide it with the information necessary to exert cognitive top-down control of behavior (Fuster, 1997; Constantinidis and Procyk, 2004). In primates, inputs from sensory systems reach the PFC via a polysynaptic pathway beginning in the primary area of each modality, which projects first to an adjacent area, beginning a sequential order of connections. Each area in this sequence projects to the next in line and also to a discrete frontal cortex area, which in turn feeds back to the projecting area (see Constantinidis and Procyk, 2004 for a recent review). Similarly, the rat medial PFC receives multimodal cortico-cortical projections from sensory, motor, and limbic cortices, which position the PFC at the nexus of several parallel networks (van Eden et al., 1992; Conde et al., 1995; Uylings et al., 2003). Highly processed sensory information is predominantly relayed to the dorsolateral PFC. In contrast, limbic system structures preferentially innervate the orbital PFC, medial PFC, and anterior cingulate cortex. In most mammals, the orbital and medial PFC (specifically PL, IL, and the agranular insular cortices in the rat) receive projections from the hippocampal complex, cingulate cortex, and from adjacent areas of the limbic cortex (Pandya and Yeterian, 1990; Conde et al., 1995; Fuster, 1997). Direct afferents also reach the orbital and medial PFC from the brainstem tegmentum, the pons, the hypothalamus, and the amygdala (Conde et al., 1995; Fuster, 1997; Uylings et al., 2003) (Fig. 8.1). The extensive connections of the medial and orbital PFC with the dorsolateral PFC may provide the neural link between WM and motivation.

Of special importance are the reciprocal connections of the PFC with the premotor and supplementary motor areas, as well as projections to the superior colliculus and motor portions of the basal ganglia. In both primates and rats the frontal lobe is the main cortical target of basal ganglia output, relayed via the thalamus. Inputs from the PFC to the dorsal and ventral striatum (including nucleus accumbens) are relayed back indirectly via the globus pallidus, thalamic nuclei, and the substantia nigra (Alexander et al., 1990; Berendse et al., 1992; Parent and Hazrati, 1995). Depending on the precise location of termination, the striatal efferents of the PFC not only control the execution of movement, but are also involved in associative and limbic aspects of behavior (Parent and Hazrati, 1995). The patterns of afferent and efferent connections and their implications for the homology of various regions within the primate and rodent frontal cortex are described in greater detail in the excellent articles by Fuster (1997), Conde et al. (1995), Preuss (1995), Uylings et al. (2003) and Gabbott et al. (2005).

8.3 Microarchitecture of the PFC

In primates, the PFC is also referred to as "granular frontal cortex" due to the presence of a layer 4, which is absent in the remaining areas of the frontal cortex. In contrast to primary visual and somatosensory cortices, layer 4 in the primate PFC is composed of small pyramidal cells instead of spiny stellate neurons, and lacks distinct sublayers (Goldman-Rakic, 1995). Thalamic inputs to the primate PFC, including the bulk of afferents from the MD, terminate in layer 4 and deep layer 3, (Giguere and Goldman-Rakic, 1988; Erickson and Lewis, 2004) suggesting that layer 3 is also involved in early processing of thalamic signals. In rodents, the entire frontal cortex is agranular and accordingly inputs from MD terminate heaviest in deep layer 3, where they contact layer 5/6 and layer 3 pyramidal cell dendrites as well as a subgroup of interneurons (Kuroda et al., 1998; Rotaru et al., 2005).

As outlined above, the PFC is characterized by its extensive connections with many other cortical regions. A large number of studies that employed anterograde and retrograde tracers have examined the cortical layers in which afferent cortico-cortical inputs terminate, as well as the laminar origin of the efferent corticocortical PFC projections. A detailed description of these laminar patterns for all the PFC areas in monkeys and rats is clearly beyond the scope of the present chapter, how-ever, differences in such patterns clearly could contribute to species differences in the properties of a hypothetical WM network.

In the primate dorsolateral PFC, layer 2/3 pyramidal cells furnish extensive axon collaterals that target other neurons within the superficial layers. Paired recordings from synaptically connected neurons in monkey PFC have been used to study the properties of unitary connections between pyramidal cells (Urban et al., 2002; Gonzalez-Burgos et al., 2004), connections from pyramidal cells to interneurons (Gonzalez-Burgos et al., 2004; Povysheva et al., 2006), and from interneurons to pyramidal cells (Gonzalez-Burgos et al., 2005a). Pyramidal cells with cell bodies in the superficial layers typically send axons down into layers 4 and 5. The axonal arbors of superficial layer interneurons also either show descending projections or are confined to the superficial layers, depending on their morphological subtype (see below) (Krimer et al., 2005; Zaitsev et al., 2005; Kroener et al., 2006). The connections between layers 2/3 and layer 5 PFC cells have not yet been characterized.

In addition to axon collaterals descending and branching in layer 5, the main axon trunk of many layer 2/3 pyramidal cells projects down to the white matter (Gonzalez-Burgos et al., unpublished observations). Thus, these cells may simultaneously send outputs locally and to other brain areas. Layer 3 pyramidal neurons in the primate PFC furnish long-range projections that travel horizontally through the PFC gray matter and terminate in stripe-like columns in the superficial layers (Levitt et al., 1993; Kritzer and Goldman-Rakic, 1995; Pucak et al., 1996; Melchitzky et al., 1998). Such long-range horizontal projections may underlie the monosynaptic excitatory synaptic currents evoked in layer 3 pyramidal neurons of monkey PFC by long-distance lateral stimulation (Gonzalez-Burgos et al., 2000). Multisite stimulation at different distances from a recorded cell, revealed the presence of "hot spots" where stimulation was much more likely to elicit a response (Gonzalez-Burgos et al., 2000). The distance between those spots was remarkably similar to the distance separating clusters of interconnected neurons revealed with anatomical techniques. These clus-tered horizontal connections appear to be reciprocal (Kritzer and Goldman-Rakic, 1995; Pucak et al., 1996), a property that may facilitate reverberative activity in neuronal ensembles, contributing to the sustained firing of PFC cells during WM tasks (Durstewitz et al., 1999). These findings suggest that clustered horizontal con-nections between superficial layer pyramidal neurons are an important feature of the functional microarchitecture of the PFC in primates. Interestingly, delay-period activity during WM tasks is most prominent in primate PFC layers 2/3 (Friedman and Goldman-Rakic, 1994).

In analogy to the columns of primary sensory cortex, clusters of nearby cells in the monkey PFC may form functional modules. Neurons within a module might share the memory for a certain location (Rao et al., 1999), display similar object selectivity (Rainer et al., 1998), encode a similar tactile stimulus frequency (Romo et al., 1999), or represent similar properties of an auditory stimulus (Fuster et al., 2000). Indeed, simultaneous electrophysiological recordings in vivo showed that nearby PFC neurons represent adjacent spatial locations (Constantinidis et al., 2001). Evidence for a comparative functional and anatomical arrangement of interconnected columns in rodents is missing (Fisk and Wyss, 1999; Gabbott et al., 2003). However, there is evidence for a lattice of distinct dendritic bundles, which may represent fundamental structural subunits that encode specific functions (Gabbott and Bacon, 1996).

In the PFC of nonprimate species most studies of local circuits have been focused on the connections between layer 5 neurons, which receive the densest innervation by DA fibers (see below). Synaptic connections between layer 5 pyramidal neurons have been described in both the PFC of rats (Hempel et al., 2000) and ferrets (Gao et al., 2001; Gao and Goldman-Rakic, 2003; Wang et al., 2006). In addition, layer 5 pyramidal neurons in ferret PFC provide excitatory input to at least two classes of layer 5 interneurons (Gao et al., 2003). These two interneurons classes in turn make GABAergic synapses onto pyramidal cells in the same layer (Gao and Goldman-Rakic, 2003). A recent study of intrinsic connections in ferret PFC provided evidence for an exceptionally high incidence of reciprocal connections between glutamatergic pyramidal neurons (Wang et al., 2006), which in theory may serve to support reverberatory activity and persistent firing. The high rate of reciprocal excitatory connections seems to be restricted to a subpopulation of layer 5 pyramidal cells characterized by the presence of "dual" apical dendrites (Wang et al., 2006). However, evidence for similar subnetworks of reciprocally connected pyramidal neurons in other mammalian species is still lacking. Pyramidal cells with dual apical dendrites have not been described in previous studies of the PFC of rats (Yang et al., 1996; Degenetais et al., 2002) or monkeys (Anderson et al., 1995; Soloway et al., 2002; Duan et al., 2003).

Comparison of the dendritic morphology of PFC neurons with that of cells from other cortical areas has revealed morphological specializations. For example, in the PFC of humans and nonhuman primates, layer 3 pyramidal cells have dendritic trees with several times more spines than primary visual cortex neurons (Elston et al., 2001; Jacobs et al., 2001), probably reflecting a high convergence of excitatory inputs onto PFC cells. In addition, PFC pyramidal neurons probably can be divided into subclasses, which, however, currently are poorly characterized. For example, contralaterally projecting neurons in PFC area 9 have dendritic trees that are larger, more complex, and more spinous than those of cells that project only ipsilaterally (Soloway et al., 2002).

In addition to morphology, the intrinsic electrical properties of single neurons can be useful to define cortical cell classes. In the monkey PFC, such properties have been examined only for superficial layer neurons. In these layers, essentially all pyramidal cells belong to the regular spiking cell class, characterized by a relatively long duration of single spikes, significant spike-frequency adaptation, and absence of bursting (Henze et al., 2000; Gonzalez-Burgos et al., 2004; Chang et al., 2005), although, in some cells, a very weakly bursting behavior leads to the presence of spike doublets. Variability in properties such as spike doublets and the time course and amplitude of the afterhyperpolarizations (AHPs) suggest the presence of electrophysiological subclasses of regular spiking pyramidal cells in monkey PFC layers

2/3 (Gonzalez-Burgos, unpublished observations). However, a detailed characterization of these properties is still missing.

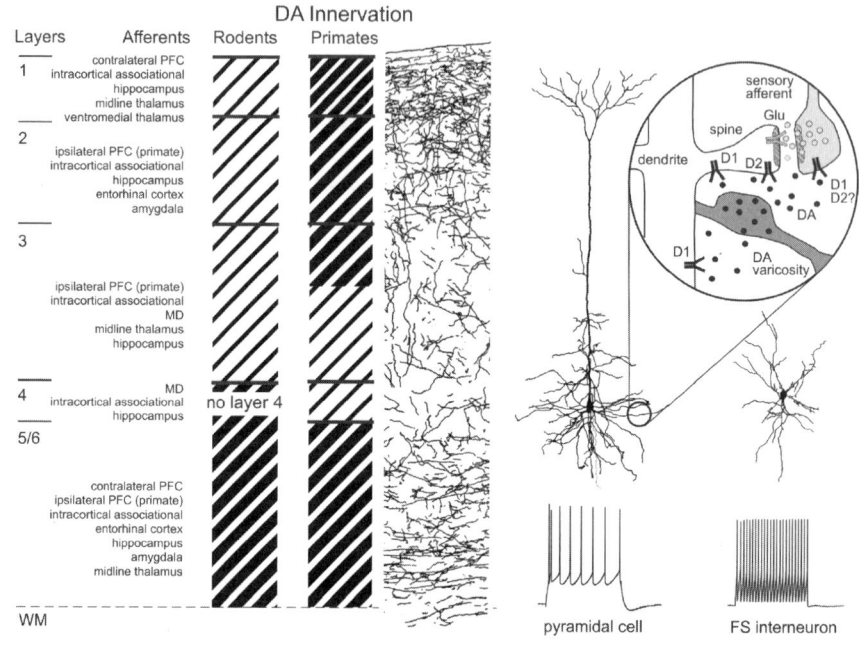

Figure 8.1 Patterns of innervation of prefrontal cortex by glutamatergic and dopaminergic afferent fibers. On the left: Laminar termination pattern of some of the main inputs to the medial PFC of rats and monkeys, as well as the dorsolateral PFC of monkeys. The shaded boxes represent the relative density of dopamine fibers in the PFC of rodents and primates, respectively. Note the absence of layer 4 in rat PFC. Shown is also an example of the bilaminar innervation pattern by dopamine fibers in area 46 of the monkey dorsolateral PFC. The diagram to the right shows the known relations at the ultrastructural level between dopamine fibers, receptors, and glutamate synapses, considering both pre- and postsynaptic components of asymmetric glutamatergic synapses. Activation of D1 and D2 receptors localized in dendritic spines and shafts, as well as in presynaptic boutons, appears to occur mainly by volume transmission, after diffusion of dopamine from the release sites. Abbreviations: MD, mediodorsal thalamic nucleus; WM, white matter; DA, dopamine; D1, D1-type dopamine receptor, D2, D2-type dopamine receptor, Glu, glutamate; FS, fast-spiking.

Only a few reports have provided exhaustive descriptions of the membrane properties of pyramidal cells in the rat PFC. In layer 5 of the rat medial PFC, Yang et al. (1996) identified four principal pyramidal cell types: regular spiking (19%), intrinsic bursting (64%), repetitive oscillatory bursting (13%), and intermediate types (4%). The class of intrinsic bursting neurons contains neurons that project to the nucleus accumbens (Yang et al., 1996). Using intracellular recordings in vivo in anesthetized rats three main classes of PFC pyramidal cells were described (Degenetais et al.,

2002). These included two types of bursting cells and also regular spiking cells, which represented the majority of all recorded neurons (81 of 115; 70%) and could be further divided into 2 subtypes based on their degree of spike frequency adaptation. Interestingly, the proportion of burst-firing versus regular-spiking neurons in these two studies varies greatly. Whether this reflects ambiguity in the categorization of individual cells; differences between in vivo versus in vitro preparations or other methodological factors, remains to be determined. In addition, in vivo the firing properties of pyramidal neurons vary greatly across the sleep–wake cycle, so that bursting neurons can behave like regular spiking cells during different states of vigilance (Steriade, 2001). Furthermore, the firing properties and adaptation rates observed with square pulse injections differ from results obtained with more realistic "noisy" current pulses that mimic synaptic bombardment (Mainen et al., 1995). Differences in intrinsic membrane properties also reflect the unique distributions and number of ion channels in individual neurons. Interestingly, the available evidence (see below) suggests that the effects of DA modulation on layer 5 pyramidal cells are independent of electrophysiological subtype (Yang and Seamans, 1996). However, differences in the various ionic mechanisms through which DA can exert its effects are likely to determine cell-type specific effects of DA.

Interneurons constitute the minority (15–25%) of PFC neurons, but are significantly more heterogeneous than pyramidal cells. For instance, at least 13 interneuron subclasses were characterized in the monkey dorsolateral PFC, based on the morphology and distribution of the axon (Lund and Lewis, 1993). In addition, the postsynaptic target sites may give insight into their functional roles within the circuit (see Fig. 8.2). Examples are the axo-axonic, or so-called chandelier cells that innervate the initial segment of the pyramidal cell axon (Howard et al., 2005). Basket cells, most of which show similar fast spiking firing characteristics as the chandelier cells (see below), make synapses onto the soma or the proximal dendrites of postsynaptic pyramidal cells. The axons of basket and chandelier cells typically project horizontally within the layer of the soma. In contrast, double bouquet neurons give rise to ascending and/or descending "cascades" of axonal branches and boutons that span several cortical layers. The axons of Martinotti cells ascend from the lamina in which the cell's soma is located. Double bouquet and Martinotti cells appear to specifically target distal dendritic compartments of postsynaptic pyramidal cells. Thus, these cells might be involved in regulation of synaptic integration along the dendritic tree of pyramidal cells. Another subclass of dendritic-targeting interneurons is the neurogliaform (or "spiderweb") cell class. In both monkey and rat PFC, neurogliaform cell axons form dense local arborizations, probably connecting with nearby postsynaptic neurons (Povysheva et al., 2007).

As in other cortical regions, PFC interneurons differ with regard to their calcium-binding protein (CBP) content, which tends to be cell-type specific, somewhat correlating with both axonal morphology and intrinsic electrical properties (see below). Chandelier neurons and most basket cells contain the CBP parvalbumin (PV); the double bouquet cells may contain either calretinin (CR) or calbindin (CB). CB seems to be present in neurogliaform and Martinotti cells as well. Neuropeptide content may also help to define interneuron subclasses. The relationship between peptide content and DA modulation is unclear, but PV-positive cells, which do not contain neuropeptides, constitute a significant target of DA modulation (Muly et al., 1998; Sesack et al., 1998b).

In experiments studying the actions of DA, for practical purposes interneurons are usually divided into two major physiological groups, fast-spiking (FS, or nonadapting) cells, and nonfast spiking (non-FS, or adapting) cells (Gorelova et al., 2002; Gao and Goldman-Rakic, 2003; Gao et al., 2003; Kroener et al., 2006; Tseng and O'Donnell, 2006). Each of these groups actually contains several morphological, and in the case of the non-FS cells also distinct physiological subgroups (Kawaguchi and Kubota, 1997; Gorelova et al., 2002; Krimer et al., 2005; Kroener et al., 2006). However, some broad rules seem to apply that correlate cells with different firing patterns with morphological classes and/or neurochemical content. Briefly, FS electrical properties correlate with PV content and with synaptic connections established onto the perisomatic compartment of the pyramidal cell membrane. On the other hand, non-FS cells that contain CB seem to target pyramidal cell dendrites and non-FS cells that contain CR seem to target predominantly, but not exclusively, other GABA neurons.

In rats, GABAergic neurons make up ~15% of all neurons, whereas in primates they represent up to 25%. These differences may reflect increased neuronal diversity, that is, the emergence of new cell types during evolution (DeFelipe et al., 2002). Between primate and nonprimate species some differences exist in the characteristics in the firing properties of interneuron subtypes (Kawaguchi, 1995; Krimer et al., 2005; Kroener, et al. 2006; Povysheva et al., 2007), the morphological characteristics of classes of interneurons (Somogyi et al., 1984; Yanez et al., 2005), and the relative number and distribution of subpopulations defined by calcium-binding proteins (Conde et al., 1994; Gabbott and Bacon, 1996; Gabbott et al., 1997; Kawaguchi and Kubota, 1997). Furthermore, cortical interneurons in rodents and primates appear to have different developmental origins (Letinic et al., 2002; Xu et al., 2004).

It has been hypothesized that the functional role of GABAergic interneurons in PFC WM processes is to enable fine-tuning of reverberating activity within a cortical module (Rao et al., 1999), and/or to provide lateral inhibition between modules that encode different features, for example, opposite spatial locations (Wilson et al., 1994; Rao et al., 1999). Computational modeling studies of delay-period activity have similarly emphasized the importance of inhibition. These studies have suggested that WM networks are characterized by bistability between a spontaneous resting state (down-state) and multiple continuous persistent activity states. In these models robust high activity states are achieved when overall inhibitory events dominate, and when excitatory synapses show a pronounced slow NMDA component (Compte et al., 2000; Durstewitz et al., 2000b; Brunel and Wang, 2001). These properties of theoretical WM networks closely match the components that govern "up-states" in vivo and in organotypic co-cultures (Timofeev et al., 2001; Seamans et al., 2003) in that the majority of unitary synaptic events that bombard the neuron during up-states are inhibitory (but see Waters and Helmchen, 2006). Thus, cortical interneurons are likely to contribute significantly to persistent activity states.

8.4 The Mesocortical Dopamine System

The dopaminergic innervation of the mammalian forebrain arises from mesencephalic cell groups designated as A8, A9, and A10. These generally correspond to the substantia nigra (SN, A9), ventral tegmental area (VTA, A10), and the retrorubral area (A8). In both primates and rodents the cortical innervation by DA fibers shows a strong rostro-caudal gradient, the innervation being densest in the frontal cortex. However, between species regional variations in density and distribution patterns exist, suggesting a heterogeneous role for DA in different neocortical areas (Berger et al., 1991; Williams and Goldman-Rakic, 1998). In primates, the widespread cortical innervation by DA-containing fibers originates in a continuum of neurons distributed across the A8-A10 cell groups (Williams and Goldman-Rakic, 1998). Input to the rodent PFC derives from largely separate populations of DA neurons in the SN and VTA, respectively (Berger et al., 1991).

The laminar pattern of DA innervation varies between rodents and primates and between PFC areas. In primate dorsolateral PFC (mostly area 46), layer 1, 2, and superficial 3 are more densely innervated than layers 5–6, whereas layers 4 and deep 3 receive weak DA input. In the motor cortices, medial PFC (area 9) and anterior cingulate cortex of monkeys (areas of the frontal cortex which all lack a distinct layer 4), dopaminergic input is similarly dense across layers (Berger et al., 1991; Lewis and Sesack 1997). In rodents, afferents from the medial VTA to the PFC provide dense DA input to the deep layers 5–6. The considerably weaker DA innervation to the superficial layer 1–3 originates in the medial SN and the lateral VTA (Fig. 8.1).

In the PFC of rats and primates, DA fibers make specialized synaptic contacts which are mostly of the symmetric type (Seguela et al., 1988; Smiley and Goldman-Rakic 1993). The remaining axonal varicosities constitute unspecialized synaptic release sites and probably contribute to the actions of DA via volume transmission. The most common synaptic target of DA terminals in the PFC of rodents and primates are the dendritic spines and shafts of pyramidal neurons (van Eden et al., 1987; Goldman-Rakic et al., 1989; Smiley and Goldman-Rakic, 1993). Many of the postsynaptic spines innervated by DA terminals also receive unlabeled asymmetric (putative excitatory) terminals. Through this "triadic" synaptic arrangement DA could modulate the excitatory afferents to pyramidal cells both pre- and postsynaptically (van Eden et al., 1987; Seguela et al., 1988; Goldman-Rakic et al., 1989; Smiley and Goldman-Rakic, 1993; Carr and Sesack, 1996; Paspalas and Goldman-Rakic, 2005; Fig. 8.1). However, immunoelectron microscopy studies suggest that such triads are relatively uncommon. For example, a substantial number of DA receptors are localized in dendritic shafts, the axon initial segment and in axonal boutons (Bergson et al., 1995; Paspalas and Goldman-Rakic, 2005). Because DA fibers rarely make appositions onto such structures, these findings suggest DA regulation of these targets by means of volume transmission (Paspalas and Goldman-Rakic, 2004). Furthermore, whereas the number of putative DA synapses per PFC pyramidal neuron is small (approximately 90 per cell) (Krimer et al., 1997), dendritic spines containing DA receptors are relatively common (Paspalas and Goldman-Rakic, 2005). This suggests that most spines that contain DA receptors do not directly receive synapses from DA fibers. Further evidence for DA volume transmission comes from the analysis of DA clearance from the extracellular space. In limbic and cortical regions, including the PFC, DA clearance rates are slower than in the striatum (Garris et al., 1993). The subcellular localization and low levels of DA transporter expression in the PFC (Sesack et al., 1998a; Lewis et al., 2001) suggest a limited ability to regulate extracellular DA levels. Indeed, the DA transporter accounts for only ~40% of DA uptake in the PFC, compared with ~95% in the striatum (Wayment et al., 2001). DA fibers in the PFC of both rats and monkeys also terminate on GABAergic neurons (Goldman-Rakic et al., 1989; Vincent et al., 1993; Smiley and Goldman-Rakic, 1993; Sesack et al., 1998b). As is true for pyramidal cells, the number of putative DA synapses per interneuron is relatively small (Krimer et al., 1997).

8.5 Dopamine Receptors in the PFC

All known DA receptors belong to the superfamily of G-protein-coupled receptors with seven transmembrane spanning domains. Five different DA receptors subtypes have been identified by molecular cloning. The pharmacological response properties to "classical" DA agonists distinguish the D1-like family (D1 and D5), and a D2-like family (D2, D3, D4) of DA receptors. Receptors of the D1 family are positively coupled to adenylyl-cyclase, while D2 receptors inhibit adenylyl-cyclase activity (Missale et al., 1998). More recent work continues to identify additional pathways through which activation of D1- or D2-type receptors can affect cellular physiology.

The laminar distribution of DA receptors in the frontal cortex largely parallels the bilaminar pattern of the innervation by dopaminergic fibers (Goldman-Rakic et al., 1990). In rodent frontal cortex there is considerable expression of mRNA for the D1 receptors in deep layers 5–6. A comparatively lower expression of D2 receptor mRNA is distributed in superficial layers 1–3 and in deep layer 5 (Gaspar et al., 1995). In general, in the frontal cortices of rats and primates, the density of D1 receptors has been estimated to be about fivefold to tenfold higher than that of D2 receptors (Lidow et al., 1991; Joyce et al., 1993; but see Vincent et al., 1993). In rodent frontal cortex there is also evidence for a considerable number of D4 receptors, but not for D3 and D5 receptors (Joyce et al., 1993; Ariano et al., 1997). In primate PFC expression of mRNA for all 5 DA receptor subtypes has been found (Lidow et al., 1998). High levels of receptor proteins are found for the D1, D4, and D5 subtypes in pyramidal neurons of superficial and deep layers (2/3 and 5–6, respectively), with layer 5 neurons showing a stronger expression of D4 and D5 receptors (Lidow et al., 1991).

Immunoelectron microscopy studies provided important information on the localization of DA receptors in the PFC at the ultrastructural level. Receptors of the D1 family were found in the membrane of both dendritic shafts and spines (Smiley et al., 1994; Bergson et al., 1995; Paspalas and Goldman-Rakic, 2004, 2005), as well as in the initial segment of the axon and in axonal boutons (Paspalas and Goldman-Rakic, 2005). Spines labeled by D1 receptor-tagging antibodies typically contain asymmetric synapses, but only rarely are targets of a symmetric synapse indicative of DA synapses (Smiley et al., 1994; Bergson et al., 1995; Paspalas and Goldman-Rakic, 2004, 2005). Similarly, D1 receptors are found in spines (Fig. 8.1), but are not localized at synapses made by DA-containing terminals labeled in the same sections (Smiley et al., 1994). Localization of D1 receptors in glutamatergic synaptic boutons, which also lack DA synapses further supports the idea that DA receptors in the PFC are commonly activated via volume transmission (Paspalas and Goldman-Rakic, 2004, 2005).

Immunoreactivity and mRNA expression for D1, D2, and D4 receptors has also been found in GABAergic interneurons of rat and primate frontal cortex (Mrzljak et al., 1996; Ariano et al., 1997; Le Moine and Gaspar, 1998; Muly et al., 1998; Vysokanov et al., 1998; Paspalas and Goldman-Rakic, 2004).

8.6 Neuromodulation of Cellular Functions by DA in Primate and Rat PFC

The complexity of the PFC network clearly limits a reductionist analysis of the effects of DA on single neurons and synaptic interactions for the purpose of elucidating the processes that are involved in PFC cell activity during behavior. Nevertheless, characterizing the effects of DA on single neurons and synaptic connections is a crucial step to develop more comprehensive models for the role of DA in PFC dependent cognitive function. Based on the selective localization of DA fibers or receptors in cortical layers, cell types or subcellular membrane compartments, several potential targets of DA neuromodulation in the PFC network can be suggested. Here we summarize results obtained in our laboratories and those of others aimed at characterizing the effects of DA on the electrical properties of single cells and on fast excitatory and inhibitory transmission studied in single cells or in small circuits.

As outlined above, all DA receptors identified so far are high affinity G-protein coupled receptors, suggesting that DA release does not trigger fast synaptic transmission but rather acts to modulate glutamatergic or GABAergic transmission and intrinsic ionic currents. DA is thought to act slowly, in low concentrations, and at a distance from the release sites, that is, via volume transmission. Therefore, bath application of agonists in vitro may reasonably well mimic the in vivo effects of DA.

DA does not produce a consistent effect on membrane potential in pyramidal neurons from primate (Gonzalez-Burgos et al., 2002) or rodent PFC (Seamans and Yang, 2004). In contrast, in both primate and rodent PFC, D1R activation consistently depolarizes FS interneurons, but not interneurons of other subtypes (Gorelova et al., 2002; Kroener et al., 2006). However, D1R-mediated depolarization of FS cells does not directly trigger action potentials (APs), yet was associated with a decrease in AP threshold to current pulses and other effects that increase excitability (Gorelova et al., 2002; Kroener et al., 2006). Similarly, in recordings from pyramidal cells of monkey and rat PFC, D1R activation produced an increase in the frequency of spontaneous, AP-dependent IPSCs (Seamans et al., 2001b; Kroener et al., 2006). This effect could result either from increased excitability and interneuron firing, or from an increased probability of AP-evoked release. The effects of DA receptor activation on AP-evoked inhibitory responses were directly tested by recording IPSCs evoked by extracellular electrical stimulation in pyramidal cells from rat PFC. Application of the endogenous agonist DA had complex effects on these IPSCs including a rapid D2 receptor-mediated depression followed by a delayed and persistent enhancement mediated by D1Rs (Seamans et al., 2001b; Trantham-Davidson et al., 2004). Pharmacological analysis suggest that DA at low concentrations (<1 μM) may preferentially activate high affinity D1Rs, while at higher concentrations D2Rs are also activated, resulting in a transient suppression of evoked IPSCs. Because extracellular stimulation activates GABAergic fibers in a nonselective manner, such a biphasic time course could also be explained by differential effects of DA on the IPSCs of different interneuron subtypes. In a study of synaptically connected pairs in the PFC of ferrets, D1R activation rapidly depressed unitary IPSPs evoked in pyramidal neurons by identified presynaptic FS cells and enhanced IPSPs elicited by non-FS neurons (Gao et al., 2003). D2 receptor activation had no significant effects on IPSPs elicited by neither interneuron type (Gao et al., 2003). Whereas FS cells have rather homogeneous electrical properties (Gorelova et al., 2002; Krimer et al., 2005; Gonzalez-Burgos et al., 2005a; Kroener et al., 2006), cells with non-FS properties comprise several subgroups (Gorelova et al., 2002; Krimer et al., 2005). Thus, it is theoretically conceivable that the cells in the relatively small sample reported by Gao and colleagues did not overlap with interneuron subtypes that made up the majority of inputs stimulated in the experiments in the rat PFC (Seamans et al., 2001b; Trantham-Davidson et al., 2004). In the rat, the cell type and receptor specificity of DA modulation of interneurons may be age-dependent, and potentially both D1R and D2R activation can increase excitability of both FS and non-FS cells in postadolescent rats (Tseng and O'Donnell, 2006); however, similar effects were not observed in young adult monkeys (Gonzalez-Burgos et al., 2005a; Kroener et al., 2006).

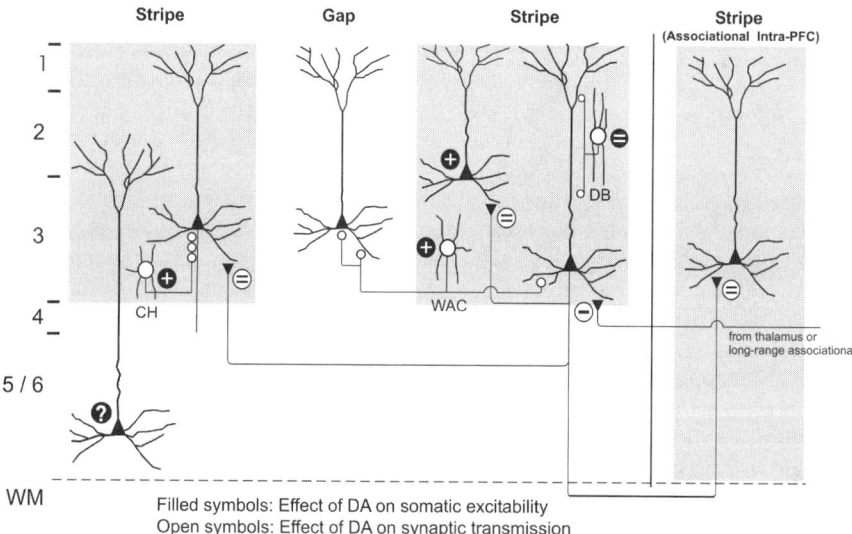

Fig. 8.2 Effects of dopamine in monkey PFC. Summary of the known effects of dopamine on different components of the PFC circuitry as described in brain slice studies in vitro. "Stripes" represent areas within the PFC that are reciprocally interconnected and which are hypothesized to form functional networks. Neurons and synaptic contacts shown in black are glutamatergic; those shown in white are GABAergic. Filled symbols indicate changes in membrane excitability, typically measured as changes in spike firing following somatic current injection. Open symbols represent changes in synaptic transmission. "=" indicates that an effect of DA or its receptor-specific agonists has been tested and no change was found; "?" indicates that effects have yet not been tested. Note that in the monkey PFC in vitro so far no specific effect of D2-receptor activation has been described; all changes noted here result from activation of receptors with either DA or D1-specific agonists. Abbreviations: CH, chandelier cell; WAC, wide arbor cell; DB, double-bouquet cell; WM, white matter; see text for details.

Recent studies also examined the effects of DA on excitatory synaptic inputs onto these cells. As in other regions of cortex, glutamate inputs onto PFC FS neurons elicit EPSPs with very fast decay time course, significantly shorter than that observed in pyramidal cells or non-FS interneurons (Gao and Goldman-Rakic, 2003; Gonzalez-Burgos et al., 2004; Povysheva et al., 2006). Such properties are observed in pyramidal cell-to-FS cell synaptic connections (Gao and Goldman-Rakic, 2003; Gonzalez-Burgos et al., 2004) as well as in spontaneous EPSPs or EPSPs evoked by extracellular electrical stimulation (Gonzalez-Burgos et al., 2004; Povysheva et al., 2006). In addition, membrane depolarization further decreases the EPSP duration in FS cells (Gonzalez-Burgos et al., 2004; Gonzalez-Burgos et al., 2005a), suggesting that NMDA receptors do not shape the duration of EPSPs at the soma of PFC FS neurons.

D1R activation significantly decreased the amplitude of EPSPs evoked by low frequency extracellular stimulation in FS cells, but affected less significantly EPSPs evoked trains of synaptic input (Gonzalez-Burgos et al., 2005b). Trains of synaptically evoked EPSPs are often used in an attempt to mimic the dynamics of a persistent input as it might occur during the elevated (~20 Hz) firing of the prefrontal network during WM. In addition, the resulting prolonged depolarization of the postsynaptic cell increases the chance of activation of NMDA currents and thus allows studying the modulation of both AMPA and NMDA components by DA (see below). In FS neurons of the ferret PFC, fast-decaying unitary EPSPs elicited by stimulation of synaptically connected pyramidal cells were unaffected by DA (Gao and Goldman-Rakic, 2003). One possibility is that DA depresses FS cell EPSPs in inputs other than those from nearby pyramidal cells. Thus, as in the case of IPSCs, the specificity of the input may determine the mode of DA modulation.

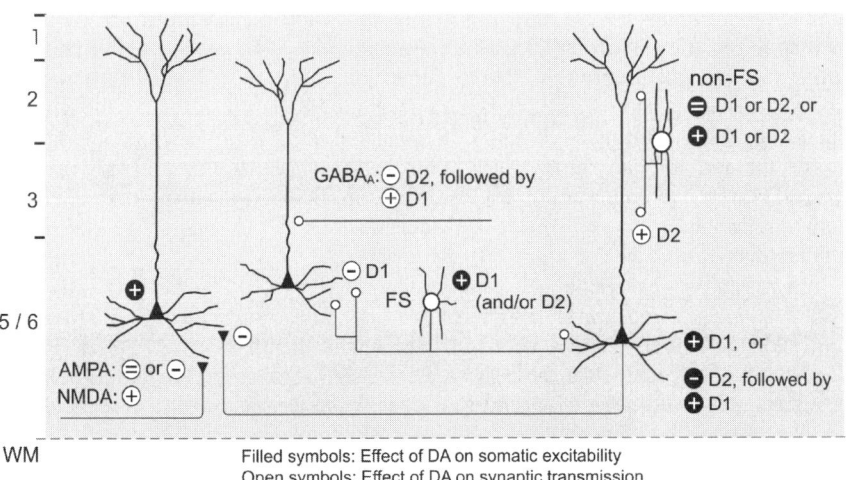

Fig. 8.3 Effects of dopamine in rat and ferret PFC. Summary of the known effects of dopamine on the PFC circuitry of rats and ferrets as described in brain slice studies in vitro. Abbreviations and symbols are identical to those in Figs. 8.1 and 8.2.

In both rat and primate D1 receptor activation increases the excitability of PFC pyramidal cells, that is, the ability of excitatory currents to elicit spike firing, without consistently changing the cells' membrane potential (Gonzalez-Burgos et al., 2002; Seamans and Yang, 2004). In slices, the nature of DA effects on PFC pyramidal cell excitability strongly depends on the time after initial application of DA agonists. Some studies have shown that early during agonist application, pyramidal cell excitability decreases (Geijo-Barrientos and Pastore, 1995; Gulledge and Jaffe, 1998; Gulledge and Jaffe, 2001; Gulledge and Stuart, 2003), or does not change (Henze et al., 2000; Seamans and Yang, 2004). In contrast, neuronal excitability increases significantly during later phases of DA application, or even during agonist washout (Gulledge and Jaffe, 1998; Henze et al., 2000; Gulledge and Jaffe, 2001; Seamans and Yang, 2004). One possibility suggested previously is that an increased IPSC frequency during agonist application could initially mask an increase in pyramidal cell excitability that persists after DA washout (Gulledge and Jaffe, 2001). However, both the increase in IPSC frequency recorded from pyramidal cells and the increase in FS cell excitability outlined above follow a delayed and persistent time course (Gorelova et al., 2002; Kroener et al., 2006).

DA, acting on D1 receptors, also modulates excitatory glutamatergic transmission onto PFC pyramidal neurons. In slices of monkey PFC, DA application did not change significantly the amplitude of unitary EPSPs recorded in synaptically connected pairs of pyramidal neurons, but depressed EPSCs evoked by extracellular electrical stimulation, suggesting that the effects are input-specific (Urban et al., 2002). EPSPs evoked by extracellular electrical stimulation in rat PFC pyramidal neurons were also modulated by D1R activation (Seamans et al., 2001a). Interestingly, whereas D1R stimulation to a varying degree decreased EPSPs evoked at low baseline frequency, it potentiated late responses in EPSP trains evoked by 20 Hz stimulation (Seamans et al., 2001a). Potentiation of late EPSPs in trains was abolished by application of an NMDA receptor antagonist (Seamans et al., 2001a), suggesting that an enhancement of NMDA currents by D1R activation is observed late during trains of synaptic input, when depolarization causes relief of the voltage dependent NMDA current Mg^{2+} block (Seamans et al., 2001a). Such depolarization may be attained only late during EPSP trains and may be absent if cells are voltage-clamped at negative potentials or are resting at significantly negative potentials, thus occluding the observation of the NMDA-dependent effect (Urban et al., 2002; Gao and Goldman-Rakic, 2003).

In pyramidal-to-pyramidal cell connections of ferret PFC, DA-receptor activation significantly depressed unitary EPSPs either early or late in trains (Gao and Goldman-Rakic, 2003), contrasting with previous results (Seamans et al., 2001a; Urban et al., 2002). The reasons for these discrepancies are not clear at this point. Some of the effects of DA in monkey and rat PFC slices reviewed above are summarized in Figs. 2 and 3.

8.7 Behavioral Physiology

As stated in the introduction, the PFC is essential to the cognitive control of behavior, in part by mediating important aspects of WM function. Moreover, DA inputs and receptors serve a critical role for these functions. In humans, WM involves multiple processes, including various buffers that transiently store information (the visuo-spatial, verbal and multimodal episodic buffers), and a central executive component that controls the manipulation of the information stored (Shallice and Burgess, 1996; Repovs and Baddeley, 2006). In animal models, recordings of neural activity may allow investigation of the cellular bases of the multiple processes that WM involves.

To investigate neuronal activity during WM, animals are typically trained to perform a delayed response task and the firing properties of PFC neurons are recorded. In a classic task, a monkey observes the experimenter bait one of two covered food wells before an opaque screen blocks the monkey's view of the wells. After a short delay, the screen is raised and the monkey must choose the previously baited well to obtain the reward. In the oculomotor delayed response task, developed more recently, the monkey must initially fixate on a dot of light presented at the center of a video screen until a cue (light flash) appears in one of eight spatial locations on the screen, equidistant from the fixation spot. The cue light is then extinguished during a delay that lasts a few seconds, and then the fixation spot is turned off, signaling the monkey to respond by performing a saccade to the spatial location where the cue was flashed, in order to obtain reward. In the absence of the cue, the response must be directed based on information stored transiently in the WM buffers.

Neurons in the primate PFC show increased (and sometimes decreased) activity during all phases (cue, delay and response) of the original (Fuster and Alexander, 1971; Kubota and Niki, 1971; Fuster, 1973) and oculomotor delayed response task (Funahashi et al., 1989). Such changes in firing rate, however, appear not to be random but to reveal the existence of populations of PFC neurons characterized by firing patterns specifically associated with particular phases of the task. One of such firing patterns was readily identified in the original studies and was characterized by maintenance of an elevated firing rate throughout the delay period (Fuster and Alexander, 1971; Kubota and Niki, 1971). Fuster and Alexander hypothesized that PFC neurons active during the cue and delay periods participated in the acquisition and temporary storage, respectively, of cue-related information implicated in the performance of the delayed response (Fuster and Alexander, 1971). At that time, Fuster and Alexander referred to this finding as the neural correlate of short-term memory. Currently, however, important distinctions are made between the short-term memory and WM processes mediated by the PFC. For example, lesions of the PFC in humans may have no significant impact in performance of tasks that require short-term memory but do not require manipulation of information in order to complete the task (Petrides, 2000). In contrast, in tasks that require response flexibility, patients with damage to their PFC commit repeated errors that they are consciously aware of, but cannot use to update behavior. These findings fit well with the idea that the PFC is necessary for the active maintenance and manipulation of information necessary to guide behavior at the end of the delay period. It is in addition important to consider that regions in the human dorsolateral and ventral PFC may be differentially involved in short-term storage versus the multiple processes involved in the central executive function. An interesting possibility suggested by lesion and neuroimaging studies in humans is that the different components of WM are localized in different, specialized regions of PFC, for example, the dorsolateral and ventral regions (Barch, 2006). In macaque monkeys and rats, the neural basis of different components of WM may be integrated into fewer and smaller frontal cortical areas.

Numerous studies have revealed systematic and stimulus- and/or response-specific changes in single-neuron firing rates during the delay periods of delayed response tasks (Quintana et al., 1988; Funahashi et al., 1989; Miller et al., 1996; Rao et al., 1997; Rainer et al., 1998; Fuster et al., 2000; Mushiake et al., 2006). The classical study of Funahashi et al. (1989) reported that 30% (69/228) of neurons exhibited tuned delay-period activity, whereas Miller et al. (1996) reported that 15–34% (7/47, 33/98) of neurons exhibited delay-period activity depending on the task. In a vibro-tactile discrimination delayed response task, large percentages of monkey PFC neurons (65–69%, in two task variants), showed modulation of their firing rate during the delay period, as a monotonic function of the stimulus properties (Romo et al., 1999). For 23–35% of these cells ("persistent" delay neurons) the cue-related modulation of firing rate was maintained throughout the entire delay period. In two other groups ("early" and "late" delay neurons, comprising 34–38% and 19–33% of the delay cells , respectively), stimulus-related modulation of firing rate occurred exclusively near the start or end of the delay (Romo et al., 1999). Thus, although cells with persistent delay activity are common in primate PFC, other populations of delay neurons show modulation of firing rate during subintervals of the delay period. Similar to the early and late delay cells recorded by Romo and colleagues, commonly the firing rate slowly and steadily ramps up or down during the delay and peaks somewhere in between (Miller et al., 1996; Chafee and Goldman-Rakic, 1998; Quintana and Fuster, 1999; Rainer et al., 1999; Rainer and Miller, 2002). A plausible interpretation is that in addition to cells with persistent delay activity (which may represent information stored in the buffers) the PFC contains cell populations whose activity contributes to the central executive function. Indeed it is evident that delay activity in a given neuron is strongly affected by the contextual task rule applied, and that the numbers of neurons exhibiting delay-period activity varies across tasks.

To assess WM function in rats, a variety of tasks have been used that differ from each other and from the delayed response tasks used in primates. Batuev et al. (1990) and Orlov et al. (1988) used a task in which a light was flashed above a food well to the right or left of the rat, and after a delay of 5 sec the rat was allowed to visit the previously lighted food well (Orlov et al., 1988; Batuev et al., 1990). In this task, the firing of 85% of the PFC cells was correlated with the response phase and approximately 54% of PFC neurons responded preferentially during the delay in a spatially selective manner. In a spatial-delayed matching to sample task, 33% (56/171) of the PFC neurons showed modulation of firing rate during the delay period (Chang et al., 2002). Remarkably, almost 30 of the 56 delay active neurons were classified as exhibiting "continuous" delay activity where their elevated firing rate lasted throughout the delay. This incidence of continuously active cells is higher than the percentage of primate neurons that usually are reported to exhibit continuous delay activity (see above), even though at 10s or longer the delay intervals were significantly greater than those typically employed in primate studies.

As noted above in the case of primate neurons, the number of delay active cells in rat PFC varied strongly with the requirements of the task. In a spatially based radial arm maze task a significant number of PFC cells encoded responses or anticipated rewards associated with responses, yet few cells displayed delay-related activity (Pratt and Mizumori, 2001). Using a spatially based Fig. 8.8 maze, Baeg et al. (2003) reported that 10.1% of rat mPFC neurons maintained continuously elevated activity throughout the delay, the duration of which depended on the time it took the rat to traverse the maze (Baeg et al., 2003). In addition, 21.3% of neurons showed significant firing rate modulation during some portion of the delay. Likewise, Jung et al. (1998) reported that 21% (40/187) of mPFC cells had a clear "WM" correlate on a radial maze task and 13% (15/113) on a Fig. 8.8 maze, although most neurons show firing rate modulation only during portions of the delay period (Jung et al., 1998). Viewed collectively, the available data support the idea that approximately similar proportions of primate and rat PFC neurons exhibit some form of delay task-related activity. In primate PFC, there are multiple forms of delay-related activity, but the finding of neurons with persistent delay-period activity is consistent across many studies since the early reports in the 1970s. In rats, mPFC neurons consistently show WM task-related activity, but the fraction of neurons with persistent modulation of firing throughout the delay appears to be variable. This is not surprising given the vastly different task employed.

Perhaps one of the clearest behavioral correlates of rat PFC cell activation is reward. Nearly one third of units showed specifically goal-related correlates in Jung et al. (1999) and 21% in Bouret and Sara (2004). For example, about 50% of the neurons with task-related firing showed increases or decreases in firing just prior to or during contact with reward in an 8 arm radial maze task (Pratt and Mizumori, 2001). Moreover, some neurons fired differentially in anticipation to high and low reward locations and others distinguished between high- and low-reward arms during reward consumption (Pratt and Mizumori, 2001). Therefore, reward-related firing in rat PFC neurons is common and may be encoded differently depending on the nature of the reward. As in the rat mPFC, considerable evidence suggests that primate PFC neurons respond to reward. In nondelayed tasks, some PFC neurons respond simply to the presentation of food rewards and this activity is abolished if the rewarding value of the stimuli is decreased by adding quinine to the food (Inoue et al., 1985). Furthermore, PFC neurons fire more vigorously to stimuli predictive of reward relative to equivalent stimuli that were irrelevant to the monkey (Yajeya et al., 1988). However, the activity of delay-active neurons was more vigorous if food itself served as a cue relative to a stimulus previously paired with food. Yet, PFC neurons do not respond directly to the presence or absence of reward, but respond similarly to different stimuli with the same behavioral significance while responding differently to identical stimuli of varying behavioral significance (Watanabe, 1996; Watanabe et al., 2002). In contrast to cells in other cortical association areas that typically code only for specific stimuli regardless of their significance (Miller et al., 1996), DLPFC activity is significantly enhanced if stimuli are of particular behavioral significance.

On a delayed response task both rewarding and aversive outcomes affect the activity of PFC neurons (Kobayashi et al., 2006). On a spatial WM task, neurons respond to expected reward, expected reward omission (Watanabe et al., 2005), as well as what type of reward will be omitted (Watanabe et al., 2002). Cues associated with more palatable rewards produce significantly greater activity in delay-active PFC neurons (Watanabe, 1996) and information about the amount of reward may be encoded parametrically by firing rate variations (Wallis and Miller, 2003). The firing rate of dorsolateral PFC neurons was not only dependent on whether the cue was predictive of the presence and magnitude of reward, but also showed a quantitatively different delay-period activity depending on what type of reward (i.e. what food item) was expected to be presented. Reward expectation appears to affect the directional selectivity of the sensory-related information held during the delay (Watanabe,

1990; Roesch and Olson, 2003; Amemori and Sawaguchi, 2006; Kobayashi et al., 2006). The timing of reward is also a strong influence on delay activity, as the level of delay-period activity depends on the predicted interval between the response and reward (Tsujimoto and Sawaguchi, 2005).

Collectively these studies demonstrate that reward modulates the firing of rat and primate PFC neurons in various ways. Specifically, the strength of PFC unit activity during delay tasks is directly related to the particular significance of the stimuli. Based on these data, one would conclude that there is a signal that provides input to PFC about specific rewards and this input drives subsets of "reward sensitive" neurons. In addition, there may be a second reward-related signal that is transferred to the PFC that modulates the memory-related firing rate of PFC neurons. A likely candidate for this signal is the mesocortical DA input (Schultz, 1997).

The DA cells in the midbrain with ascending projections to striatum and cortex have clear reward-related correlates. Midbrain dopaminergic neurons exhibit a phasic burst of action potentials immediately following unexpected salient events, which include sudden novel stimuli (Schultz and Romo, 1990; Ljungberg et al., 1992), and primary rewards (Ljungberg et al., 1992). With experience, DA neurons respond to conditioned stimuli that predict potential rewards. This advanced signal does not provide information about the reward itself, but rather provides a prediction error signal that may encode the discrepancy between the predicted and the actual reward (Schultz, 1997). Alternatively, the signal emanating from DA neurons may not encode reward per se, but the ability or will to engage in effortful responding to obtain reward (Salamone et al., 2005). This is related to the construct of incentive salience or wanting (Berridge, 2006). Therefore, DA appears to signal expected reward or serves as a motivational substrate to obtain them.

This signal may then bias the manner in which PFC neurons and networks process information (Durstewitz et al., 2000a; Gonzalez-Burgos et al., 2002; Seamans and Yang, 2004). Based on many of the biophysical characteristics outlined in this chapter, the DA signal could make WM networks in the PFC either more or less robust to distractors, or more or less flexible in incorporating new information. In regards specifically to delay-period activity, the prediction is that the optimal amount of DA release in PFC should increase the signal-to-noise ratios of delay-period activity. This indeed has been reported in some studies that investigated the effects of iontophoretic application of DAergic drugs on delay-period activity in primate DLPFC (Sawaguchi, 2001). Therefore, DA, in a manner consistent with the modulation of delay-period activity by natural rewards described above, provides no specific information, but rather modulates the way in which specific WM information is processed. However, low doses of DA D1 receptor antagonists have also been shown to strengthen delay-period activity (Williams and Goldman-Rakic, 1995). The reason for this is that different levels of activation of D1 receptors may have different physiological effects at different concentrations. It was suggested that the facilitatory effect of D1 antagonists on delay activity may be related to removal of DA-mediated activation of inhibitory interneurons. This would be counteracted at different concentrations of DA by the modulation of NMDA and other intrinsic currents that enhanced delay-period activity (Durstewitz et al., 2000a).

As concluded above, in many cases the DA modulation of interneurons and pyramidal cells at the single-neuron level is conserved across species. Therefore, the possibility exists that the reward-related modulation of WM information may also be conserved. Currently, there is no clear data to address this issue, because so far no studies have examined the effects of manipulations of DA receptor activation on the firing of rat PFC neurons in vivo, during tasks that require WM.

8.8 Conclusions

In this chapter we focused on the importance of DA inputs to the functioning of the cellular networks of the PFC. Activity in the PFC contributes to multiple components of the WM process, including short-term storage and manipulation of the stored information to guide future behavior. Proper WM requires DA input to the PFC, but what information this DA signal conveys to the PFC is still a matter of debate. Reward presentation is a strong modulator of PFC and DA cell firing, suggesting that the role of DA input to PFC may be related to the importance of reward for the cognitive control of behavior. A comprehensive characterization of the effects of DA modulation on the firing patterns of PFC neurons during WM tasks is still lacking. Importantly, several of the effects of reward magnitude on the firing of PFC neurons during different phases of WM tasks may be expected to depend on signals provided by midbrain DA neurons. Without doubt, future studies in rodents will be of importance to our understanding of how DA modulation of single-neuron activity translates to its effects on large-scale networks that encode WM. At the cellular level, there are both similarities and differences in the effects of DA on PFC neurons and synaptic connections examined so far appear to be conserved across species. The most obvious difference between species appears to relate to the intrinsic organization of the PFC, that is, how the individual components of the network (e.g., various cell types, or the DAergic innervation) are combined to form functional modules. Thus, while the rat serves as a good model system to address many questions about DA modulation of WM and reward, the validity of these findings for primate and ultimately human cognition remains an issue for future experiments.

Acknowledgments

The authors are greatly indebted to Drs. David Lewis and German Barrionuevo in whose laboratories all monkey slice physiology experiments were conducted. Support provided by grants from NARSAD, NIH MH41456 (G.G.B), MH065924 (J.K.S.), and NIDA P50DA015369-04 (S.K.).

References

Alexander, G.E., Crutcher, M.D., and DeLong, M.R. (1990). Basal ganglia-thalamocortical circuits: parallel substrates for motor, oculomotor, "prefrontal" and "limbic" functions. Prog Brain Res 85, 119–146.
Amemori, K. and Sawaguchi, T. (2006) Contrasting effects of reward expectation on sensory and motor memories in primate prefrontal neurons. Cereb Cortex 16, 1002–1015.
Anderson, S.A., Classey, J.D., Conde, F., Lund, J.S., and Lewis, D.A. (1995) Synchronous development of pyramidal neuron dendritic spines and parvalbumin-immunoreactive

chandelier neuron axon terminals in layer III of monkey prefrontal cortex. Neuroscience 67, 7–22.

Ariano, M.A., Wang, J., Noblett, K.L., Larson, E.R., and Sibley, D.R. (1997) Cellular distribution of the rat D4 dopamine receptor protein in the CNS using anti-receptor antisera. Brain Res 752, 26–34.

Baeg, E.H., Kim, Y.B., Huh, K., Mook-Jung, I., Kim, H.T., and Jung, M.W. (2003) Dynamics of population code for working memory in the prefrontal cortex. Neuron 40, 177–188.

Barch, D.M. (2006) What can research on schizophrenia tell us about the cognitive neuroscience of working memory? Neuroscience 139, 73–84.

Batuev, A.S., Kursina, N.P., and Shutov, A.P. (1990) Unit activity of the medial wall of the frontal cortex during delayed performance in rats. Behav Brain Res 41, 95–102.

Berendse, H.W., Galis -de Graaf, Y., and Groenewegen, H.J. (1992) Topographical organization and relationship with ventral striatal compartments of prefrontal corticostriatal projections in the rat. J Comp Neurol 316, 314–347.

Berger, B., Gaspar, P., and Verney, C. (1991) Dopaminergic innervation of the cerebral cortex: unexpected differences between rodents and primates. Trends Neurosci 14, 21–27.

Bergson, C., Mrzljak, L., Smiley, J.F., Pappy, M., and Goldman-Rakic, P.S. (1995) Regional, cellular, and subcellular variations in the distribution of D1 and D5 dopamine receptors in primate brain. J Neurosci 15, 7821–7836.

Berridge, K.C. (2006) The debate over dopamine's role in reward: the case for incentive salience. Psychopharmacol (Berl) 191 (3): 391–431.

Brozoski, T.J., Brown, R.M., Rosvold, H.E., and Goldman, P.S. (1979) Cognitive deficit caused by regional depletion of dopamine in prefrontal cortex of rhesus monkey. Science 205, 929–932.

Brunel, N. and Wang, X.J. (2001) Effects of neuromodulation in a cortical network model of object working memory dominated by recurrent inhibition. J Comp Neurosci 11, 63–85.

Carr, D.B. and Sesack, S.R. (1996) Hippocampal afferents to the rat prefrontal cortex: synaptic targets and relation to dopamine terminals. J Comp Neurol 369, 1–15.

Chafee, M.V. and Goldman-Rakic, P.S. (1998) Matching patterns of activity in primate prefrontal area 8a and parietal area 7ip neurons during a spatial working memory task. J Neurophysiol 79, 2919–2940.

Chang, J.Y., Chen, L., Luo, F., Shi, L.H., and Woodward, D.J. (2002) Neuronal responses in the frontal cortico-basal ganglia system during delayed matching-to-sample task: ensemble recording in freely moving rats. Exp Brain Res 142, 67–80.

Chang, Y.M., Rosene, D.L., Killiany, R.J., Mangiamele, L.A., and Luebke, J.I. (2005) Increased action potential firing rates of layer 2/3 pyramidal cells in the prefrontal cortex are significantly related to cognitive performance in aged monkeys. Cereb Cortex 15, 409–418.

Compte, A., Brunel, N., Goldman-Rakic, P.S., and Wang, X.J. (2000) Synaptic mechanisms and network dynamics underlying spatial working memory in a cortical network model. Cereb Cortex 10, 910–923.

Conde, F., Lund, J.S., Jacobowitz, D.M., Baimbridge, K.G., and Lewis, D.A. (1994) Local circuit neurons immunoreactive for calretinin, calbindin D- 28k or parvalbumin in monkey prefrontal cortex: distribution and morphology. J Comp Neurol 341, 95–116.

Conde, F., Maire-Lepoivre, E., Audinat, E., and Crepel, F. (1995) Afferent connections of the medial frontal cortex of the rat. II. Cortical and subcortical afferents. J Comp Neurol 352, 567–593.

Constantinidis, C. and Procyk, E. (2004) The primate working memory networks. Cogn Affect. Behav Neurosci 4, 444–465.

Constantinidis, C. and Steinmetz, M.A. (1996) Neuronal activity in posterior parietal area 7a during the delay periods of a spatial memory task. J Neurophysiol 76, 1352–1355.

Constantinidis, C., Franowicz, M.N., and Goldman-Rakic, P.S. (2001) Coding specificity in cortical microcircuits: a multiple -electrode analysis of primate prefrontal cortex. J Neurosci 21, 3646–3655.

DeFelipe, J., Alonso-Nanclares, L., and Arellano, J.I. (2002) Microstructure of the neocortex: comparative aspects. J Neurocytol 31, 299–316.

Degenetais, E., Thierry, A.M., Glowinski, J., and Gioanni, Y. (2002) Electrophysiological properties of pyramidal neurons in the rat prefrontal cortex: an in vivo intracellular recording study. Cereb Cortex 12, 1–16.

Duan, H., Wearne, S.L., Rocher, A.B., Macedo, A., Morrison, J.H., and Hof, P.R. (2003) Age-related dendritic and spine changes in corticocortically projecting neurons in macaque monkeys. Cereb Cortex 13, 950–961.

Durstewitz, D., Kelc, M., and Gunturkun, O. (1999) A neurocomputational theory of the dopaminergic modulation of working memory functions. J Neurosci 19, 2807–2822.

Durstewitz, D., Seamans, J.K., and Sejnowski, T.J. (2000a) Dopamine-mediated stabilization of delay-period activity in a network model of prefrontal cortex. J Neurophysiol 83, 1733–1750.

Durstewitz, D., Seamans, J.K., and Sejnowski, T.J. (2000b) Neurocomputational models of working memory. Nature Neurosci 3, 1184–1191.

Elston, G.N., Benavides-Piccione, R., and DeFelipe, J. (2001) The pyramidal cell in cognition: a comparative study in human and monkey. J Neurosci 21, RC163.

Erickson, S.L. and Lewis, D.A. (2004) Cortical connections of the lateral mediodorsal thalamus in cynomolgus monkeys. J Comp Neurol 473, 107–127.

Fisk, G.D. and Wyss, J.M. (1999) Associational projections of the anterior midline cortex in the rat: intracingulate and retrosplenial connections. Brain Res 825, 1–13.

Friedman, H.R. and Goldman-Rakic, P.S. (1994) Coactivation of prefrontal cortex and inferior parietal cortex in working memory tasks revealed by 2DG functional mapping in the rhesus monkey. J Neurosci 14, 2775–2788.

Funahashi, S., Bruce, C.J., and Goldman-Rakic, P.S. (1989) Mnemonic coding of visual space in the monkey's dorsolateral prefrontal cortex. J Neurophysiol 61, 331–349.

Fuster, J.M. (1973) Unit activity in prefrontal cortex during delayed-response performance: neuronal correlates of transient memory. J Neurophysiol 36, 61–78.

Fuster, J.M. (1997) The Prefrontal Cortex: anatomy, physiology and neuropsychology of the frontal lobe, 3rd ed. Raven Press, New York, NY.

Fuster, J.M. and Alexander, G.E. (1971) Neuron activity related to short-term memory. Science 173, 652–654.

Fuster, J.M., Bodner, M , and Kroger, J.K. (2000) Cross-modal and cross-temporal association in neurons of frontal cortex. Nature 405, 347–351.

Gabbott, P.L. and Bacon, S.J. (1996) Local circuit neurons in the medial prefrontal cortex (areas 24a,b,c, 25 and 32) in the monkey: I. Cell morphology and morphometrics. J Comp Neurol 364, 567–608.

Gabbott, P.L., Jays, P.R., and Bacon, S.J. (1997) Calretinin neurons in human medial prefrontal cortex (areas 24a,b,c, 32', and 25) J Comp Neurol 381, 389–410.

Gabbott, P.L., Warner, T.A., Jays, P.R., and Bacon, S.J. (2003) Areal and synaptic interconnectivity of prelimbic (area 32), infralimbic (area 25) and insular cortices in the rat. Brain Res 993, 59–71.

Gabbott, P.L., Warner, T.A., Jays, P.R., Salway, P., and Busby, S.J. (2005) Prefrontal cortex in the rat: projections to subcortical autonomic, motor, and limbic centers. J Comp Neurol 492, 145–177.

Gao, W.J. and Goldman-Rakic, P.S. (2003) Selective modulation of excitatory and inhibitory microcircuits by dopamine. Proc Natl Acad Sci USA 100, 2836–2841.

Gao, W.J., Krimer, L.S., and Goldman-Rakic, P.S. (2001) Presynaptic regulation of recurrent excitation by D1 receptors in prefrontal circuits. Proc Natl Acad Sci USA 98, 295–300.

Gao, W.J., Wang, Y., and Goldman-Rakic, P.S. (2003) Dopamine modulation of perisomatic and peridendritic inhibition in prefrontal cortex. J Neurosci 23, 1622–1630.

Garris, P.A., Collins, L.B., Jones, S.R., and Wightman, R.M. (1993) Evoked extracellular dopamine in vivo in the medial prefrontal cortex. J Neurochem 61, 637–647.

Gaspar, P., Bloch, B., and Le Moine, C. (1995) D1 and D2 receptor gene expression in the rat frontal cortex: cellular localization in different classes of efferent neurons. Eur J Neurosci 7, 1050–1063.

Geijo-Barrientos, E. and Pastore, C. (1995) The effects of dopamine on the subthreshold electrophysiological responses of rat prefrontal cortex neurons in vitro. Eur J Neurosci 7, 358–366.

Giguere, M. and Goldman-Rakic, P.S. (1988) Mediodorsal nucleus: areal, laminar, and tangential distribution of afferents and efferents in the frontal lobe of rhesus monkeys. J Comp Neurol 277, 195–213.

Goldman-Rakic, P.S. (1995) Cellular basis of working memory. Neuron 14, 477–485.

Goldman-Rakic, P.S., Leranth, C., Williams, S.M., Mons, N., and Geffard, M. (1989) Dopamine synaptic complex with pyramidal neurons in primate cerebral cortex. Proc Natl Acad Sci USA 86, 9015–9019.

Goldman-Rakic, P.S., Lidow, M.S., and Gallager, D.W. (1990) Overlap of dopaminergic, adrenergic, and serotoninergic receptors and complementarity of their subtypes in primate prefrontal cortex. J Neurosci 10, 2125–2138.

Gonzalez-Burgos, G., Barrionuevo, G., and Lewis, D.A. (2000) Horizontal synaptic connections in monkey prefrontal cortex: an in vitro electrophysiological study. Cereb Cortex 10, 82–92.

Gonzalez-Burgos, G., Krimer, L.S., Povysheva, N.V., Barrionuevo, G., and Lewis, D.A. (2005a) Functional properties of fast spiking interneurons and their synaptic connections with pyramidal cells in primate dorsolateral prefrontal cortex. J Neurophysiol 93, 942–953.

Gonzalez-Burgos, G., Krimer, L.S., Urban, N.N., Barrionuevo, G., and Lewis, D.A. (2004) Synaptic efficacy during repetitive activation of excitatory inputs in primate dorsolateral prefrontal cortex. Cereb Cortex 14, 530–542.

Gonzalez-Burgos, G., Kroener, S., Krimer, L.S., Seamans, J.K., Urban, N.N., Henze, D.A., Lewis, D.A., and Barrionuevo, G. (2002) Dopamine modulation of neuronal function in the monkey prefrontal cortex. Physiol Behav 77, 537–543.

Gonzalez-Burgos, G., Kroener, S., Seamans, J.K., Lewis, D.A., and Barrionuevo, G. (2005b) Dopaminergic modulation of short-term synaptic plasticity in fast-spiking interneurons of primate dorsolateral prefrontal cortex. J Neurophysiol 94, 4168–4177.

Gorelova, N., Seamans, J.K., and Yang, C.R. (2002) Mechanisms of dopamine activation of fast-spiking interneurons that exert inhibition in rat prefrontal cortex. J Neurophysiol 88, 3150–3166.

Gulledge, A.T. and Jaffe, D.B. (2001) Multiple effects of dopamine on layer V pyramidal cell excitability in rat prefrontal cortex. J Neurophysiol 86, 586–595.

Gulledge, A.T. and Stuart, G.J. (2003) Action potential initiation and propagation in layer 5 pyramidal neurons of the rat prefrontal cortex: absence of dopamine modulation. J Neurosci 23, 11363–11372.

Gulledge, A.T., and Jaffe, D.B. (1998) Dopamine decreases the excitability of layer V pyramidal cells in the rat prefrontal cortex. J Neurosci 18, 9131–9151.

Hempel, C.M., Hartman, K.H., Wang, X.J., and Turrigiano, G.G. (2000) Multiple forms of short-term plasticity at excitatory synapses in rat medial prefrontal cortex. J Neurophysiol 83, 3031–3041.

Henze, D.A., Gonzalez-Burgos, G.R., Urban, N.N., Lewis, D.A., and Barrionuevo, G. (2000) Dopamine increases excitability of pyramidal neurons in primate prefrontal cortex. J Neurophysiol 84, 2799–2809.

Howard, A., Tamas, G., and Soltesz, I. (2005) Lighting the chandelier: new vistas for axo-axonic cells. Trends Neurosci 28, 310–316.

Inoue, M., Oomura, Y., Aou, S., Nishino, H., and Sikdar, S.K. (1985) Reward related neuronal activity in monkey dorsolateral pre frontal cortex during feeding behavior. Brain Res 326, 307–312.

Jacobs, B., Schall, M., Prather, M., Kapler, E., Driscoll, L., Baca, S., Jacobs, J., Ford, K., Wainwright, M., and Treml, M. (2001) Regional dendritic and spine variation in human cerebral cortex: a quantitative golgi study. Cereb Cortex 11, 558–571.

Joyce, J.N., Goldsmith, S., and Murray, A. (1993) Neuroanatomical localization of D1 versus D2 receptors: Similar organization in the basal ganglia of the rat, cat and human and disparate organization in the cortex and limbic system. In D1:D2 Dopamine Receptor Interactions, ed. Waddington, J.L., pp. 23–49. Academic press, London.

Jung, M.W., Qin, Y., McNaughton, B.L., and Barnes, C.A. (1998) Firing characteristics of deep layer neurons in prefrontal cortex in rats performing spatial working memory tasks. Cereb Cortex 8, 437–450.

Kawaguchi, Y. (1995) Physiological subgroups of nonpyramidal cells with specific morphological characteristics in layer II/III of rat frontal cortex. J Neurosci 15, 2638–2655.

Kawaguchi, Y. and Kubota, Y. (1997) GABAergic cell subtypes and their synaptic connections in rat frontal cortex. Cereb Cortex 7, 476–486.

Kobayashi, S., Nomoto, K., Watanabe, M., Hikosaka, O., Schultz, W., and Sakagami, M. (2006) Influences of rewarding and aversive outcomes on activity in macaque lateral prefrontal cortex. Neuron 51, 861–870.

Krimer, L.S., Jakab, R.L., and Goldman-Rakic, P.S. (1997) Quantitative three-dimensional analysis of the catecholaminergic innervation of identified neurons in the macaque prefrontal cortex. J Neurosci 17, 7150–7461.

Krimer, L.S., Zaitsev, A.V., Czanner, G., Kroener, S., Gonzalez-Burgos, G., Povysheva, N.V., Iyengar, S., Barrionuevo, G., and Lewis, D.A. (2005) Cluster analysis-based physiological classification and morphological properties of inhibitory neurons in layers 2-3 of monkey dorsolateral prefrontal cortex. J Neurophysiol 94, 3009–3022.

Kritzer, M.F. and Goldman-Rakic, P.S. (1995) Intrinsic circuit organization of the major layers and sublayers of the dorsolateral prefrontal cortex in the rhesus monkey. J Comp Neurol 359, 131–143.

Kroener, S., Krimer, L.S., Lewis, D.A., and Barrionuevo, G. (2006) Dopamine increases inhibition in the monkey dorsolateral prefrontal cortex through cell type-specific modulation of interneurons. Cereb Cortex Epub June 13, 2006. doi:10.1093/cercor/bhl012.

Kubota, K. and Niki, H. (1971) Prefrontal cortical unit activity and delayed alternation performance in monkeys. J Neurophysiol 34, 337–347.

Kuroda, M., Yokofujita, J., and Murakami, K. (1998) An ultrastructural study of the neural circuit between the prefrontal cortex and the mediodorsal nucleus of the thalamus. Prog Neurobiol 54, 417–458.

Le Moine, C. and Gaspar, P. (1998) Subpopulations of cortical GABAergic interneurons differ by their expression of D1 and D2 dopamine receptor subtypes. Brain Res Mol Brain Res 58, 231–236.

Letinic, K., Zoncu, R., and Rakic P. (2002) Origin of GABAergic neurons in the human neo-cortex. Nature 417, 645–649.

Levitt, J.B., Lewis, D.A., Yoshioka, T., and Lund, J.S. (1993) Topography of pyramidal neuron intrinsic connections in macaque monkey prefrontal cortex (areas 9 and 46) J Comp Neurol 338, 360–376.

Lewis, D.A. and Sesack, S.R. (1997) Dopamine systems in the primate brain. In Handbook of Chemical Neuroanatomy, Vol. 13 The Primate Nervous System, eds. Bloom, F.E., Bjork-lund, A., and Hokfelt, T., pp. 263–375. Elsevier, New York.

Lewis, D.A., Melchitzky, D.S., Sesack, S.R., Whitehead, R.E., Auh, S., and Sampson, A.R. (2001) Dopamine transporter immunoreactivity in monkey cerebral cortex: regional, laminar, and ultrastructural localization. J Comp Neurol 432, 119–136.

Lidow, M.S., Goldman-Rakic, P.S., Gallager, D.W., Rakic P. (1991) Distribution of dopa-minergic receptors in the primate cerebral cortex: quantitative autoradiographic analysis using [3H]raclopride, [3H]spiperone and [3H]SCH23390. Neuroscience 40, 657–671.

Lidow, M.S., Wang, F., Cao, Y., and Goldman-Rakic, P.S. (1998) Layer V neurons bear the majority of mRNAs encoding the five distinct dopamine receptor subtypes in the primate prefrontal cortex. Synapse 28, 10–20.

Ljungberg, T., Apicella, P., and Schultz, W. (1992) Responses of monkey dopamine neurons during learning of behavioral reactions. J Neurophysiol 67, 145–163.

Lund, J.S. and Lewis, D.A. (1993) Local circuit neurons of developing and mature macaque prefrontal cortex: Golgi and immunocytochemical characteristics. J Comp Neurol 328, 282–312.

Mainen, Z.F., Joerges, J., Huguenard, J.R., and Sejnowski, T.J. (1995) A model of spike initia-tion in neocortical pyramidal neurons. Neuron 15, 1427–1439.

Melchitzky, D.S., Sesack, S.R., Pucak, M.L., and Lewis, D.A. (1998) Synaptic targets of pyramidal neurons providing intrinsic horizontal connections in monkey prefrontal cor-tex. J Comp Neurol 390, 221–224.

Miller, E.K., Erickson, C.A., and Desimone, R. (1996) Neural mechanisms of visual working memory in prefrontal cortex of the macaque. J Neurosci 16, 5154–5167.

Missale, C., Nass, S.R., Robinson, S.W., Jaber, M., and Caron, M.G. (1998) Dopamine recep-tors: From structure to function. Physiol Rev 78, 189–225.

Mrzljak, L., Bergson, C., Pappy, M., Huff, R., Levenson, R., and Goldman-Rakic, P.S. (1996) Localization of dopamine D4 receptors in GABAergic neurons of the primate brain. Nature 381, 245–248.

Muly, E.C., Szigeti, K., and Goldman-Rakic, P.S. (1998) D1 receptor in interneurons of ma-caque prefrontal cortex: distribution and subcellular localization. J Neurosci 18, 10553–10565.

Mushiake, H., Saito, N., Sakamoto, K., Itoyama, Y., and Tanji, J. (2006) Activity in the lateral prefrontal cortex reflects multiple steps of future events in action plans. Neuron 50, 631–641.

Orlov, A.A., Kurzina, N.P., and Shutov, A.P. (1988) Activity of medial wall neurons in frontal cortex of rat brain during delayed response reactions. Neurosci and Behav Physiol 18, 31–37.

Pandya, D.N. and Yeterian, E.H. (1990) Prefrontal cortex in relation to other cortical areas in rhesus monkey: architecture and connections. Prog Brain Res 85, 63–94.

Parent, A. and Hazrati, L.N. (1995) Functional anatomy of the basal ganglia. I. The cortico-basal ganglia-thalamo -cortical loop. Brain Res Brain Res Rev 20, 91–127.

Paspalas, C.D. and Goldman-Rakic, P.S. (2004) Microdomains for dopamine volume neuro-transmission in primate prefrontal cortex. J Neurosci 24, 5292–5300.

Paspalas, C.D. and Goldman-Rakic, P.S. (2005) Presynaptic D1 dopamine receptors in pri-mate prefrontal cortex: target-specific expression in the glutamatergic synapse. J Neuro-sci 25, 1260–1267.

Petrides, M. (2000) Impairments in working memory after frontal cortical excisions. Adv Neurol 84, 111–118.

Petrides, M. and Pandya, D.N. (1999) Dorsolateral prefrontal cortex: comparative cytoarchi-tectonic analysis in the human and the macaque brain and corticocortical connection pat-terns. Eur J Neurosci 11, 1011–1036.

Povysheva, N.V., Gonzalez-Burgos, G., Zaitsev, A.V., Kroener, S., Barrionuevo, G., Lewis, D.A., and Krimer, L.S. (2006) Properties of excitatory synaptic responses in fast-spiking interneurons and pyramidal cells from monkey and rat prefrontal cortex. Cereb Cortex 16, 541–552.

Povysheva, N.V., Zaitsev, A.V., Kroener, S., Krimer, O.A., Rotaru, D.C., Gonzalez-Burgos, G., Lewis, D.A., and Krimer, L.S. (2007) Electrophysiological differences between neurogliaform cells from monkey and rat prefrontal cortex. J Neurophysiol 97(2): 1030–9.

Pratt, W.E. and Mizumori, S.J. (2001) Neurons in rat medial prefrontal cortex show anticipatory rate changes to predictable differential rewards in a spatial memory task. Behav Brain Res 123, 165–183.

Preuss, T.M. (1995) Do rats have prefrontal cortex? The Rose-Woolsey-Akert program reconsidered. J Cog Neurosci 7, 1–24.

Pucak, M.L., Levitt, J.B., Lund, J.S., and Lewis, D.A. (1996) Patterns of intrinsic and associational circuitry in monkey prefrontal cortex. J Comp Neurol 376, 614–630.

Quintana, J. and Fuster, J.M. (1999) From perception to action: temporal integrative functions of prefrontal and parietal neurons. Cereb Cortex 9, 213–221.

Quintana, J., Yajeya, J., and Fuster, J.M. (1988) Prefrontal representation of stimulus attributes during delay tasks. I. Unit activity in cross-temporal integration of sensory and sensory-motor information. Brain Res 474, 211–221.

Rainer, G. and Miller, E.K. (2002) Timecourse of object-related neural activity in the primate prefrontal cortex during a short-term memory task. Eur J Neurosci 15, 1244–1254.

Rainer, G., Asaad, W.F., and Miller, E.K. (1998) Memory fields of neurons in the primate prefrontal cortex. Proc Natl Acad Sci USA 95, 15008–15013.

Rainer, G., Rao, S.C., and Miller, E.K. (1999) Prospective coding for objects in primate prefrontal cortex. J Neurosci 19, 5493–5505.

Rao, S.C., Rainer, G., and Miller, E.K. (1997) Integration of what and where in the primate prefrontal cortex. Science 276, 821–824.

Rao, S.G., Williams, G.V., and Goldman-Rakic, P.S. (1999) Isodirectional tuning of adjacent interneurons and pyramidal cells during working memory: evidence for microcolumnar organization in PFC. J Neurophysiol 81, 1903–1916.

Repovs, G. and Baddeley, A. (2006) The multi-component model of working memory: explorations in experimental cognitive psychology. Neuroscience 139, 5–21.

Roesch, M.R. and Olson, C.R. (2003) Impact of expected reward on neuronal activity in prefrontal cortex, frontal and supplementary eye fields and premotor cortex. J Neurophysiol 90, 1766–1789.

Romo, R., Brody, C.D., Hernandez, A., and Lemus, L. (1999) Neuronal correlates of parametric working memory in the prefrontal cortex. Nature 399, 470–473.

Rotaru, D.C., Barrionuevo, G., and Sesack, S.R. (2005) Mediodorsal thalamic afferents to layer III of the rat prefrontal cortex: synaptic relationships to subclasses of interneurons. J Comp Neurol 490, 220–238.

Salamone, J.D., Correa, M., Mingote, S.M., and Weber, S.M. (2005) Beyond the reward hypothesis: alternative functions of nucleus accumbens dopamine. Curr Opin Pharmacol 5, 34–41.

Sawaguchi, T. (2001) The effects of dopamine and its antagonists on directional delay-period activity of prefrontal neurons in monkeys during an oculomotor delayed-response task. Neurosci Res 41, 115–128.

Sawaguchi, T. and Goldman-Rakic, P.S. (1991) D1 dopamine receptors in prefrontal cortex: involvement in working memory. Science 251, 947–950.

Schultz, W. (1997) Dopamine neurons and their role in reward mechanisms. Curr Opin Neurobiol 7, 191–197.

Schultz, W. and Romo, R. (1990) Dopamine neurons of the monkey midbrain: contingencies of responses to stimuli eliciting immediate behavioral reactions. J Neurophysiol 63, 607–624.

Seamans, J.K. and Yang, C.R. (2004) The principal features and mechanisms of dopamine modulation in the prefrontal cortex. Prog Neurobiol 74, 1–58.

Seamans, J.K., Durstewitz, D., Christie, B., Stevens, C.F., and Sejnowski, T.J. (2001a) Dopamine D1/D5 receptor modulation of excitatory synaptic inputs to layer V prefrontal cortex neurons. Proc Natl Acad Sci USA 98, 301–306.

Seamans, J.K., Gorelova, N., Durstewitz, D., and Yang, C.R. (2001b) Bidirectional dopamine modulation of GABAergic inhibition in prefrontal cortical pyramidal neurons. J Neurosci 21, 3628–3638.

Seamans, J.K., Nogueira, L., and Lavin, A. (2003) Synaptic basis of persistent activity in prefrontal cortex in vivo and in organotypic cultures. Cereb Cortex 13, 1242–1250.

Seguela, P., Watkins, K.C., and Descarries, L. (1988) Ultrastructural features of dopamine axon terminals in the anteromedial and the suprarhinal cortex of adult rat. Brain Res 442, 11–22.

Sesack, S.R., Hawrylak, V.A., Matus, C., Guido, M.A., and Levey, A.I. (1998a) Dopamine axon varicosities in the prelimbic division of the rat prefrontal cortex exhibit sparse immunoreactivity for the dopamine transporter. J Neurosci 18, 2697–2708.

Sesack, S.R., Hawrylak, V.A., Melchitzky, D.S., and Lewis, D.A. (1998b) Dopamine innervation of a subclass of local circuit neurons in monkey prefrontal cortex: ultrastructural analysis of tyrosine hydroxylase and parvalbumin immunoreactive structures. Cereb Cortex 8, 614–622.

Shallice, T. and Burgess, P. (1996) The domain of supervisory processes and temporal organization of behaviour. Philos Trans R Soc Lond B Biol Sci 351, 1405–1411.

Smiley, J.F. and Goldman-Rakic, P.S. (1993) Heterogeneous targets of dopamine synapses in monkey prefrontal cortex demonstrated by serial section electron microscopy: a laminar analysis using the silver-enhanced diaminobenzidine sulfide (SEDS) immunolabeling technique. Cereb Cortex 3, 223–238.

Smiley, J.F., Levey, A.I., Ciliax, B.J., and Goldman-Rakic, P.S. (1994) D1 dopamine receptor immunoreactivity in human and monkey cerebral cortex: predominant and extrasynaptic localization in dendritic spines. Proc Natl Acad Sci USA 91, 5720–5724.

Soloway, A.S., Pucak, M.L., Melchitzky, D.S., and Lewis, D.A. (2002) Dendritic morphology of callosal and ipsilateral projection neurons in monkey prefrontal cortex. Neuroscience 109, 461–471.

Somogyi, P., Kisvarday, Z.F., Freund, T.F., and Cowey, A. (1984) Characterization by Golgi impregnation of neurons that accumulate 3H-GABA in the visual cortex of monkey. Exp Brain Res 53, 295–303.

Steriade, M. (2001) Impact of network activities on neuronal properties in corticothalamic systems. J Neurophysiol 86, 1–39.

Trantham-Davidson, H., Neely, L.C., Lavin, A., and Seamans, J.K. (2004) Mechanisms underlying differential D1 versus D2 dopamine receptor regulation of inhibition in prefrontal cortex. J Neurosci 24, 10652–10659.

Tseng, K.Y. and O'Donnell, P. (2006) Dopamine modulation of prefrontal cortical interneurons changes during adolescence. Cereb Cortex 17 (5): 1235–40.

Tsujimoto, S. and Sawaguchi, T. (2005) Neuronal activity representing temporal prediction of reward in the primate prefrontal cortex. J Neurophysiol 93, 3687–3692.

Urban, N.N., Gonzalez-Burgos, G., Henze, D.A., Lewis, D.A., and Barrionuevo, G. (2002) Selective reduction by dopamine of excitatory synaptic inputs to pyramidal neurons in primate prefrontal cortex. J Physiol 539, 707–712.

Uylings, H.B., Groenewegen, H.J., and Kolb, B. (2003) Do rats have a prefrontal cortex? Behav Brain Res 146, 3–17.

van Eden, C.G., Hoorneman, E.M., Buijs, R.M., Matthijssen, M.A., Geffard, M., and Uylings, H.B. (1987) Immunocytochemical localization of dopamine in the prefrontal cortex of the rat at the light and electron microscopical level. Neuroscience 22, 849–862.

van Eden, C.G., Lamme, V.A., and Uylings, H.B. (1992) Heterotopic cortical afferents to the medial prefrontal cortex in the rat. A combined retrograde and anterograde tracer study. Eur J Neurosci 4, 77–97.

Vincent, S.L., Khan, Y., and Benes, F.M. (1993) Cellular distribution of dopamine D1 and D2 receptors in rat medial prefrontal cortex. J Neurosci 13, 2551–2564.

Vysokanov, A., Flores-Hernandez, J., and Surmeier, D.J. (1998) mRNAs for clozapine-sensitive receptors co-localize in rat prefrontal cortex neurons. Neurosci Lett 258, 179–182.

Wallis, J.D. and Miller, E.K. (2003) Neuronal activity in primate dorsolateral and orbital prefrontal cortex during performance of a reward preference task. Eur J Neurosci 18, 2069–2081.

Wang, Y., Markram, H., Goodman, P.H., Berger, T.K., Ma, J., and Goldman-Rakic, P.S. (2006) Heterogeneity in the pyramidal network of the medial prefrontal cortex. Nat Neurosci 9, 534–542.

Watanabe, M. (1990) Prefrontal unit activity during associative learning in the monkey. Exp Brain Res 80, 296–309.

Watanabe, M. (1996) Reward expectancy in primate prefrontal neurons. Nature 382, 629–632.

Watanabe, M., Hikosaka, K., Sakagami, M., and Shirakawa, S. (2002) Coding and monitoring of motivational context in the primate prefrontal cortex. J Neurosci 22, 2391–2400.

Watanabe, M., Hikosaka, K., Sakagami, M., and Shirakawa, S. (2005) Functional significance of delay-period activity of primate prefrontal neurons in relation to spatial working memory and reward/omission-of-reward expectancy. Exp Brain Res 166, 263–276.

Waters, J. and Helmchen, F. (2006) Background synaptic activity is sparse in neocortex. J Neurosci 26, 8267–8277.

Wayment, H.K., Schenk, J.O., and Sorg, B.A. (2001) Characterization of extracellular dopamine clearance in the medial prefrontal cortex: role of monoamine uptake and monoamine oxidase inhibition. J Neurosci 21, 35–44.

Williams, G.V. and Goldman-Rakic, P.S. (1995) Modulation of memory fields by dopamine D1 receptors in prefrontal cortex. Nature 376, 572–575.

Williams, S.M. and Goldman-Rakic, P.S. (1998) Widespread origin of the primate mesofrontal dopamine system. Cereb Cortex 8, 321–345.

Wilson, F.A., O'Scalaidhe, S.P., and Goldman-Rakic, P.S. (1994) Functional synergism between putative gamma-aminobutyrate- containing neurons and pyramidal neurons in prefrontal cortex. Proc Natl Acad Sci USA 91, 4009–4013.

Xu, Q., Cobos, I., De La, C.E., Rubenstein, J.L., and Anders on, S.A. (2004) Origins of cortical interneuron subtypes. J Neurosci 24, 2612–2622.

Yajeya, J., Quintana, J., and Fuster, J.M. (1988) Prefrontal representation of stimulus attributes during delay tasks. II. The role of behavioral significance. Brain Res 474, 222–230.

Yanez, I.B., Munoz, A., Contreras, J., Gonzalez, J., Rodriguez-Veiga, E., and DeFelipe, J. (2005) Double bouquet cell in the human cerebral cortex and a comparison with other mammals. J Comp Neurol 486, 344–360.

Yang, C.R. and Seamans, J.K. (1996) Dopamine D1 receptor actions in layers V-VI rat prefrontal cortex neurons in vitro: modulation of dendritic-somatic signal integration. J Neurosci 16, 1922–1935.

Yang, C.R., Seamans, J.K., and Gorelova, N. (1996) Electrophysiological and morphological properties of layers V-VI principal pyramidal cells in rat prefrontal cortex in vitro. J Neurosci 16, 1904–1921.

Zaitsev, A.V., Gonzalez-Burgos, G., Povysheva, N.V., Kroener, S., Lewis, D.A., and Krimer, L.S. (2005) Localization of calcium-binding proteins in physiologically and morphologically characterized interneurons of monkey dorsolateral prefrontal cortex. Cereb Cortex 15, 1178–1186.

Chapter 9

Dopamine D1 and Glutamate N-Methyl-D-Aspartate Receptors: *An Essential Interplay in Prefrontal Cortex Synaptic Plasticity*

María Sol Kruse and Thérèse M. Jay
INSERM U796-Univ Paris Descartes, Physiopathologie des Maladies Psychiatriques,
2ter rue d'Alesia, 75014 Paris, France

9.1 Introduction

In recent years, it has become clear that dopamine (DA) is able to modulate directly dendritic excitability of prefrontal neurons and control higher cognitive functions. These data have surged interest on the role of DA on long-term potentiation (LTP) and long-term depression (LTD) induced in the prefrontal cortex and the mechanism by which DA could modulate synaptic plasticity.

In the prefrontal cortex, DA and glutamate terminals converge on pyramidal neurons located mostly in deep layers (V and VI) of the prelimbic and anterior cingulate areas. Spines of pyramidal neurons are often the targets of paired dopamine and excitatory terminals forming synaptic triads (Carr and Sesack, 2000). D1 family (D1R) of dopamine receptors is the most common dopamine receptor subtype found in prefrontal cortex. Ultrastructural analyses in monkey and rat prefrontal cortex have revealed the expression of D1 receptors in distal dendrites and spines of pyramidal cells. This receptor is found in close proximity to putative glutamatergic axon terminals giving rise to asymmetric synapses on the same spine (Bergson et al., 1995; Muly et al., 2001). The preferential localization of D1 receptors to the spines and shafts of pyramidal cells supports postsynaptic mechanisms for a potential role of D1 receptors in synaptic plasticity and memory processes.

More recently it was shown that D1 receptors and the ionotropic glutamate receptor, NMDA physically interact at postsynaptic sites in the hippocampus and striatum (Lee et al., 2002; Fiorentini et al., 2003; Pei et al., 2004). Through this interaction D1 receptor is enabled to modulate NMDA-current independently of the generation of second messengers. Very recent studies from our laboratory indicate that D1 and NMDA receptors are also forming a heterocomplex in the prefrontal cortex (Kruse and Jay, 2005).

The aim of the present chapter is to provide a brief overview on DA (mostly via D1 receptors) contribution to synaptic plasticity in the prefrontal cortex and consider new insights into the possible cellular mechanisms underlying these plastic changes. The data presented herein, significant for current understanding prefrontal memory mechanisms, could help finding new strategies for psychiatric disorders associated to cognitive deficits such as schizophrenia.

9.2 D1 Modulation of Synaptic Plasticity in the Prefrontal Cortex

Pharmacological and genetic evidence for a role of D1 receptors in the bidirectional modulation of synaptic plasticity in the prefrontal cortex has been demonstrated in the past few years by different groups including ours. We conducted in vivo studies in the intact brain of anesthetized rats and found that DA durably modulates the efficiency of the NMDA receptor-dependent LTP induced at hippocampal to prefrontal cortex synapses (Jay et al., 1995; Gurden et al., 1999). Our data have provided strong support to demonstrate that DA exerts a facilitatory role on the induction of LTP and that the integrity of the mesocortical DA system is necessary for the expression of this hippocampal-prefrontal cortex synaptic plasticity. Transient stimulation of the mesocortical DA neurons (ventral tegmental area, VTA) shown to produce an increase of their DA release capacity (Garris et al., 1993) is sufficient to produce a long-lasting increase in the magnitude of LTP and a significant correlation between the level of cortical DA and the magnitude of hippocampal-prefrontal LTP indicates that a minimum level of DA in the prefrontal cortex is necessary to induce LTP (Gurden et al., 1999). Further investigation into the role of specific D1 and D2 prefrontal DA receptors showed that LTP at hippocampal to prefrontal cortex synapses is significantly higher when the D1 agonist, SKF81297, is locally infused (reverse dialysis) in the prefrontal cortex prior to tetanus. The increase in LTP amplitude is significantly larger at certain doses tested when compared to ACSF-controls, demonstrating that an optimal range of D1 receptor activation is necessary to induce sustained enhancement of prefrontal LTP. Conversely, application of the D1 receptor antagonist SCH23390 at different doses in the prefrontal cortex dose-dependently impaired LTP at hippocampal to prefrontal cortex synapses. The D2 receptor antagonist sulpiride did not affect cortical LTP. Thus, these studies showing a clear facilitating effect of a D1 agonist on LTP induction and maintenance demonstrated that D1 but not D2 prefrontal DA receptors play an essential role on the expression of this NMDA-dependent LTP at hippocampal-prefrontal synapses (Gurden et al., 2000).

Using prefrontal cortex slices from adult rats and mice, it was also found that D1 receptor plays a critical role in the intracortical prefrontal LTP as it does in the hippocampal-prefrontal pathway in the intact animal (Huang et al., 2004). The D1 receptor agonist (6-APB, 6-chloro-7,8-dihydroxy-3allyl-1phenyl-2,3,4,5-tetrahydro-1H-3-benzazepine hydrobromide) facilitates the maintenance of LTP which is dependent on NMDA receptors. Conversely, the D1 antagonist SCH23390 depressed LTP. In the same paper, the authors show that this facilitating effect of DA via D1 receptor activation is absent in heterozygous D1 receptor knockout mice, that is, a 50% reduction of D1 receptor has a strong influence on synaptic plasticity in the prefrontal cortex. Not only LTP but also LTD, the flip side of synaptic plasticity is facilitated by DA in wild type mice but not in heterozygous D1 mice and this DA facilitation depends on both D1 and D2 receptors. In wild type mice, LTD is blocked by SCH23390 and sulpiride. Thus, these pharmacological and genetic results strengthen the action of DA and D1 receptors in the modulation of synaptic plasticity in the medial prefrontal cortex.

Other studies reported a more complex scheme on DA modulation of synaptic plasticity in prefrontal cortex slices prepared from juvenile rats where LTD appears easier to induce than LTP. Tetanic stimulation in layer I-II afferents-layer V neuron glutamatergic synapses induces an NMDA receptor-independent LTD when the stimulation is coupled with a bath-application of DA and this facilitating effect on LTD is mediated by both D1 and D2 receptors (Otani et al., 1999). However, if DA is applied, washed out, and then applied again while coupled to high frequency stimuli, LTP is induced instead of LTD (Blond et al., 2002). Synaptic activation of NMDA receptors concurrent to activation of D1 and D2 receptors are necessary for the induction of this novel form of LTP (Matsuda et al., 2006). Recently, Young and Yang (2005) observed that LTP within layers V → layer V input in the prefrontal cortex could be obtained by switching tetanic stimulation to a theta frequency pattern. However, the NMDA and D1 dependency of this theta-dependent LTP needs further investigation.

Altogether these data show a close interplay between NMDA receptor function and D1 receptor activity that is required to induce bidirectional synaptic plasticity in the medial prefrontal cortex.

9.3 D1 and NMDA Receptors: Mechanisms of Action

At the postsynaptic level, interactions between the D1 and NMDA receptors appear to be particularly relevant. Several mechanisms underlying this interaction, could contribute to the D1 potentiation of NMDA responses: (i) a coordinated regulation of receptor trafficking at the synaptic sites; (ii) the formation of a heteromeric D1/NMDA receptor complex; (iii) a D1R-dependent second messengers mediated phosphorylation. In this part we will review these different forms or regulation and their importance in synaptic plasticity of the prefrontal cortex.

9.3.1 RECEPTOR TRAFFICKING

DA via D1 receptors activates the adenylil cyclase (AC)-cAMP protein kinase A (PKA) signaling pathway. cAMP, PKA activity, and Ca^{2+} are key regulators of synaptic plasticity in the hippocampus, striatum, and prefrontal cortex (Frey et al., 1993; Jay et al., 1998; Spencer and Murphy, 2002). As already pointed out by several works, DA through D1 receptors potentiates NMDA currents (Cepeda and Levine, 1998) and this synergism which occurs at the postsynaptic level appears to be mediated through both a PKA and Ca^{2+}-dependent mechanisms (Wang and O'Donnell, 2001; Tseng and O'Donnell, 2003, 2004).

LTP induction increases phosphorylation of the PKA site of the AMPA receptor GluR1 subunit, whereas LTD induction dephosphorylates this site (Lee et al., 2000). During LTP there is an increase in the number of AMPA receptor in the postsynaptic membrane. In prefrontal neurons it has been recently shown that D1 receptor stimulation or PKA activation is sufficient to produce GluR1 externalization onto the cell membrane and a subsequent activation of synaptic NMDA receptors can drive these newly externalized AMPA receptors into synapses (Sun et al., 2005). Both AMPA and NMDA receptors could be modified during plasticity. Recent studies performed in developmentally mature rat CA1 show that LTP promotes surface expression of NR1 and NR2A subunits of the NMDA receptor (Grosshans et al., 2002; Goebel et al., 2005).

In the striatum, D1 receptor activation increases NMDA receptor subunits NR1, NR2A and NR2B in the synaptosomal membrane fraction (Dunah and Standaert 2001). On the other hand, NMDA receptor activation increases the D1 receptor availability in the plasma membrane and dendritic spines and enhances D1 receptor-mediated cAMP accumulation (Scott et al., 2002). Thus, the modulation of receptor trafficking at postsynaptic sites could be one of the mechanisms by which the D1-NMDA receptor interactions enhance synaptic plasticity and potentiate NMDA responses. Although there is no evidence linking D1 receptors and trafficking of NMDA receptors in neocortical pyramidal neurons, it is possible that a similar mechanism also plays a role in enhancing NMDA responses in the prefrontal cortex.

9.3.2 D1 AND NMDA RECEPTORS FORM A HETEROCOMPLEX

A number of intracellular events contribute to the D1-mediated enhancement of NMDA function. Recently, a novel mechanism of interaction between D1 and NMDA receptors was found in the striatum where D1 receptors via its intracellular C-terminus directly interact with the NR1 subunit (Fiorentini et al., 2003) and in the hippocampus where the D1 interaction occurs via the NR1 NR2A-NMDA receptor subunits (Lee et al., 2002). The functional properties of both receptors are then modified by this heteromerization (Fig. 9.1) (Cepeda and Levine, 2006). In striatal medial spiny neurons, D1 receptor selectively interacts with the NR1 subunit of the NMDA receptor to form a constitutive oligomeric complex that is recruited to the plasma membrane by the NR2B subunit. A direct protein-protein interaction with the NMDA receptor is thus one of the mechanisms directing the trafficking of D1 receptors to specific subcellular compartments in the striatum. This interaction was also shown to abolish D1 receptor internalization, a crucial adaptive response that normally occurs upon agonist stimulation (Fiorentini et al., 2003).

In hippocampal neurons, D1R-NR1 physical interaction enables NMDA receptors to increase plasma membrane insertion of D1 receptors and enhances D1 receptor-mediated cAMP accumulation (Pei et al., 2004). The heterocomplex is also implicated in the attenuation of NMDA receptor-mediated excitotoxicity and the inhibition of NMDA receptor-gated current (Lee et al., 2002), possibly as a protective mechanism to prevent the deleterious effect of excessive D1-NMDA receptor stimulation. Studies from our laboratory have shown that D1 and NMDA receptors co-localize and co-precipitates together in the rat prefrontal cortex (Kruse and Jay,

2005). However, how this interaction modulates the functional properties of the D1-NMDA receptor heterocomplex in the prefrontal cortex still remain to be determined.

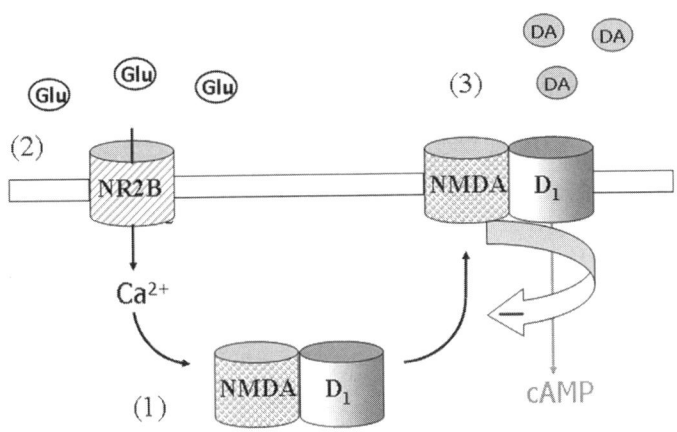

Figure 9.1 Diagram showing the subcellular distributions of D1-NR1 (NR2A) heterocomplex and its physiological responses. (1) D1 and NMDA receptors are assembled as dimers in the endoplasmic reticulum. The C-terminal tail of D1 receptors physically interacts with either the NR1 or NR2A NMDA receptor subunits. (2) The presence of NR2B subunit (striatum) and/or the activation of NMDA receptors (hippocampus) induce the translocation of the heterocomplex to the plasma membrane. (3) At the cellular membrane, D1 receptor stimulation would lead to cAMP accumulation and reduce NMDA-current by decreasing the number of NMDA receptor subunits available in the surface (open arrow).

9.3.3 MODULATION OF PHOSPHORYLATION: D1-NMDA RECEPTORS CROSSTALK

Activation of PKA by D1 receptors can directly phosphorylate a target protein or initiate a cascade of phosphorylation events by activating the DA and cAMP-regulated phosphoprotein (DARPP-32). Once activated, DARPP-32 becomes a potent inhibitor of protein phosphatase 1 (PP1) (Hemmings et al., 1984). Conversely, PP1 can be activated by the calcium/calmodulin-dependent phosphatase calcineurin, activated by calcium influx through NMDA receptors. Calcineurin dephosphorylates DARPP-32 and then through inactivation of DARPP-32 blocks the inhibition of PP1. On the other hand, CaMKII, a key enzyme in LTP is a substrate for PP1 and the inhibition of PP1 seems to promote the activation of CaMKII (Blitzer et al., 1998) (Fig. 9.2). Thus, the control of PP1 through DARPP-32, a key regulator of DA transmission, is likely to have a significant effect on the regulation of the synaptic strength in plasticity. Studies performed on DARPP-32 knock-out mice revealed that both corticostriatal LTP and LTD are absent in these animals, and that these two forms of synaptic plasticity can be restored after application of okadaic acid, an inhibitor of PP-1 (Calabresi et al., 2000). DA through D1 receptors is also able to control the rate of phosphorylation/dephosphorylation of the NR1 subunit which is required for a functional NMDA receptor, and these mechanisms occur through the DA/D1/PKA/DARPP-32/PP1 pathway (Snyder et al., 1998). In contrast, D2 receptors are able to block the D1 receptor-mediated increase in NR1 phosphorylation (Edwards et al., 2002).

The importance of the second messenger Ca^{2+} has also been questioned in the DA modulation of synaptic plasticity. A rise in postsynaptic Ca^{2+} appears necessary for triggering the D1/D5-induced effect in the hippocampus, although the source of Ca^{2+} is still unclear (Gorelova and Yang, 2000). It could result from voltage-gated Ca^{2+} channels and from Ca^{2+} entry through the NMDA or Ca^{2+} permeable AMPA receptor (Surmeier et al., 1995). It has also been reported recently that D1 receptor stimulation releases Ca^{2+} from intracellular stores in cultured neocortical neurons after priming with Gq-coupled receptors agonists (Lezcano and Bergson, 2002). Moreover, recent studies from our laboratory suggest that D1 receptor activation enhances the iCa^{2+} signal induced by NMDA in prefrontal primary cultures via PKA activation (unpublished data). The initiation of a Ca^{2+}-dependent signaling cascade involves PKA and CaMKII, two critical mediators of NMDA receptor-dependent LTP. In principle, changes in intracellular Ca^{2+} can have profound effects on cellular concentrations of cAMP if appropriate isoforms of adenylyl cyclase are present. Thus, DA receptors could integrate multiple signals to produce maximal cAMP signals which play a critical role in LTP. PKA activity could serve as a gate for synaptic plasticity by modulating calmodulin-dependent protein kinase II (CaMKII) through the PP1, the phosphatase that primarily regulates LTP expression (Blitzer et al., 1998). Furthermore, PKA stimulates the transcription of a number of genes by catalyzing the phosphorylation of the cAMP regulatory element binding protein (CREB), and PP1 retains the ability of CREB to stimulate transcription. Here is another step where, DA receptors could control the kinetics and duration of phosphorylation of CREB through the PKA/PP1 signaling complex. DA as a critical regulator of CRE-mediated gene expression has already been shown in striatal neurons (Konradi et al., 1996; Liu and Graybiel, 1996).

Very recent findings from our laboratory (Hotte et al., 2007) have shown that the late phase (2-h post-tetanus) of LTP at hippocampal to prefrontal cortex synapses leads to activation of DARPP-32 and CREB in the prefrontal cortex. Given the role of the transcription factor CREB in long lasting forms of synaptic plasticity, these interactions could explain the strong impact of DA through D1 receptors on the duration of LTP that was observed in prefrontal LTP (Gurden et al., 1999, 2000; Huang et al., 2004). D1 receptors could potentially lead to phosphorylation-dependent activation of two major cAMP-regulated proteins involved in the regulation of protein synthesis: the transcription factor CREB and DARPP-32.

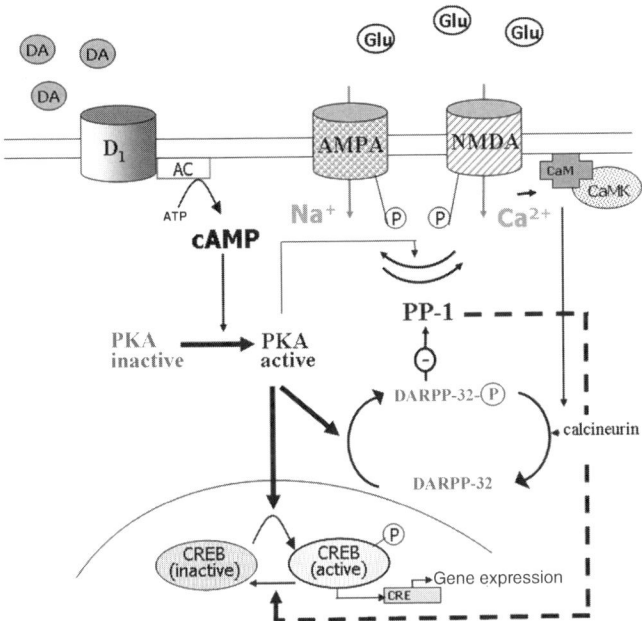

Figure 9.2 Role of D1/cAMP/PKA/PP1 signaling pathway mediating the DA modulation of LTP. The D1 receptor positively coupled to adenylate cyclase (AC) increases AC activity leading to the formation of cAMP which subsequently activates the cAMP-dependent PKA by binding to the regulatory subunits. The release of the active catalytic subunit permits phosphorylation of specific target substrates localized in different cellular compartments (cytosol, membrane, nucleus). (1) Activated PKA phosphorylates both AMPA and NMDA receptors and DARPP-32. Once phosphorylated, DARPP-32 is converted to a potent inhibitor of PP-1 that will promote phosphorylation of CaMKII. (2) Concurrently, activation of NMDA receptors increases the influx of Ca^{2+} into the cell which activates through the Ca^{2+}/calmodulin complex the CaM kinases (II and IV) and also the protein phosphatase 2B, calcineurin (PP2B). Calcineurin is able to dephosphorylate DARPP-32 which then promotes dephosphorylation through a desinhibition of PP1. (3) PKA also phosphorylates CREB, switching this transcription factor from its inactive to its active state. Conversely, PP-1 specifically dephosphorylates CREB. The control of PP1 through DARPP-32, a key regulator of DA transmission and NMDA receptors activity, is likely to have a significant effect on the DA regulation of LTP (adapted from Jay, 2003).

9.4 D1-NMDA Mediated Responses: A Multiple Point Regulation

In the brain, the outcome response to activation of interacting receptors depends on the order of neurotransmitters release. For example, activation of D1 receptors due to DA release by unexpected reward can prime cortical pyramidal neurons. Timing is also an important requirement for DA when LTP is induced experimentally in the prefrontal cortex, since DA release evoked by VTA stimulation should occur before excitatory afferents are activated in order to induce potentiation (Jay, 2003; Wolf et al., 2004). In contrast, DA release when occurring after tetanus-induced LTP does not produce any further increase in the potentiation. Thus, it is a sequential application of the two paradigms which is able to induce the synaptic enhancement. In a similar way, infusion of the D1 agonist has to be infused or applied in the prefrontal cortex on in vivo or in vitro preparations prior to tetanus to get a larger LTP and/or a longer lasting LTP.

DA concentration and the mode of release have also a significant effect on synaptic plasticity. Dopaminergic neurons have been shown to fire in two firing modes, tonic and bursts at higher frequency. Phasic DA release is a critical mediator of reward-related processes but tonic background DA concentration is believed to regulate cognitive functions (Goto and Grace, 2005). Phasic and tonic DA release may produce different effects on LTP. Background endogenous DA in the prefrontal cortex is a determining factor in the magnitude of LTP expression in vivo (Gurden et al., 1999). When using prefrontal cortex slices, which are mostly, deprived of tonic DA inputs, pairing DA with a weak tetanic stimulation, a protocol usually inducing LTD induces LTP when "primed" with DA (Matsuda et al., 2006). This different background level in DA may also explain the difference in the NMDA- versus non NMDA-dependent prefrontal LTP recorded in vivo at hippocampal-prefrontal synapses and in vitro in prefrontal slices. During tetanic stimulation of the ventral hippocampus DA is phasically released. DA is increased in the prefrontal cortex immediately after stimulation of the glutamate afferents (Gurden et al., 2000) which suggests that phasic activation of DA neurons may induce long-term synaptic changes. Some of these effects which depend on the exact timing of the DA activation relative to the event to be conditioned are in analogy to behavioral learning, and are thus compatible with the notion of DA responses acting as neurochemical reinforcement (Goldman-Rakic 1998; Seamans and Yang, 2004). DA inputs that regulate prefrontal synaptic plasticity may influence the subsequent storage of information. It was recently shown that activation of the D1/D5 receptors during memory encoding is necessary for the persistence of a memory trace in the hippocampus (O'Carroll et al., 2006). The downstream consequences of this appropriate co-activation of glutamatergic and dopaminergic activity might include the up-regulation of protein synthesis, and the subsequent stabilization of synaptic changes at synapses "tagged" by glutamatergic stimulation.

Pyramidal cells in the prefrontal cortex show membrane potential shifts between a negative down state and a depolarized up state (O'Donnell and Grace, 1995; Lewis and O'Donnell, 2000). D1 receptors stabilize prefrontal cortex pyramidal neurons into up states through both intrinsic and synaptic mechanisms (Lewis and O'Donnell 2000; Tseng and O'Donnell, 2005). Therefore, during tetanic stimulation of glutamate afferents to pyramidal neurons where LTP is induced, DA probably maintains those neurons to a depolarization level and facilitates the induction of LTP.

It is possible to speculate that once NMDA receptors are activated, oligomerization of the D1-NMDA receptor occurs, providing a regulated delivery to the plasma membrane, dendritic spines or both (Scott et al., 2002; Fiorentini et al., 2003) (Fig. 9.1).

Once in the plasma membrane, the traditional pathway involving D1 receptor activation and the cAMP-PKA-DARPP-32 cascade could produce various effects that enhance NMDA receptor function and LTP (Cepeda and Levine, 1998; Jay, 2003) (Fig. 9.2).

In summary, there are many steps by which the D1-NMDA receptor in the prefrontal cortex could regulate synaptic plasticity, memory mechanisms and cognition. D1-NMDA receptor interactions together with other neurotransmitters are able to fine-tune normal neuronal functions in the prefrontal cortex but the characteristics of these interactions have to be defined. The potential for postsynaptic cross-talk between D1 and NMDA receptors in prefrontal cortex projection neurons could have important functional implications and may offer a promising target for development of novel pharmacological tools. There is now increasing evidence that concurrent alterations of dopamine and glutamate function may play a central role in the pathophysiology of psychiatric but also neurological disorders.

Acknowledgments

This work is supported by INSERM postdoctoral fellowship (MSK) and ACI French Research Minister Grant (TMJ).

References

Bergson C., Mrzljak L., Smiley J.F., Pappy M., Levenson R., Goldman-Rakic P.S. (1995) Regional, cellular, and subcellular variations in the distribution of D1 and D5 dopamine receptors in primate brain. J. Neurosci. 15:7821–7836.

Blitzer R.D., Connor J.H., Brown G.P., Wong T., Shenolikar S., Iyengar R., Landau E.M. (1998) Gating of CaMKII by cAMP-regulated protein phosphatase activity during LTP. Science 280:1940–1942.

Blond O., Crepel F., Otani S. (2002) Long-term potentiation in rat prefrontal slices facilitated by phased application of dopamine. Eur. J. Pharmacol. 438:115–116.

Calabresi P., Gubellini P., Centonze D., Picconi B., Bernardi G., Chergui K., Svenningsson P., Fienberg A.A, Greengard P. (2000) Dopamine and cAMP-regulated phosphoprotein 32 kDa controls both striatal long-term depression and long-term potentiation, opposing forms of synaptic plasticity. J. Neurosci. 20:8443–8451.

Carr D.B., Sesack S.R. (2000) Dopamine terminals synapse on callosal projection neurons in the rat prefrontal cortex. J. Comp. Neurol. 425:275–283.

Cepeda C., Levine M.S. (1998) Dopamine and N-methyl-D-aspartate receptor interactions in the neostriatum. Dev. Neurosci. 20:1–18.

Cepeda C., Levine M.S. (2006) Where do you think you are going? The NMDA-D1 receptor trap. Sci. STKE. 2006:pe20.

Dunah A.W., Standaert D.G. (2001) Dopamine D1 receptor-dependent trafficking of striatal NMDA glutamate receptors to the postsynaptic membrane. J. Neurosci. 21:5546–5558.

Edwards S., Simmons D.L., Galindo D.G., Doherty J.M., Scott A.M., Hughes P.D., Wilcox R.E. (2002) Antagonistic effects of dopaminergic signaling and ethanol on protein kinase A-mediated phosphorylation of DARPP-32 and the NR1 subunit of the NMDA receptor. Alcohol Clin. Exp. Res. 26:173–180.

Fiorentini C., Gardoni F., Spano P., Di Luca M., Missale C. (2003) Regulation of dopamine D1 receptor trafficking and desensitization by oligomerization with glutamate *N*-methyl-D-aspartate receptors. J. Biol. Chem. 278:20196–20202.

Frey U., Huang Y.Y., Kandel E.R. (1993) Effects of cAMP simulate a late stage of LTP in hippocampal CA1 neurons. Science 260:1661–1664.

Garris P.A., Collins L.B., Jones S.R., Wightman R.M. (1993) Evoked extracellular dopamine in vivo in the medial prefrontal cortex. J. Neurochem. 61:637–647.

Goebel S.M., Alvestad R.M., Coultrap S.J., Browning M.D. (2005) Tyrosine phosphorylation of the *N*-methyl-D-aspartate receptor is enhanced in synaptic membrane fractions of the adult rat hippocampus. Brain Res. Mol. Brain Res. 142:65–79.

Goldman-Rakic P.S. (1998) The cortical dopamine system: role in memory and cognition. Adv. Pharmacol. 42:707–711.

Gorelova N.A., Yang C.R. (2000) Dopamine D1/D5 receptor activation modulates a persistent sodium current in rat prefrontal cortical neurons in vitro. J. Neurophysiol. 84:75–87.

Goto Y., Grace A.A. (2005) Dopamine-dependent interactions between limbic and prefrontal cortical plasticity in the nucleus accumbens: disruption by cocaine sensitization. Neuron 47:255–266.

Grosshans D.R., Clayton D.A., Coultrap S.J., Browning M.D. (2002) LTP leads to rapid surface expression of NMDA but not AMPA receptors in adult rat CA1. Nat. Neurosci. 5: 27–33.

Gurden H., Tassin J.P., Jay T.M. (1999) Integrity of the mesocortical dopaminergic system is necessary for complete expression of in vivo hippocampal-prefrontal cortex long-term potentiation. Neuroscience 94:1019–1027.

Gurden H., Takita M., Jay T.M. (2000) Essential role of D1 but not D2 receptors in the NMDA receptor-dependent long-term potentiation at hippocampal-prefrontal cortex synapses in vivo. J. Neurosci. 20:RC106.

Hemmings Jr., H.C., Nairn A.C., Greengard P. (1984) DARPP-32, a dopamine- and adenosine 3':5'-monophosphate-regulated neuronal phosphoprotein. II. Comparison of the kinetics of phosphorylation of DARPP-32 and phosphatase inhibitor 1. J. Biol. Chem. 259:14491–14497.

Hotte M., Thuault S., Dineley K.T., Hemmings H.C., Jr., Nairn A.C., Jay T.M. (2007) Phosphorylation of CREB and DARPP-32 during late LTP at hippocampal to prefrontal cortex synapses in vivo. Synapse 61:24–28.

Huang Y.Y., Simpson E., Kellendonk C., Kandel E.R. (2004) Genetic evidence for the bidirectional modulation of synaptic plasticity in the prefrontal cortex by D1 receptors. Proc. Natl. Acad. Sci. U.S.A. 101:3236–3241.

Jay T.M. (2003) Dopamine: a potential substrate for synaptic plasticity and memory mechanisms. Prog. Neurobiol. 69:375–390.

Jay T.M., Burette F., Laroche S. (1995) NMDA receptor-dependent long-term potentiation in the hippocampal afferent fibre system to the prefrontal cortex in the rat. Eur. J. Neurosci. 7:247–250.

Jay T.M., Gurden H., Yamaguchi T. (1998) Rapid increase in PKA activity during long-term potentiation in the hippocampal afferent fibre system to the prefrontal cortex in vivo. Eur. J. Neurosci. 10:3302–3306.

Konradi C., Leveque J.C., Hyman S.E. (1996) Amphetamine and dopamine-induced immediate early gene expression in striatal neurons depends on postsynaptic NMDA receptors and calcium. J. Neurosci. 16:4231–4239.

Kruse M.S., Jay T.M. (2005) Direct interaction of dopamine D1 with NMDA NR1 receptors in rat prefrontal cortex. Society for Neuroscience, 34.4.

Lee F.J., Xue S., Pei L., Vukusic B., Chery N., Wang Y., Wang Y.T., Niznik H.B., Yu X.M., Liu F. (2002) Dual regulation of NMDA receptor functions by direct protein-protein interactions with the dopamine D1 receptor. Cell 111:219–230.

Lee H.K., Barbarosie M., Kameyama K., Bear M.F., Huganir R.L. (2000) Regulation of distinct AMPA receptor phosphorylation sites during bidirectional synaptic plasticity. Nature 405:955–959.

Lewis B.L., O'Donnell P. (2000) Ventral tegmental area afferents to the prefrontal cortex maintain membrane potential 'up' states in pyramidal neurons via D(1) dopamine receptors. Cereb. Cortex. 10:1168–1175.

Lezcano N., Bergson C. (2002) D1/D5 dopamine receptors stimulate intracellular calcium release in primary cultures of neocortical and hippocampal neurons. J. Neurophysiol. 87:2167–2175.

Liu F.C., Graybiel A.M. (1996) Spatiotemporal dynamics of CREB phosphorylation: transient versus sustained phosphorylation in the developing striatum. Neuron 17:1133–1144.

Matsuda Y., Marzo A., Otani S. (2006) The presence of background dopamine signal converts long-term synaptic depression to potentiation in rat prefrontal cortex. J. Neurosci. 26:4803–4810.

Muly E.C., Greengard P., Goldman-Rakic P.S. (2001) Distribution of protein phosphatases-1 alpha and -1 gamma 1 and the D(1) dopamine receptor in primate prefrontal cortex: evidence for discrete populations of spines. J. Comp. Neurol. 440:261–270.

O'Carroll C.M., Martin S.J., Sandin J., Frenguelli B., Morris R.G. (2006) Dopaminergic modulation of the persistence of one-trial hippocampus-dependent memory. Learn. Mem. 13:760–769.

O'Donnell P., Grace A.A. (1995) Different effects of subchronic clozapine and haloperidol on dye-coupling between neurons in the rat striatal complex. Neuroscience 66:763–767.

Otani S., Auclair N., Desce J.M., Roisin M.P., Crepel F. (1999) Dopamine receptors and groups I and II mGluRs cooperate for long-term depression induction in rat prefrontal cortex through converging postsynaptic activation of MAP kinases. J. Neurosci. 19:9788–9802.

Pei L., Lee F.J., Moszczynska A., Vukusic B., Liu F. (2004) Regulation of dopamine D1 receptor function by physical interaction with the NMDA receptors. J. Neurosci. 24:1149–1158.

Scott L., Kruse M.S., Forssberg H., Brismar H., Greengard P., Aperia A. (2002) Selective up-regulation of dopamine D1 receptors in dendritic spines by NMDA receptor activation. Proc. Natl. Acad. Sci. U.S.A. 99:1661–1664.

Seamans J.K., Yang C.R. (2004) The principal features and mechanisms of dopamine modulation in the prefrontal cortex. Prog. Neurobiol. 74:1–58.

Snyder G.L., Fienberg A.A., Huganir R.L., Greengard P. (1998) A dopamine/D1 receptor/protein kinase A/dopamine- and cAMP-regulated phosphoprotein (Mr 32 kDa)/protein phosphatase-1 pathway regulates dephosphorylation of the NMDA receptor. J. Neurosci. 18:10297–10303.

Spencer J.P., Murphy K.P. (2002) Activation of cyclic AMP-dependent protein kinase is required for long-term enhancement at corticostriatal synapses in rats. Neurosci. Lett. 329:217–221.

Sun X., Zhao Y., Wolf M.E. (2005) Dopamine receptor stimulation modulates AMPA receptor synaptic insertion in prefrontal cortex neurons. J. Neurosci. 25:7342–7351.

Surmeier D.J., Bargas J., Hemmings H.C., Jr., Nairn A.C., Greengard P. (1995) Modulation of calcium currents by a D1 dopaminergic protein kinase/phosphatase cascade in rat neostriatal neurons. Neuron 14:385–397.

Tseng K.Y., O'Donnell P. (2003) Dopamine-glutamate interactions in the control of cell excitability in medial prefrontal cortical pyramidal neurons from adult rats. Ann. N. Y. Acad. Sci. 1003:476–478.

Tseng K.Y., O'Donnell P. (2004) Dopamine-glutamate interactions controlling prefrontal cortical pyramidal cell excitability involve multiple signaling mechanisms. J. Neurosci. 24:5131–5139.

Tseng K.Y., O'Donnell P. (2005) Post-pubertal emergence of prefrontal cortical up states induced by D1-NMDA co-activation. Cereb. Cortex 15:49–57.

Wang J., O'Donnell P. (2001) D(1) dopamine receptors potentiate nmda-mediated excitability increase in layer V prefrontal cortical pyramidal neurons. Cereb. Cortex 11:452–462.

Wolf M.E., Sun X., Mangiavacchi S., Chao S.Z. (2004) Psychomotor stimulants and neuronal plasticity. Neuropharmacology 47 Suppl 1:61–79.

Young C.E., Yang C.R. (2005) Dopamine D1-like receptor modulates layer- and frequency-specific short-term synaptic plasticity in rat prefrontal cortical neurons. Eur. J. Neurosci. 21:3310–3320.

Chapter 10

Prefrontal Cortical Synaptic Plasticity: *The Roles of Dopamine and Implication for Schizophrenia*

Yukiori Goto[1] and Satoru Otani[2]
[1]Department of Psychiatry, McGill University, Montreal, QC, Canada H3A 1A1;
[2]Department of Neurobiology, University of Paris VI-CNRS, Paris 75005, France

10.1 Introduction

The prefrontal cortex (PFC) is central in mediating executive functions in goal-directed behavior, for which proper dopamine (DA) actions of information processing modulation is essential in this area. It is now evident that, as in the case of the hippocampus, the PFC undergoes neuronal adaptation processes in its networks with induction of synaptic plasticity such as long-term potentiation (LTP) and short-term potentiation (STP). A prominent characteristic of synaptic plasticity in the PFC is that its induction mechanisms involve DA as an essential modulatory molecule. As such, DA-dependent plastic changes occurring in PFC network have important roles for PFC-mediated cognitive functions. Nevertheless, little attempt has been made to characterize the nature of PFC neuronal adaptation by synaptic plasticity, given that the PFC is thought to be the area of temporary storage and manipulation of information, known as working memory. However, accumulating evidences now indicate that the functions of the PFC cannot be fully explained just as the region of an online representation and handling of information. Importance of DA-dependent synaptic plasticity is further encouraged by possible disruption of synaptic plasticity mechanism in the PFC in psychiatric disorders such as schizophrenia, drug addiction, and depression.

In this chapter, we describe the underlying cellular mechanisms of DA action on synaptic plasticity induction in the PFC, and possible roles of PFC synaptic plasticity in executive functions as well as its disruptions in the pathophysiology of schizophrenia.

10.2 Dopamine Systems in the PFC

DA neurons in the ventral tegmental area (VTA) innervates into the PFC (Thierry et al., 1973; Lewis et al., 1987). PFC neurons in turn projects into the VTA, yielding glutamatergic excitatory drive on DA neurons (Murase et al., 1993; Au-Young et al., 1999). As such, the PFC and VTA organize a reciprocal loop, with which greater PFC activity induces more DA release in this area (Carr and Sesack, 2000; Laruelle et al., 2003).

Both D1 and D2 receptors are expressed in PFC neurons of rodents and primates (Lidow et al., 1991; Weiner et al., 1991). It seems that DA release differently activates these DA receptors; lower DA release primary stimulates D1, whereas higher DA release stimulates D2, receptors (Trantham-Davidson et al., 2004). Thus, DA release could modulate PFC neuronal activity specifically associated with the mechanisms mediated by D1 and D2 receptor stimulations. D1 activation is known to facilitate calcium influx into the cells by NMDA channel phosphorylation via cAMP-PKA second messenger cascade (Greengard et al., 1999). Thus, PFC neuronal excitability is enhanced by this D1–NMDA interaction (Tseng and O'Donnell, 2004). Nevertheless, presynaptic D1 receptor activation is also shown to attenuate glutamate release from excitatory projections in the PFC (Seamans et al., 2001). The impacts of D2 activation on PFC neurons are less characterized. One recent study has shown that D2 stimulation facilitates burst spike firing in PFC neurons without changing their excitability (Wang and Goldman-Rakic, 2004). Other studies have described D2 effects on GABA interneurons excitability in the PFC of adult (Tseng and O'Donnell, 2004, 2006), but not prepubertal (Gorelova et al., 2002; Tseng and O'Donnell, 2006) animals. Although the roles of DA on cellular effects are still in controversy, the mutual agreement could be made on which DA actions maintains the balance of excitation and inhibitions on PFC neuronal activity.

10.3 Cellular Mechanisms of Synaptic Plasticity

Synaptic plasticity such as LTP and long-term depression (LTD) and their underlying cellular mechanisms are extensively studied in the hippocampus (HPC) as it relates to their roles in learning and long-term memory (Bear and Malenka, 1994). LTP and LTD have been also reported in the PFC (Otani et al., 1998; Gurden et al., 1999; Takita et al., 1999). However, less attention has been paid to these forms of synaptic plasticity in this region because the main function of the PFC is thought to be "temporal" storage and manipulation of information (i.e., working memory) that does not require long-term neuronal adaptation of the network (Goldman-Rakic, 1995; Miller and Cohen, 2001). Nevertheless, DA dependence for induction of LTP and LTD (Otani et al., 1998; Gurden et al., 1999) suggests important, yet undetermined, functional roles of long-term synaptic plasticity in the PFC as DA deficits are described to impair PFC functions and implicated in some psychiatric disorders.

Under in vivo condition, high-frequency stimulation in the HPC induces LTP in their afferents into the PFC (Gurden et al., 1999). This LTP requires D1, but not D2, activation in the PFC, since the presence of D1 antagonist blocks the induction of LTP in the PFC, whereas D2 antagonist does not (Gurden et al., 2000). In addition, LTP in the HPC–PFC pathway can be enhanced by increased release of dopamine in the PFC by VTA stimulation (Gurden et al., 1999). Furthermore, this LTP is prevented by pretreatment of NMDA receptor antagonist (Gurden et al., 1999), suggesting that the LTP induction in limbic afferents involves D1–NMDA synergism in the PFC. Interestingly, the D1–NMDA receptor synergism in the PFC is critical for normal acquisition of operant lever pressing behavior (Baldwin et al., 2002). In both cases, a critical second messenger activated by concurrent D1 and NMDA receptor stimulation may be protein kinase A (Gurden et al., 1999; Baldwin et al., 2002).

Different from in vivo condition, high-frequency stimulation does not reliably induce long-term synaptic plasticity under in vitro slice preparation (Otani et al., 1998, 1999). But if DA is present at the time of the stimulation, it induces LTD, but not LTP (Otani et al., 1998, 1999). More recently, however, we found in PFC slices that the same LTD protocol induces LTP if D1 and D2 receptors are prestimulated ("primed") by DA (Matsuda et al., 2006). Particularly, a continuous background presence of low-concentration of DA, mimicking tonic DA release in intact brain, facilitated LTP induction and converted the above LTD into LTP (Matsuda et al., 2006). In PFC slices where DA fibers are severed, DA receptors are thought to be left largely unstimulated during the incubation period (at least 4 h before an exogenous DA application). Under in vivo condition, in contrast, DA neurons exhibit tonic spontaneous firing that provides background DA concentrations in target areas. Our results therefore suggest that this tonic DA secures the induction of LTP in the PFC. Without this tonic DA, high-frequency synaptic input fails to induce LTP and, rather, facilitates LTD. This LTP–LTD conversion in the absence of tonic background DA represents pathological replacement of LTP by LTD. We of course do not exclude possible roles of physiologically induced forms of LTD in PFC network, notably by prolonged low-frequency input (Takita et al., 1999; Huang et al., 2004).

Cellular mechanisms of the LTP and LTD induced in PFC slices differ in some respects but share certain features. A largest difference is that during induction by high-frequency stimulation, LTP requires stimulation of both D1 and D2 receptors (Matsuda et al., 2006) whereas LTD appears to be induced even when only one type of receptors is stimulated (Otani et al., 1998). Another clear difference is that while LTD is induced in an NMDA receptor-independent manner, LTP requires synaptic activation of NMDA receptors particularly during the "priming" phase. But both forms of plasticity are dependent on postsynaptic depolarization, postsynaptic calcium rises, and metabotropic glutamate receptor (mGluR) activation (Otani et al., 1999). We are currently trying to identify critical biochemical factors involved in the "priming" process induced by tonic DA that converts the LTD to LTP.

There is an interesting difference in LTP induction between in vivo and in vitro conditions. Although LTP induced in limbic afferents in vivo is solely dependent on D1 receptor activation, LTP induced within the PFC circuitry in slice preparation requires both D1 and D2 receptors [but see Huang et al. (2004) showing LTP induction with D1, but not D2, activation in PFC slices]. This difference may be due to synaptic plasticity induction in different circuitries. Thus, the limbic afferents mainly project into the deep layers (layers V–VI) of the PFC (Jay and Witter, 1991), whereas tetanic stimulation to induce LTP in PFC slices is applied to the superficial layers (layers I–II) that mainly contain cortico-cortical fibers (Levitt et al., 1993). Our recent study supports this idea (manuscript in preparation). Thus, tetanic stimulation at the frequency of theta cycle given in the HPC could induce STP of synaptic responses in the PFC that depends on D1, but not D2, activation. In contrast, another tetanic stimulation at gamma frequency applied to the superficial layers of the PFC induced similar STP, which depends on both D1 and D2 receptor activation. As the theta (~7 Hz) and the gamma (~40 Hz) frequencies were chosen to mimic neuronal activities physiologically occurring in the hippocampus (Klemm, 1976) and the cortico-cortical (Tallon-Baudry et al., 1998) inputs, respectively, it appears that pathway specific variations of DA receptor dependency (D1 receptor alone vs. D1 + D2 receptors) exist when physiological ranges of neuronal activities in a given pathways are approximately respected.

10.4 Functional Roles of Synaptic Plasticity

Long-term synaptic plasticity in the PFC has been relatively ignored due to a general consensus that a major function of the PFC is a temporary storage of retrieved or acquired information for behavioral guidance, that is, working memory (Goldman-Rakic, 1995). According to this view, PFC neurons and their synaptic connections serve as a devise analogous to a computer buffer (working memory buffer) for which the PFC neuronal network appears to be unrelated to long-term plastic changes. Thus, the PFC network temporarily represents and compares multiple sources of information, which is emptied after the completion of ongoing behavior for a next use. Such a mechanism may involve short-term rather than long-term plasticity in PFC neurons mediated by rapid modulation of ion channels by DA to transiently alter firing properties of PFC neurons (Yang and Seamans, 1996). Nonetheless, recent studies suggest the roles of the PFC in certain forms of long-term memory that may involve lasting synaptic and neuronal plasticity in its network such as STP, LTP, LTD, and LTP of intrinsic excitability (EPSP-spike potentiation).

First, inhibition of molecules essential for long-term plasticity induction in the PFC was shown to affect emotional memory involving temporal association of events. Thus, intra-PFC injection of extracellular signal-regulated kinase (Erk) inhibitor prior to training session of trace fear conditioning disrupted fear memory tested 2–3 days later without affecting initial task acquisition (Runyan et al., 2004). Furthermore, administration of NMDA receptor antagonist into the PFC immediately after training of trace eyeblink conditioning impaired consolidation of such associative memory (Takehara-Nishiuchi et al., 2006). Similarly, a PFC lesion given after training sessions of a similar trace eyeblink task also caused deficits on retention of the associative memory (Simon et al., 2005). These studies suggest that formation and retention of long-term memory requires neuronal substrates involving NMDA- and kinase-activation in the PFC. Indeed, LTP induction in the PFC also depends on NMDA receptors (Gurden et al., 1999; Matsuda et al., 2006) and Erk activation (our unpublished preliminary data). More direct evidence of LTP induction in the PFC network as a neuronal substrate for memory formation came from the studies of Garcia and coworkers (Herry and Garcia, 2002). Using fear-conditioning paradigm, they have shown a correlation between persistent facilitation of synaptic responses (i.e., LTP) at thalamic and limbic inputs into the PFC and a successful extinction of conditioned fear. Moreover, they also found that mechanical induction of LTD in the PFC by low-frequency tetanic stimulation blocked the fear extinction. A study by

Mulder et al. (2000) further suggested the presence of lasting neuronal changes engaged during memory retention; they found stable increases and decreases of PFC unit activity in response to conditioned stimulus that evolved during visual discrimination learning.

What kind of long-term memory could be mediated by these lasting neuronal changes in the PFC? Recent studies in the PFC of rodents (Birrell and Brown, 2000; Egerton et al., 2005) and primates (Wallis et al., 2001) indicate the function of the PFC for mediating behavioral performance based on "rules" (Miller and Cohen, 2001) that depends on context-dependent temporal association between stimulus and action (Fuster et al., 2000). PFC lesion (Birrell and Brown, 2000) and chronic treatment of the NMDA antagonist, phencyclidine that attenuates PFC activity (Egerton et al., 2005) are shown to impair behavioral adaptation to rule changes. These studies suggest that the PFC networks indeed serve as a buffer-like devise in which perceptual inputs are consciously coupled to an action repertory in order to construct adapted behaviors. However, importantly, we are tempted to suggest that this working memory buffer itself is composed of *stable* neuronal connections that rely on the induction of long-term synaptic plasticity. Such "PFC neuronal traces" could carry long-term information of how behaviors are structured in time appropriate for situations, without which no one can effectively adapt its behavior to a challenging context. Thus, concrete perceptual information temporarily retrieved or newly acquired (i.e., working memory content) would be processed through these stable neuronal connections representing a context or a rule at the given time, so that the perceptual information is combined to an appropriate action repertory. This idea explains the deficits in long-term memory formation and retention induced after the inhibition of PFC activity (Runyan et al., 2004; Takehara-Nishiuchi et al., 2006) or lesion of the PFC (Simon et al., 2005), because these manipulations cause inability of the animals to acquire new temporal structure of behavior as well as to use it during memory retention.

10.5 Implications for Schizophrenia

There are accumulating evidences that the pathophysiology of schizophrenia may involve disruption of synaptic plasticity in the PFC.

First, schizophrenia brains exhibit abnormalities in the PFC (Weinberger et al., 1994) and associated DA system (Knable and Weinberger, 1997), the HPC (Benes, 1999), and functional connectivity between the HPC and PFC (Meyer-Lindenberg et al., 2005). As such, the patients exhibit deficits in cognitive functions that are mediated by the PFC and HPC such as working memory (Goldman-Rakic and Selemon, 1997), behavioral flexibility (Haut et al., 1996), and episodic memory (Heckers et al., 1998). Recent human imaging studies reported that D1 receptor expression in the PFC is altered in schizophrenia brains, although these are still controversial; one study have shown increased (Abi-Dargham et al., 2002), whereas another study have reported decreased (Okubo et al., 1997), D1 receptors. Furthermore, given that NMDA antagonists such as phencyclidine are known to induce schizophrenia-like symptoms in normal subjects, NMDA hypofunction has been proposed in the pathophysiology of schizophrenia (Javitt and Zukin, 1991; Jentsch and Roth, 1999). As such, synaptic plasticity induced at HPC afferents into the PFC that depends on D1–NMDA interactions are possibly disrupted in schizophrenia.

To test this possibility, we have examined LTP induction at the HPC–PFC pathway in an animal model of schizophrenia in which cortical development is transiently disrupted with prenatal exposure to the antimitotic agent, methylazoxymethanol acetate (MAM) (Goto and Grace, 2006). Animals with prenatal MAM exposure exhibit abnormal behaviors that are consistent with those observed in schizophrenia patients. We found that abnormal augmentation of LTP induced at HPC afferents into the PFC that was more severely disrupted by acute stress exposure than in normal animals. Furthermore, these deficits could be reversed by pretreatment of D1 antagonist and D2 agonist, suggesting that imbalance of D1 and D2 systems in the PFC may be responsible for such altered synaptic plasticity in the animal model of schizophrenia. Stress is a negative environmental factor affecting synaptic plasticity and can be a trigger of relapse in schizophrenia symptoms. As consequence of abnormally augmented synaptic plasticity in the PFC of the animal model, information processing mediated by HPC–PFC interaction appears to be greater vulnerability to stress exposure. We observed that acute, weaker stress could facilitate LTP induction at HPC inputs into the PFC in normal animals, whereas longer stress disrupted LTP induction. In normal animals, the relationship between stress exposure and LTP induction exhibit an inverted U-shape (Yerkes and Dodson, 1908). In animal models of schizophrenia, even weaker stress exposure is sufficient to disrupt LTP induction, resulting in the shift of the inverted U-shape toward greater vulnerability for stress exposure. Similar U-shape relationship has been described between working memory and D1 stimulation in the PFC (Lidow et al., 1998; Yang and Chen, 2005; Williams and Castner, 2006) as well as between working memory and PFC activation (Callicott et al., 2003; Manoach, 2003). Of particular interest is that inverted U-shape relationship between working memory and PFC activation and its shift toward lower working memory load has been observed in schizophrenia patients (Manoach, 2003).

10.6 Conclusions

The PFC networks express their neuronal adaptation processes with induction of DA-dependent long-term synaptic plasticity such as LTP and LTD. Recent studies including ours suggest that such plasticity mechanisms involve DA stimulation on specific subtypes of DA receptors depending on the inputs with which the PFC interacts. Although the exact roles of such DA-dependent synaptic plasticity changes occurring in the PFC networks remains to be determined, it is likely that synaptic plasticity in the PFC networks has critical roles for its functions such as a "memory trace" mediating context or rule-dependent information processing within the PFC networks. Unveiling the cellular mechanisms of synaptic plasticity in the PFC is of particular importance for the pathophysiology of schizophrenia. Indeed, many candidate genes suggested in schizophrenia encode molecules associated with such synaptic plasticity in the PFC (Harrison and Weinberger, 2005). Therefore, future studies in this topic promises bringing advancements of our knowledge and clinical treatments of the disorder.

Acknowledgments

We would like to thank Dr Charles Yang for helpful discussion. Supported by NARSAD Young Investigator Award (Y.G.), the French Minister of Research and the Centre National de la Recherche Scientifique (S.O.).

References

Abi-Dargham, A., Mawlawi, O., Lombardo, I., Gil, R., Martinez, D., Huang, Y., Hwang, D. R., Keilp, J., Kochan, L., Van Heertum, R., Gorman, J. M., and Laruelle, M. (2002). Prefrontal dopamine D1 receptors and working memory in schizophrenia. *J Neurosci* **22,** 3708–19.

Au-Young, S. M., Shen, H., and Yang, C. R. (1999). Medial prefrontal cortical output neurons to the ventral tegmental area (VTA) and their responses to burst-patterned stimulation of the VTA: neuroanatomical and in vivo electrophysiological analyses. *Synapse* **34,** 245–55.

Baldwin, A. E., Sadeghian, K., and Kelley, A. E. (2002). Appetitive instrumental learning requires coincident activation of NMDA and dopamine D1 receptors within the medial prefrontal cortex. *J Neurosci* **22,** 1063–71.

Bear, M. F., and Malenka, R. C. (1994). Synaptic plasticity: LTP and LTD. *Curr Opin Neurobiol* **4,** 389–99.

Benes, F. M. (1999). Evidence for altered trisynaptic circuitry in schizophrenic hippocampus. *Biol Psychiatry* **46,** 589–99.

Birrell, J. M., and Brown, V. J. (2000). Medial frontal cortex mediates perceptual attentional set shifting in the rat. *J Neurosci* **20,** 4320–4.

Callicott, J. H., Mattay, V. S., Verchinski, B. A., Marenco, S., Egan, M. F., and Weinberger, D. R. (2003). Complexity of prefrontal cortical dysfunction in schizophrenia: more than up or down. *Am J Psychiatry* **160,** 2209–15.

Carr, D. B., and Sesack, S. R. (2000). Projections from the rat prefrontal cortex to the ventral tegmental area: target specificity in the synaptic associations with mesoaccumbens and mesocortical neurons. *J Neurosci* **20,** 3864–73.

Egerton, A., Reid, L., McKerchar, C. E., Morris, B. J., and Pratt, J. A. (2005). Impairment in perceptual attentional set-shifting following PCP administration: a rodent model of set-shifting deficits in schizophrenia. *Psychopharmacology (Berl)* **179,** 77–84.

Fuster, J. M., Bodner, M., and Kroger, J. K. (2000). Cross-modal and cross-temporal association in neurons of frontal cortex. *Nature* **405,** 347–51.

Goldman-Rakic, P. S. (1995) Cellular basis of working memory. *Neuron* **14,** 477–85.

Goldman-Rakic, P. S., and Selemon, L. D. (1997). Functional and anatomical aspects of prefrontal pathology in schizophrenia. *Schizophr Bull* **23,** 437–58.

Gorelova, N., Seamans, J. K., and Yang, C. R. (2002). Mechanisms of dopamine activation of fast-spiking interneurons that exert inhibition in rat prefrontal cortex. *J Neurophysiol* **88,** 3150–66.

Goto, Y., and Grace, A. A. (2006). Alterations in medial prefrontal cortical activity and plasticity in rats with disruption of cortical development. *Biol Psychiatry*, **60,** 1259–67.

Greengard, P., Allen, P. B., and Nairn, A. C. (1999). Beyond the dopamine receptor: the DARPP-32/protein phosphatase-1 cascade. *Neuron* **23,** 435–47.

Gurden, H., Takita, M., and Jay, T. M. (2000). Essential role of D1 but not D2 receptors in the NMDA receptor-dependent long-term potentiation at hippocampal-prefrontal cortex synapses in vivo. *J Neurosci* **20,** RC106.

Gurden, H., Tassin, J.-P., and Jay, T. M. (1999). Integrity of the mesocortical dopaminergic system is necessary for complete expression of in vivo hippocampal-prefrontal cortex long-term potentiation. *Neuroscience* **94**, 1019–27.

Harrison, P. J., and Weinberger, D. R. (2005). Schizophrenia genes, gene expression, and neuropathology: on the matter of their convergence. *Mol Psychiatry* **10**, 40–68.

Haut, M. W., Cahill, J., Cutlip, W. D., Stevenson, J. M., Makela, E. H., and Bloomfield, S. M. (1996). On the nature of Wisconsin Card Sorting Test performance in schizophrenia. *Psychiatry Res* **65**, 15–22.

Heckers, S., Rauch, S. L., Goff, D., Savage, C. R., Schacter, D. L., Fischman, A. J., and Alpert, N. M. (1998). Impaired recruitment of the hippocampus during conscious recollection in schizophrenia. *Nat Neurosci* **1**, 318–23.

Herry, C., and Garcia, R. (2002). Prefrontal cortex long-term potentiation, but not long-term depression, is associated with the maintenance of extinction of learned fear in mice. *J Neurosci* **22**, 577–83.

Huang, Y. Y., Simpson, E., Kellendonk, C., and Kandel, E. R. (2004). Genetic evidence for the bidirectional modulation of synaptic plasticity in the prefrontal cortex by D1 receptors. *Proc Natl Acad Sci U S A* **101**, 3236–41.

Javitt, D. C., and Zukin, S. R. (1991). Recent advances in the phencyclidine model of schizophrenia. *Am J Psychiatry* **148**, 1301–8.

Jay, T. M., and Witter, M. P. (1991). Distribution of hippocampal CA1 and subicular efferents in the prefrontal cortex of the rat studied by means of anterograde transport of *Phaseolus vulgaris*-leucoagglutinin. *J Comp Neurol* **313**, 574–86.

Jentsch, J. D., and Roth, R. H. (1999). The neuropsychopharmacology of phencyclidine: from NMDA receptor hypofunction to the dopamine hypothesis of schizophrenia. *Neuropsychopharmacology* **20**, 201–25.

Klemm, W. R. (1976). Physiological and behavioral significance of hippocampal rhythmic, slow activity ("theta rhythm"). *Prog Neurobiol* **6**, 23–47.

Knable, M. B., and Weinberger, D. R. (1997). Dopamine, the prefrontal cortex and schizophrenia. *J Psychopharmacol* **11**, 123–31.

Laruelle, M., Kegeles, L. S., and Abi-Dargham, A. (2003). Glutamate, dopamine, and schizophrenia: from pathophysiology to treatment. *Ann N Y Acad Sci* **1003**, 138–58.

Levitt, J. B., Lewis, D. A., Yoshioka, T., and Lund, J. S. (1993). Topography of pyramidal neuron intrinsic connections in macaque monkey prefrontal cortex (areas 9 and 46). *J Comp Neurol* **338**, 360–76.

Lewis, D. A., Campbell, M. J., Foote, S. L., Goldstein, M., and Morrison, J. H. (1987). The distribution of tyrosine hydroxylase-immunoreactive fibers in primate neocortex is widespread but regionally specific. *J Neurosci* **7**, 279–90.

Lidow, M. S., Goldman-Rakic, P. S., Gallager, D. W., and Rakic, P. (1991). Distribution of dopaminergic receptors in the primate cerebral cortex: quantitative audioradiographic analysis using [3H]raclopride, [3H]spiperone, and [3H]SCH23390. *Neuroscience* **40**, 657–71.

Lidow, M. S., Williams, G. V., and Goldman-Rakic, P. S. (1998). The cerebral cortex: a case for a common site of action of antipsychotics. *Trends Pharmacol Sci* **19**, 136–40.

Manoach, D. S. (2003). Prefrontal cortex dysfunction during working memory performance in schizophrenia: reconciling discrepant findings. *Schizophr Res* **60**, 285–98.

Matsuda, Y., Marzo, A., and Otani, S. (2006). The presence of background dopamine signal converts long-term synaptic depression to potentiation in rat prefrontal cortex. *J Neurosci* **26**, 4803–10.

Meyer-Lindenberg, A. S., Olsen, R. K., Kohn, P. D., Brown, T., Egan, M. F., Weinberger, D. R., and Berman, K. F. (2005). Regionally specific disturbance of dorsolateral prefrontal-hippocampal functional connectivity in schizophrenia. *Arch Gen Psychiatry* **62**, 379–86.

Miller, E. K., and Cohen, J. D. (2001). An integrative theory of prefrontal cortex function. *Annu Rev Neurosci* **24,** 167–202.

Mulder, A. B., Nordquist, R., Orgut, O., and Pennartz, C. M. (2000). Plasticity of neuronal firing in deep layers of the medial prefrontal cortex in rats engaged in operant conditioning. *Prog Brain Res* **126,** 287–301.

Murase, S., Grenhoff, J., Chouvet, G., Gonon, F. G., and Svensson, T. H. (1993). Prefrontal cortex regulates burst firing and transmitter release in rat mesolimbic dopamine neurons studied in vivo. *Neurosci Lett* **157,** 53–6.

Okubo, Y., Suhara, T., Suzuki, K., Kobayashi, K., Inoue, O., Terasaki, O., Someya, Y., Sassa, T., Sudo, Y., Matsushima, E., Iyo, M., Tateno, Y., and Toru, M. (1997). Decreased prefrontal dopamine D1 receptors in schizophrenia revealed by PET. *Nature* **385,** 634–6.

Otani, S., Auclair, N., Desce, J.-M., Roisin, M.-P., and Crepel, F. (1999). Dopamine receptors and groups I and II mGluRs cooperate for long-term depression induction in rat prefrontal cortex through converging postsynaptic activation of MAP kinases. *J Neurosci* **19,** 9788–02.

Otani, S., Blond, O., Desce, J. M., and Crepel, F. (1998). Dopamine facilitates long-term depression of glutamatergic transmission in rat prefrontal cortex. *Neuroscience* **85,** 669–76.

Runyan, J. D., Moore, A. N., and Dash, P. K. (2004). A role for prefrontal cortex in memory storage for trace fear conditioning. *J Neurosci* **24,** 1288–95.

Seamans, J. K., Durstewitz, D., Christie, B. R., Stevens, C. F., and Sejnowski, T. J. (2001). Dopamine D1/D5 receptor modulation of excitatory synaptic inputs to layer V prefrontal cortex neurons. *Proc Natl Acad Sci U S A* **98,** 301–6.

Simon, B., Knuckley, B., Churchwell, J., and Powell, D. A. (2005). Post-training lesions of the medial prefrontal cortex interfere with subsequent performance of trace eyeblink conditioning. *J Neurosci* **25,** 10740–6.

Takehara-Nishiuchi, K., Nakao, K., Kawahara, S., Matsuki, N., and Kirino, Y. (2006). Systems consolidation requires postlearning activation of NMDA receptors in the medial prefrontal cortex in trace eyeblink conditioning. *J Neurosci* **26,** 5049–58.

Takita, M., Izaki, Y., Jay, T. M., Kaneko, H., and Suzuki, S. S. (1999). Induction of stable long-term depression in vivo in the hippocampal-prefrontal cortex pathway. *Eur J Neurosci* **11,** 4145–8.

Tallon-Baudry, C., Bertrand, O., Peronnet, F., and Pernier, J. (1998). Induced gamma-band activity during the delay of a visual short-term memory task in humans. *J Neurosci* **18,** 4244–54.

Thierry, A. M., Blanc, G., Sobel, A., Stinus, L., and Golwinski, J. (1973). Dopaminergic terminals in the rat cortex. *Science* **182,** 499–501.

Trantham-Davidson, H., Neely, L. C., Lavin, A., and Seamans, J. K. (2004). Mechanisms underlying differential D1 versus D2 dopamine receptor regulation of inhibition in prefrontal cortex. *J Neurosci* **24,** 10652–9.

Tseng, K. Y., and O'Donnell, P. (2004). Dopamine-glutamate interactions controlling prefrontal cortical pyramidal cell excitability involve multiple signaling mechanisms. *J Neurosci* **24,** 5131–9.

Tseng, K. Y., and O'Donnell, P. (2007). Dopamine modulation of prefrontal cortical interneurons changes during adolescence. *Cereb Cortex* 17, 1235–40.

Wallis, J. D., Anderson, K. C., and Miller, E. K. (2001). Single neurons in prefrontal cortex encode abstract rules. *Nature* **411,** 953–6.

Wang, Y., and Goldman-Rakic, P. S. (2004). D2 receptor regulation of synaptic burst firing in prefrontal cortical pyramidal neurons. *Proc Natl Acad Sci U S A* **101,** 5093–8.

Weinberger, D. R., Aloia, M. S., Goldberg, T. E., and Berman, K. F. (1994). The frontal lobes and schizophrenia. *J Neuropsychiatry Clin Neurosci* **6,** 419–27.

Weiner, D. M., Levey, A. I., Sunahara, R. K., Niznik, H. B., O'Dowd, B. F., Seeman, P., and Brann, M. R. (1991). D1 and D2 dopamine receptor mRNA in rat brain. *Proc Natl Acad Sci U S A* **88,** 1859–63.

Williams, G. V., and Castner, S. A. (2006). Under the curve: critical issues for elucidating D1 receptor function in working memory. *Neuroscience* **139,** 263–76.

Yang, C. R., and Chen, L. (2005). Targeting prefrontal cortical dopamine D1 and N-methyl-D-aspartate receptor interactions in schizophrenia treatment. *Neuroscientist* **11,** 452–70.

Yang, C. R., and Seamans, J. K. (1996). Dopamine D1 receptor actions in layers V-VI rat prefrontal cortex neurons in vitro: modulation of dendritic-somatic signal integration. *J Neurosci* **16,** 1922–35.

Yerkes, R. M., and Dodson, J. D. (1908). The relation of strength of stimulus to rapidity of habit-formation. *J Comp Neurol Psychol* **18,** 459–82.

Chapter 11

Acquiring the Excitatory and the Inhibitory Action of Dopamine in the Prefrontal Cortex During Postnatal Development

Kuei Y. Tseng

Department of Cellular & Molecular Pharmacology, Rosalind Franklin University of Medicine and Science, The Chicago Medical School, North Chicago, IL 60064, USA

11.1 Introduction

Mesocortical dopamine (DA) transmission and its regulation of prefrontal cortical (PFC) firing (also known as persistent activity) have been correlated with several cognitive task-related, decision processes including working memory, reward, and attention (Goldman-Rakic et al., 2000; Horvitz, 2000; Jay, 2003; O'Donnell, 2003). More importantly, it has been proposed that one of the major roles of mesocortical DA is to increase the signal detection ratio by reducing the impact of irrelevant inputs to the PFC. An enhancement of the mesocortical DA signal in the PFC would therefore contribute to optimize cognitive task performance and the response to motivationally salient stimuli (Horvitz, 2000; Cohen et al., 2002; Schultz, 2002; O'Donnell, 2003). We also know that cognitive functions that depend on PFC DA, such as decision making and working memory, do change with the transition to adulthood (Funahashi and Inoue, 2000; Segalowitz and Davies, 2004). They are typically acquired and refined during puberty and late adolescence when widespread structure refinement and functional neuronal maturation occurs in cortical circuits (Giedd et al., 1999; Casey et al., 2000; Spear, 2000; Gogtay et al., 2004). Although the cellular mechanisms underlying this DA-PFC interaction are not fully identified and often contradictory, disruption of this DA modulation has been associated with the characteristic postpubertal/late adolescence emergence of cognitive deficits in schizophrenia (Lipska and Weinberger, 1998, 2000).

In this chapter I will first review recent electrophysiological findings on how DA D1 and D2 receptors influence excitatory and inhibitory neurotransmission in the PFC during the peripubertal transition to adulthood, the cellular mechanism underlying this DA modulation as well as its functional significance. Secondly, I will discuss how a developmental disruption of PFC DA modulation may underlie some of the cortical pathophysiological changes observed in schizophrenia and related psychiatry disorders.

11.2 Dopamine-Glutamate Interactions in the Prefrontal Cortex Changes During the Peripubertal Transition to Adulthood

11.2.1 POSTPUBERTAL ENHANCEMENT OF D1 FACILITATION OF NMDA FUNCTION IN THE PREFRONTAL CORTEX

DA modulation of excitatory and inhibitory neurotransmission in the prefrontal Cortex (PFC) undergoes several changes during postnatal development (Benes et al., 2000; Tseng and O'Donnell, 2005a; Tseng and O'Donnell, 2006), probably due to changes that occur at both cellular and subcellular levels during cortical maturation (Spear, 2000; Tarazi and Baldessarini, 2000; Zhu, 2000). Although the cellular mechanisms by which these processes occur are not entirely identified, evidences indicate that DA may facilitate the acquisition of mature cognitive abilities by virtue of a D1-dependent enhancement of NMDA functions in the PFC (Gurden et al., 1999, 2000; Wang and O'Donnell, 2001; Tseng and O'Donnell, 2005a). It is well documented that hippocampal-PFC long-term potentiation (LTP) is enhanced by activation of D1, but not D2, receptors (Gurden et al., 2000). The dependence of LTP on an intact mesocortical projection is evidenced by the lack of LTP formation in animals with DA depletion (Gurden et al., 1999). A role for mesocortical DA on cognitive functioning is further supported by behavioral studies showing that PFC D1 receptors improve memory retrieval and working memory performance (Seamans et al., 1998; Floresco and Phillips, 2001), and D1-NMDA coactivation in the PFC is required for appetitive instrumental learning in adult rats (Baldwin et al., 2002).

At the cellular level, D1-NMDA coactivation can elicit recurrent plateau depolarizations resembling in vivo "UP states" in PFC pyramidal neurons, an effect that becomes evident only after puberty (Tseng and O'Donnell, 2005a). Despite these observations, there is still some debate on whether UP and DOWN transitions are present in awake animals (reviewed by Tseng and O'Donnell, 2005b), evidences indicates that UP states are important cellular elements of information processing, particularly by enabling ensembles of active neurons to synchronize at any given moment (O'Donnell, 2003). UP states or plateau depolarizations could last from a few hundred microseconds and repeat every second under anesthesia and sleep, to several seconds or minutes as observed in the awake state during alert conditions (reviewed by Tseng and O'Donnell, 2005b). Therefore, a D1 facilitation of plateau depolarization in the PFC would provide a temporal window during which context relevant inputs can drive pyramidal neurons into synchronized firing and NMDA-dependent synaptic plasticity would be enabled. Because activation of mesocortical DA is context-dependent and related to attention and salient stimuli (Horvitz, 2000; Cohen et al., 2002; Schultz, 2002), the relevant ongoing activity in the PFC, that is, that mediated by AMPA and NMDA receptors, could become enhanced and reinforced by setting and maintaining a population of pyramidal neurons into the "UP state" via activation of local D1 receptors (O'Donnell, 2003; Tseng and O'Donnell,

2005b). These events may depend on potentiation of NMDA function by a number of mechanisms, including, but probably not limited to, D1 receptor activation, that ultimately enhances protein kinase A activity and postsynaptic Ca^{2+} signaling (Tseng and O'Donnell, 2005b). Although most cortical synapses have been formed at prepubertal (PD 28–35) ages, many of the mature morphological and physiological properties of neocortical neurons do not emerge until puberty/adolescence (Petit et al., 1988; Zhu, 2000). In particular, pyramidal neurons dendritic Na^+ and Ca^{2+} regenerative potentials, which are important players in synaptic plasticity, become effective in coupling distal apical dendrites with somata at PD 42 (Zhu, 2000). Furthermore, both D1 and NMDA receptors levels change during postnatal development, and thus, D1-NMDA interactions may regulate PFC synaptic function differentially, depending on the stage of cortical maturation, with prepubertal PFC neurons being less likely to exhibit strong potentiation of NMDA-dependent plasticity by D1 receptors. Therefore, an important role in converting prepubertal cortical synaptic organization into a mature network includes the delayed acquisition of adult levels of D1 dopamine receptors (Leslie et al., 1991; Tarazi et al., 1999; Tarazi and Baldessarini, 2000) and NMDA receptor subunits (Williams et al., 1993; Monyer et al., 1994) combined with a late maturation of the intrinsic physiological properties in pyramidal neurons (Zhu, 2000). The emergence of D1 enhancement of NMDA functions (Tseng and O'Donnell, 2005) would therefore have a strong influence on PFC plasticity during puberty/adolescence, which in turn may be critical for developing and acquiring mature cognitive abilities.

11.2.2 D2 ATTENUATION OF EXCITATORY SYNAPTIC TRANSMISSION IN THE PREFRONTAL CORTEX INCREASES AFTER PUBERTY

The impact of D2 receptors on PFC function is less clear when compared to the D1 effects. There is evidence for a D2 modulation of working memory (Druzin et al., 2000) and age-related PFC cognitive functions (Arnsten et al., 1995), yet demonstration of the mechanisms by which D2 receptors influence PFC neuronal activity have been inconclusive. A D2 attenuation of inhibitory postsynaptic currents in pyramidal neurons has been observed in the prepubertal PFC (Scamans et al., 2001; Trantham-Davidson et al., 2004) whereas in other studies, D2 activation decreases pyramidal neurons excitability (Gulledge and Jaffe, 1998; Tseng and O'Donnell, 2004) and downregulates AMPA and NMDA-mediated excitatory responses (Tseng and O'Donnell, 2004). This latter observation is consistent with the effect of D2 receptors on PFC excitatory response induced by local synaptic activation. D2 activation clearly exerts a powerful downregulation of intracortical AMPA/kainite-dependent synaptic response as indicated by the ability of the D2 agonist quinpirole to attenuate pyramidal neurons excitatory postsynaptic potential (EPSP) elicited from layers I–II stimulation (Tseng and O'Donnell, 2007). Time course analyses revealed that the evoked EPSP remained attenuated even after quinpirole was removed. This effect was not observed in presence of the D2 antagonist eticlopride, confirming that the quinpirole-induced synaptic depression is indeed D2 mediated. More importantly, the impact of this D2 modulation was more pronounced in the adult PFC as revealed by magnitude and duration of the synaptic inhibition. A period of at least 25 min was required to partially recover the EPSP amplitude to baseline. Interestingly, the

GABA-A antagonist picrotoxin significantly reduced the duration of the post-quinpirole effect without affecting the degree of the initial D2 inhibition. This distinctive D2-glutamate-GABA interaction was not observed in the prepubertal PFC, probably because GABAergic interneurons in the PFC acquire their mature response to D2 during adolescence (see Section 11.3; Tseng and O'Donnell, 2006). Thus, D2 attenuation of PFC fast excitatory synaptic transmission could be sustained by virtue of a D2-dependent enhancement of local GABAergic tone.

Several pre- and postsynaptic mechanisms could mediate the D2-dependent synaptic inhibition of AMPA/kainite-mediated transmission in the adult PFC. These include a direct postsynaptic activation of phospholipase lipase C-IP$_3$ and inhibition of protein kinase A signaling pathways by D2 receptors in deep-layer pyramidal neurons (Tseng and O'Donnell, 2004). This inhibition could also occur indirectly, through an enhancement of local GABAergic tone (Grobin and Deutch, 1998; Gulledge and Jaffe, 1998; Tseng and O'Donnell, 2004), most likely by virtue of a D2-dependent increase of inhibitory interneurons excitability (Tseng and O'Donnell, 2006). Although the mechanisms by which this characteristic D2-GABA-dependent modulation of intracortical synaptic transmission regulate the overall PFC outcome and functions remain to be determined, it becomes apparent that a D2 enhancement of GABAergic interneuron excitability underlies the D2-dependent-GABA-mediated long-lasting EPSP depression in pyramidal neurons as these two events become evident only in the adult PFC. In fact, a similar D2-dependent long-lasting enhancement of deep-layer GABAergic interneuron excitability to 5–7 min quinpirole was observed in the PFC, but only after puberty (Tseng and O'Donnell, 2006). Thus, a direct modulation of postsynaptic signaling pathways by D2 receptors (Tseng and O'Donnell, 2004) may initiate the attenuation of excitatory inputs to deep-layer pyramidal neurons, a synaptic inhibition that could be prolonged by recruiting local GABA interneuron via a D2-mediated mechanism, in the adult PFC (Tseng and O'Donnell, 2006).

From a functional standpoint, the inhibitory D2 effect on fast excitatory neurotransmission observed in the PFC may increase signal detection by attenuating weak signals, whereas the concurrent dampening of NMDA responses by D2 receptors via enhancement of GABA transmission (Tseng and O'Donnell, 2004) may be critical to control the timing as well as the spatial selectivity of pyramidal neurons firing (Rao et al., 2000). Therefore, it seems plausible that by acquiring the distinctive mature D2-dependent regulation of PFC synaptic transmission in pyramidal neurons during the peripubertal/periadolescence transition to adulthood throughout the modulation of local GABAergic tone should enhance the saliency of output neurons in the PFC and contribute to the development of an adult cognitive profile.

11.3 Differential Impacts of DA, D1, and D2 Receptors on Prefrontal Cortical Interneurons During Adolescence

It is widely recognized that cortical inhibitory interneurons, by interacting with the glutamatergic inputs, play an important role in determining the timing and spatial selectivity of pyramidal neurons firing (Traub et al., 1996; Rao et al., 2000; Szabadics et al., 2001; Markram et al., 2004). GABAergic interneurons in the PFC also receive DA innervation from the VTA (Benes et al., 1993; Sesack et al., 1998a, b), and express both D1 and D2 receptors (Vincent et al., 1993, 1995; Smiley et al., 1994; Gaspar et al., 1995; Mrzljak et al., 1996; Muly et al., 1998). Although we know little about how DA–GABA interactions influence PFC function, there is evidence suggesting that part of the inhibitory action of DA in the PFC is due to an enhancement of local GABAergic tone (Pirot et al., 1992; Gulledge and Jaffe, 1998; Gorelova et al., 2002; Tseng and O'Donnell, 2004, 2006). For instance, D2 (not D1) activation in the adult PFC enhance local GABA release (Grobin and Deutch, 1998) and mesocortical stimulation with trains of pulses mimicking DA cell burst firing (5 pulses delivered at 20 Hz) typically elicits prolonged plateau depolarizations (UP state) along with suppression of action potential firing in deep-layer pyramidal neurons (Lewis and O'Donnell, 2000). This characteristic firing suppression in pyramidal neurons occurs frequently in the absence of coincident excitatory inputs to the PFC (Lewis and O'Donnell, 2000; Tseng et al., 2006a) and it matches the temporal course of fast-spiking (FS) interneuron excitation (Tseng et al., 2006a).

On the other hand, in vitro electrophysiological recordings of PFC GABAergic interneurons from juvenile rats (brain slices obtained from PD 14-28) revealed that only D1, not D2, receptors play a role in regulating FS interneurons excitability (Gorelova et al., 2002). A similar selective D1-dependent upregulation of PFC GABAergic function was observed in prepubertal (PD 28-35) animals whereas the D2 influence becomes evident only after puberty (Tseng and O'Donnell, 2006). Perhaps the most interesting, yet surprising observation is the powerful excitatory action exerted by D2 activation on deep-layer GABAergic interneurons in the developmentally mature (PD 50-80) PFC (Tseng and O'Donnell, 2006). Bath application of the D2 agonist quinpirole increases the excitability of FS and non-FS (NFS) interneurons, an effect that was completely blocked by the selective D2 receptor antagonist eticlopride. Thus, the net effect of mesocortical DA in the adult PFC is to drive GABAergic interneuron firing by both D1- and D2-dependent mechanisms (Tseng and O'Donnell, 2006).

The involvement of both D1 and D2 DA receptors in facilitating local GABAergic function in the adult PFC (Fig. 11.1) presumably causes a powerful inhibition of pyramidal neurons firing in the presence of phasic DA, as shown in vivo in response to electrical (Lewis and O'Donnell, 2000) and chemical (Tseng et al., 2006a) stimulation of the mesocortical projection neurons. This modulation could in turn influence the timing and spatial selectivity of ensembles of output neurons in the PFC, probably by the D2- and GABA-mediated attenuation of excitatory inputs to deep-layer pyramidal neurons (Tseng and O'Donnell, 2007a, b). In addition to the postpubertal emergence of a D1 enhancement of NMDA function (Tseng and O'Donnell, 2005), an increased impact of D2-dependent inhibition with the postpubertal acquisition of DA (particularly D2) modulation of local GABAergic interneurons (Tseng and O'Donnell, 2004; Tseng and O'Donnell, 2006) may represent another important functional characteristic of the adult mesocortical system. These postpubertal changes may ultimately enhance the detection of relevant and salient signals through two concurrent events. A mature D2-GABA interaction may provide a more efficient mechanism to limit neuronal firing originated from asynchronous inputs, which in turn would facilitate relevant signals to drive specific ensembles of PFC pyramidal neurons into UP states by virtue of a D1 enhancement of NMDA function (Lewis and O'Donnell, 2000; Tseng and O'Donnell, 2005). If these events were coordinated with the arrival of strong coincident excitatory inputs (e.g., from the hippocampus), the representation encoded in this ensemble of PFC neurons would be reinforced and synchronized neuronal firing during prolonged UP states (i.e., persistent activity) would be enabled. Acquisition of postpubertal DA control of excitatory and inhibitory neurotransmission could be critical to fine-tuning PFC output activity responsible for mature cognitive processes that are refined during puberty and adolescence. Thus, DA modulation of PFC interneuronal firing may be a critical step to engage pyramidal neuron activity during working memory and goal-directed behaviors.

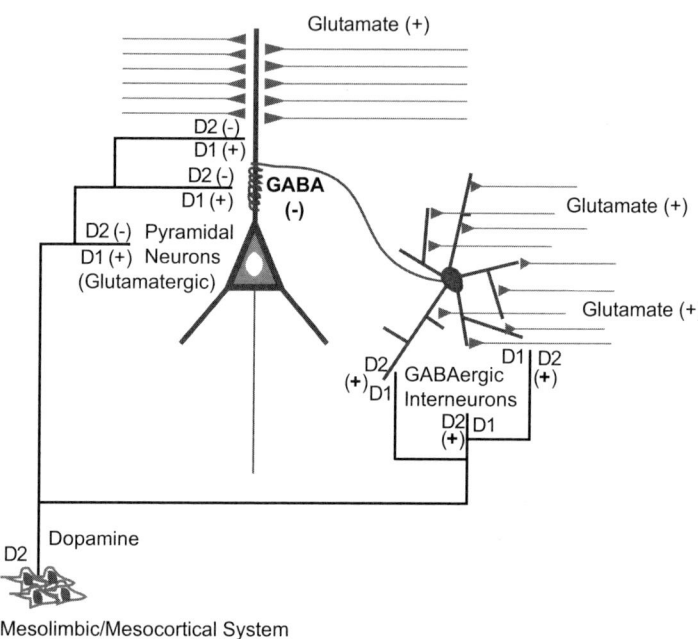

Figure 11.1 Simplified diagram showing how dopamine D1 and D2 receptors regulate the interaction between deep-layer pyramidal neurons (output neurons) and GABAergic interneurons in the medial prefrontal cortex (prelimbic and infralimbic). Note that D2 receptor activation exerts different effects on pyramidal neuron and GABAergic interneuron excitability (see also Tseng and O'Donnell, 2004, 2007).

11.4 Developmental Disruptions of Prefrontal Cortical Dopamine Glutamate Interactions and Schizophrenia: Lessons from an Animal Model

Prefrontal disruption in animals with a neonatal ventral hippocampal lesion (NVHL) has been proposed to model the cortical deficits observed in schizophrenia (O'Donnell et al., 2002). This neonatal manipulation yields a variety of behavioral and neurochemical abnormalities resembling those found in patients with schizophrenia (Lipska and Weinberger, 1998, 2000). Hyperlocomotion, excessive reactivity to stress (i.e., measured with repeated intraperitoneal saline injection, inescapable foot shock, or a swimming test), sensorimotor gating deficits, reduced social interactions and working memory impairments are typically observed in NVHL animals (Lipska et al., 1993, 1995; Chambers et al., 1996; Becker et al., 1999; Le Pen et al., 2000; Lipska et al., 2002; Chrapusta et al., 2003). More importantly, these behavioral alterations do not emerge until the animals reach to puberty and young adulthood (Lipska and Weinberger, 1998, 2000). Although the mechanisms underlying these changes are not fully understood, treatment with antipsychotic drugs has been shown to effectively reverse some of the abnormal behavioral and physiological responses associated with the NVHL (Lipska and Weinberger, 1994; Goto and O'Donnell, 2002; Le Pen and Moreau, 2002), indicating that the mesocortical and mesolimbic DA systems are compromised in these animals. Accordingly, PFC pyramidal neurons in NVHL rats respond with an abnormal increase in firing instead of the classically observed decrease (Lewis and O'Donnell, 2000) to mesocortical activation (O'Donnell et al., 2002). This effect was observed in postpubertal NVHL animals but not with analogous lesions performed in adulthood (O'Donnell et al., 2002). This latter observation indicates that the altered PFC response may reflect abnormal postnatal developmental changes within the mesocortical-PFC pathway. Furthermore, PFC lesions can reverse NVHL-induced behavioral deficits (Lipska et al., 1998) as well as the abnormal mesolimbic responses in the nucleus accumbens (Goto and O'Donnell, 2004).

DA–glutamate interactions become compromised in the PFC of NVHL animals after puberty (Tseng et al., 2007). As discussed earlier in this chapter, DA modulation of PFC glutamatergic transmission matures after puberty (Tseng and O'Donnell, 2005a) and involves multiple cellular mechanisms including activation of GABAergic interneurons and postsynaptic signaling cascades (Tseng and O'Donnell, 2004, 2006). Therefore, an early inactivation of the hippocampal formation may alter the postnatal development of the mesocortical/mesolimbic system as well as the maturation local GABAergic interneurons in the PFC, in particular the parvalbumin-containing cells. Consequently, the abnormal/hyperreactive PFC response to mesocortical stimulation observed in NVHL animals (O'Donnell et al., 2002; Tseng et al., 2006b) could reflect a postnatal disruption of DA regulation of PFC GABAergic function that normally matures during adolescence (Tseng and O'Donnell, 2006). Although this hypothesis has yet to be examined, recent studies show that the NVHL selectively downregulates the expression of glutamate decarboxylase-67 mRNA in the PFC, the rate-limiting enzyme for GABA synthesis (Lipska et al., 2003a, c). Thus, functional attenuation of local GABAergic transmission could contribute to alter the balance of appropriate synaptic events in the PFC leading to the cortical deficits observed in these animals.

In summary, DA modulation of PFC inhibition and excitation become developmentally compromised in the PFC of NVHL animals. These changes could lead to the abnormal cognitive performance in NVHL animals by setting inappropriate coordination between pyramidal neurons and GABAergic interneurons, which in turn would alter the spatial selectivity of PFC neuronal response to excitatory inputs. A similar cortical disruption has been associated in schizophrenia (Lewis et al., 2004), a disorder with a clear prefrontal dysfunction that becomes evident during late adolescence and early adulthood. Thus, the postpubertal/late adolescence emergence of PFC anomalies after an early inactivation of the ventral hippocampus coincides with the timing of symptom emergence in schizophrenia.

11.5 Conclusions

Proper development of postpubertal mesocortical DA control of excitatory and inhibitory neurotransmission (Fig. 11.2) is critical for the acquisition of mature cognitive processes that emerge during late adolescence. It can be proposed that the progressive maturation of synaptic connections combined with the distinctive physiological changes in the PFC circuit during the peripubertal transition to adulthood described here are critical to acquire the mature cognitive profile by establishing appropriate information processing mechanisms that can support more complex behavioral outcomes.

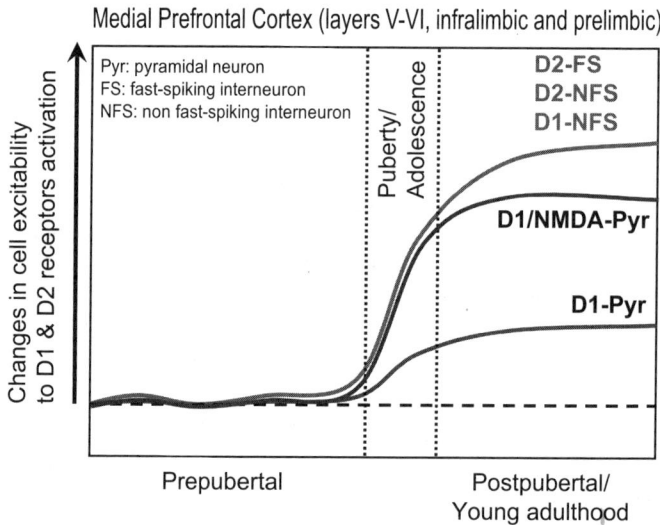

Figure 11.2 Diagram summarizing the impact of dopamine D1 and D2 receptors on medial prefrontal cortex deep-layer pyramidal neuron and GABAergic interneuron excitability during the peripubertal transition to young adulthood in rats (see also Tseng and O'Donnell, 2004, 2005, 2007a,b Tseng et al., 2007).

This process includes the periadolescent maturation of PFC interneurons response to mesocortical DA, which in turn has important implications for the refinement of PFC functions as interneurons are essential for proper tuning of cortical network dynamics (Kondo and Kawaguchi, 2001; Zhu et al., 2004), including the coordination of oscillatory activity (Fellous et al., 2001; Shu et al., 2003). Therefore, postnatal disruption of these developmental changes may lead to inappropriate PFC functioning that would become evident after puberty or late in adolescence. These dysfunctions maybe expressed as working memory and attention deficits, two major cognitive impairments typically observed in schizophrenia (reviewed by Harrison and Weinberger, 2005). We can further speculate that early developmental disruption would lead to abnormal assembly of PFC circuits which would not result in behavioral anomalies until adolescence when these connections acquire an adult profile, particularly in relation to their modulation by DA (Tseng et al., 2006b, 2007), and perhaps by other monoamines as well. Thus, DA and other monoamine systems may differentially impact PFC neuronal and synaptic plasticity depending on the developmental stage of cortical maturation.

Acknowledgments

I would like to thank Dr Anthony West for thoughtful comments on a previous version of the chapter. A special thank to Prof Patricio O'Donnell for his unconditional support, training, and friendship during my postdoctoral fellowship in his laboratory at Albany Medical College (2001–2006).

References

Arnsten, A.F., Cai, J.X., Steere, J.C. and Goldman-Rakic, P.S. (1995) Dopamine D2 receptor mechanisms contribute to age-related cognitive decline: the effects of quinpirole on memory and motor performance in monkeys. J. Neurosci. 15, 3429–3439.

Baldwin, A.E., Sadeghian, K. and Kelley, A.E. (2002) Appetitive instrumental learning requires coincident activation of NMDA and dopamine D1 receptors within the medial prefrontal cortex. J. Neurosci. 22, 1063–1071.

Becker, A., Grecksch, G., Bernstein, H.G., Hollt, V. and Bogerts, B. (1999) Social behaviour in rats lesioned with ibotenic acid in the hippocampus: quantitative and qualitative analysis. Psychopharmacology (Berl). 144, 333–338.

Benes, F.M., Vincent, S.L. and Molloy, R. (1993) Dopamine-immunoreactive axon varicosities form nonrandom contacts with GABA-immunoreactive neurons of rat medial prefrontal cortex. Synapse. 15, 285–295.

Benes, F.M., Taylor, J.B. and Cunningham, M.C. (2000) Convergence and plasticity of monoaminergic systems in the medial prefrontal cortex during the postnatal period: implications for the development of psychopathology. Cereb. Cortex. 10, 1014–1027.

Casey, B.J., Giedd, J.N. and Thomas, K.M. (2000) Structural and functional brain development and its relation to cognitive development. Biol. Psychol. 54, 241–257.

Chambers, R.A., Moore, J., McEvoy, J.P. and Levin, E.D. (1996) Cognitive effects of neonatal hippocampal lesions in a rat model of schizophrenia. Neuropsychopharmacology. 15, 587–594.

Chrapusta, S.J., Egan, M.F., Wyatt, R.J., Weinberger, D.R. and Lipska, B.K. (2003) Neonatal ventral hippocampal damage modifies serum corticosterone and dopamine release responses to acute footshock in adult Sprague-Dawley rats. Synapse. 47, 270–277.

Cohen, J.D., Braver, T.S. and Brown, J.W. (2002) Computational perspectives on dopamine function in prefrontal cortex. Curr. Opin. Neurobiol. 12, 223–229.

Druzin, M.Y., Kurzina, N.P., Malinina, E.P. and Kozlov, A.P. (2000) The effects of local application of D2 selective dopaminergic drugs into the medial prefrontal cortex of rats in a delayed spatial choice task. Behav. Brain. Res. 109, 99–111.

Fellous, J.M., Houweling, A.R., Modi, R.H., Rao, R.P., Tiesinga, P.H. and Sejnowski, T.J. (2001) Frequency dependence of spike timing reliability in cortical pyramidal cells and interneurons. J. Neurophysiol. 85, 1782–1787.

Floresco, S.B. and Phillips, A.G. (2001) Delay-dependent modulation of memory retrieval by infusion of a dopamine D1 agonist into the rat medial prefrontal cortex. Behav. Neurosci. 115, 934–939.

Funahashi, S. and Inoue, M. (2000) Neuronal interactions related to working memory processes in the primate prefrontal cortex revealed by cross-correlation analysis. Cereb. Cortex. 10, 535–551.

Gaspar, P., Bloch, B. and Le Moine, C. (1995) D1 and D2 receptor gene expression in the rat frontal cortex: cellular localization in different classes of efferent neurons. Eur. J. Neurosci. 7, 1050–1063.

Giedd, J.N., Blumenthal, J., Jeffries, N.O., Rajapakse, J.C., Vaituzis, A.C., Liu, H., Berry, Y.C., Tobin, M., Nelson, J. and Castellanos, F.X. (1999) Development of the human corpus callosum during childhood and adolescence: a longitudinal MRI study. Prog Neyropsychopharmacol Biol. Psychiatry. 23, 571–588.

Gogtay, N., Giedd, J.N., Lusk, L., Hayashi, K.M., Greenstein, D., Vaituzis, A.C., Nugent, T.F., 3rd, Herman, D.H., Clasen, L.S., Toga, A.W., Rapoport, J.L. and Thompson, P.M. (2004) Dynamic mapping of human cortical development during childhood through early adulthood. Proc. Natl. Acad. Sci. U S A. 101, 8174–8179.

Goldman-Rakic, P.S., Muly, E.C., 3rd. and Williams, G.V. (2000) D(1) receptors in prefrontal cells and circuits. Brain Res Brain Res Rev. 31, 295–301.

Gorelova, N., Seamans, J.K. and Yang, C.R. (2002) Mechanisms of dopamine activation of fast-spiking interneurons that exert inhibition in rat prefrontal cortex. J. Neurophysiol. 88, 3150–3166.

Goto, Y. and O'Donnell, P. (2002) Delayed mesolimbic system alteration in a developmental animal model of schizophrenia. J. Neurosci. 22, 9070–9077.

Grobin, A.C. and Deutch, A.Y. (1998) Dopaminergic regulation of extracellular gamma-aminobutyric acid levels in the prefrontal cortex of the rat. J. Pharmacol. Exp. Ther. 285, 350–357.

Gulledge, A.T. and Jaffe, D.B. (1998) Dopamine decreases the excitability of layer V pyramidal cells in the rat prefrontal cortex. J. Neurosci. 18, 9139–9151.

Gurden, H., Takita, M. and Jay, T.M. (2000) Essential role of D1 but not D2 receptors in the NMDA receptor-dependent long-term potentiation at hippocampal-prefrontal cortex synapses in vivo. J. Neurosci. 20, RC106.

Gurden, H., Tassin, J.P. and Jay, T.M. (1999) Integrity of the mesocortical dopaminergic system is necessary for complete expression of in vivo hippocampal-prefrontal cortex long-term potentiation. Neuroscience. 94, 1019–1027.

Harrison, P.J. and Weinberger, D.R. (2005) Schizophrenia genes, gene expression, and neuropathology: on the matter of their convergence. Mol. Psychiatry. 10, 40–68; image 45.

Horvitz, J.C. (2000) Mesolimbocortical and nigrostriatal dopamine responses to salient nonreward events. Neuroscience. 96, 651–656.

Jay, T.M. (2003) Dopamine: a potential substrate for synaptic plasticity and memory mechanisms. Prog. Neurobiol. 69, 375–390.

Kondo, S. and Kawaguchi, Y. (2001) Slow synchronized bursts of inhibitory postsynaptic currents (0.1-0.3 Hz) by cholinergic stimulation in the rat frontal cortex in vitro. Neuroscience. 107, 551–560.

Le Pen, G., Grottick, A.J., Higgins, G.A., Martin, J.R., Jenck, F. and Moreau, J.L. (2000) Spatial and associative learning deficits induced by neonatal excitotoxic hippocampal damage in rats: further evaluation of an animal model of schizophrenia. Behav. Pharmacol. 11, 257–268.

Le Pen, G. and Moreau, J.L. (2002) Disruption of prepulse inhibition of startle reflex in a neurodevelopmental model of schizophrenia: reversal by clozapine, olanzapine and risperidone but not by haloperidol. Neuropsychopharmacology. 27, 1–11.

Leslie, C.A., Robertson, M.W., Cutler, A.J. and Bennett, J.P., Jr. (1991) Postnatal development of D1 dopamine receptors in the medial prefrontal cortex, striatum and nucleus accumbens of normal and neonatal 6-hydroxydopamine treated rats: a quantitative autoradiographic analysis. Brain. Res. Dev. Brain. Res. 62, 109–114.

Lewis, D.A., Volk, D.W. and Hashimoto, T. (2004) Selective alterations in prefrontal cortical GABA neurotransmission in schizophrenia: a novel target for the treatment of working memory dysfunction. Psychopharmacology (Berl). 174, 143–150.

Lewis, B.L. and O'Donnell, P. (2000) Ventral tegmental area afferents to the prefrontal cortex maintain membrane potential 'up' states in pyramidal neurons via D(1) dopamine receptors. Cereb. Cortex. 10, 1168–1175.

Lipska, B.K., Jaskiw, G.E. and Weinberger, D.R. (1993) Postpubertal emergence of hyperresponsiveness to stress and to amphetamine after neonatal excitotoxic hippocampal damage: a potential animal model of schizophrenia. Neuropsychopharmacology. 9, 67–75.

Lipska, B.K. and Weinberger, D.R. (1994) Subchronic treatment with haloperidol and clozapine in rats with neonatal excitotoxic hippocampal damage. Neuropsychopharmacology. 10, 199–205.

Lipska, B.K., Chrapusta, S.J., Egan, M.F. and Weinberger, D.R. (1995) Neonatal excitotoxic ventral hippocampal damage alters dopamine response to mild repeated stress and to chronic haloperidol. Synapse. 20, 125–130.

Lipska, B.K. and Weinberger, D.R. (1998) Prefrontal cortical and hippocampal modulation of dopamine-mediated effects. Adv. Pharmacol. 42, 806–809.

Lipska, B.K. and Weinberger, D.R. (2000) To model a psychiatric disorder in animals: schizophrenia as a reality test. Neuropsychopharmacology. 23, 223–239.

Lipska, B.K., Aultman, J.M., Verma, A., Weinberger, D.R. and Moghaddam, B. (2002) Neonatal damage of the ventral hippocampus impairs working memory in the rat. Neuropsychopharmacology. 27, 47–54.

Lipska, B.K., Lerman, D.N., Khaing, Z.Z. and Weinberger, D.R. (2003a) The neonatal ventral hippocampal lesion model of schizophrenia: effects on dopamine and GABA mRNA markers in the rat midbrain. Eur. J. Neurosci. 18, 3097–3104.

Lipska, B.K., Lerman, D.N., Khaing, Z.Z., Weickert, C.S. and Weinberger, D.R. (2003c) Gene expression in dopamine and GABA systems in an animal model of schizophrenia: effects of antipsychotic drugs. Eur. J. Neurosci. 18, 391–402.

Markram, H., Toledo-Rodriguez, M., Wang, Y., Gupta, A., Silberberg, G. and Wu, C. (2004) Interneurons of the neocortical inhibitory system. Nat. Rev. Neurosci. 5, 793–807.

Monyer, H., Burnashev, N., Laurie, D.J., Sakmann, B. and Seeburg, P.H. (1994) Developmental and regional expression in the rat brain and functional properties of four NMDA receptors. Neuron. 12, 529–540.

Mrzljak, L., Bergson, C., Pappy, M., Huff, R., Levenson, R. and Goldman-Rakic, P.S. (1996) Localization of dopamine D4 receptors in GABAergic neurons of the primate brain. Nature. 381, 245–248.

Muly, E.C., 3rd, Szigeti, K. and Goldman-Rakic, P.S. (1998) D1 receptor in interneurons of macaque prefrontal cortex: distribution and subcellular localization. J. Neurosci. 18, 10553–10565.

O'Donnell, P., Lewis, B.L., Weinberger, D.R. and Lipska, B.K. (2002) Neonatal hippocampal damage alters electrophysiological properties of prefrontal cortical neurons in adult rats. Cereb Cortex. 12, 975–982.

O'Donnell, P. (2003) Dopamine gating of forebrain neural ensembles. Eur. J. Neurosci. 17, 429–435.

Petit, T.L., LeBoutillier, J.C., Gregorio, A. and Libstug, H. (1988) The pattern of dendritic development in the cerebral cortex of the rat. Brain. Res. 469, 209–219.

Pirot, S., Godbout, R., Mantz, J., Tassin, J.P., Glowinski, J. and Thierry, A.M. (1992) Inhibitory effects of ventral tegmental area stimulation on the activity of prefrontal cortical neurons: evidence for the involvement of both dopaminergic and GABAergic components. Neuroscience. 49, 857–865.

Rao, S.G., Williams, G.V. and Goldman-Rakic, P.S. (2000) Destruction and creation of spatial tuning by disinhibition: GABA(A) blockade of prefrontal cortical neurons engaged by working memory. J. Neurosci. 20, 485–494.

Schultz, W. (2002) Getting formal with dopamine and reward. Neuron. 36, 241–263.

Seamans, J.K., Floresco, S.B. and Phillips, A.G. (1998) D1 receptor modulation of hippocampal-prefrontal cortical circuits integrating spatial memory with executive functions in the rat. J. Neurosci. 18, 1613–1621.

Seamans, J.K., Gorelova, N., Durstewitz, D. and Yang, C.R. (2001) Bidirectional dopamine modulation of GABAergic inhibition in prefrontal cortical pyramidal neurons. J. Neurosci. 21, 3628–3638.

Segalowitz, S.J. and Davies, P.L. (2004) Charting the maturation of the frontal lobe: an electrophysiological strategy. Brain. Cogn. 55, 116–133.

Sesack, S.R., Hawrylak, V.A., Matus, C., Guido, M.A. and Levey, A.I. (1998a) Dopamine axon varicosities in the prelimbic division of the rat prefrontal cortex exhibit sparse immunoreactivity for the dopamine transporter. J. Neurosci. 18, 2697–2708.

Sesack, S.R., Hawrylak, V.A., Melchitzky, D.S. and Lewis, D.A. (1998b) Dopamine innervation of a subclass of local circuit neurons in monkey prefrontal cortex: ultrastructural analysis of tyrosine hydroxylase and parvalbumin immunoreactive structures. Cereb. Cortex. 8, 614–622.

Shu, Y., Hasenstaub, A., Badoual, M., Bal, T. and McCormick, D.A. (2003) Barrages of synaptic activity control the gain and sensitivity of cortical neurons. J. Neurosci. 23, 10388–10401.

Smiley, J.F., Levey, A.I., Ciliax, B.J. and Goldman-Rakic, P.S. (1994) D1 dopamine receptor immunoreactivity in human and monkey cerebral cortex: predominant and extrasynaptic localization in dendritic spines. Proc. Natl. Acad. Sci. U S A. 91, 5720–5724.

Spear, L.P. (2000) The adolescent brain and age-related behavioral manifestations. Neurosci. Biobehav. Rev. 24, 417–463.

Szabadics, J., Lorincz, A. and Tamas, G. (2001) Beta and gamma frequency synchronization by dendritic GABAergic synapses and gap junctions in a network of cortical interneurons. J. Neurosci. 21, 5824–5831.

Tarazi, F.I., Tomasini, E.C. and Baldessarini, R.J. (1999) Postnatal development of dopamine D1-like receptors in rat cortical and striatolimbic brain regions: An autoradiographic study. Dev. Neurosci. 21, 43–49.

Tarazi, F.I. and Baldessarini, R.J. (2000) Comparative postnatal development of dopamine D(1), D(2) and D(4) receptors in rat forebrain. Int. J. Dev. Neurosci. 18, 29–37.

Trantham-Davidson, H., Neely, L.C., Lavin, A. and Seamans, J.K. (2004) Mechanisms underlying differential D1 versus D2 dopamine receptor regulation of inhibition in prefrontal cortex. J. Neurosci. 24, 10652–10659.

Traub, R.D., Whittington, M.A., Stanford, I.M. and Jefferys, J.G. (1996) A mechanism for generation of long-range synchronous fast oscillations in the cortex. Nature. 383, 621–624.

Tseng, K.Y. and O'Donnell, P. (2004) Dopamine-glutamate interactions controlling prefrontal cortical pyramidal cell excitability involve multiple signaling mechanisms. J. Neurosci. 24, 5131–5139.

Tseng, K.Y. and O'Donnell, P. (2005a) Post-pubertal emergence of prefrontal cortical up states induced by D1-NMDA co-activation. Cereb. Cortex. 15, 49–57.

Tseng, KY. and O'Donnell, P. (2005b) Dopaminergic modulation of cortical and striatal up states. In: J.P. Bolam, C.A. Ingham and P.J. Magill (Eds.), *The Basal Ganglia VIII, SECTION: Physiological and Anatomical Studies of the Functional Organization of the Basal Ganglia*. Springer Science and Business Media, New York, pp. 467–474.

Tseng, K.Y., Mallet, N., Toreson, K.L., Le Moine, C., Gonon, F. and O'Donnell, P. (2006a) Excitatory response of prefrontal cortical fast-spiking interneurons to ventral tegmental area stimulation in vivo. Synapse. 59, 412–417.

Tseng, K.Y., Amin, F., Lewis, B. L. and O'Donnell, P. (2006b) Altered prefrontal cortical metabolic response to mesocortical activation in adult animals with a neonatal ventral hippocampal lesion. Biol. Psychiatry. 60, 585–590.

Tseng, K.Y. and O'Donnell, P. (2007a) Dopamine modulation of prefrontal cortical interneurons changes during adolescence. Cereb. Cortex. 17, 1235–1240.

Tseng, K.Y. and O'Donnell, P. (2007b) Dopamine D2 receptors recruit a GABA component for their attenuation of excitatory synaptic transmission in the adult rat prefrontal cortex. Synapse. (In press).

Tseng, K.Y., Lewis, B.L., Lipska, B.K. and O'Donnell, P. (2007) Post-pubertal disruption of medial prefrontal cortical dopamine-glutamate interactions in a developmental animal model of schizophrenia. Biol. Psychiatry. Jan 2; [Epub ahead of print]

Vincent, S.L., Khan, Y. and Benes, F.M. (1993) Cellular distribution of dopamine D1 and D2 receptors in rat medial prefrontal cortex. J. Neurosci. 13, 2551–2564.

Vincent, S.L., Pabreza, L. and Benes, F.M. (1995) Postnatal maturation of GABA-immunoreactive neurons of rat medial prefrontal cortex. J. Comp. Neurol. 355, 81–92.

Wang, J. and O'Donnell, P. (2001) D(1) dopamine receptors potentiate NMDA-mediated excitability increase in layer V prefrontal cortical pyramidal neurons. Cereb. Cortex. 11, 452–462.

Williams, K., Russell, S.L., Shen, Y.M. and Molinoff, P.B. (1993) Developmental switch in the expression of NMDA receptors occurs in vivo and in vitro. Neuron. 10, 267–278.

Zhu, J.J. (2000) Maturation of layer 5 neocortical pyramidal neurons: amplifying salient layer 1 and layer 4 inputs by Ca^{2+} action potentials in adult rat tuft dendrites. J. Physiol. 526 Pt 3, 571–587.

Zhu, Y., Stornetta, R.L. and Zhu, J.J. (2004) Chandelier cells control excessive cortical excitation: characteristics of whisker-evoked synaptic responses of layer 2/3 nonpyramidal and pyramidal neurons. J Neurosci. 24, 5101–5108.

Chapter 12

Dopamine and Noradrenaline Coupling in the Cerebral Cortex

Paola Devoto and Giovanna Flore
B.B. Brodie, Department of Neuroscience, University of Cagliari, Italy

12.1 Anatomical Consideration

The catecholaminergic innervation of rodent cerebral cortex differs in distribution of areas of dopaminergic and noradrenergic afferents. Noradrenergic innervation, originating from the locus coeruleus, is widespread and uniformly distributed throughout the cortex, predominantly in layer I (Audet et al., 1988). On the contrary, dopaminergic innervation, originating from ventral tegmental (VTA), and to a lesser degree from dorsal part of substantia nigra-pars compacta, is mainly concentrated in specific cortical areas, namely infralimbic-prelimbic, anterior cingulate and entorhinal cortices and innervates layers III, V and VI (Descarries et al., 1987). However, a consistent mismatch exists between DA innervation and its receptors. Indeed, the latter are widely spread throughout the cerebral cortex and not confined to DA innervated areas (Martres et al., 1985; Lidow et al., 1989; Richfield et al., 1989; Goldsmith and Joyce, 1994; Wedzony et al., 2000).

DA is present in noradrenergic neurons, being the biochemical precursor of NA. Furthermore, there is increasing evidence that DA may act as a physiological ligand at noradrenergic receptors (Zhang et al., 2004), and dopaminergic stimulation of noradrenergic receptors has been documented in brain stem (Crochet and Sakai, 2003), preoptic area (Cornill et al., 2002) and hippocampus (Malenka and Nicoll, 1986).

At variance with subcortical areas, the medial prefrontal cortex is characterised by an extremely scarce expression of DA transporter (DAT) (Ciliax et al., 1995; Sesak et al., 1998), while NA transporter (NAT) is well represented (Miner et al., 2000; Sanders et al., 2005). Furthermore, DA is a high affinity substrate for NAT (Horn, 1973; Raiteri et al., 1977). As a consequence of these observations, it is likely that in the cerebral cortex a consistent fraction of extracellular DA is recaptured into noradrenergic terminals by NAT. In this regard, it has been suggested that in medial prefrontal cortex extracellular DA increase might be the passive consequence of augmented extracellular NA, the parallel increases being due to competition for the same transporter (Carboni et al., 1990; Pozzi et al., 1994; Yamamoto and Novotney, 1998; Moron et al., 2002).

The previously expounded anatomical and physiological findings together with other experimental observations (see below) led us to postulate that in some circumstances DA is not only recaptured, but also released from noradrenergic neurons, where it acts as a co-transmitter. We also suggested that this phenomenon takes place not only in the richly DA innervated medial prefrontal cortex, but also in other cortical areas that have a significantly less amount of DA terminals (Devoto and Flore, 2006).

12.2 Tissue and Extracellular Catecholamine Content

A difference in dopaminergic innervation is reflected in a contrasting DA and DOPAC tissue content, as shown in Table 12.1. The nucleus accumbens, which receives its abundant dopaminergic input from VTA, has a DA and DOPAC content measurable in thousands of picograms per milligram of tissue.

Mesocortical dopaminergic neurons, which are believed to be less than 10% of VTA dopaminergic neurons, terminate in the medial prefrontal cortex, which has a DA content four times higher than the scarcely, if any, innervated parietal cortex.

The differences in DA innervation are also evidenced by DOPAC tissue levels, equaling DA values. On the other hand, in line with the homogeneous noradrenergic innervation of cerebral cortex, NA content is the same in the two cortical areas and higher in nucleus accumbens, which receives its noradrenergic innervation not only from locus coeruleus but also from nucleus of tractus solitarius (Delfs et al., 1988).

Table 12.1 Tissue NA, DA and DOPAC Concentrations in Different Rat Brain Areas

Cerebral area	NA	DA	DOPAC
Medial prefrontal cortex	363 ± 11.3	72.3 ± 3.6	36.4 ± 3.2
Parietal cortex	329 ± 11.6	18.7 ± 1.5	18.4 ± 2.3
Nucleus accumbens	814 ± 39.0	5402.6 ± 232.4	1573.7 ± 94.5

Values are expressed as picograms per milligram tissue, and are the means $\pm$ SEM of data obtained from 65 rats.

At difference with tissue contents, extracellular DA, as measured by means of microdialysis in freely moving rats, is found at the same concentration in three cortical areas profoundly distinctive as to dopaminergic innervation: medial prefrontal, occipital and parietal cortex (Table 12.2). In contrast, extracellular DOPAC concentrations are significantly different among cortical areas, being highest in medial prefrontal cortex and lowest in occipital cortex, therefore reflecting dopaminergic innervation intensity.

Table 12.2 Extracellular NA, DA and DOPAC Concentrations in Different Rat Brain Areas

Cerebral area (n of rats)	NA	DA	DOPAC
Medial prefrontal cortex (173)	3.10 ± 0.15	2.72 ± 0.14	190.68 ± 10.80
Parietal cortex (64)	2.96 ± 0.18	2.26 ± 0.14	122.71 ± 17.71
Occipital cortex (28)	3.80 ± 0.42	2.24 ± 0.17	23.48 ± 2.84
Nucleus accumbens (33)	8.18 ± 0.90	40.21 ± 7.12	2902.82 ± 263.56

Data collected from animals implanted with vertical microdialysis probes. Extracellular concentration of NA, DA and DOPAC are expressed in picograms per sample. All values are presented as mean ± SEM.

For comparison, Table 12.1 also reports extracellular NA, DA and DOPAC values in nucleus accumbens, a cerebral area particularly rich in dopaminergic afferents that also receives NA innervation, in which DOPAC values are far higher than in the cerebral cortex. Thus, extracellular DOPAC seems to be a better index of dopaminergic innervation intensity than extracellular DA concentration.

On the other hand, NA levels were very similar in the three cortices, as expected from its homogeneous innervation, and highest in nucleus accumbens, which also has higher NA tissue level than cortices.

12.3 Locus Coeruleus or Ventral Tegmental Area Modulation: Effects on Extracellular Catecholamine Content in Terminal Areas

By means of multi-probe microdialysis, extracellular neurotransmitter levels can be recorded in neuronal terminal areas during local administration of drugs into their nuclei of origin. Such kind of experiments demonstrated that drug perfusion into the locus coeruleus profoundly affected both NA and DA levels in cerebral cortex, but only NA in other dopaminergic projection areas. Indeed, locus coeruleus LC perfusion with the α_2 adrenergic agonist clonidine decreased NA and DA levels in the medial prefrontal cortex as well as in the occipital cortex, whilst affecting NA but not DA levels in the ventral striatum (Kawahara et al., 2001; Devoto et al., 2003a). The decrease elicited by clonidine perfusion into the locus coeruleus was also evident during cerebral cortex perfusion with desipramine, which blocks NAT activity (Devoto et al., 2003a). This finding implicates that the decrease of DA observed following clonidine infusion into locus coeruleus, is not due to augmented reuptake of DA by NAT, but to DA diminished release from neuronal terminals. Furthermore, drugs that stimulate neuronal activity such as NMDA, kainate, carbachol or bicuculline, locally applied into locus coeruleus, elicited parallel increases in both NA and DA in medial prefrontal cortex, but not in nucleus accumbens, where only NA levels were increased (Kawahara et al., 2001). Neuronal stimulation can moreover be accomplished by direct administration of electrical impulses. This treatment, delivered into locus coeruleus, increased extracellular levels of both catecholamines in cortical areas (medial prefrontal, occipital and parietal cortices) but only NA, and not DA, in caudate nucleus (Devoto et al., 2005a, b).On the other hand, the D_2 dopaminergic agonist quinpirole, locally perfused into the VTA, failed to affect medial prefrontal cortex extracellular DA levels (Fig. 12.1), whereas it lowered DA release in the nucleus accumbens to about 40% of baseline (Devoto and Flore, unpublished observations). It should be stressed that DA release in the medial prefrontal cortex from dopaminergic neurons arising from VTA cannot be excluded by this observation. Indeed, infusion of glutamatergic, GABAergic and cholinergic drugs into the VTA has been demonstrated to modulate extracellular DA levels in the medial prefrontal cortex (Westerink et al., 1998a). Similarly, VTA stimulation with local infusion of the muscarinic cholinergic agonist oxotremorine increases DA, but not NA levels in the medial prefrontal cortex (Gronier et al., 2000).

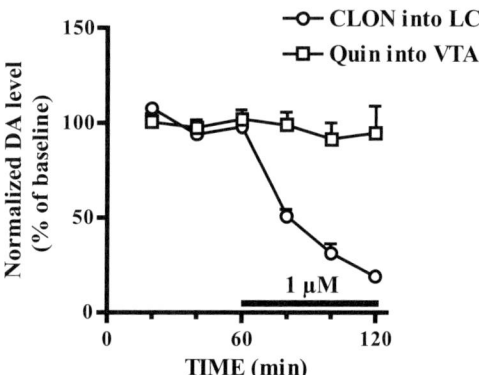

Figure 12.1 Extracellular dopamine levels in the medial prefrontal cortex during perfusion of 1 μM clonidine (CLON) into the locus coeruleus (LC) or 1 μM quinpirole (Quin) into the ventral tegmental area (VTA). All drugs were perfused during 1 h as indicated by the horizontal bar. Data are expressed as percent of basal mean value (mean ± SEM; n=4 rats).

12.4 D$_2$- or α$_2$-Receptor Modulations of Extracellular Catecholamines

Quinpirole also proved ineffective on cortical DA release after either systemic administration or local perfusion into the cerebral cortex (Devoto et al., 2001, 2003b). Consistent with a lack of D$_2$ receptor control on DA release in the medial prefrontal cortex, haloperidol, despite of its action on the mesocortical DA system (Gessa et al., 2000), failed to modify DA concentrations in the medial prefrontal cortex and in other cortical areas (Pehek and Yamamoto, 1994; Li et al., 1998; Kuroki et al., 1999; Devoto et al., 2001, 2004a). However, systemic administration of haloperidol increases DOPAC levels in the medial prefrontal cortex, but not in the occipital cortex (Gessa et al., 2000; Devoto et al., 2004a). This effect seems to be the consequence of haloperidol-induced activation of dopaminergic neurons firing, known to stimulate tyrosine hydroxylase activity (Zigmond et al., 1989; Nakahara et al., 2000). On the other hand, drugs that stimulate noradrenergic system affect extracellular DA and DOPAC levels in the cerebral cortex. A pronounced decrease of both NA and DA was found in the medial prefrontal cortex following local perfusion of the α_2 agonist clonidine (Devoto et al., 2001). Similar decreases of DA and NA in both the medial prefrontal and the occipital cortices were observed after systemic administration of clonidine (Devoto et al., 2001, 2003a, 2004b). Conversely, selective α_2-adrenoceptor antagonists, such as RS 79948 and idazoxan, systemically or locally administered in medial prefrontal, parietal and occipital cortex, increased not only extracellular NA concentrations, but also those of DA and DOPAC (Devoto et al., 2001; Kawahara et al., 2001; Devoto et al., 2004a). Furthermore, systemic administration of idazoxan failed to increase extra-neuronal DA and DOPAC in caudate and accumbens nuclei, its effect being selective for cerebral cortex (Hertel et al., 1999). Idazoxan efficacy in increasing prefrontal cortex DA was still present in animals in which the VTA was perfused with the sodium channel blocker tetrodotoxin. Indeed, tetrodotoxin infusion into VTA blocked mesocortical dopaminergic cell neuronal activity, as demonstrated by the decrease of extracellular DA levels in medial prefrontal cortex (Westerink et al., 1998a; Hertel et al., 1999), nevertheless it did not prevent idazoxan-induced increase in DA (Hertel et al., 1999). This observation implies that a fraction of medial prefrontal cortex DA does not originate from the VTA. This observation supports the hypothesis of a DA-NA co-release in the prefrontal cortex, probably originated from the noradrenergic terminals from the locus coeruleus.

In line with an α_2-adrenoceptor mediated effect on catecholamine levels in the cerebral cortex, drugs with α_2-adrenoceptor antagonistic properties, such as the anti-depressants mianserine and mirtazapine or atypical antipsychotics, such as clozapine, olanzapine and zotepine, have been shown to elicit parallel increases of NA and DA in the cerebral cortex, without affecting subcortical DA levels (Tanda et al., 1996; Li et al., 1998; Devoto et al., 2003b, 2004b; Nakamura et al., 2005). The selective in-crease in dopaminergic release in the medial prefrontal cortex has been claimed to subserve the efficacy of atypical antipsychotic drugs to treat negative symptoms in schizophrenia (Meltzer, 1994; Lidow et al., 1998). For example, clozapine not only enhance DA and NA levels in the medial prefrontal cortex but also in the occipital cortex (Li et al., 1998; Westerink et al., 2003b; Devoto et al., 2004a). Interestingly, this effect was counteracted by clonidine, but not by quinpirole, suggesting that the catecholamine increases induced by clozapine might be due to its antagonistic effect on α_2-adrenoceptors (Devoto et al., 2003b). In fact, clozapine is known to bind to a variety of different receptors including DA and NA receptors (Bymaster et al., 1996).

12.5 Conclusions

The data reported here highlight the functional significance of DA-NA co-release in medial prefrontal cortex. NA and DA are considered to play a central role in arousal, anxiety, stress, fear, reward as well as in the pathophysiological changes observed in schizophrenia, depression and in the action of psychostimulants, antipsychotics and antidepressants. Extracellular DA and NA concentrations have been shown to covary with many of the above conditions. Our results raise the issue as to whether DA and

NA neurons are distinctly influenced in these conditions or whether the observed co-variations of the two amines may depend, at least in part, on the fact that they are co-released by the same NA neurons. Furthermore, spontaneous and drug-induced release of cortical DA is not restricted within the medial prefrontal cortical circuit. The potential physiological or pathophysiological impact of DA in other cortical areas should be investigated.

References

Audet, M.A., Doucet, G., Oleskevich, S., Descarries, L. (1988) Quantified regional and laminar distribution of the noradrenaline innervation in the anterior half of the adult rat cerebral cortex. J. Comp. Neurol. 273, 224–240.

Bymaster, F.P., Calligaro, D.O., Falcone, J.F., Marsh, R.D., Moore, N.A., Tye, N.C., Seeman, P., Wong, D.T. (1996) Radioreceptor binding profile of the atypical antipsychotic olanzapine. Neuropsychopharmacology 14, 87–96.

Carboni, E., Tanda, G.L., Frau, R., Di Chiara, G. (1990) Blockade of the noradrenaline carrier increases extracellular dopamine concentrations in the prefrontal cortex: evidence that dopamine is taken up in vivo by noradrenergic terminals. J. Neurochem. 55, 1067–1070.

Ciliax, B.J., Heilman, C., Demchyshyn, L.L., Pristupa, Z.B., Ince, E., Hersch, S.M., Niznik, H.B., Levey, A.I. (1995) The dopamine transporter: immunochemical characterization and localization in brain. J. Neurosci. 15, 1714–1723.

Cornill, C.A., Balthazart, J., Motte, P., Massotte, L., Seutin, V. (2002) Dopamine activates noradrenergic receptors in the preoptic area. J. Neurosci. 24, 5410–5419.

Crochet, S., Sakai, K, (2003) Dopaminergic modulation of behavioural states in mesopontine tegmentum: a reverse microdialysis study in freely moving cats. Sleep 26, 801–806.

Delfs, J.M., Zhu, Y., Druhan, J.P., Aston-Jones, G.S. (1998) Origin of noradrenergic afferents to the shell subregion of the nucleus accumbens: anterograde and retrograde tract-tracing studies in the rat. Brain Res. 806, 127–140.

Descarries, L., Lemay, B., Doucet, G., Berger, B. (1987) Regional and laminar density of the dopamine innervation in adult rat cerebral cortex. Neuroscience 21, 807–824.

Devoto, P., and Flore, G. (2006) On the origin of cortical dopamine: is it a co-transmitter in noradrenergic neurons? Curr. Neuropharmacol. 4, 115–125.

Devoto, P., Flore, G., Pani, L., Gessa, G.L. (2001) Evidence for co-release of noradrenaline and dopamine from noradrenergic neurons in the cerebral cortex. Mol. Psychiatry 6, 657–664.

Devoto, P., Flore, G., Longu, G., Pira, L., Gessa, G.L. (2003a) Origin of extracellular dopamine from dopamine and noradrenaline neurons in the medial prefrontal and occipital cortex. Synapse 50, 200–205.

Devoto, P., Flore, G., Vacca, G., Pira, L., Arca, A., Casu, M.A., Pani, L., Gessa, G.L. (2003b) Co-release of noradrenaline and dopamine from noradrenergic neurons in the cerebral cortex induced by clozapine, the prototype atypical antipsychotic. Psychopharmacology 167, 79–84.

Devoto, P., Flore, G., Longu, G., Pira, L., Gessa, G.L. (2004a) Alpha2-adrenoceptor mediated co-release of dopamine and noradrenaline from noradrenergic neurons in the cerebral cortex. J. Neurochem. 88, 1003–1009.

Devoto, P., Flore, G., Pira, L., Longu, G., Gessa, G.L. (2004b) Mirtazapine-induced co-release of dopamine and noradrenaline from noradrenergic neurons in the medial prefrontal and occipital cortex. Eur. J. Pharmacol. 487, 105–111.

Devoto, P., Flore, G., Saba, P., Fà, M., Gessa, G.L. (2005a) Co-release of noradrenaline and dopamine in the cerebral cortex elicited by single train and repeated train stimulation of the locus coeruleus. BMC Neuroscience 6, 31. http://www.biomedcentral.com/1471-2202/6/31.

Devoto, P., Flore, G., Saba, P., Fà, M., Gessa, G.L. (2005b) Stimulation of the locus coeruleus elicits noradrenaline and dopamine release in the medial prefrontal and parietal cortex. J. Neurochem. 92, 368–374.

Gessa, G.L., Devoto, P., Diana, M., Flore, G., Melis, M., Pistis, M. (2000) Dissociation of haloperidol, clozapine and olanzapine effects on electrical activity of mesocortical dopamine neurons and dopamine release in the prefrontal cortex. Neuropsychopharmacology 22, 642–649.

Goldsmith, S.K., Joyce, J.N. (1994) Dopamine D2 receptors in hippocampus and parahippocampal cortex of rat, cat and human in relation to tyrosine hydroxylase-immunoreactive fibers. Hippocampus 4, 354–373.

Gronier, B., Perry, K.W., Rasmussen, K. (2000) Activation of the mesocorticolimbic dopaminergic system by stimulation of muscarinic cholinergic receptors in the ventral tegmental area. Psychopharmacology 147, 347–355.

Hertel, P., Nomikos, G.G., Svensson, T.H. (1999) Idazoxan preferentially increases dopamine output in the rat medial prefrontal cortex at the nerve terminal level. Eur. J. Pharmacol. 371, 153–158.

Horn, A.S. (1973) Structure-activity relations for the inhibition of catecholamine uptake into synaptosomes from noradrenergic and dopaminergic neurons in rat brain homogenates. Br. J. Pharmacol. 47, 332–338.

Kawahara, H., Kawahara, Y., Westerink, B.H.C. (2001) The noradrenaline-dopamine interaction in the rat medial prefrontal cortex studied by multi-probe microdialysis. Eur. J. Pharmacol. 418, 177–186.

Kuroki, T., Meltzer, H.Y., Ichikawa, J. (1999) Effect of antipsychotic drugs on extracellular dopamine levels in rat medial prefrontal cortex and nucleus accumbens. J. Pharmacol. Exp. Ther. 288, 774–781.

Li, X.-M., Perry, K.W., Wong, D.T., Bymaster, F.P. (1998) Olanzapine increases in vivo dopamine and norepinephrine release in rat prefrontal cortex, nucleus accumbens and striatum. Psychopharmacology 136, 153–161.

Lidow, M.S., Goldman-Rakic, P.S., Rakic, P., Innis, R.B. (1989) Dopamine D2 receptors in the cerebral cortex: distribution and pharmacological characterization with [3H]raclopride. Proc. Natl. Acad. Sci. U.S.A. 86, 6412–6416.

Lidow, M.S., Williams, G.V., Goldman-Rakic, P.S. (1998) The cerebral cortex: a case for a common site of action of antipsychotics. Trends Pharmacol. Sci. 19, 136–140.

Malenka, R.C., Nicoll, R.A. (1986) Dopamine decreases the calcium-activated afterhyperpolarization in hippocampal CA1 pyramidal cells. Brain Res. 379, 210–215.

Martres, M.P., Bouthenet, M.L., Sales, N., Sokoloff, P., Schwartz, J.C. (1985) Widespread distribution of brain dopamine receptors evidenced with [125I]iodosulpiride, a highly selective ligand. Science 228, 752–755.

Meltzer, H.Y. (1994) An overview of the mechanism of action of clozapine. J. Clin. Psychiatry, 55, 47–52.

Miner, L.H., Schroeter, S., Blakely, R.D., Sesack, S.R. (2000) Ultrastructural localization of the serotonin transporter in superficial and deep layers of the rat prelimbic prefrontal cortex and its spatial relationship to dopamine terminals. J. Comp. Neurol. 427, 220–234.

Moron, J.A., Brockington, A., Wise, R.A., Rocha, B.A., Hope, B.T. (2002) Dopamine uptake through the norepinephrine transporter in brain regions with low levels of the dopamine transporter: evidence from knock-out mouse lines. J. Neurosci. 22, 389–395.

Nakahara, D., Nakamura, M., Furukawa, H., Furuno, N. (2000) Intracranial self-stimulation increases differentially in vivo hydroxylation of tyrosine but similarly in vivo hydroxylation of tryptophan in rat medial prefrontal cortex, nucleus accumbens and striatum. Brain Res. 864, 124–129.

Nakamura, S., Ago, Y., Itoh, S., Koyama, Y., Baba, A., Matsuda, T. (2005) Effect of zotepine on dopamine, serotonin and noradrenaline release in rat prefrontal cortex. Eur. J. Pharmacol. 528, 95–98.

Pehek, E.A., Yamamoto, B.K. (1994) Differential effects of locally administered clozapine and haloperidol on dopamine efflux in the rat prefrontal cortex and caudate-putamen. J. Neurochem. 63, 2118–2124.

Pozzi, L., Invernizzi, R., Cervo, L., Vallebuona, F., Samanin, R. (1994) Evidence that extracellular concentrations of dopamine are regulated by noradrenergic neurons in the frontal cortex of rats. J. Neurochem. 63, 195–200.

Raiteri, M., Del Carmine, R., Bertollini, A., Levi, G. (1977) Effects of sympaticomimetic amines on the synaptosomal transport of noradrenaline, dopamine, 5-hydroxytryptamine. Eur. J. Pharmacol. 41, 133–143.

Richfield, E.K., Young, A.B., Penney, J.B. (1989) Comparative distribution of dopamine D1 and D2 receptors in the cerebral cortex of rats, cats and monkeys. J. Comp. Neurol. 286, 409–426.

Sanders, J.D., Happe, H.K., Bylund, D.B., Murrin, L.C. (2005) Development of the norepinephrine transporter in the rat CNS. Neuroscience 130, 107–117.

Sesak, S.R., Hawrylak, V.A., Matus, C., Guido, M.A., Levey, A.I. (1998) Dopamine axon varicosities in the prelimbic division of the rat prefrontal cortex exhibit sparse immunoreactivity for the dopamine transporter. J. Neurosci. 18, 2696–2708.

Tanda, G., Bassareo, V., Di Chiara, G. (1996) Mianserin markedly and selectively increases extracellular dopamine in the prefrontal cortex as compared to the nucleus accumbens of the rat. Psychopharmacology 123, 127–130.

Wedzony, K., Chocyk, A., Mackowiak, M., Fijal, K., Czyrak, A. (2000) Cortical localization of dopamine D4 receptors in the rat brain – immunocytochemical study. J. Physiol. Pharmacol. 51, 205–221.

Westerink, B.H.C., Enrico, P., Feimann, J., De Vries, J.B. (1998a). The pharmacology of mesocortical dopamine neurons: a dual-probe microdialysis study in the ventral tegmental area and prefrontal cortex of the rat brain. J. Pharmacol. Exp. Ther. 285, 143–154.

Westerink, B.H.C., de Boer, P., de Vries, J.B., Kruse, C.G., Long, S.K. (1998b) Antipsychotic drugs induce similar effects on the release of dopamine and noradrenaline in the medial prefrontal cortex of the rat brain. Eur. J. Pharmacol. 361, 27–33.

Yamamoto, B.K., Novotney, S. (1998) Regulation of extracellular dopamine by the norepinephrine transporter. J. Neurochem. 71, 274–280.

Zhang, W.P., Ouyang, M., Thomas, S.A. (2004) Potency of catecholamines and other L-tyrosine derivatives at the cloned mouse adrenergic receptors. Neuropharmacology 47, 438–449.

Zigmond, R.E., Schwarzschild, M.A., Rittenhouse, A.R. (1989) Acute regulation of tyrosine hydroxylase by nerve activity and by neurotransmitters via phosphorilation. Ann. Rev. Neurosci. 12, 415–461.

Chapter 13

Regulation of Cortical Functions by the Central Noradrenergic System: *Emerging Properties from an Old Friend*

Marco Atzori[1], Humberto Salgado[1] and Kuei Y. Tseng[2]

[1]School for Behavioral & Brain Sciences, University of Texas at Dallas, Richardson, TX 75080, USA; [2]Department of Cellular & Molecular Pharmacology, Rosalind Franklin University of Medicine and Science/The Chicago Medical School, North Chicago, IL 60064, USA

13.1 Introduction

The locus coeruleus (LC)-norepinephrine (NE) and the ventral tegmental area (VTA)-dopamine (DA) systems are two important brainstem neuromodulatory ascending pathways with a widespread cortical distribution. Historically, the NE system has been implicated in arousal whereas DA signals have been linked to reward and motivation. In addition to this early interpretation, recent findings indicate that the NE system also play an important role in the control of complex behaviors (Devilbiss and Waterhouse, 2004; Aston-Jones and Cohen, 2005a; Chamberlain et al., 2006). For example, neuronal activity in the LC, particularly the phasic firing mode, has been associated to the outcome of certain task-related decision processes, and it has been proposed that this enhancement of NE signal (presumably in the cortex) helps to optimize task performance (Aston-Jones and Cohen, 2005a). A similar pattern of firing response to task-related events has also been observed in DA neurons (Lidow et al., 1998). These data suggest that both NE and DA systems are responsive to motivationally salient events such as reward predictors.

The central NE system also plays a crucial role in determining the outcome of brain function in response to acute and chronic stress. Many neurochemical studies, in fact, have shown that NE is able to produce a stress response resulting from activation of the hypothalamus-pituitary-adrenal axis (Morilak et al., 1987; Valentino et al., 1993; Pacak et al., 1995; Cecchi et al., 2002).

In the past 20 years, extensive studies have been conducted to elucidate the role of LC NE during complex and specific behavioral performances and stress. Although the global effect of NE activation seems to lead to overall increases in neural responsiveness, alertness, and a temporary refinement of perceptual receptive fields, little is known about how NE receptors interact with other neural systems, in particular in brain regions involved in executive functions and cognition. In this chapter we will first review data from animal studies reporting the effects of NE on cortical neurons, and, secondly, we will summarize how NE-single cell interactions impact cortical functions by changing the behavior of cortical circuits.

13.2 Cortical Norepinephrine Receptors

Both NE and DA are catecholamines synthesized from the aromatic amino acid tyrosine. The synthesis of NE requires three steps: tyrosine is first transformed in L-DOPA by the enzyme tyrosine hydroxylase (TH). L-DOPA is then decarboxylated by aromatic 1-amino acid decarboxylase producing dopamine (DA). LC neurons are identified by their content of the enzyme dopamine-β-hydroxylase, which converts DA in NE.

NE exerts its modulatory action by binding to membrane receptors belonging to the seven transmembrane domains-, G-protein-coupled receptors superfamily. Nine distinct adrenergic receptors have been characterized and subdivided into three types of receptors (α_1, α_{12} and β). The three subtypes of α_1 receptors α_{1A}, α_{1B} and α_{1D}, activate a phospholipase C (PLC) cascade leading to the synthesis of phospholipids and the activation of protein kinase C (PKC). α_2 type receptors comprise also three different subtypes: α_{2A}, α_{2B} and α_{2C}. Activation of α_2 adrenoceptors inhibits adenylyl cyclase activation decreasing the levels of cAMP (see Table 13.1). On the contrary, activation of any of the three different types of β-adrenoceptors (β_1, β_2, and β_3, increases the synthesis of cAMP, increasing PKA activation. The results are summarized and reported in Table 13.1. Because of the similarity of the chemical structure of DA and NE, adrenoceptors possess a high affinity also for DA. Several studies reported slightly different values of the affinity constant of NE for adrenergic receptors (Table 13.1) (Hieble and Ruffolo, 1996; Rey et al., 2001; Cornil et al., 2002; Swaminath et al., 2004; Ramos and Arnsten, 2007).

Table 13.1 Properties of Adrenoceptors.

Receptor family	α_1	α_2	β
Affinity of NE for NE receptors	(EC_{50}) 330 nM (Ramos and Arnsten, 2007); (E_{C50}) 1.7 μM (Hieble and Ruffolo, 1996)	(EC_{50}) 56 nM (Ramos and Arnsten, 2007); (E_{C50}) 9.3 μM (Cornil et al., 2002)	(EC_{50}) 740 nM (Ramos and Arnsten, 2007); (E_{C50}) 300 nM (Swaminath et al., 2004)
Affinity of DA for NE receptors	(EC_{50}) 5 μM (Rey et al., 2001)	(EC_{50}) 40 μM (Cornil et al., 2002)	(EC_{50}) 10 μM (Swaminath et al., 2004)
Agonists	Phenylephrine	Clonidine	Isoproterenol, Isoprenaline
Antagonists	Prazosin	Yohimbine	Propanolol
G-protein coupled	Gq	Gi/o	Gs
Actions	(+) PLC	(+) Adenylyl cyclase	(−) Adenylyl cyclase

Affinity for NE and DA, Prototypical Agonists and Antagonists, G-protein Coupling and Intracellular Effects. (+) Stimulation, (−) Inhibition.

13.2.1 MEMBRANE EFFECTS OF NE

Application of NE can induce usually modest changes in membrane resting potential. Pralong and Magistretti (Pralong and Magistretti, 1994) found that in the entorhinal cortex, a brain area richly innervated by noradrenergic fibers integrating high-level cortical inputs with limbic signals, more than half of the neurons (presumably pyramidal) recorded from LII displayed a hyperpolarizing response to 50 μM NE, while a smaller fraction (~10%) of neurons showed hyperpolarization. On the contrary, most types of GABAergic frontal cortex interneurons respond to NE with a depolarization, sometimes preceded by a short-lasting hyperpolarization (Kawaguchi and Shindou, 1998).

Arguably, the single most important cellular effect of NE -shared with several other neurotransmitters- is the decrease of the apamine-sensitive, voltage-independent, slow component of the Ca^{2+}-independent after-hyperpolarization current (sAHP). The effect, discovered first in the hippocampus (Charpak et al., 1990) and later confirmed in voltage-clamp recordings from layer V pyramidal neurons in the sensory-motor rat cortex (Lorenzon and Foehring, 1992, 1993), is mediated by β-receptors (reviewed in McCormick, 1992). The block of the sAHP is most remarkable because it induces a strong increase in single cell excitability measurable in current clamp recordings by the increase in firing frequency in response to a depolarizing current pulse. The block of the sAHP accounts possibly for most of the results from current-clamp studies showing an increase in firing frequency following NE application (Waterhouse et al., 1988; Dodt et al., 1991; Mouradian et al., 1991; Law-Tho et al., 1993; Lorenzon and Foehring, 1993; Devilbiss and Waterhouse, 2000) (Table 13.2).

Table 13.2 Differential Effects of NE on Pyramidal Neurons Firing at Different Cortical Regions and Layers. NER: NE Receptors.

Cortical region	Changes in firing frequency	NER	Remarks	Reference
Prefrontal cortex	↑	β	No change in V_h LII/III	Dodt et al. (1991)
	↑	β	LV	Law-Tho et al. (1993)
	↑ (heterogeneity, age dependence?)	α	Non-fast-spiking GABA interneurons; LV, PN 18–22 mostly ↑ in V_h	Kawaguchi and Shindou (1998)
Prelimbic area, layers V/VI	↑	$α_2$, β	Synaptic cause (?) no change in V_h	Andrews and Lavin (2006)
Sensory-motor cortex	↑	β	PN < 22	Lorenzon and Foehring (1993)
Somatosensory cortex	↑	$α_1$	Ach-induced firing L IV–V	Mouradian et al. (1991)
	(↓) ↑	$α_1$	Glutamate evoke from LII/III, LV biphasic in LII/III	Devilbiss and Waterhouse (2000)
	↑	?		Waterhouse et al. (1988)

PN: Postnatal; ↓ Decrease; ↑ Increase.

The NE-induced increase in pyramidal cell excitability following methylphenidate applications in the prefrontal cortex can also be explained with the block of the SHP current (Andrews and Lavin, 2006). In the latter study α_2 -but also β antagonists- effectively blocked methylphenidate-induced increase in firing frequency following a depolarizing square pulse of current injection. It is not clear whether the methyl- phenidate effect would be mediated by direct activation of adrenergic receptors or rather by block of the NE reuptaker. A similar mechanism has not been shown for GABAergic interneurons, although the tetrodotoxin (TTX) -dependent increase in GABAergic spontaneous synaptic currents displayed in at least two cortical studies (Gellman and Aghajanian, 1993; Kawaguchi and Shindou, 1998) is suggestive of a similar mechanism, demonstrated in non cortical cells (Aghajanian, 1985; McCormick and Wang, 1991).

Although NE is probably not affecting directly the sAHP by modulating Ca^{2+} channels, NE also directly blocks both N- and P/Q-type Ca^{2+} channels, by activation of α_2 receptors (Timmons et al., 2004). NE can also affect neuronal excitability by increasing other two types of voltage-gated conductances: the cationic, hyperpolari- zation-activated current I_h, (Pralong and Magistretti, 1994) and the persistent Na^+ current (Foehring et al., 1989). Data are summarized in Table 13.3.

Table 13.3 Neuronal Effects of NE on Voltage- and Ca^{2+}-Gated Conductances.

NE receptors	Brain region	Current	References
α_2	Sensorimotor cortex	$\downarrow$ N- and P/Q-type Ca^{2+} channels	Timmons et al. (2004)
α_2	Entorhinal cortex	$\uparrow I_h$	Pralong and Magistretti (1994)
β	Sensorimotor cortex	$\downarrow$ sAHP	Lorenzon and Foehring (1993); Foehring et al. (1989)
β	Sensorimotor cortex	$\downarrow$ mAHP	Lorenzon and Foehring (1992)
?	Sensorimotor cortex	$\uparrow$ persistent $I(Na^+)$	Foehring et al. (1989)

I: Current; AHP: After-hyperpolarization; $\downarrow$ Decrease; $\uparrow$ Increase.

13.2.2 EFFECTS OF NE ON SYNAPTIC TRANSMISSION

Studies investigating the effects of NE on synaptic transmission mediated by iono- tropic ligand-gated channels are remarkably consistent in showing a decrease in glutamate-evoked currents and an increase in GABAergic responses. The two ef- fects, separately or in combination, induce a decrease in global excitability that may explain why LC lesions can be epileptogenic, while LC stimulation has antiepileptic consequences (Altman and Corcoran, 1983; Neuman, 1986; Stanton et al., 1992).

NE Block of AMPAR- and NMDAR-Mediated Currents

While several studies converge in concluding that ionotropic glutamatergic currents are blocked postsynaptically by NE, the extent and the pharmacological properties of the NE-induced block largely vary among different studies. For instance, an early study on LV neurons of the prefrontal cortex reported that NE application decreases both alpha-amino-3-hydroxy-5-methyl-4-isoxazolepropionic acid receptor (AMPAR) and *N*-methyl-D-aspartic acid receptor (NMDAR)-mediated currents by more than 70% (Law-Tho et al., 1993), while the activation of α_2Rs decreases AMPAR-mediated currents in the entorhinal cortex by 40% (Pralong and Magistretti, 1994).AMPAR-mediated currents are blocked by only ~20% in LII/III GABAergic interneurons of the visual cortex (Kobayashi et al., 2000). The latter study also reports that NE applications reduce NMDAR-mediated current by approximately 60% through activation of α_1 receptors (Kobayashi et al., 2000). N-methyl-D-aspartate-receptor (NMDAR)-mediated current is also blocked by 40% by the activation of α_1 and/or α_2 receptors in the prefrontal cortex (Liu et al., 2006). NE decreases also glutamatergic responses from the olfactory cortex (Hasselmo et al., 1997), by differentially affecting transmission in different tiers of layer I. All previous studies agree in a postsynaptic origin for NE-induced decrease of both the AMPAR-mediated and the NMDAR-mediated synaptic currents.

NE Increases GABAergic Responses

Most if not all reports on the effect of NE on γ-amino butyric A receptor (GABA$_A$R)-mediated signaling show an increase in the size of synaptic currents, mediated mostly (but not exclusively) by α_1 receptors (Table 13.4).

Table 13.4 Effect of NE on Synaptic Transmission in Different Cortical Regions, their Possible Synaptic Locus and NE Receptor (NER) Subtypes Involved.

Receptors	Cortical area	Locus	NER	Remarks	References
NMDAR	Prefrontal	↓ Pos	α_1/α_2	↓32%	Liu et al. (2006)
	Visual	↓ Pos	α_1	↓43%	Kobayashi et al. (2000)
AMPAR	Visual	↓ Pos	α_1	↓82%	Kobayashi et al. (2000)
	Entorhinal	↓ Pos	α_2	~PN30	Pralong and Magistretti (1994)
	Prefrontal	↓ Pos?	α_1	>70%	Law-Tho et al. (1993)
	Olfactory (piriform)	↓ Pos	?	differential effect within layers Ia versus. Ib	Hasselmo et al. (1997)
	Somatosensory	↓ Pos?	α	LII/III	Dodt et al. (1991)
GABA$_A$R	Visual	↑ ?	α_1	PN 13-35	Kobayashi et al. (2000)
	Sensorimotor	↑ Pre	α_1	developmentally regulated	Bennett et al. (1998)
	Frontal	↑ Pre	α_1	P18-22	Kawaguchi and Shindou (1998)
	Somatosensory	↑ Pre?	β		Waterhouse et al. (1980)

Pre: Presynaptic; Pos: Postsynaptic; ↓ Decrease; ↑ Increase.

Kobayashi and co-authors (Kobayashi et al. 2000), report that activation of α_1 receptors increases GABA$_A$R-mediated currents, similar to the results in the study by Bennett and coworkers (Bennett et al., 1998), reporting a presynaptically-, α_1-mediated increase of IPSCs amplitude in the sensory-motor cortex of the rat.

Waterhouse and colleagues propose a postsynaptic origin of the increase of GABA release induced by the activation of β receptors in the somatosensory cortex of the rat (Waterhouse et al., 1980).

Kawaguchi and Shindou (1998) show that both amplitude and frequency of GABA$_A$R-mediated inhibitory postsynaptic currents are greatly increased after application of 50 µM NE in a tetrodotoxin-sensitive manner. This very detailed study continues by systematically exploring the effect of NE on a variety of GABAergic interneuronal populations, demonstrating that in spite of large functional and immunohistochemical differences, most types of GABAergic interneurons respond to NE or to the α agonist FNA with a postsynaptic depolarization. Table 13.4 summarizes the effects of NE on synaptic transmission.

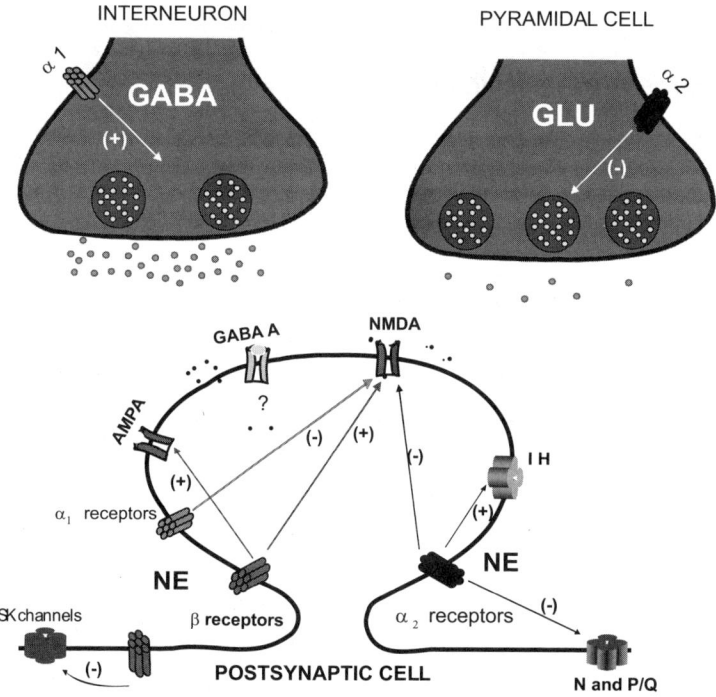

Figure 13.1 Alpha receptors in presynaptic terminals modulate synaptic release. Alpha receptors in postsynaptic terminals modulate multiple targets including SK channels responsible for the afterhyperpolarization current, different types of Ca^{2+} channels, I$_h$, AMPA, NMDA, and GABA$_A$ receptors. Plus and minus signs alongside the arrows indicate functional increases and decreases, respectively. Arrows only represent the final target of modulation, which is mediated by activation of G-proteins (Table 13.1).

In summary, three main effects stand out from our review on the cellular effects of NE: (i) NE induces a generalized increase in excitability due to a block of the mAHP and sAHP current; (ii) NE produces a mild but global enhancement of inhibition by simultaneously hyperpolarizing excitatory neurons and depolarizing GABAergic interneurons; (iii) NE produces a net synaptic enhancement of inhibition by adding postsynaptic inhibition of glutamatergic ionotropic currents to enhancement of GABAergic synaptic currents. Figure 13.1 illustrates a few representative pathway associated with cellular functions of NE.

13.3 Noradrenergic Regulation of Cortical Information Processing

It is well known that the central NE system is critical in general arousal and environmental responsiveness as revealed by changes in LC neuronal firing in association with salient stimuli, in particular those that elicit behavioral responses (Foote et al., 1980; Aston-Jones and Bloom, 1981a, b). In addition to the complexity of the effects of NE on different cell types across cortical circuits (see above), one of the distinctive action of NE is to increase the ratio of excitatory or inhibitory synaptically evoked activity, while decreasing spontaneous activity in the same neuron (Waterhouse and Woodward, 1980; Waterhouse et al., 1980, 1984). Although this modulatory action depends on the different NE receptor subtypes involved, it has been proposed that the major role of the central NE system is to provide appropriate gain control of sensory network information (Foote et al., 1983; Hurley et al., 2004; Aston-Jones and Cohen 2005a, b). This could reflect the ability of NE to interact with other monoamines (i.e., serotonin, dopamine) to regulate specific receptive field properties in individual cells by altering the representation of sensory information in ensemble of sensory neurons in the neocortex (Devilbiss and Waterhouse, 2000, 2004; Hurley et al., 2004).

Recent electrophysiological findings also indicate that the cortical NE system may be implicated in the regulation of cognitive performance (for review see Aston-Jones and Cohen, 2005a, b; Chamberlain et al., 2006). Among the many cortical regions that receive NE and dopamine innervation, the prefrontal cortex stands out as one of the main cortical areas critically involved in a number of cognitive functions, such as working memory, reward and attention (Aston-Jones and Cohen, 2005a, b), particularly in its double role of activator and recipient of modulation of brainstem nuclei, with which it shares a great deal of reciprocal anatomical connections. Although the cellular and synaptic mechanisms involved in these modulations remain controversial and often contradictory, the relationship between working memory performance and prefrontal cortical NE and DA levels are often nonlinear (Lidow et al., 1998; Devilbiss and Waterhouse, 2000, 2004; Aston-Jones and Cohen, 2005a, b). In fact, moderate levels of NE strengthen working memory and increase delay-related neuronal activity through a postsynaptic action on α_2 receptors, while high levels of NE, released during stress, impair working memory and reduce delay-related firing via activation of α_1 receptors (Arnsten et al., 1988; Arnsten and Li, 2005; Birnbaum et al., 2004). Similarly, the effect of DA (D1 receptors) on working memory and delay-related activity showed an inverted U dose/response relationship (Lidow et al., 1998). Furthermore, multiple experimental evidence indicate that disturbances of both systems are implicated in several cognitive deficits observed in many psychiatry disorders including schizophrenia, depression as well as attention deficits disorders

(Goldman-Rakic et al., 2004; Castner and Goldman-Rakic, 2004; Castner et al., 2004).

It is interesting to compare the cortical effects of NE with those of the acetylcholine, whose fibers also display a high density throughout the cortical mantle, and whose release is also triggered by similarly salient stimuli (Sarter et al., 2005). Acetylcholine and NE share two important avenues of cellular effects: they both enhance sAHP currents and decrease glutamatergic signaling (acetylcholine with presynaptic muscarinic receptors, and NE by postsynaptic depression of glutamatergic ionotropic channels). On the other hand, the two neuromodulators have opposite effects on GABAergic synaptic signals: while acetylcholine decreases GABAergic IPSCs (Hasselmo and Bower, 1992; Fukudome et al., 2004), NE increases GABAergic synaptic signaling (see Table 13.3). One possibility is that acetylcholine and NE are invested of complementary roles such that the combined effect of NE suppresses horizontal to the advantage of vertical (intracolumnar) signal propagation, as reported in the visual cortex (Kobayashi et al., 2000), while acetylcholine would inhibit intracolumnar and promoting laminar signal transmission by activating nicotinic receptors and muscarinic receptors on different types of GABAergic interneurons (Xiang et al., 1998).

The involvement of NE in developmental regulation is confirmed by the discovery that NE β receptors are present in different classes of neocortical LI neurons at least during the first postnatal week (Schwartz et al., 1998). The simultaneous, concerted, activation of a main type of layer I neurons (the non-Cajal-Retzius cells) following NE application could indicate a neurotrophic role of NE in early perinatal life (Schwartz et al., 1998).

The data reviewed here underscore the importance of gaining a more thorough knowledge of cortical interneuronal responses as well as on the interplay among NE and the other modulatory neurotransmitters for understanding the nature of cortical working modes and how the brain switches from one cortical operation to the other.

Acknowledgments

This work was in part supported by grant NIDCD 1R01-DC005986-01A1 and NARSAD foundation/Sidney Baer Trust (M.A.).

References

Aghajanian, G. K., 1985. Modulation of a transient outward current in serotonergic neurones by alpha 1-adrenoceptors. Nature. 315, 501–503.

Altman, I. M. and Corcoran, M. E., 1983. Facilitation of neocortical kindling by depletion of forebrain noradrenaline. Brain Res. 270, 174–177.

Andrews, G. D. and Lavin, A., 2006. Methylphenidate increases cortical excitability via activation of alpha-2 noradrenergic receptors. Neuropsychopharmacology. 31, 594–601.

Arnsten, A. F., Cai, J. X. and Goldman-Rakic, P. S., 1988. The alpha-2 adrenergic agonist guanfacine improves memory in aged monkeys without sedative or hypotensive side effects: evidence for alpha-2 receptor subtypes. J Neurosci. 8, 4287–4298.

Arnsten, A. F. and Li, B. M., 2005. Neurobiology of executive functions: catecholamine influences on prefrontal cortical functions. Biol Psychiatry. 57, 1377–1384.

Aston-Jones, G. and Bloom, F. E., 1981a. Activity of norepinephrine-containing locus coeruleus neurons in behaving rats anticipates fluctuations in the sleep-waking cycle. J Neurosci. 1, 876–886.

Aston-Jones, G. and Bloom, F. E., 1981b. Nonrepinephrine-containing locus coeruleus neurons in behaving rats exhibit pronounced responses to non-noxious environmental stimuli. J Neurosci. 1, 887–900.

Aston-Jones, G. and Cohen, J. D., 2005a. Adaptive gain and the role of the locus coeruleus-norepinephrine system in optimal performance. J Comp Neurol. 493, 99–110.

Aston-Jones, G. and Cohen, J. D., 2005b. An integrative theory of locus coeruleus-norepinephrine function: adaptive gain and optimal performance. Annu Rev Neurosci. 28, 403–450.

Bennett, B. D., Huguenard, J. R. and Prince, D. A., 1998. Adrenergic modulation of GABAA receptor-mediated inhibition in rat sensorimotor cortex. J Neurophysiol. 79, 937–946.

Birnbaum, S. G., Yuan, P. X., Wang, M., Vijayraghavan, S., Bloom, A. K., Davis, D. J., Gobeske, K. T., Sweatt, J. D., Manji, H. K. and Arnsten, A. F., 2004. Protein kinase C overactivity impairs prefrontal cortical regulation of working memory. Science. 306, 882–884.

Castner, S. A. and Goldman-Rakic, P. S., 2004. Enhancement of working memory in aged monkeys by a sensitizing regimen of dopamine D1 receptor stimulation. J Neurosci. 24, 1446–1450.

Castner, S. A., Goldman-Rakic, P. S. and Williams, G. V., 2004. Animal models of working memory: insights for targeting cognitive dysfunction in schizophrenia. Psychopharmacology (Berl). 174, 111–125.

Cecchi, M., Khoshbouei, H. and Morilak, D. A., 2002. Modulatory effects of norepinephrine, acting on alpha 1 receptors in the central nucleus of the amygdala, on behavioral and neuroendocrine responses to acute immobilization stress. Neuropharmacology. 43, 1139–1147.

Chamberlain, S. R., Muller, U., Blackwell, A. D., Robbins, T. W. and Sahakian, B. J., 2006. Noradrenergic modulation of working memory and emotional memory in humans. Psychopharmacology (Berl). 188, 397–407.

Charpak, S., Gahwiler, B. H., Do, K. Q. and Knopfel, T., 1990. Potassium conductances in hippocampal neurons blocked by excitatory amino-acid transmitters. Nature. 347, 765–767.

Cornil, C. A., Balthazart, J., Motte, P., Massotte, L. and Seutin, V., 2002. Dopamine activates noradrenergic receptors in the preoptic area. J Neurosci. 22, 9320–9330.

Devilbiss, D. M. and Waterhouse, B. D., 2000. Norepinephrine exhibits two distinct profiles of action on sensory cortical neuron responses to excitatory synaptic stimuli. Synapse. 37, 273–282.

Devilbiss, D. M. and Waterhouse, B. D., 2004. The effects of tonic locus ceruleus output on sensory-evoked responses of ventral posterior medial thalamic and barrel field cortical neurons in the awake rat. J Neurosci. 24, 10773–10785.

Dodt, H. U., Pawelzik, H. and Zieglgansberger, W., 1991. Actions of noradrenaline on neocortical neurons in vitro. Brain Res. 545, 307–311.

Foehring, R. C., Schwindt, P. C. and Crill, W. E., 1989. Norepinephrine selectively reduces slow Ca2+- and Na+-mediated K+ currents in cat neocortical neurons. J Neurophysiol. 61, 245–256.

Foote, S. L., Aston-Jones, G. and Bloom, F. E., 1980. Impulse activity of locus coeruleus neurons in awake rats and monkeys is a function of sensory stimulation and arousal. Proc Natl Acad Sci U S A. 77, 3033–3037.

Foote, S. L., Bloom, F. E. and Aston-Jones, G., 1983. Nucleus locus ceruleus: new evidence of anatomical and physiological specificity. Physiol Rev. 63, 844–914.

Fukudome, Y., Ohno-Shosaku, T., Matsui, M., Omori, Y., Fukaya, M., Tsubokawa, H., Taketo, M. M., Watanabe, M., Manabe, T. and Kano, M., 2004. Two distinct classes of mus-carinic action on hippocampal inhibitory synapses: M2-mediated direct suppression and M1/M3-mediated indirect suppression through endocannabinoid signalling. Eur J Neuro-sci. 19, 2682–2692.

Gellman, R. L. and Aghajanian, G. K., 1993. Pyramidal cells in piriform cortex receive a convergence of inputs from monoamine activated GABAergic interneurons. Brain Res. 600, 63–73.

Goldman-Rakic, P. S., Castner, S. A., Svensson, T. H., Siever, L. J. and Williams, G. V., 2004. Targeting the dopamine D1 receptor in schizophrenia: insights for cognitive dys-function. Psychopharmacology (Berl). 174, 3–16.

Hasselmo, M. E. and Bower, J. M., 1992. Cholinergic suppression specific to intrinsic not afferent fiber synapses in rat piriform (olfactory) cortex. J Neurophysiol. 67, 1222–1229.

Hasselmo, M. E., Linster, C., Patil, M., Ma, D. and Cekic, M., 1997. Noradrenergic suppres-sion of synaptic transmission may influence cortical signal-to-noise ratio. J Neurophysiol. 77, 3326–3339.

Hieble, J. P. and Ruffolo, Jr., R. R., 1996. Subclassification and nomenclature of alpha 1- and alpha 2-adrenoceptors. Prog Drug Res. 47, 81–130.

Hurley, L. M., Devilbiss, D. M. and Waterhouse, B. D., 2004. A matter of focus: mono-aminergic modulation of stimulus coding in mammalian sensory networks. Curr Opin Neurobiol. 14, 488–495.

Kawaguchi, Y. and Shindou, T., 1998. Noradrenergic excitation and inhibition of GABAergic cell types in rat frontal cortex. J Neurosci. 18, 6963–6976.

Kobayashi, M., Imamura, K., Sugai, T., Onoda, N., Yamamoto, M., Komai, S. and Watanabe, Y., 2000. Selective suppression of horizontal propagation in rat visual cortex by norepi-nephrine. Eur J Neurosci. 12, 264–272.

Law-Tho, D., Crepel, F. and Hirsch, J. C., 1993. Noradrenaline decreases transmission of NMDA- and non-NMDA-receptor mediated monosynaptic EPSPs in rat prefrontal neu-rons in vitro. Eur J Neurosci. 5, 1494–1500.

Lidow, M. S., Williams, G. V. and Goldman-Rakic, P. S., 1998. The cerebral cortex: a case for a common site of action of antipsychotics. Trends Pharmacol Sci. 19, 136–140.

Liu, W., Yuen, E. Y., Allen, P. B., Feng, J., Greengard, P. and Yan, Z., 2006. Adrenergic modulation of NMDA receptors in prefrontal cortex is differentially regulated by RGS proteins and spinophilin. Proc Natl Acad Sci U S A. 103, 18338–18343.

Lorenzon, N. M. and Foehring, R. C., 1992. Relationship between repetitive firing and after-hyperpolarizations in human neocortical neurons. J Neurophysiol. 67, 350–363.

Lorenzon, N. M. and Foehring, R. C., 1993. The ontogeny of repetitive firing and its modula-tion by norepinephrine in rat neocortical neurons. Brain Res Dev Brain Res. 73, 213–223.

McCormick, D. A., 1992. Neurotransmitter actions in the thalamus and cerebral cortex. J Clin Neurophysiol. 9, 212–223.

McCormick, D. A. and Wang, Z., 1991. Serotonin and noradrenaline excite GABAergic neu-rones of the guinea-pig and cat nucleus reticularis thalami. J Physiol. 442, 235–255.

Morilak, D. A., Fornal, C. A. and Jacobs, B. L., 1987. Effects of physiological manipulations on locus coeruleus neuronal activity in freely moving cats. III. Glucoregulatory challenge. Brain Res. 422, 32–39.

Mouradian, R. D., Sessler, F. M. and Waterhouse, B. D., 1991. Noradrenergic potentiation of excitatory transmitter action in cerebrocortical slices: evidence for mediation by an alpha 1 receptor-linked second messenger pathway. Brain Res. 546, 83–95.

Neuman, R. S., 1986. Suppression of penicillin-induced focal epileptiform activity by locus ceruleus stimulation: mediation by an alpha 1-adrenoceptor. Epilepsia. 27, 359–366.

Pacak, K., Palkovits, M., Kopin, I. J. and Goldstein, D. S., 1995. Stress-induced norepineph-rine release in the hypothalamic paraventricular nucleus and pituitary-adrenocortical and sympathoadrenal activity: in vivo microdialysis studies. Front Neuroendocrinol. 16, 89–150.

Pralong, E. and Magistretti, P. J., 1994. Noradrenaline reduces synaptic responses in normal and tottering mouse entorhinal cortex via alpha 2 receptors. Neurosci Lett. 179, 145–148.

Ramos, B. P. and Arnsten, A. F., 2007. Adrenergic pharmacology and cognition: focus on the prefrontal cortex. Pharmacol Ther. 2007, 113(3):523–36.

Rey E, Hernandez-iaz FJ, Abreu P, Alonso R, Tabares L. (2001) Dopamine induces intracellular Ca^{2+} signals mediated by alpha 1B- adrenoceptors in rat pineal cells. Eur J Pharmacol 26; 430(1):9–17.

Sarter, M., Hasselmo, M. E., Bruno, J. P. and Givens, B., 2005. Unraveling the attentional functions of cortical cholinergic inputs: interactions between signal-driven and cognitive modulation of signal detection. Brain Res Brain Res Rev. 48, 98–111.

Schwartz, T. H., Rabinowitz, D., Unni, V., Kumar, V. S., Smetters, D. K., Tsiola, A. and Yuste, R., 1998. Networks of coactive neurons in developing layer 1. Neuron. 20, 541–552.

Stanton, P. K., Mody, I., Zigmond, D., Sejnowski, T. and Heinemann, U., 1992. Noradrenergic modulation of excitability in acute and chronic model epilepsies. Epilepsy Res Suppl. 8, 321–334.

Swaminath, G., Xiang, Y., Lee, T. W., Steenhuis, J., Parnot, C. and Kobilka, B. K., 2004. Sequential binding of agonists to the beta2 adrenoceptor. Kinetic evidence for intermediate conformational states. J Biol Chem. 279, 686–691.

Timmons, S. D., Geisert, E., Stewart, A. E., Lorenzon, N. M. and Foehring, R. C., 2004. alpha2-Adrenergic receptor-mediated modulation of calcium current in neocortical pyramidal neurons. Brain Res. 1014, 184–196.

Valentino, R. J., Foote, S. L. and Page, M. E., 1993. The locus coeruleus as a site for integrating corticotropin-releasing factor and noradrenergic mediation of stress responses. Ann N Y Acad Sci. 697, 173–188.

Waterhouse, B. D. and Woodward, D. J., 1980. Interaction of norepinephrine with cerebrocortical activity evoked by stimulation of somatosensory afferent pathways in the rat. Exp Neurol. 67, 11–34.

Waterhouse, B. D., Moises, H. C. and Woodward, D. J., 1980. Noradrenergic modulation of somatosensory cortical neuronal responses to iontophoretically applied putative neurotransmitters. Exp Neurol. 69, 30-49.

Waterhouse, B. D., Moises, H. C., Yeh, H. H., Geller, H. M. and Woodward, D. J., 1984. Comparison of norepinephrine- and benzodiazepine-induced augmentation of Purkinje cell responses to gamma-aminobutyric acid (GABA). J Pharmacol Exp Ther. 228, 257-267.

Waterhouse, B. D., Sessler, F. M., Cheng, J. T., Woodward, D. J., Azizi, S. A. and Moises, H. C., 1988. New evidence for a gating action of norepinephrine in central neuronal circuits of mammalian brain. Brain Res Bull. 21, 425-432.

Xiang, Z., Huguenard, J. R. and Prince, D. A., 1998. Cholinergic switching within neocortical inhibitory networks. Science. 281, 985-988.

Chapter 14

Neuromodulation of Cortical Synaptic Plasticity

Alfredo Kirkwood
Mind Brain Institute, Johns Hopkins University, Baltimore, MD 21218, USA

14.1 Introduction

Sensory experience can shape the synaptic connectivity of the sensory cortices during its maturation in infants, and during learning in adults. Besides sensory inputs, cortical plasticity also depends on neuromodulatory inputs conveying information on the individual's behavioral state. As a result, experience-dependent cortical plasticity requires the individual to be awake and attentive. In contrast, passive experience usually does not leave a permanent trace on cortical connectivity. The traditional view on the neuromodulatory control of cortical plasticity has emphasized their role in increasing neural excitability, improving the signal to noise ratio, and their role in controlling the propagation of activity through the cortex. This Chapter will review the notion that neuromodulators control sensory-induced plasticity at a synaptic level by setting the gain of mechanisms like long-term potentiation (LTP) and long-term depression (LTD). We will discuss how intracellular signal generated by different neuromodulator receptors can interact with the biochemical machinery responsible for LTP and LTD. Finally, we will discuss the emerging notion that different neuromodulator receptors can specifically control LTP and LTD by regulating the trafficking of glutamate AMPA receptors in and out of the synapse.

14.2 Neuromodulators as Permissive Factors in Cortical Plasticity

Experience-induced cortical plasticity was firmly established by Hubel and Wiesel (Wiesel and Hubel, 1963), who showed that brief monocular deprivation (MD) at early age, shifts the visual response in visual cortical cells toward the non-deprived eye. Since then, comparable forms of developmental plasticity induced by altered sensory experience have also been demonstrated in the somatosensory (Fox, 1992) and auditory cortices (Zhang et al., 2002). The role of neuromodulatory inputs in cortical plasticity was recognized early on in visual cortex, where manipulations that disrupt adrenergic function prevent the effects of monocular deprivation (Kasamatsu and Pettigrew, 1979a; Gordon et al., 1988). These findings were subsequently extended to other neuromodulatory systems like acetylcholine (Bear and Singer, 1986; Gu and Singer, 1993) serotonin, (Gu and Singer, 1995), and also to plasticity in other sensory cortices (Osterheld-Haas et al., 1994; Zhu and Waite, 1998). On the other hand, studies in visual somatosensory and auditory cortices indicate that the reduced plasticity in adult cortex can be enhanced by activation of noradrenergic, cholinergic, and dopaminergic centers that project to the cortex (Kasamatsu et al., 1985; Kilgard and Merzenich 1998; Shulz et al., 2000; Ego-Stengel et al., 2001). The general implication of these results is that neuromodulators act as enabling factors that permit or gate experience-dependent plasticity. It has also been proposed that reduction in the supply of neuromodulators with age might account for the reduced capacity for plasticity in adults (Kasamatsu and Pettigrew, 1979b).

It is now well accepted that the remodeling of cortical connectivity by sensory experience requires activity-dependent mechanisms that allow strengthening and weakening of synaptic connections. Although experiments conducted in vivo underscored the role of neuromodulators in plasticity, this approach has provided limited insight on the underlying cellular mechanism due to the difficulty of manipulating neural activity and the levels of neuromodulators in a controlled manner. Therefore, we will focus on reviewing experiments conducted in vitro, in the slice preparation that allow detailed analysis of the cellular mechanisms of neuromodulation.

14.3 Neuromodulatory Control of Synaptic Plasticity

Currently, the most comprehensive models of bidirectional synaptic modification are NMDA receptor dependent forms of LTP and LTD. Before reviewing the ample literature documenting effects of neuromodulators in the induction of LTP and LTD, it is worth recapitulating the essentials of the biochemical mechanisms of LTP and LTP. According to the prevailing view, the state of correlation between pre- and postsynaptic activity is converted by voltage-dependent NMDA receptor channels into a graded postsynaptic Ca^{2+} signal that triggers LTP when it exceeds some critical value (the "modification threshold") and triggers LTD when it falls below this level (Lisman, 1989; Cummings et al., 1996; Malenka and Bear, 2004). Ultimately, these different Ca^{2+} signals alter the balance of protein kinase/phosphatase activity and the phosphorylation state of synaptic proteins involved in LTP and LTD (Barria et al., 1997; Lee et al., 2000). Recent evidence suggests that the insertion and removal of AMPA receptors at the synapse are important downstream events for LTP and LTD, respectively (Malinow and Malenka, 2002).

Activation of neuromodulatory receptors might control plasticity at various levels. Because the NMDA receptor is voltage-dependent, neuromodulation of plasticity could occur indirectly by regulating the excitability of the postsynaptic neuron or by altering the properties of inhibition. In addition, the NMDA receptor itself is a phosphoprotein that is subject to regulation (Roche et al., 1994; Lee, 2006). Regulation of plasticity could also occurs downstream of NMDA receptor activation, for example via release of Ca^{2+} from intracellular stores or by altering the activity of the protein kinases and phosphatases that control synaptic efficacy. Finally, neuromodulators could also affect the phosphorylation state of AMPA that controls the trafficking of the receptors in and out of the synapse (Esteban et al., 2003; Lee et al., 2003; Boehm et al., 2006).

14.4 Neuromodulation of Cortical LTP

Neuromodulation of the induction of cortical LTP was first demonstrated in rat visual cortical slices, in the white matter to layer II/III pathway. LTP in this pathway is lost in adults (Kato et al., 1991) due to a developmental increase in synaptic inhibition (Kirkwood et al., 1995) and is an attractive model to study critical periods of plasticity. The initial observation was that activation of beta adrenergic receptors, alone or in conjunction with muscarinic receptors, permits the induction of this form of LTP in adults (Brocher et al., 1992; Kato, 1993). Subsequently it was shown that LTP in this pathway is also facilitated by 5-HT$_{2C}$ serotonergic receptors in visual cortical slices from cats (Kojic et al., 1997, 2000, 2001). The finding that the activation of adrenergic, cholinergic, and serotonergic receptors can rescue LTP is in harmony with the idea that these neuromodulators play a permissive role in experience-dependent plasticity, and also with the notion that reduced neuromodulatory input contributes to the loss of plasticity after the critical period. However, more recently other investigators have reported a suppressive role for 5-HT$_{2C}$ and 5-HT1in rat visual cortical slices (Edagawa et al., 1998, 2000, 2001; Kim et al., 2006), and suggested that developmental increases in serotonin levels contribute to the loss of LTP.

The seemingly conflicting results regarding the role of serotonergic receptors might relate to the complexity of circuits studied, and the multiple actions of serotonin receptors. Layer II/III synaptic responses evoked by white matter stimulation have a substantial polysynaptic component reflecting the relay of activity via layer IV cells (Aizenman et al., 1996), which in turn is limited by the recruitment of feed-forward inhibition (Rozas et al., 2001). Thus, 5-HT$_{2C}$ can reduce LTP by enhancing the excitability GABAergic interneurons (Edagawa et al., 2001). On the other hand, 5-HT$_{2C}$ receptors might facilitate LTP by boosting the Ca^{2+} in layer II/III pyramidal cells via release of Ca^{2+} from intracellular stores (Kojic et al., 2001). In a similar fashion, multiple mechanisms could account for the β-noradrenergic and muscarinic modulation of LTP. Initially, the facilitation of LTP was attributed to the enhanced excitability of layer II/III cells (Brocher et al., 1992). However, M1 receptors can also reduce evoked GABA release by inhibitory interneurons (Valentino and Dingle-dine, 1981; Behrends and ten Bruggencate, 1993; Murakoshi, 1995; Kimura and Baughman, 1997). Finally, besides increasing excitability, β-adrenergic receptor activation might also inhibit protein phosphatase 1 (Thomas et al., 1996), which has been implicated in the mechanism for LTD induction (Bear and Malenka, 1994).

The regulation of LTP by dopamine, the other major neuromodulator, has been examined in vitro in the layer III inputs to layer V cells in slices from rat prefrontal cortex. Although in vivo LTP in this pathway can be readily induced by tetanic stimulation (Gurden et al., 2000), in vitro LTP induction requires prior activation of dopamine receptors (Blond et al., 2002). Interestingly, LTP could not be induced during application of dopamine, but only after the removal of the neuromodulator (Blond et al., 2002; Otani et al., 2003). The mechanisms underlying this peculiar dopaminergic priming of LTP remain unknown. However, available evidence suggests a plausible scenario. Activation of dopamine receptors results in an acute depression of synaptic transmission that could transiently impair the induction of LTP (Otani et al., 2003). On the other hand, studies on cultured cells indicate that activation of D1 receptors, which stimulates cAMP production, produces a lasting increase in the surface expression of GluR1 containing AMPA receptors (Sun et al., 2005). These newly surfaced AMPA receptors could then be subsequently incorporated into synapses upon appropriate activation of NMDA receptors during LTP induction (Derkach et al., 2007). Our unpublished observations indicate (Seol et al., submitted) that β-adrenergic receptors, which stimulate cAMP production, produce a lasting increases the phosphorylation of the GluR1 subunit at s845, a phosphorylation sites necessary for the surface expression of AMPA receptors and LTP. Interestingly, similar actions of β-adrenergic agonists have been described in the hippocampus, suggesting that regulating phosphorylation of AMPA receptors is a common mechanism of neuromodulation.

14.5 Neuromodulation of Cortical LTD

The role of neuromodulators in cortical LTD is relatively well understood. Indeed recent studies suggest that neuromodulators do not facilitate LTD but rather are an obligatory step in the induction of this form of plasticity. The first indication of an essential role came from the observation that norepinephrine and cholinergic agonists reduce the activity requirement for the induction of LTD on the layer IV inputs to layer II/II by at least an order of magnitude (Kirkwood et al., 1999). In the presence of norepinephrine or acetylcholine LTD can be induced with 40 stimulation pulses, whereas under standard experimental conditions LTD induction typically requires 900 pulses. This dramatic facilitation of LTD by acetylcholine is mediated by muscarinic receptors of the M1 subtype, and the effect of norepinephrine is mediated by α1 adrenoreceptors. Both M1 and α1 receptors are coupled to the phospholipase C (PLC) signaling pathway and stimulates phosphoinositide turnover, suggesting that this could be a final common biochemical pathway for the modulation of LTD. A similar cholinergic and adrenergic facilitation of LTD has been described in the hippocampus (Kirkwood et al., 1999; Scheiderer et al., 2004; Etkin et al., 2006) and the induction of LTD in prefrontal cortex depends on the activation of PLC-coupled dopamine receptors (Otani et al., 1998, 2002). Thus, PLC-coupled receptors appear to play a common role in LTD across different synapses.

Further studies revealed that the induction of LTD in visual cortex requires the activation of the PLC pathway along with NMDAR's (Choi et al., 2005). Importantly, multiple receptors coupled to PLC, including adrenergic, serotonergic, muscarinic, and metabotropic glutamate receptors can independently support the induction of LTP. This functional redundancy might account for the observation that different neuromodulatory systems can substitute each other in supporting naturally occurring experience-dependent plasticity. Another important point is that synaptic modifications

(LTD) will occur only when NMDA receptors are activated in conjunction with activation of PLC-linked neuromodulators. In this context, the activation of NMDA receptors will convey information about specific external inputs, while neuromodulators could provide information about the behavioral state of the organism, reflecting such variables as attention and arousal.

14.6 Summary and Conclusions

Studies conducted in vitro in slice preparations indicate that neuromodulators act as permissive factors for the induction of NMDAR dependent LTP and LTD. The exact mechanisms of neuromodulation on plasticity are not fully understood and multiple explanations have been proposed, ranging from changes in the propagation of neural activity through cortical circuits to changes in the intracellular machinery responsible for LTP and LTD. The proliferation of proposed mechanisms might have resulted from the complexity of circuits studied and also from the use of the natural agonists, which can activate multiple types of receptors. However, insightful common themes have emerged from studying select receptors on simpler monosynaptic circuits. An emerging pattern is that neuromodulation of receptors that stimulate cAMP production promote LTP, while receptors coupled to PLC promote LTD. Importantly, these selective effects on LTD and LTP occur downstream from the activation of NMDAR, likely at the level of AMPA receptor phosphorylation and trafficking. These findings call for a revision of our views on synaptic plasticity to incorporate the notion that neuromodulators and the patterns of NMDAR activation during conditioning stimulation co-determine the polarity and magnitude of synaptic modification.

Acknowledgments

We thank Dr HK Lee for critically reviewing the manuscript.

References

Aizenman C, Kirkwood A, Bear M (1996) Current source density analysis of evoked responses in visual cortex in vitro: implications for the regulation of long-term potentiation. Cereb Cortex 6:751–758.

Barria A, Muller D, Derkach V, Griffith LC, Soderling TR (1997) Regulatory phosphorylation of AMPA-type glutamate receptors by CaM-KII during long-term potentiation. Science 276:2042–2045.

Bear MF, Singer W (1986) Modulation of visual cortical plasticity by acetylcholine and noradrenaline. Nature 320:172–176.

Bear MF, Malenka RC (1994) Synaptic plasticity: LTP and LTD. Curr Opin Neurobiol 4: 389–399.

Behrends JC, ten Bruggencate G (1993) Cholinergic modulation of synaptic inhibition in the guinea pig hippocampus in vitro: excitation of GABAergic interneurons and inhibition of GABA release. J Neurophysiol 69:626–629.

Blond O, Crepel F, Otani S (2002) Long-term potentiation in rat prefrontal slices facilitated by phased application of dopamine. Eur J Pharmacol 438:115–116.

Boehm J, Kang MG, Johnson RC, Esteban J, Huganir RL, Malinow R (2006) Synaptic incorporation of AMPA receptors during LTP is controlled by a PKC phosphorylation site on GluR1. Neuron 51:213–225.

Brocher S, Artola A, Singer W (1992) Agonists of cholinergic and noradrenergic receptors facilitate synergistically the induction of long-term potentiation in slices of rat visual cortex. Brain Res 573:27–36.

Choi S, J C, Jiang B, GH S, SS M, JS H, Shin HS, Gallagher M, Kirkwood A (2005) Multiple receptors coupled to PLC gate LTD in visual cortex. J Neurosci 25:11433–11443.

Cummings JA, Mulkey RM, Nicoll RA, Malenka RC (1996) Ca2+ signalling requirements for long-term depression in the hippocampus. Neuron 16:825–833.

Derkach VA, Oh MC, Guire ES, Soderling TR (2007) Regulatory mechanisms of AMPA receptors in synaptic plasticity. Nat Rev Neurosci 8:101–113.

Edagawa Y, Saito H, Abe K (1998) 5-HT1A receptor-mediated inhibition of long-term potentiation in rat visual cortex. Eur J Pharmacol 349:221–224.

Edagawa Y, Saito H, Abe K (2000) The serotonin 5-HT2 receptor-phospholipase C system inhibits the induction of long-term potentiation in the rat visual cortex. Eur J Neurosci 12:1391–1396.

Edagawa Y, Saito H, Abe K (2001) Endogenous serotonin contributes to a developmental decrease in long-term potentiation in the rat visual cortex. J Neurosci 21:1532–1537.

Ego-Stengel V, Shulz DE, Haidarliu S, Sosnik R, Ahissar E (2001) Acetylcholine-dependent induction and ixpression of functional plasticity in the barrel cortex of the adult rat. J Neurophysiol 86:422–437.

Esteban JA, Shi SH, Wilson C, Nuriya M, Huganir RL, Malinow R (2003) PKA phosphorylation of AMPA receptor subunits controls synaptic trafficking underlying plasticity. Nat Neurosci 6:136–143.

Etkin A, Alarcon JM, Weisberg SP, Touzani K, Huang YY, Nordheim A, Kandel ER (2006) A role in learning for SRF: deletion in the adult forebrain disrupts LTD and the formation of an immediate memory of a novel context. Neuron 50:127–143.

Fox K (1992) A critical period for experience-dependent synaptic plasticity in rat barrel cortex. J Neurosci 12:1826–1838.

Gordon B, Allen EE, Trombley PQ (1988) The role of norepinephrine in plasticity of visual cortex. Prog Neurobiol 30:171–191.

Gu Q, Singer W (1993) Effects of intracortical infusion of anticholinergic drugs on neuronal plasticity in kitten striate cortex. Eur J Neurosci, 5:475–485.

Gu Q, Singer W (1995) Involvement of serotonin in developmental plasticity in kitten visual cortex. Eur J Neurosci 7:1146–1153.

Gurden H, Takita M, Jay TM (2000) Essential role of D1 but not D2 receptors in the NMDA receptor-dependent long-term potentiation at hippocampal-prefrontal cortex synapses in vivo. J Neurosci 20:RC106.

Kasamatsu T, Pettigrew JD (1979a) Preservation of binocularity after monocular deprivation in the striate cortex of kittens treated with 6-OHDA. J Comp Neurol 185:139–162.

Kasamatsu T, Pettigrew JD (1979b) Restoration of visual cortical plasticity by local microperfusion of norepinephrine. J Comp Neurol 185:163–182.

Kasamatsu T, Watabe K, Heggelund P, Scholler E (1985) Plasticity in cat visual cortex restored by electrical stimulation of the locus coeruleus. Neurosci Res 2:365–386.

Kato N (1993) Mechanisms of beta-adrenergic facilitation in rat visual cortex. Neuroreport 4:1087–1090.

Kato N, Artola A, Singer W (1991) Developmental changes in the susceptibility to long-term potentiation of neurones in rat visual cortex slices. Dev Brain Res 60:43–50.

Kilgard MP, Merzenich MM (1998) Cortical map reorganization enabled by nucleus basalis activity. Science 279:1714–1718.

Kim HS, Jang HJ, Cho KH, Hahn SJ, Kim MJ, Yoon SH, Jo YH, Kim MS, Rhie DJ (2006) Serotonin inhibits the induction of NMDA receptor-dependent long-term potentiation in the rat primary visual cortex. Brain Res 1103:49–55.

Kimura F, Baughman RW (1997) Distinct muscarinic receptor subtypes suppress excitatory and inhibitory synaptic responses in cortical neurons. J Neurophysiol 77:709–716.

Kirkwood A, Lee H-K, Bear MF (1995) Co-regulation of long-term potentiation and experience-dependent plasticity in visual cortex by age and experience. Nature 375:328–331.

Kirkwood A, Rozas C, Kirkwood J, Perez F, Bear MF (1999) Modulation of long-term depression in visual cortex by acetylcholine and norepinephrine. J Neurosci 19:1599–1609.

Kojic L, Gu Q, Douglas. RM, Cynader MX (1997) Serotonin facilitates synaptic plasticity in kitten visual cortex: an in vitro study. Dev Brain Res 101:299–304.

Kojic L, Gu Q, Douglas RM, Cynader MS (2001) Laminar distribution of cholinergic- and serotonergic-dependent plasticity within kitten visual cortex. Dev Brain Res 126:157–162.

Kojic L, Dyck RH, Gu Q, Douglas. RM, Matsuraba J, Cynader MS (2000) Columnar distribution of serotonin-dependent plasticity within kitten striate cortex. Proc Natl Acad Sci U S A 97:1841–1844.

Lee HK (2006) Synaptic plasticity and phosphorylation. Pharmacol Ther 112:810–832.

Lee HK, Barbarosie M, Kameyama K, Bear MF, Huganir RL (2000) Regulation of distinct AMPA receptor phosphorylation sites during bidirectional synaptic plasticity. Nature 405:955–959.

Lee HK, Takamiya K, Han JS, Man H, Kim CH, Rumbaugh G, Yu S, Ding L, He C, Petralia RS, Wenthold RJ, Gallagher M, Huganir RL (2003) Phosphorylation of the AMPA receptor GluR1 subunit is required for synaptic plasticity and retention of spatial memory. Cell 112:631–643.

Lisman J (1989) A mechanism for the Hebb and the anti-Hebb processes underlying learning and memory. Proc Natl Acad Sci U S A 86:9574–9578.

Malenka RC, Bear MF (2004) LTP and LTD: an embarrassment of riches. Neuron 44:5–21.

Malinow R, Malenka RC (2002) AMPA receptor trafficking and synaptic plasticity. Annu Rev Neurosci 25.103–126.

Murakoshi T (1995) Cholinergic modulation of synaptic transmission in the rat visual cortex in vitro. Vision Res 35:25–35.

Osterheld-Haas MC, Van der Loos H, Hornung JP (1994) Monoaminergic afferents to cortex modulate structural plasticity in the barrelfield of the mouse. Dev Brain Res 77:189–202.

Otani S, Blond O, Desce JM, Crepel F (1998) Dopamine facilitates long-term depression of glutamatergic transmission in rat prefrontal cortex. Neuroscience 85:669–676.

Otani S, Daniel H, Takita M, Crepel F (2002) Long-term depression induced by postsynaptic group II metabotropic glutamate receptors linked to phospholipase C and intracellular calcium rises in rat prefrontal cortex. J Neurosci 22:3434–3444.

Otani S, Daniel H, Roisin MP, Crepel F (2003) Dopaminergic modulation of long-term synaptic plasticity in rat prefrontal neurons. Cereb Cortex 13:1251–1256.

Roche KW, Tingley WG, Huganir RL (1994) Glutamate receptor phosphorylation and synaptic plasticity. Curr Opin Neurobiol 4:383–388.

Rozas C, Frank H, J. HA, Morales B, Bear MF, Kirkwood A (2001) Developmental inhibitory gate controls the relay of activity to the superficial layers of the visual cortex. J Neuroscience 21:6791–6801.

Scheiderer CL, Dobrunz LE, McMahon LL (2004) Novel form of long-term synaptic depression in rat hippocampus induced by activation of alpha 1 adrenergic receptors. J Neurophysiol 91:1071–1077.

Shulz DE, Sosnik R, Ego V, Haidarliu S, Ahissar E (2000) A neuronal analogue of state-dependent learning. Nature 403:549–553.

Sun X, Zhao Y, Wolf ME (2005) Dopamine receptor stimulation modulates AMPA receptor synaptic insertion in prefrontal cortex neurons. J Neurosci 25:7342–7351.

Thomas MJ, Moody TD, Makinson M, O'Dell TJ (1996) Activity-dependent beta-adrenergic modulation of low frequency stimulation induced LTP in the hippocampal Ca1 region. Neuron 17:475–482.

Valentino RJ, Dingledine R (1981) Presynaptic inhibitory effect of acetylcholine in the hippocampus. J Neurosci 1:784–792.

Wiesel TN, Hubel DH (1963) Single cell responses in striate cortex of kittens deprived of vision in one eye. J Neurophysiol 26:1003–1017.

Zhang LI, Bao S, Merzenich MM (2002) Disruption of primary auditory cortex by synchronous auditory inputs during a critical period. Proc Natl Acad Sci U S A 99:2309–2314.

Zhu XO, Waite PM (1998) Cholinergic depletion reduces plasticity of barrel field cortex. Cereb Cortex 8:63–72.

Chapter 15

Dopaminergic Modulation of Prefrontal Cortex Network Dynamics

Daniel Durstewitz
Centre for Theoretical and Computational Neuroscience, Faculty of Science, Portland Square A220, Drake Circus, University of Plymouth, Plymouth, PL4 8AA, UK

15.1 Introduction

Our knowledge about how the neocortex fulfils its diverse computational duties is at present very rudimentary. As knowledge about biochemical, physiological, and anatomical processes and structures accumulates, we seem to be even more and more puzzled by the unveiling layers of mind-blowing complexity and dense networks of interaction between so many biophysical and biochemical variables. It seems clear that a full understanding of nervous system function cannot be gained solely by dissection of physiological processes into simple cause-effect chains, as these are blurred and lost within the multitude of highly non-linear reciprocal interactions and feedback loops governing the dynamics of the whole nervous system, yielding often completely counter-intuitive results. Although simple cause-effect chains are the building blocks of higher order dynamics and an in depth understanding of them therefore remains important, additional approaches are required that put these building blocks back together again, and study the emergent dynamics of the whole system with its implications for computation and cognition. Computational Neuroscience tries to provide some of these tools, and the present article will give an account of the dopaminergic modulation of prefrontal cortex (PFC) from this perspective.

15.2 What is the Prefrontal Cortex Doing?

The PFC is the area of neocortex that has most tightly been linked to 'higher level' cognitive processes such as planning, thinking, problem solving, and working memory (Milner and Petrides, 1984; Fuster, 1997; Dehaene et al., 1999; Miller and Cohen, 2001; Procyk and Goldman-Rakic, 2006). Surely many of these processes are not carried out by the PFC in isolation but by an interconnected network of brain areas within which the PFC seems to be a key player (Dagher et al., 1999; Duncan and Owen, 2000), organizing behaviour in time based on the recent history of sensory inputs and information retrieved from long-term memory (Fuster, 2001; Miller and Cohen 2001). Many of the cognitive processes and tasks the PFC is involved in have two key elements in common: (i) A temporal sequence of events, not just the current sensory input, has to be processed to guide behaviour, and (ii) a prediction about a future state of affairs can be derived based on that temporal sequence of events. These two properties are present in a nutshell in the simplest type of working-memory tasks, delayed-response (DR) tasks. For instance, in one such type of task, the delayed-non-matching to sample (D-N-MTS) task, an animal would first briefly see a cue stimulus which then disappears for a delay period of usually ≥ 1 sec, after which a choice situation would come up where the animal would be required to pick the one of two stimuli that does not match the previously presented cue. The cue is trial-unique in these tasks, implying it is not sufficient to retrieve just from long-term memory the correct response (like in operant conditioning) but the previously presented cue has to be actively maintained for a brief interval. Hence one core feature of these tasks is that the correct response in the choice situation depends on the recent history or temporal sequence of events, not just on the present sensory input. More generally, such tasks have also been termed 'non-Markovian decision problems' in the machine learning literature (Sutton and Barto, 1998) to capture the fact that – unlike in a Markov process – prediction of the next state of the environment depends not solely on the present state of observables but on their recent past (the recent sequence of events).

15.3 Computational Requirements and Basic Electrophysiological Correlates

So how does the PFC solve such tasks, and which role could dopamine (DA) possibly play in it? Mainly through in vivo electrophysiological recordings we have gained some insight into how working memory at least in its simplest form, as assessed in DR tasks, could potentially be implemented at the neural level. Close to nothing is known about the neurodynamical implementation of still higher-level PFC-associated mental activities like planning and problem solving (but see Dehaene et al., 1999; Procyk and Goldman-Rakic, 2006; Unterrainer and Owen, 2006). A major observation from in vivo recordings is that many PFC neurons increase their average firing rate throughout the delay period, during which no external cues are present, and then drop back to their baseline firing level after termination of the delay or trial (Kubota and Niki, 1971; Fuster, 1973; Quintana et al., 1988; Funahashi et al., 1989; Miller et al., 1996; Rao et al., 1997; Rainer et al., 1998; Fuster et al., 2000). The elevated firing activity of these cells (termed delay-active cells in the following) is often stimulus (or response) selective, and its spontaneous or evoked cessation during the delay often provokes behavioural errors on the task (Fuster, 1973; Quintana et al., 1988; Funahashi et al., 1989; Sobotka et al., 2005). These findings led to the conclusion that these cells probably encode an active (online) memory in their 'persistently' elevated firing rates that allows to solve the non-Markovian decision problem by bridging the temporal gaps between previously and currently presented information.

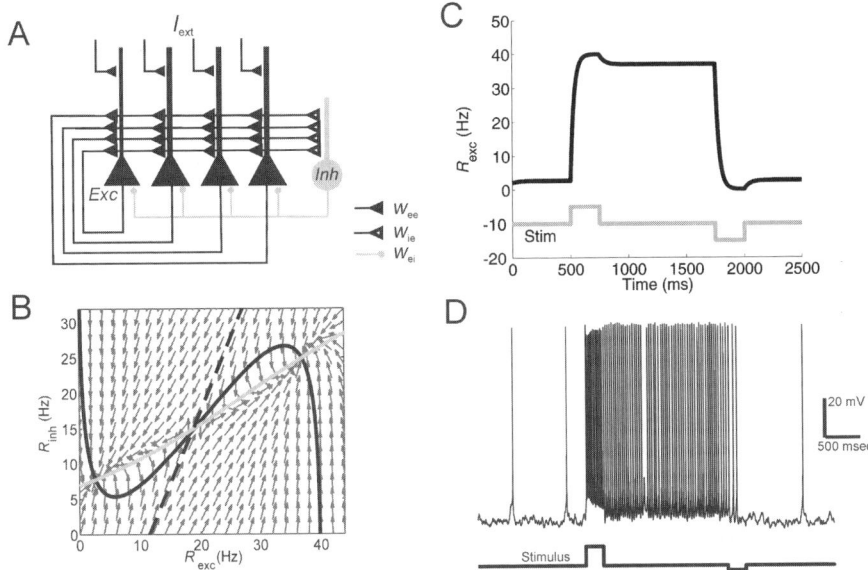

Figure 15.1 Firing rate bistability through interaction of positive and negative network feedback loops. (A) Simple model network architecture that may yield bistable behaviour: pyramidal cells (triangles) are recurrently coupled through excitatory synapses, and also excite W_{ie} a population of interneurons (circle) which feeds back inhibition W_{ei} to the pyramidal cells. (B) State space of such a model system, depicting the average firing rate of interneurons (R_{inh}) versus average firing rate of pyramidal cells (R_{exc}). Grey arrows indicate the flow field, that is, the direction of change of the average firing rates at various points in the state space. The solid-black and solid light-grey lines indicate the points at which the rate of change of firing rate of the pyramidal cells and interneurons, respectively, becomes zero. Intersections of these lines yield three fixed points at which neither the firing rate of the pyramidal cells nor that of the interneurons changes anymore, hence the whole system is in a steady state. The upper and lower fixed points are stable against small perturbations and hence are attractors of the system dynamics, while the middle one is unstable and divides the state space into two basins of attraction (black dashed line). (C) An excitatory stimulus switches the system into its high attractor state which is maintained even after offset of the stimulus, and an inhibitory stimulus causes the trajectory to cross the border between the basins of attraction again in the other direction, thus turning persistent activity off. (D) Bistability in a biophysical model network (from a simulation as in Durstewitz and Seamans, 2002). [Reproduced from Durstewitz and Seamans (2006) with permission from Elsevier (copyright 2005 IBRO)].

Already Joaquin Fuster in his seminal 1973 paper pointed out that there is considerable variance in the firing rate profiles of the delay-active cells. But partly for reasons of theoretical parsimony it was often assumed that these variations could mainly be considered as noise evolving around a rather plateau-like elevated firing state, as exhibited by the 'prototypical' stimulus-selective delay cells reported in, for example, Funahashi et al. (1989). Indeed, from a computational point of view, cells (or cellular populations) that have two stable firing rate states (=attractors; Fig. 15.1) between which they can be moved back and forth, are all that is needed to accomplish most DR tasks employed experimentally so far (Fig. 15.2; Durstewitz et al., 2000b). The cue would switch a cue-selective population from its low into its high firing rate state, maintained for instance by recurrent excitation (Fig. 15.1). Then, given the right network connectivity established within the many training trials used in DR tasks, at the time of the decision the choice stimuli plus the active cue-encoding cell assembly would conjointly determine the correct response (Fig. 15.2). A simple and straightforward explanation based on in vivo electrophysiological findings.

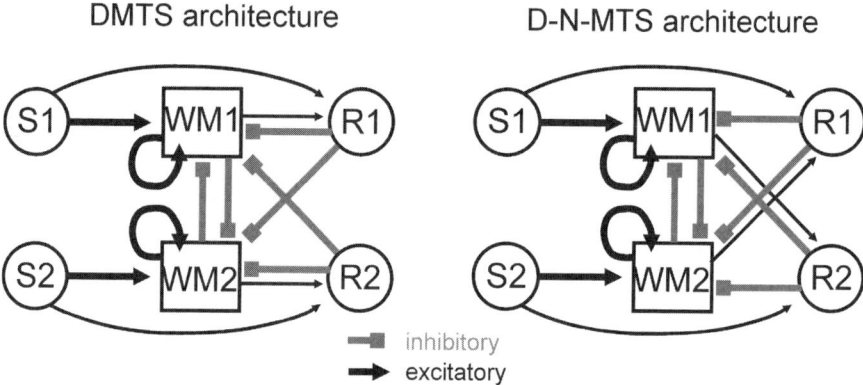

Figure 15.2 Neural network architectures for solving delayed-matching-to-sample (DMTS, left) and delayed-non-matching-to-sample (D-N-MTS, right) tasks. Two stimulus-specific neural populations (S1, S2) feed into recurrently connected working-memory circuits that could maintain stimulus-specific information over temporal delays (WM1, WM2). Response-selective neural populations could trigger two different types of behavioural output (R1, R2) and receive input both from the stimulus-specific and working-memory populations. Since input from either one of these two types of populations is too weak to trigger response-population firing (indicated by the thin black arrowed lines), conjoint activity within a stimulus-specific and working-memory population is necessary to cross the threshold and trigger a response. After a response, all working-memory networks are shut down by inhibitory feedback (grey lines) from the response neurons, terminating delay activity. Moreover, different working-memory representations should be mutually exclusive for these tasks and hence WM populations inhibit each other. In the left circuit, the temporal sequence S1 – delay – S1+S2 will trigger R1 while S2 – delay – S1+S2 will trigger R2 (assuming S-activity habituates before WM-activity is fully turned on during the presentation). In the right D-N-MTS circuit, response contingencies are reversed.

15.4 Dopamine as a Control Parameter of PFC Attractor Dynamics

What would be the role of DA in this scenario? One way to find out is to start with dopamine's experimentally established effects on voltage-gated and synaptic ion channels and implement them into a biophysically sufficiently realistic network model based on the ideas outlined in the previous section. This is the basic strategy we followed for a number of years, using neural models consisting of several somato-dendritic compartments and equipped with a variety of synaptic and voltage-gated conductances based on Hodgkin–Huxley-like gating kinetics that reproduced basic patch-clamp electrophysiological data (Durstewitz et al., 1999, 2000a; Durstewitz and Seamans, 2002; see also Brunel and Wang, 2001; Cohen et al., 2002; Dreher et al., 2002 for similar approaches, ideas, and results).

DA exerts its effects by modulating a variety of voltage- and ligand-gated ion channels, including, through D1-receptors, an enhancement of NMDA, $GABA_A$ and persistent Na^+ channels, and excitability of interneurons, while down-regulating presynaptic release probability and slow K^+ channels (Yang and Seamans, 1996; Zheng et al., 1999; Gorelova and Yang, 2000; Gao et al., 2001; Seamans et al., 2001a; Seamans et al., 2001b; Gorelova et al., 2002; Dong and White, 2003). DAs effects on various high-voltage gated Ca^{2+} channels are more complex (Yang and Seamans 1996; Young and Yang, 2004), including both facilitating and depressing effects. Hence, DA through D1 receptors pulls on a plethora of different biophysical parameters of pyramidal and interneuronal cells, with partly 'excitatory' and partly 'inhibitory' consequences. Many of these effects are reversed through activation of DA D2-receptors (Nishi et al., 1997; Zheng et al., 1999; Seamans et al., 2001b) present with D1 receptors on the same cells, further complicating the interpretation of DA's functional role. So DA's effects at the biophysical level are highly complex, and the only firm conclusion that can be drawn so far is that it would be utterly wrong to conceive DA simply as an excitatory or inhibitory neurotransmitter (see also Lapish et al., 2007) – it is a modulator modifies a large repertoire of biophysical properties rather than including fast de-or hyperpolarizing effects.

DA receptors are metabotropic, exerting their effects through G-protein-coupled cascades, and accordingly many of the DA-mediated effects seem to develop on a much slower time scale than the fast synaptic interactions through AMPA, NMDA, and GABA currents, in the range of at least seconds to minutes (Greengard, 2001a; Greengard, 2001b; Nishi et al., 2002; Nishi et al., 2005; Lapish et al., 2006). Another notable feature of the DA system is its spatially broad and diffuse action – DA neurons in the midbrain often respond like a rather homogeneous population (Schultz and Romo, 1990; Ljungberg et al., 1992), their fibres may form more than 100,000 varicosities (an order of magnitude more than the typical neocortical pyramidal cell; Fallon and Loughlin, 1995; Descarries et al., 1987), exhibiting the properties of volume transmission (reviewed in Lapish et al., 2006). Taken together, from the viewpoint of dynamical systems theory, dopamine may hence be seen as a control parameter: Rather than being an integral part of the fast information processing dynamics in neocortical networks, DA changes a diverse set of biophysical parameters of PFC neurons and synapses on a slow time scale and in a spatially unspecific manner.

When we integrated many of the DA-mediated cellular and synaptic effects into biophysical network models that exhibited memory attractor dynamics along the lines paved in Figs. 15.1 and 15.2, we observed that D1 stimulation increased the gap between firing rate attractor states belonging to different items, while it can be inferred that D2 stimulation, based on its often opposite cellular actions, would diminish the gap (Fig. 15.3A, B; Durstewitz et al., 1999, 2000a; Durstewitz and Seamans, 2002).

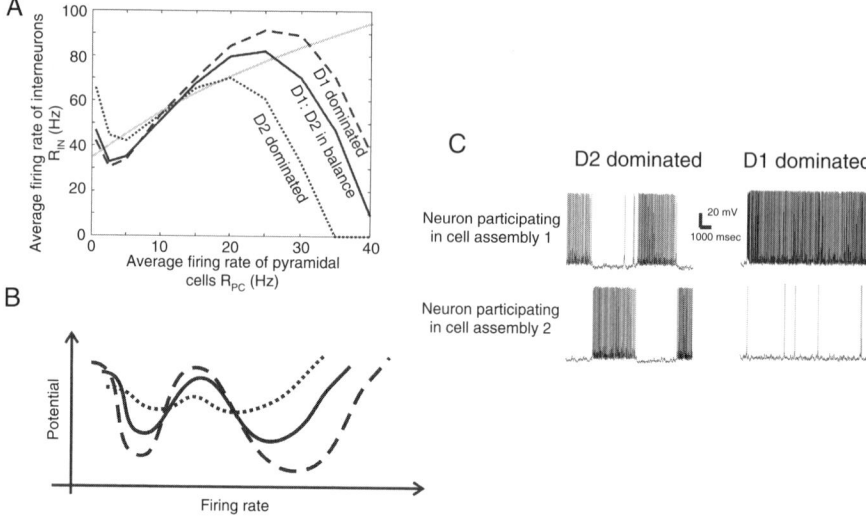

Figure 15.3 Dopaminergic modulation of activity states and attractor basins. (A) State space representation of a system of recurrently coupled pyramidal cells and interneurons, spanned by the average firing rates of these two neural populations (see legend of Fig. 15.1B for more explanation on this type of graph). This state space was derived by numerical simulations of a network of biophysically realistic conductance-based neurons as described in Durstewitz and Seamans (2002). In addition to the curve indicating where the average firing rate of the interneurons becomes stationary (grey line), there are three curves showing where the average firing rate of the pyramidal cells would remain stationary under conditions of high D1 (dashed-black line), high D2 (dotted-black line), and balanced D1 versus D2 (solid-black line) stimulation. Recall from Fig. 15.1B that intersections of the grey with any of the black curves indicate fixed points of the system (where the whole network attains stationary activity), with the uppermost and lowermost being stable and the middle ones being unstable fixed points. Note that stable fixed points move inward and the stationarity curve for the pyramidal cells flattens out along the curve for the interneurons as the network is pushed from a D1-dominated into a D2-dominated regime. This indicates that low and high firing attractor states approach each other in firing rate and that their regions of robustness shrink. (B) Representation of the same information as in (A) (with corresponding line styles) in terms of a potential or 'energy landscape' (note that this graph is just a schema; strictly speaking the simulated system does not have a true potential function). Minima of the potential correspond to the fixed point attractors in (A), and the state of the system may be envisioned as a ball rolling down into the nearest minimum. The graph makes clear that it becomes much harder to switch between different attractor states in the D1 dominated regime as the minima move apart and the 'valleys' become much steeper. Conversely, in the D2 dominated regime the valleys become so flat and nearby that noise may easily push the system from one state into the other. Also note that the ease of 'attractor hopping' could in principle be regulated purely by the steepness of the valley slopes, without any change in the position of the minima, hence without any change in average firing rates. (C) Network simulation illustrating the fact that the system spontaneously switches (or may cycle) among different attractors (neural representations) in the D2 dominated regime, while robustly maintaining a once elicited attractor in the D1 dominated regime.

In other words, D1-mediated effects increased the 'energy barrier', the amount of stimulation necessary to hop between different memory states, by enlarging and deepening the basins of attraction of active memory states. This means it becomes harder to disturb currently active delay cells under D1 influence, increasing their robustness to internal or external noise and distracters and salvaging the active memory contents across the delay (Durstewitz et al., 1999, 2000a; Durstewitz and Seamans, 2002). On contrary, D2 activation would make it easier to switch among different memory states, enhancing prefrontal network flexibility by allowing much more representations to hop in and out of the working-memory buffer in brief succession (Fig. 15.3C). In summary, since evidence suggests that the relative degree of D1 versus D2 stimulation could be regulated through DA concentration (Zheng et al., 1999; Trantham-Davidson et al., 2004), these modelling results suggest that DA may set the right balance for different task requirements. At medium concentrations it would foster the robust maintenance of working-memory contents during highly goal-directed behaviour, while at high concentrations it may allow sifting through the memory base in a highly flexible manner when solutions to novel behavioural problems are required.

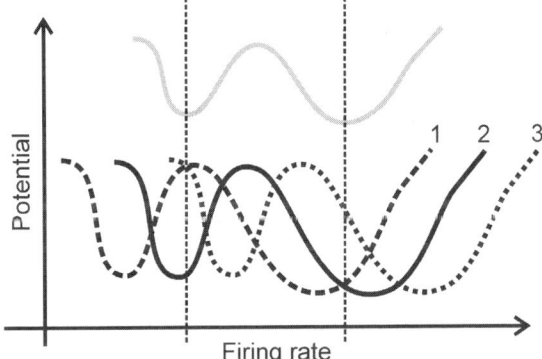

Figure 15.4 Three different scenarios for moving the system from a neutral (grey potential function) into a D1 dominated regime (black curves). For scenario 1 (dashed-black) both low and high attractor states are reduced in firing rate, for scenario 3 (dotted-black) both are increased, and for scenario 2 (solid-black) the low state is further reduced while the high state is further enhanced in firing rate. Note that despite these differences in change of firing rates, the overall shape of the black curve stays the same, implying that all three scenarios result in the same changes of robustness of attractor states.

We emphasize that it is quite remarkable that the variety of different DA-induced biophysical changes may ultimately pull on the same strings, moving network dynamics into one or the other direction (Durstewitz et al., 1999, 2000a). Taking for instance the D1-induced strengthening of both excitatory NMDA and inhibitory GABA$_A$ currents, these will lead to a fiercer competition among different cell assemblies while at the same time boosting recurrent activity within the currently active assembly. On top of this, NMDA currents increase non-linearly with membrane voltage which is usually higher in the dendrites of currently active neurons, in contrast to the decrease in AMPA and linear behaviour of GABA currents. This could further expand the gap between foreground and background neuron firing.

The results summarized above could explain the decreased performance and behavioural failures in working-memory tasks at too high or too low levels of DA receptor stimulation (Arnsten et al., 1994; Murphy et al., 1996a; Murphy et al., 1996b; Cai and Arnsten, 1997; Zahrt et al., 1997; Tost et al., 2006; Meyer-Lindenberg and Weinberger, 2007) as they imply an optimal DA level for the robust maintenance of goal-related information. Moreover, they can account for a number of different in vivo electrophysiological observations. In a series of experiments Sawaguchi and co-workers (Sawaguchi et al., 1988, 1990a, b) found that DA enhances task-related activity much more than spontaneous firing of PFC neurons, an effect that is susceptible mainly to D1 but not D2 receptor blockade. On the other hand, Vijayraghavan and co-workers (Vijayraghavan et al., 2007) in a very recent study observed that although D1 receptor agonists decrease both delay activity related to the preferred as well as to a non-preferred stimulus of a neuron, they reduce the latter much more than the former. Hence, although the overall direction of effects is different in these two sets of studies, they both predict that DA via D1 receptors magnifies the gap between target-selective delay activity and non-target related activity. In another direct electrophysiological test of our theory, Lavin and co-workers (Lavin et al., 2005) observed that after VTA stimulation spontaneous firing of PFC neurons in vivo was suppressed while stimulus-evoked firing was actually enhanced. Thus, while the overall effect of D1 stimulation on firing rates may depend on the balance between different synaptic components, the dynamical implication may still be an enlarging and deepening of basins of attraction resulting in an increased barrier for hopping between attractor states (Fig. 15.4). In fact, although the experimentally identified differential regulation of target versus non-target delay activity in general supports our model, strictly speaking DA could achieve its functional effects without any apparent changes in firing rates at all (Fig. 15.3). This is an important point as it makes clear that functionally highly important effects of neuromodulators may not always be identifiable at the level of neural firing rates.

15.5 Is this all There is – Puzzle Solved?

The models outlined above seem to capture some core aspects of PFC-dependent behaviour as well as of PFC electrophysiology, and the dopaminergic modulation of both these aspects. However, it seems clear that Fig. 15.2 and the related behavioural paradigms provide a yet very much over-simplified and limited view of PFC function. Both at the behavioural-cognitive and the neurodynamical level PFC networks seem to be much more powerful and intricate (see Durstewitz and Seamans, 2006 for a review). In fact, stimulus-dependent bi- or multi-stable neural behaviour as required in typical DR tasks has been observed in many other brain areas, including midbrain nuclei, spinal motoneurons, and invertebrate systems (Marder et al., 1996; Kiehn and Eken, 1998; Aksay et al., 2001), suggesting that basically all of them may in principle be capable of solving 'DR tasks', provided input and output conditions are casted in appropriate physiological terms.

In Section 15.2 we already pointed out that the PFC may be involved in non-Markovian decision problems in a much wider sense than implied by simple DR tasks, including prediction of future states of affairs based on extracted abstract rules (Wallis et al., 2001) and 'mental simulations' of situation-action chains (Averbeck et al., 2006; Averbeck and Lee, 2007). But even in the apparently rather simple setting of a DR task, animal brains seem to be doing much more than simply maintaining cue information and integrating it with the choice information following the delay: They seem to actively predict both the forthcoming choice situation as well as its time of occurrence, as indicated by behavioural and electrophysiological evidence (Quintana and Fuster, 1999; Rainer et al., 1999; Sargisson and White, 2001; Brody et al., 2003; Durstewitz, 2004). At the electrophysiological level, these processes are reflected by the fact that neural activity associated with previously presented cues usually decays throughout the delay (rather than maintaining a constantly elevated level) while at the same time activity associated with the upcoming choice slowly ramps up (Quintana and Fuster, 1999; Rainer et al., 1999). Since the slope of these ramping activity profiles adjusts to the expected time of occurrence (Komura et al., 2001; Brody et al., 2003; Durstewitz, 2003, 2004; Reutimann et al., 2004), it has been proposed that these neurons encode information about the temporal interval between the cue and choice situation as well. In fact, there seems to be a rather wide spectrum of different temporal activity profiles throughout the delay (Rainer and Miller, 2002; first indicated by Fuster, 1973), indicating that the network dynamics in PFC even in apparently simple DR tasks, and even at the level of firing rate changes, may be much more complex than previously realized.

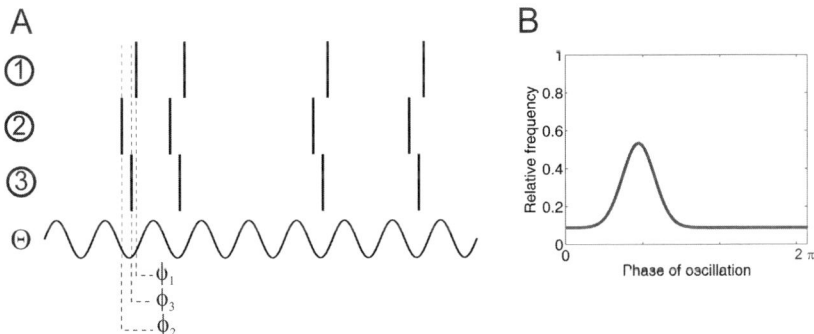

Figure 15.5 Phase-coding. (A) Spiking pattern of three neurons (circles labelled 1–3) that stereotypically repeats itself several times at irregular intervals, but always phase-locked to an underlying theta population oscillation (as measured, e.g. by the local field potential). Information may be encoded in this spiking pattern as a phase-vector (ϕ_1, ϕ_2, ϕ_3), where the ϕ_i denote the relative position (phase) within the theta cycle. This pattern could be read out by a coincidence detector if the axonal delays $\Delta ax_2 > \Delta ax_3 > \Delta ax_1$ are such that they compensate for the time differences within the theta cycle [reproduced from Durstewitz and Seamans (2006) with permission from Elsevier (copyright 2005 IBRO)]. (B) Phase-histogram that allows assessing the possibility of phase-locking in experimental recordings. It charts the relative frequency of spike occurences as a function of the phase of the underlying network rhythm (as, e.g. extracted from the LFP). The location of the peak indicates where most spikes occur relative to the phase of the underlying network oscillation, while its width indicates the temporal precision of the spike-time locking.

But there are yet other dynamical phenomena that seem to bear a relationship to working-memory processing. Oscillations in various frequency bands, predominantly in the theta and gamma range, accompany working memory in a manner that suggests a specific involvement in information processing by (i) often appearing or increasing in amplitude selectively during the delay period (Raghavachari et al., 2001; Lutzenberger et al., 2002; Pesaran et al., 2002), (ii) increasing in amplitude with working-memory load (Howard et al., 2003), and (iii) being stimulus-specific in some reports (Pesaran et al., 2002; Lee et al., 2005; but see Compte et al., 2003). Oscillations can both be a signature and a carrier signal for the temporal organization of spiking activity among multiple neurons which can be exploited as an alternative scheme for maintaining and processing information (Fig. 15.5). Such phase- or temporal-coding schemes often come with the benefit of allowing much faster and more powerful computational operations than firing rate-based computations (Hopfield, 1995; Hopfield and Brody, 2001; Brody and Hopfield, 2003). Indeed, precise temporal relations between the spiking times of multiple neurons associated with significant task events (Riehle et al., 1997), or between the spiking or bursting times of single neurons and the population oscillation (phase-locking; Lee et al., 2005) have been reported during working-memory tasks. Lee et al. (2005) provided evidence that such phase-locking occurs in a stimulus-specific manner and is sustained throughout the delay-period, suggesting that information can be maintained in working or short-term memory even in the absence of alterations in firing rate.

15.6 Dopaminergic Modulation of Oscillations and Dynamical Regimes

Since dopamine has such a profound effect on all major synaptic currents as well as on many voltage-gated intrinsic conductances, it will also somehow influence the more complex computational dynamics described above. Indeed, a couple of studies indicate that dopamine may alter the power of oscillatory activity in both slow- and high-frequency ranges in various telencephalic regions including the PFC (Montaron et al., 1982; Ruskin et al., 2003; Peters et al., 2004; Gao, 2007). Recently we observed in our computational models that dopamine via D1-receptor activation may enhance slow delta-frequency (0.5–2 Hz) oscillations by boosting peak and diminishing trough activity in line with the D1-mediated reduction in low and amplification of high firing rate states we described previously (Durstewitz et al., 200a; Durstewitz and Seamans, 2002), and that has been supported experimentally (Lavin et al., 2005). In our model, these slow-frequency oscillations are mediated by NMDA synaptic currents which can both shift single neurons into an oscillatory regime and synchronize burst-like activity among many neurons (Tseng and O'Donnell, 2005; Durstewitz and Gabriel, 2007; Golomb et al., 2006). Since D1 receptors enhance both NMDA and $GABA_A$ synaptic currents, this may lead to the amplification of slow network oscillations as described above. These findings also fit with the recent observation by Tseng and O'Donnell (2005) that D1 activation allows for the emergence of slow-frequency down-/up-state cycles in a PFC slice preparation, and that this phenomenon depends on NMDA input as well (see also Beggs and Plenz, 2003; Durstewitz and Gabriel, 2007).

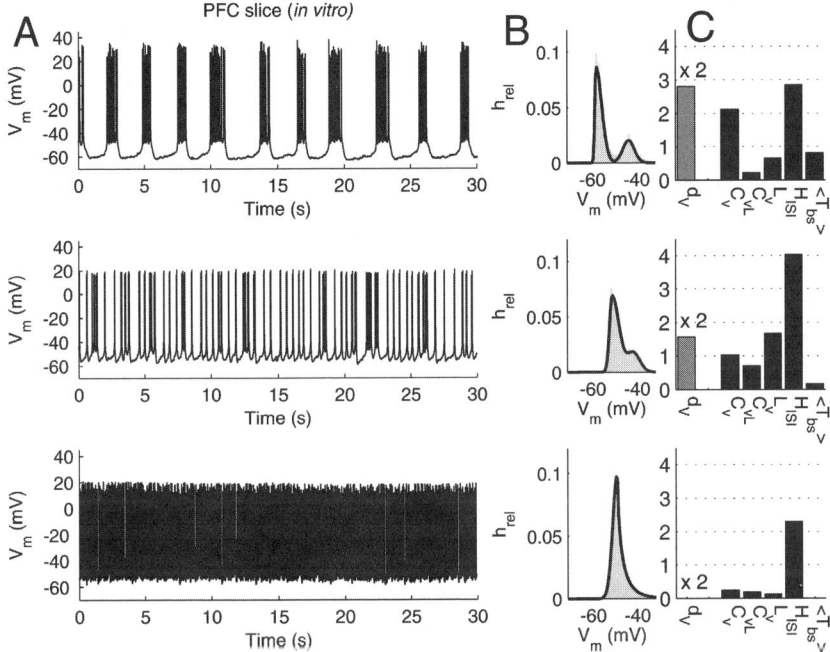

Figure 15.6 NMDA-induced spiking patterns in an adult rat prefrontal cortex slice. (A) Example traces from three different prefrontal neurons with long-duration bursts (top), highly irregular spiking (middle), and continuously regular spiking (bottom). Pharmacological conditions were similar in all three cases (NMDA: 8 μM; SKF 38393: 3 μM (top) or 4 μM (middle and bottom). (B) Membrane potential distributions of these neurons indicating a shift from a clearly bimodal (top), through an intermediate (middle), to a clearly unimodal (bottom) distribution (grey = V_m distribution, black line = function fit). (C) Index for the amount of bimodality within the membrane potential distribution (d_V), various statistical indices for the amount of spiking irregularity (C_V, C_{VL}, L_V, H_{ISI}), and average duration of bursts ($<T_{bs}>$ in seconds) The C_V is closest to 1 (as in a Poisson process) while the other 'true' irregularity indices (C_{VL}, L_V, H_{ISI}) peak for the intermediated regime. see Durstewitz and Gabriel (2007) for details. [reproduced from Durstewitz and Gabriel (2007) by permission of oxford university press].

How these changes in oscillatory dynamics, if confirmed in vivo, will impact on the potential for phase-coding is still an open question. One possible hypothesis one may put forward is that a D1-dominated regime, characterized by strong network oscillations, will allow for the stable maintenance of a few phase-coding patterns, while a D2-dominated regime, characterized by weak and spatially less coherent oscillations, will allow for the simultaneous representation of many phase-coding patterns, yet in a much less robust manner. Note that this idea recapitulates our model for dopamine's functional impact on firing rate attractors (see above), now for the phase-coding domain.

But dopamine may alter network dynamics in a yet more profound way. Recently we observed that NMDA currents could switch pyramidal cells into different dynamical regimes in a PFC slice preparation and a PFC network model based on these in vitro data (Fig. 15.6; Durstewitz and Gabriel, 2007). Neurons in a PFC slice bathed in NMDA and a low concentration of a D1-receptor agonist [a preparation similar to the one employed by Tseng and O'Donnell (2005)] exhibited different types of NMDA-induced behaviour, ranging from strong slow-frequency oscillatory burst-like activity [as reported before by Tseng and O'Donnell (2005)], highly irregular mixtures of single spikes, doublets, and 'mini-bursts', to highly regular continuous single spiking activity (Fig. 15.6). Non-linear spike-train analysis of the in vitro recordings and biophysical network modelling suggested that the irregular intermediate regime exhibits features of deterministic chaos (Strogatz, 1994), and that the overall strength of NMDA synaptic inputs could move neurons back and forth among these different regimes. One very important point to note here is that the level of NMDA input, rather than just changing the average firing rate of a neuron, changes the temporal patterning of spiking activity! Since both D1- and D2-receptor activation modulates NMDA currents with opposite sign (Zheng et al., 1999; Seamans et al., 2001a) it is conceivable that dopaminergic input does not only affect the strength of oscillatory activity but may move PFC neurons among different dynamical regimes characterized by more or less ordered or disordered spike-train dynamics, a possibility we are currently exploring). Such a finding would be highly important from a computational point of view in the light of recent simulation studies that suggest that the computational power and ability to process long temporal sequences in recurrent neural networks may be optimized in certain dynamical regimes (Bertschinger and Natschläger, 2004; Legenstein and Maass, 2007).

15.7 Conclusions

We have started with the idea that the PFC of mammals may process non-Markovian decision problems in a very general sense, that is, may implement the cognitive ability to forecast future states of affairs based on the recent history and temporal sequence of environmental events. One component of this ability is working memory as captured in its simplest form by DR tasks employed in primates. DR tasks could basically be solved based on a very simple computational architecture (Fig. 15.2) which rests on the ability of recurrent networks to exhibit multi-stability of different firing rate attractor states, as seemingly supported by PFC in vivo electrophysiology (Fuster, 1973; Funahashi et al., 1989). In this scenario, dopamine based on its cellular and synaptic actions would through D1-receptors increase and through D2-receptors lower the 'energy barrier' between different attractor states, making it harder/easier to jump between them. This could both explain a number of diverse behavioural findings on working-memory failures and set shifting abilities upon manipulations of the prefrontal dopamine input (Sawaguchi and Goldman-Rakic, 1994; Dias et al., 1997; Zahrt et al., 1997; Seamans et al., 1998; Mehta et al., 2004;

Floresco et al., 2006; Tost et al., 2006), and various electrophysiological findings suggesting an increased dissociation of target and non-target related firing rates by D1 activation (Sawaguchi et al., 1990a, b; Lavin et al., 2005; Vijayraghavan et al., 2007).

However, both at the cognitive-behavioural and the neurodynamical level PFC functionality seems to be much more complex. In particular, various studies suggest a prominent role of oscillatory activity and phase-locking phenomena in information processing and maintenance during working-memory tasks (Pesaran et al., 2002; Howard et al., 2003; Lee et al., 2005). Again, dopamine based on its numerous synaptic and cellular actions may foster oscillations and robust phase-coding schemes through D1 receptors but enhance the variety of phase-locked patterns while diminishing their stability through D2-receptors. Hence, at the computational level, dopamine's function both with regards to firing rate attractors and to phase-coding schemes may be to balance the ability to robustly maintain items in working memory and shield them from distraction with the ability to flexibly integrate and represent new information.

Finally, dopamine specifically through its actions on NMDA and GABA$_A$ synaptic currents may move PFC neurons into different dynamical regimes, characterized by different amounts of order in the temporal patterning of spiking activity, optimized for the task of deriving predictions from the processing of temporal sequences (Bertschinger and Natschläger, 2004).

Many of these hypotheses are still preliminary and require much more in depth investigation at the computational, in vivo electrophysiological, and behavioural levels. Should some of these ideas turn out to be right, this could lead to major breakthroughs both in our understanding of dopamine modulation of PFC functionality as well as malfunctioning states such as schizophrenia or ADHD (Winterer and Weinberger, 2004).

References

Aksay, E., Gamkrelidze, G., Seung, H.S., Baker, R. and Tank, D.W. (2001) In vivo intracellular recording and perturbation of persistent activity in a neural integrator. Nat. Neurosci. 4(2), 184–193.

Arnsten, A.F., Cai, J.X., Murpy, B.L. and Goldman-Rakic, P.S. (1994) Dopamine D1 receptor mechanisms in the cognitive performance of young and aged monkeys. Psychopharmacology (Berl.) 116, 143–151.

Averbeck, B.B. and Lee, D. (2007) Prefrontal neural correlates of memory for sequences. J. Neurosci. 27(9), 2204–2211.

Averbeck, B.B., Sohn, J.W. and Lee, D. (2006) Activity in prefrontal cortex during dynamic selection of action sequences. Nat. Neurosci. 276–282.

Beggs, J.M. and Plenz, D. (2003) Neuronal avalanches in neocortical circuits. J. Neurosci. 23, 11167–11177.

Bertschinger, N. and Natschläger, T. (2004) Real-time computation at the edge of chaos in recurrent neural networks. Neural Comput. 16, 1413–1436.

Brody, C.D. and Hopfield, J.J. (2003) Simple networks for spike-timing-based computation, with application to olfactory processing. Neuron 37, 843–852.

Brody, C.D., Hernandez, A., Zainos, A. and Romo, R. (2003) Timing and neural encoding of somatosensory parametric working memory in macaque prefrontal cortex. Cereb. Cortex 13, 1196–1207.

Brunel, N. and Wang, X.J. (2001) Effects of neuromodulation in a cortical network model of object working memory dominated by recurrent inhibition. J. Comput. Neurosci. 11(1), 63–85.

Cai, J.X. and Arnsten, A.F. (1997) Dose-dependent effects of the dopamine D1 receptor agonists A77636 or SKF81297 on spatial working memory in aged monkeys. J. Pharmacol. Exp. Ther. 283, 183–189.

Cohen, J.D., Braver, T.S. and Brown, J.W. (2002) Computational perspectives on dopamine function in prefrontal cortex. Curr. Opin. Neurobiol. 12(2), 223–229.

Compte, A., Constantinidis, C., Tegner, J., Raghavachari, S., Chafee, M.V., Goldman-Rakic, P.S. and Wang, X.J. (2003) Temporally irregular mnemonic persistent activity in prefrontal neurons of monkeys during a delayed response task. J. Neurophysiol. 90, 3441–3454.

Dagher, A., Owen, A.M., Boecker, H. and Brooks, D.J. (1999) Mapping the network for planning: a correlational PET activation study with the Tower of London task. Brain 122, 1973–1987.

Dehaene, S., Jonides, J., Smith, E.E. and Spitzer, M. (1999) Thinking and problem solving. In: Fundamental Neuroscience, M.J. Zigmond, F.E. Bloom, S.C. Landis, J.L. Roberts and L.R. Squire (eds.). San Diego: Academic Press, pp. 1543–1564.

Descarries, L., Lemay, B., Doucet, G. and Berger, B. (1987) Regional and laminar density of the dopamine innervation in adult rat cerebral cortex. Neuroscience 21, 807–824.

Dias, R., Robbins, T.W. and Roberts, A.C. (1997) Dissociable forms of inhibitory control within prefrontal cortex with an analog of the Wisconsin card sort test: restriction to novel situations and independence from "on-line" processing. J. Neurosci. 17, 9285–9297.

Dong, Y. and White, F.J. (2003) Dopamine D1-class receptors selectively modulate a slowly inactivating potassium current in rat medial prefrontal cortex pyramidal neurons. J. Neurosci. 23, 2686–2695.

Dreher, J.C., Guigon, E. and Burnod, Y. (2002) A model of prefrontal cortex dopaminergic modulation during the delayed alternation task. J. Cogn. Neurosci. 14, 853–865.

Duncan, J. and Owen, A.M. (2000) Common regions of the human frontal lobe recruited by diverse cognitive demands. Trends Neurosci. 23, 475–483.

Durstewitz, D. (2003) Self-organizing neural integrator predicts interval times through climbing activity. J. Neurosci. 23, 5342–5353.

Durstewitz, D. (2004) Neural representation of interval time. Neuroreport 15, 745–749.

Durstewitz, D. and Seamans, J.K. (2002) The computational role of dopamine D1 receptors in working memory. Neural Networks 15, 561–572.

Durstewitz, D. and Seamans, J.K. (2006) Beyond bistability: biophysics and temporal dynamics of working memory. Neuroscience 139, 119–133.

Durstewitz, D. and Gabriel, T. (2007) Dynamical basis of irregular spiking in NMDA-driven prefrontal cortex neurons. Cereb. Cortex 17, 894–908. [Epub 2006].

Durstewitz, D., Kelc, M. and Güntürkün, O. (1999) A neurocomputational theory of the dopaminergic modulation of working memory functions. J. Neurosci. 19, 2807–2822.

Durstewitz, D., Seamans, J.K. and Sejnowski, T.J. (2000a) Dopamine-mediated stabilization of delay-period activity in a network model of prefrontal cortex. J. Neurophysiol. 83, 1733–1750.

Durstewitz, D., Seamans, J.K. and Sejnowski, T.J. (2000b) Neurocomputational models of working memory. Nat. Neurosci. 3(Suppl.) 1184–1191.

Fallon, J.H. and Loughlin, S.E. (1995) Substantia nigra. In: Paxinos, G (ed.) The Rat Nervous System. Academic, San Diego, pp. 215–237.

Floresco, S.B., Magyar, O., Ghods-Sharifi, S., Vexelman, C. and Tse, M.T. (2006) Multiple dopamine receptor subtypes in the medial prefrontal cortex of the rat regulate set-shifting. Neuropsychopharmacology. 31, 297–309.

Funahashi, S., Bruce, C.J. and Goldman-Rakic, P.S. (1989) Mnemonic coding of visual space in the monkey's dorsolateral prefrontal cortex. J. Neurophysiol. 61, 331–349.

Fuster, J. (1997) The prefrontal cortex: anatomy, physiology, and neuropsychology of the frontal lobe. Philadelphia: Lippincott-Raven.

Fuster, J.M. (1973) Unit activity in prefrontal cortex during delayed-response performance: neuronal correlates of transient memory. J. Neurophysiol. 36, 61–78.

Fuster, J.M. (2001) The prefrontal cortex – an update: time is of the essence. Neuron 30, 319–333.

Fuster, J.M., Bodner, M. and Kroger, J.K. (2000) Cross-modal and cross-temporal association in neurons of frontal cortex. Nature 405, 347–351.

Gao, W.J. (2007) Acute clozapine suppresses synchronized pyramidal synaptic network activity by increasing inhibition in the ferret prefrontal cortex. J. Neurophysiol. 97(2), 1196–1208.

Gao, W.J., Krimer, L.S. and Goldman-Rakic, P.S. (2001) Presynaptic regulation of recurrent excitation by D1 receptors in prefrontal circuits. Proc. Natl. Acad. Sci. U.S.A. 98, 295–300.

Golomb, D., Shedmi, A., Curtu, R. and Ermentrout, G.B. (2006) Persistent synchronized bursting in cortical tissues with low magnesium concentration: a modeling study. J. Neurophysiol. 95, 1049–1067.

Gorelova, N.A. and Yang, C.R. (2000) Dopamine D1/D5 receptor activation modulates a persistent sodium current in rat prefrontal cortical neurons in vitro. J. Neurophysiol. 84, 75–87.

Gorelova, N., Seamans, J.K. and Yang, C.R. (2002) Mechanisms of dopamine activation of fast-spiking interneurons that exert inhibition in rat prefrontal cortex. J. Neurophysiol. 88, 3150–3166.

Greengard, P. (2001a) The neurobiology of dopamine signaling. Biosci. Rep. 21(3), 247–269.

Greengard, P. (2001b) The neurobiology of slow synaptic transmission. Science 294(5544), 1024–1030.

Hopfield, J.J. (1995) Pattern recognition computation using action potential timing for stimulus representation. Nature 376, 33–36.

Hopfield, J.J. and Brody, C.D. (2001) What is a moment? Transient synchrony as a collective mechanism for spatiotemporal integration. Proc. Natl. Acad. Sci. U.S.A. 98(3), 1282–1287.

Howard, M.W., Rizzuto, D.S., Caplan, J.B., Madsen, J.R., Lisman, J., Aschenbrenner-Scheibe, R., Schulze-Bonhage, A. and Kahana, M.J. (2003) Gamma oscillations correlate with working memory load in humans. Cereb. Cortex 13, 1369–1374.

Kiehn, O. and Eken, T. (1998) Functional role of plateau potentials in vertebrate motor neurons. Curr. Opin. Neurobiol. 8(6), 746–752. Review.

Komura, Y., Tamura, R., Uwano, T., Nishijo, H., Kaga, K. and Ono, T. (2001) Retrospective and prospective coding for predicted reward in the sensory thalamus. Nature 412, 546–549.

Kubota, K. and Niki, H. (1971) Prefrontal cortical unit activity and delayed alternation performance in monkeys. J. Neurophysiol. 34, 337–347.

Lapish, C.C., Kroener, S., Durstewitz, D., Lavin, A. and Seamans, J.K. (2007) The ability of the mesocortical dopamine system to operate in distinct temporal modes. Psychopharmacology 191, 609–625. [Epub 2006].

Lavin, A., Nogueira, L., Lapish, C.C., Wightman, R.M., Phillips P.E. and Seamans J.K. (2005) Mesocortical dopamine neurons operate in distinct temporal domains using multimodal signaling. J. Neurosci. 25, 5013–5023.

Lee, H., Simpson, G.V., Logothetis, N.K. and Rainer, G. (2005) Phase locking of single neuron activity to theta oscillations during working memory in monkey extrastriate visual cortex. Neuron 45, 147–156.

Legenstein, R.A. and Maass, W. (2007) Edge of chaos and prediction of computational performance for neural microcircuit models. Neural Networks, in press.

Ljungberg, T., Apicella, P. and Schultz, W. (1992) Responses of monkey dopamine neurons during learning of behavioral reactions. J. Neurophysiol. 67, 145–163.

Lutzenberger, W., Ripper, B., Busse, L., Birbaumer, N. and Kaiser J. (2002) Dynamics of gamma-band activity during an audiospatial working memory task in humans. J. Neurosci. 22, 5630–5638.

Marder, E., Abbott, L.F., Turrigiano, G.G., Liu, Z. and Golowasch, J. (1996) Memory from the dynamics of intrinsic membrane currents. Proc. Natl. Acad. Sci. U.S.A. 93, 13481–13486.

Mehta, M.A., Manes, F.F., Magnolfi, G., Sahakian, B.J. and Robbins, T.W. (2004) Impaired set-shifting and dissociable effects on tests of spatial working memory following the dopamine D2 receptor antagonist sulpiride in human volunteers. Psychopharmacology (Berl). 176, 331–342.

Meyer-Lindenberg, A. and Weinberger, D.R. (2007) Intermediate phenotypes and genetic mechanisms of psychiatric disorders. Nat. Rev. Neurosci. 7, 818–827.

Miller, E.K. and Cohen, J.D. (2001) An integrative theory of prefrontal cortex function. Annu. Rev. Neurosci. 24, 167–202.

Miller, E.K., Erickson, C.A. and Desimone, R. (1996) Neural mechanisms of visual working memory in prefrontal cortex of the macaque. J. Neurosci. 16, 5154–5167.

Milner, B. and Petrides, M. (1984) Behavioural effects of frontal-lobe lesions in man. Trends Neurosci. 7, 403–407.

Montaron, M.F., Bouyer, J.J., Rougeul, A. and Buser, P. (1982) Ventral mesencephalic tegmentum (VMT) controls electrocortical beta rhythms and associated attentive behaviour in the cat. Behav. Brain Res. 6, 129–145.

Murphy, B.L., Arnsten, A.F.T., Goldman-Rakic, P.S. and Roth, R.H. (1996a) Increased dopamine turnover in the prefrontal cortex impairs spatial working memory performance in rats and monkeys. Proc. Natl. Acad. Sci. U.S.A. 93, 1325–1329.

Murphy, B.L., Arnsten, A.F.T., Jentsch, J.D. and Roth, R.H. (1996b) Dopamine and spatial working memory in rats and monkeys: pharmacological reversal of stress-induced impairment. J. Neurosci. 16, 7768–7775.

Nishi, A., Snyder, G.L. and Greengard, P. (1997) Bidirectional regulation of DARPP-32 phosphorylation by dopamine. J. Neurosci. 17(21), 8147–8155.

Nishi, A., Bibb, J.A., Matsuyama, S., Hamada, M., Higashi, H., Nairn, A.C. and Greengard, P. (2002) Regulation of DARPP-32 dephosphorylation at PKA- and Cdk5-sites by NMDA and AMPA receptors: distinct roles of calcineurin and protein phosphatase-2A. J. Neurochem. 81(4), 832–841.

Nishi, A., Watanabe, Y., Higashi, H., Tanaka, M., Nairn, A.C. and Greengard, P. (2005) Glutamate regulation of DARPP-32 phosphorylation in neostriatal neurons involves activation of multiple signaling cascades. Proc. Natl. Acad. Sci. U.S.A. 102(4), 1199–1204.

Pesaran, B., Pezaris, J.S., Sahani, M., Mitra, P.P. and Andersen, R.A. (2002) Temporal structure in neuronal activity during working memory in macaque parietal cortex. Nat. Neurosci. 5, 805–811.

Peters, Y., Barnhardt, N.E. and O'Donnell, P. (2004) Prefrontal cortical up states are synchronized with ventral tegmental area activity. Synapse 52, 143–152.

Procyk, E. and Goldman-Rakic, P.S. (2006) Modulation of dorsolateral prefrontal delay activity during self-organized behavior. J. Neurosci. 26(44), 11313–11323.

Quintana, J. and Fuster, J.M. (1999) From perception to action: temporal integrative functions of prefrontal and parietal neurons. Cereb. Cortex 9, 213–221.

Quintana, J., Yajeya, J. and Fuster, J.M. (1988) Prefrontal representation of stimulus attributes during delay tasks. I. Unit activity in cross-temporal integration of sensory and sensory-motor information. Brain Res. 474, 211–221.

Raghavachari, S., Kahana, M.J., Rizzuto, D.S., Caplan, J.B., Kirschen, M.P., Bourgeois, B., Madsen, J.R. and Lisman, J.E. (2001) Gating of human theta oscillations by a working memory task. J. Neurosci. 21, 3175–3183.

Rainer, G. and Miller, E.K. (2002) Timecourse of object-related neural activity in the primate prefrontal cortex during a short-term memory task. Eur. J. Neurosci. 15, 1244–1254.

Rainer, G., Asaad, W.F. and Miller, E.K. (1998) Selective representation of relevant information by neurons in the primate prefrontal cortex. Nature 393, 577–579.

Rainer, G., Rao, S.C. and Miller, E.K. (1999) Prospective coding for objects in primate prefrontal cortex. J. Neurosci. 19, 5493–5505.

Rao, S.C., Rainer, G. and Miller, E.K. (1997) Integration of what and where in the primate prefrontal cortex. Science 276, 821–824.

Reutimann, J., Yakovlev, V., Fusi, S. and Senn, W. (2004) Climbing neuronal activity as an event-based cortical representation of time. J. Neurosci. 24, 3295–3303.

Riehle, A., Grün, S., Diesmann, M. and Aertsen, A. (1997) Spike synchronization and rate modulation differentially involved in motor cortical function. Science 278, 1950–1953.

Ruskin, D.N., Bergstrom, D.A., Tierney, P.L. and Walters, J.R. (2003) Correlated multisecond oscillations in firing rate in the basal ganglia: modulation by dopamine and the subthalamic nucleus. Neuroscience 117, 427–438.

Sargisson, R.J. and White, K.G. (2001) Generalization of delayed matching to sample following training at different delays. J. Exp. Anal. Behav. 75, 1–14.

Sawaguchi, T. and Goldman-Rakic, P.S. (1994) The role of D1-dopamine receptor in working memory: local injections of dopamine antagonists into the prefrontal cortex of rhesus monkeys performing an oculomotor delayed-response task. J. Neurophysiol. 71, 515–528.

Sawaguchi, T., Matsumara, M. and Kubota, K. (1988) Dopamine enhances the neuronal activity of spatial short-term memory task in the primate prefrontal cortex. Neurosci. Res. 5, 465–473.

Sawaguchi, T., Matsumara, M. and Kubota, K. (1990a) Catecholaminergic effects on neuronal activity related to a delayed response task in monkey prefrontal cortex. J. Neurophysiol. 63, 1385–1400.

Sawaguchi, T., Matsumara, M. and Kubota, K. (1990b) Effects of dopamine antagonists on neuronal activity related to a delayed response task in monkey prefrontal cortex. J. Neurophysiol. 63, 1401–1412.

Schultz, W. and Romo, R. (1990) Dopamine neurons of the monkey midbrain: contingencies of responses to stimuli eliciting immediate behavioral reactions. J. Neurophysiol. 63, 607–624.

Seamans, J.K., Floresco, S.B. and Phillips, A.G. (1998) D1 receptor modulation of hippocampal-prefrontal cortical circuits integrating spatial memory with executive functions in the rat. J. Neurosci. 18, 1613–1621.

Seamans, J.K., Durstewitz, D., Christie, B.R., Stevens, C.F. and Sejnowski, T.J. (2001a) Dopamine D1/D5 receptor modulation of excitatory synaptic inputs to layer V prefrontal cortex neurons. Proc. Nat. Acad. Sci. U.S.A. 98, 301–306.

Seamans, J.K., Gorelova, N., Durstewitz, D. and Yang, C.R. (2001b) Bidirectional dopamine modulation of GABAergic inhibition in prefrontal cortical pyramidal neurons. J. Neurosci. 21, 3628–3638.

Sobotka, S., Diltz, M.D. and Ringo, J.L. (2005) Can delay-period activity explain working memory? J. Neurophysiol. 93, 128–136.

Strogatz, S.H. (1994) Nonlinear dynamic and chaos. Reading, MA: Addison-Wesley.

Sutton, R.S. and Barto, A.G. (1998) Reinforcement learning. Cambridge: MIT Press.

Tost, H., Meyer-Lindenberg, A., Klein, S., Schmitt, A., Hohn, F., Tenckhoff, A., Ruf, M., Ende, G., Rietschel, M., Henn, F.A. and Braus, D.F. (2006) D2 antidopaminergic modulation of frontal lobe function in healthy human subjects. Biol. Psychiatry 60, 1196–1205.

Trantham-Davidson, H., Neely, L.C., Lavin, A. and Seamans, J.K. (2004) Mechanisms underlying differential D1 versus D2 dopamine receptor regulation of inhibition in prefrontal cortex. J. Neurosci. 24, 10652–10659.

Tseng, K.Y. and O'Donnell, P. (2005) Post-pubertal emergence of prefrontal cortical up states induced by D1-NMDA co-activation. Cereb. Cortex 15, 49–57.

Unterrainer, J.M. and Owen, A.M. (2006) Planning and problem solving: from neuropsychology to functional neuroimaging. J. Physiol. (Paris) 99(4–6), 308–317.

Vijayraghavan, S., Wang, M., Birnbaum, S.G., Williams, G.V. and Arnsten, A.F. (2007) Inverted-U dopamine D1 receptor actions on prefrontal neurons engaged in working memory. Nat. Neurosci. 10(3), 376–384.

Wallis, J.D., Anderson, K.C. and Miller, E.K. (2001) Single neurons in prefrontal cortex encode abstract rules. Nature 411, 953–956.

Winterer, G. and Weinberger, D.R. (2004) Genes, dopamine and cortical signal-to-noise ratio in schizophrenia. Trends Neurosci. 27, 683–690.

Yang, C.R. and Seamans, J.K. (1996) Dopamine D1 receptor actions in layers V-VI rat prefrontal cortex neurons in vitro: modulation of dendritic-somatic signal integration. J. Neurosci. 16, 1922–1935.

Young, C.E. and Yang, C.R. (2004) Dopamine D1/D5 receptor modulates state-dependent switching of soma-dendritic Ca2+ potentials via differential protein kinase A and C activation in rat prefrontal cortical neurons. J. Neurosci. 24, 8–23.

Zahrt, J., Taylor, J.R., Mathew, R.G. and Arnsten, A.F. (1997) Supranormal stimulation of D1 dopamine receptors in the rodent prefrontal cortex impairs spatial working memory performance. J. Neurosci. 17, 8528–8535.

Zheng, P., Zhang, X.X., Bunney, B.S. and Shi, W.X. (1999) Opposite modulation of cortical N-methyl-D-aspartate receptor-mediated responses by low and high concentrations of dopamine. Neuroscience 91, 527–535.

Chapter 16

Stable and Unstable Activation of the Prefrontal Cortex with Dopaminergic Modulation

Shoji Tanaka

Department of Electrical and Electronics Engineering, Sophia University, 7-1 Kioicho, Chiyoda-ku, Tokyo 102-8554, Japan

Abstract. The so-called inverted-U shaped profile of the dependence of working memory performance on D1 receptor activation in the prefrontal cortex (PFC) illustrates how the cognitive function declines with the deviation of the dopamine (DA) level in the PFC from the optimum point. A recent computational study, however, suggests that the PFC circuit is not always stable along the inverted-U shaped curve. This is due to effectively positive feedback control of the prefronto-mesoprefrontal system under hypodopaminergic conditions. The instability of the PFC circuit makes the activity of the PFC largely fluctuating. Because of this, even a slight increase in DA releasability causes a catastrophic jump of the activity of the PFC from a low to a high level. Under hyperdopaminergic conditions, in contrast, the PFC circuit is stable by means of negative feedback control. This chapter reviews these computational studies and argues the stable and unstable activation of the PFC in connection with cognitive deficits observed in schizophrenia.

16.1 Introduction

16.1.1 COGNITIVE FUNCTIONS OF THE PREFRONTAL CORTEX

The prefrontal cortical mechanisms for cognition have been studied intensively for decades (e.g., Goldman-Rakic, 1987; Fuster, 1997). The prefrontal cortex (PFC) is the center for the maintenance and manipulation of working memory. If a large-scale network in the brain mediates executive functions, the PFC is also considered to play a central role in the operation of the network of the executive system.

16.1.2 A CIRCUIT VIEW

The maintenance of working memory is evidenced by persistent spike trains with selectivity that was recorded from neurons in the primate PFC (e.g., Funahashi et al., 1989). It requires modest activation of both pyramidal neurons and GABAergic interneurons in the PFC because the excitation and inhibition are to be balanced in the circuit so that the neurons exhibit persistent activity with rather low frequency. Because this balance is subtle, working memory representation is susceptible to distracters and noise, and, at the same time, it would be readily manipulated. The nature of working memory is, therefore, largely dependent on the dynamics of the PFC circuit. However, the dynamical characteristics of the PFC circuit are mostly unknown yet.

16.1.3 PSYCHIATRY

The impairment of working memory and other cognitive functions is one of the core symptoms of schizophrenia (e.g., Park and Holzman, 1992; Goldman-Rakic, 1994). The notion that cognitive functions are generated by the dynamics of relevant circuits in the brain leads to the hypothesis that psychiatric patients with cognitive deficits would have PFC circuit dynamics that are different from that of healthy subjects [the "*circuit dynamics hypothesis of schizophrenia*" (Tanaka, 2006)]. Knowing the difference would be crucial not only for understanding the mechanisms of the diseases but for designing pharmacological treatment strategies.

16.2 Dopaminergic Modulation of the Prefrontal Cortical Dynamics

16.2.1 D1 RECEPTOR CONTRIBUTION TO DOPAMINERGIC MODULATION

The pioneering work by Brozoski et al. (1979) and the succeeding studies have suggested the involvement of dopamine (DA) in cognitive functions. Dopaminergic modulation of signal transmission and neuronal excitability alters the circuit dynamics for cognition.

A computational study by Durstewitz and coworkers suggests that D1 receptor stimulation increases the robustness of working memory representation (Durstewitz et al., 2000; Durstewitz and Seamans, 2002). Such kind of qualitative change in working memory representation through D1 receptor activation would be a basis for cognitive operations. Actually, a computer simulation of dopaminergic modulation of the PFC circuit for working memory has shown that the same cue input can form different working memory when the activation level of D1 receptors is slightly different (Tanaka, 2002a). Low D1 receptor activation makes working memory readily replaceable by a newly loading item. With slightly higher activation of D1 receptors, both old and new working memory items are represented simultaneously in the same circuit. When the D1 receptor activation is increased further, working memory that is stored in the circuit prevents loading of additional working memory items. These results has lead Tanaka (2002a) to propose the hypothesis that DA can control fundamental cognitive operations by changing the D1 receptor activation in the PFC (the "*operational control hypothesis of dopamine*"). It is interesting to note that the robustness of working memory representation and the switching of the operations of working memory essentially depend on the glutamatergic transmission efficacy through the NMDA receptor on the neurons in the circuit. In computer simulations, NMDA hypofunction results in the failure of maintenance of working memory

(Durstewitz et al., 2000; Brunel and Wang, 2001; Durstewitz and Seamans, 2002; Tanaka 2002a, b).

16.3 The Prefronto-Mesoprefrontal System

16.3.1 CLOSED-LOOP ARCHITECTURE

Closed-loop circuitry is one of its notable characteristics of the mesocortical DA system (Yamashita and Tanaka, 2003, 2005). Because of this, the PFC can have influence on the activity of DA neurons in the midbrain, thereby regulating itself.

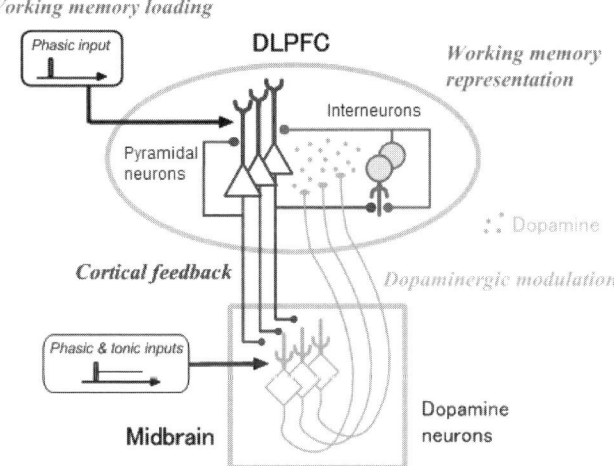

Figure 16.1 A simplified diagram of the prefronto-mesoprefrontal system. The DLPFC circuit is composed of pyramidal neurons and interneurons. The pyramidal neurons send glutamatergic projections, monosynaptically and/or polysynaptically (not shown in the diagram), to the midbrain DA neurons, which send axons to the DLPFC. DA modulates the activities of both the pyramidal neurons and the interneurons. The phasic input to the DLPFC pyramidal neurons triggers the dynamics for working memory formation. The phasic and tonic inputs to the DA unit evokes the activity of the DA unit, and then the DA release, which mimics that observed in experiments (Grace, 1995). Note, however, that it is the glutamatergic input through the cortical feedback projection that contributes to the regulation of the DA release from the axon terminals in this closed-loop circuit model.

There is evidence in rodents showing that mesocortical DA neurons send axons to the cortex and receive feedback projections from the cortex (Tzschentke, 2001; Sesack and Carr, 2002; Sesack et al., 2003). In primates, the DA afferents to the dorsolateral PFC (DLPFC) originate from a widespread continuum of neurons in the substantia nigra (SN), ventral tegmental area (VTA), and the retrorubral area (RRA) (Williams and Goldman-Rakic, 1998). Among these, the SN/RRA sends the major afferents to the DLPFC. A recent anatomical tracing study in monkeys suggests that, in contrast to the mPFC, which projects to primarily the VTA and RRA, the DLPFC sends projections to primarily the SN/RRA with fewer projections to the VTA (Frankle et al., 2006). The cortical projections to the midbrain DA neurons in primates appear to be limited compared to those of rodents (Frankle et al., 2006). However, because midbrain DA neurons are enriched in glutamate receptors (Paquet et al., 1997), there is also a possibility that the DLPFC influences the midbrain DA neurons polysynaptically. Therefore, the DLPFC could control the activity of midbrain DA neurons monosynaptically and/or polysynaptically. These results suggest reciprocal connections between DLPFC neurons and the midbrain DA neurons (Fig. 16.1).

16.4 The Prefrontal Cortex

16.4.1 PREFRONTAL CORTICAL CIRCUIT

The model DLPFC circuit is composed of a pyramidal neuron unit (a population of pyramidal neurons) and an inhibitory interneuron unit (a population of GABAergic interneurons), as shown in Fig. 16.1. These units have mutual and recurrent connections. The pyramidal neuron unit receives a transient external input, which triggers the dynamics of the circuit. The state equations of these units are given by

$$\frac{dx_p}{dt} = -\frac{x_p}{\tau_p} + W_{pp}(z) f(x_p) - W_{np} f(x_n) + I_{cue}$$

$$\frac{dx_n}{dt} = -\frac{x_n}{\tau_n(z)} + W_{pn}(z) f(x_p) - W_{nn} f(x_n)$$

(16.1)

where x_p and x_n are the population activities of the pyramidal neuron unit and the interneuron unit, respectively, τ_p and τ_n are the time constants of these units, W_{ij} $(i, j = p, n)$ is the synaptic efficacy from unit i to j, and I_{cue} is the transient external input that mimics the cue input (Tanaka, 2006; Yamashita and Tanaka, 2005). The activation function is given by

$$f(x) = \begin{cases} f_{max} \tan h (0.15x) & x \geq 0 \\ 0 & x < 0 \end{cases}$$

(16.2)

where f_{max} is the maximum firing rate (100 sp/s). This activation function is assumed to be common to both neuron units. The parameters that depend on z are subject to dopaminergic modulation, where z is the D1 receptor activation level (see Section 16.5.2).

16.4.2 DOPAMINERGIC MODULATION VIA D1 RECEPTORS

The activation of D1 receptors affects the conductances of ion channels, such as NMDA, AMPA, and potassium, thereby changing the efficacy of glutamatergic signal transmission (W_{pp} and W_{pn}) and the excitability of the interneuron (τ_n) (Yamashita and Tanaka, 2005; Tanaka, 2006). These are described by

$$W_{pp}(z) = W_{pp}(0)(1 + az)$$

$$W_{pn}(z) = W_{pn}(0)(1 + bz) \tag{16.3}$$

$$\tau_n(z) = \tau_n(0)(1 + cz)$$

where a, b, and c are constants.

16.4.3 THE EQUILIBRIUM STATE OF THE PREFRONTAL CORTEX

The equilibrium state of this model circuit is obtained from Eq. 16.1 (by setting $d/dt = 0$) to be

$$x_p = \tau_p \left[W_{pp}(z)f(x_p) - W_{np}f(x_n) \right]$$

$$x_n = \tau_n(z) \left[W_{pn}(z)f(x_p) - W_{nn}f(x_n) \right] \tag{16.4}$$

From the above equations, we have

$$x_p = \tau_p \left[W_{pp}(z)f(x_p) - W_{np}f \left[\tau_n(z) \left\{ \left[W_{pn}(z) - \frac{W_{nn}}{W_{np}} W_{pp}(z) \right] f(x_p) + \frac{W_{nn}}{\tau_p W_{np}} x_p \right\} \right] \right]$$

$$x_n = \tau_n(z) \left[W_{pn}(z)f \left[\tau_p \left\{ \frac{W_{pp}(z)}{\tau_n(z)W_{pn}} x_n + \left(\frac{W_{nn}}{W_{pn}} W_{pp}(z) - W_{np} \right) f(x_n) \right\} \right] - W_{nn}f(x_n) \right] \tag{16.5}$$

These equations specify the equilibrium states of the pyramidal neuron unit and the interneuron unit, respectively.

16.5 The Dopaminergic System

16.5.1 THE MESOPREFRONTAL DOPAMINERGIC SYSTEM

In the closed-loop system, the glutamatergic input from the PFC controls the activity of the DA unit. The dynamics of the dopaminergic system obeys the following set of equations (Tanaka, 2006):

$$\frac{dx_d}{dt} = -\frac{x_d}{\tau_d} + W_{pd}(z)\, f(x_p) + I_{phasic} + I_{tonic}$$

$$\frac{dy}{dt} = -\frac{y}{\tau_y} + W_{dy}\, f(x_d)$$

(16.6)

where x_d is the population activity of the DA unit, τ_d is the time constant, W_{pd} is the synaptic efficacy of the corticomesencephalic projection, I_{phasic} and I_{tonic} are the phasic and tonic inputs to the DA unit, y is the DA release in the PFC, τ_y is the time constant, and W_{dy} is the DA releasability, which is the efficacy of the DA release by the activation of the DA neurons.

16.5.2 THE EQUILIBRIUM STATE OF THE MESOPREFRONTAL DOPAMINERGIC SYSTEM

From Eq. 16.6, we obtain the equilibrium state of the dopaminergic system as

$$x_d = \tau_{max}\left[W_{pd}\, f(x_p) + I_{tonic} \right]$$

$$y = \tau_y W_{dy}\, f(x_d)$$

(16.7)

The D1 receptor activation depends on the DA release: $z = z_{max} f(y)$, where z_{max} is a constant. Substituting Eq. 16.7 into this equation, we have the D1 receptor activation as

$$z = z_{max} f\left[\tau_y W_{dy}\, f\left\{ \tau_d \left[W_{pd}\, f(x_p) + I_{tonic} \right] \right\} \right]$$

(16.8)

This equation specifies the equilibrium state of the dopaminergic system.

16.6. The Operating Point of the Prefrontal Cortical Circuit

The operating point of the PFC circuit is determined by the interaction between the PFC and the mesoprefrontal dopaminergic system as follows (Fig. 16.2): The inverted-U shaped curve represents the equilibrium condition of the PFC circuit, which is given by Eq. 16.5. This curve depends only on the parameters of the PFC circuit, including the activity of the D1 receptors on the PFC neurons. The monotonically increasing curve represents the equilibrium condition of the mesocortical dopaminergic system, which is given by Eq. 16.8. This curve covaries with the parameters of the dopaminergic system: increasing z_{max}, W_{dy}, W_{pd}, τ_y, or τ_d decreases the slope of the curve. It intersects with the inverted-U shaped curve with zero, one, or two intersections, depending on the DA releasability, for example. The intersections, which are fixed points, are indicated by circles or triangles. The intersections with circles are stable and those with triangles are unstable, which are judged mathematically with a conventional method (e.g., Khalil, 2000). The origin is always stable, which is indicated by a circle. Figure 16.2 also shows trajectories of state transition from various initial states, converging to one of the stable fixed points. This becomes the operating point of the system. When the DA releasability is higher, the monotonically increasing curve is less steep. In this case, the stable fixed point has higher stability (Fig. 16.2C). When the DA releasability decreases, the stable fixed point approaches the unstable fixed point (Fig. 16.2A). Both fixed points merge at a certain value of the DA releasability, and disappear for the DA releasability lower than this critical value.

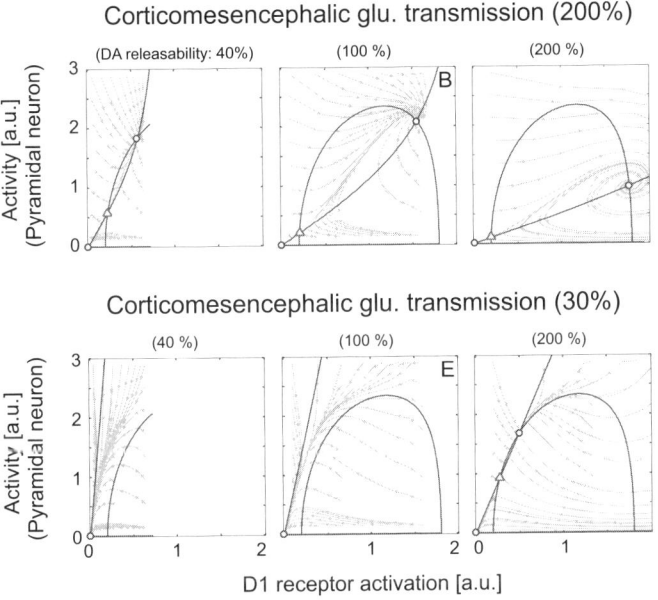

Figure 16.2 Fixed points of the system under the dopaminergic modulation of the PFC. The curves show the inverted-U shaped curve and the monotonically increasing curve. The inverted-U shaped curve represents the equilibrium condition of the PFC circuit, while the monotonically increasing curve represents the equilibrium condition of the mesocortical dopaminergic system. The gray trajectories with arrows show how the states with different initial conditions converge to one of the stable fixed points. The circles indicate the stable fixed points. The triangles indicate the unstable fixed points.

Figure 16.3 illustrates these dynamical characteristics of the prefronto-mesoprefrontal system. The inverted-U shaped curve is independent of the DA release. The DA release changes the slope of the monotonically increasing curve, which is simply depicted by a straight line in this figure. Between the two intersections of these two curves, the right one is always a stable fixed point and the left one is always an unstable fixed point. The unstable region is the range of D1 receptor activation in which the unstable fixed point exists on the inverted-U shaped curve.

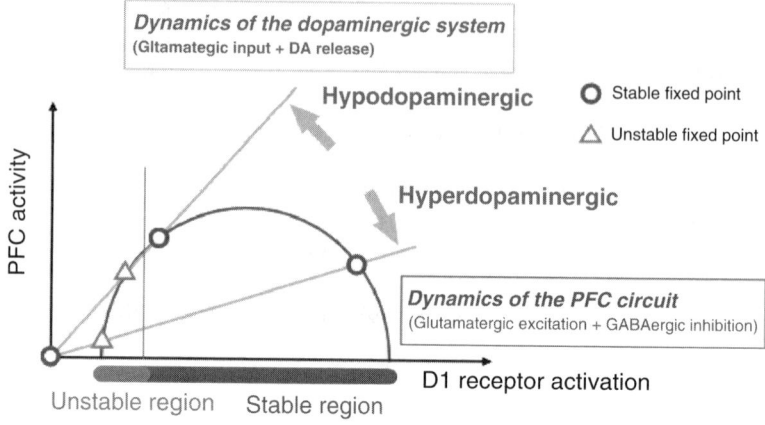

Figure 16.3 An illustration showing how the operating point of the PFC circuit could be determined. The slope of the line varies with the DA releasability. The operating point of the PFC circuit is one of the stable fixed points. Under hypodopaminergic conditions, the operating point is close to the unstable fixed point. In contrast, hyperdopaminergic conditions bring the operating point toward right, being away from the unstable fixed point.

16.7. Instability of the Prefrontal Cortical Circuit

16.7.1 HYPODOPAMINERGIC STATES OF THE PREFRONTAL CORTEX

That hypodopaminergic states are unstable is rich in implications. The PFC of patients with schizophrenia has lower baseline dopaminergic transmission compared to normal subjects (Davis et al., 1991; Okubo et al., 1997; Laruelle, 2000; Abi-Dargham et al., 2002; Weinberger and Laruelle, 2002; Guo et al., 2003). This hypodopaminergic state is associated with negative symptoms and cognitive impairment in schizophrenia (Kahn and Davis, 2000; Abi-Dargham et al., 2002). Figure 16.4 shows the relationship between the PFC activity and the DA releasability as well as the relationship between the D1 receptor activation and the DA releasability. When the glutamatergic transmission in the cortical feedback projection is low (Fig. 16.4A), the PFC activity suddenly jumps to a high level from a very low level by an increase in the DA releasability, called *catastrophic jump* (Tanaka, 2006).

16.7.2 OVERACTIVATION OF THE PREFRONTAL CORTEX

Many studies have reported that the PFC of patients with schizophrenia are hypoactive when the subjects perform working memory tasks relative to healthy subjects (Paulman et al., 1990; Andreasen et al., 1992; Carter et al., 1998; Meyer-Lindenberg et al., 2002; Ramsey et al., 2002). Recent studies, however, report overactivation during working memory performance (Manoach et al., 1999, 2000; Callicott et al., 2000, 2003; Weinberger et al., 2001; Manoach, 2003). Both are associated with the impairment of cognitive functions including working memory, which are commonly seen in schizophrenia (for reviews: Elvevag and Goldberg, 2000; Kuperberg and Heckers, 2000; Goldberg et al., 2003; Lewis, 2004; Winterer and Weinberger, 2004). The model reviewed in this chapter suggests that both overactivation and underactivation can occur under hypodopaminergic conditions. The bifurcation to hypoactive and hyperactive states is due to the *instability* of the system. Even a slight difference in the DA releasability can cause a big difference in the activity of the PFC. These are formulated as the "*instability theory of schizophrenia*" (Tanaka, 2006).

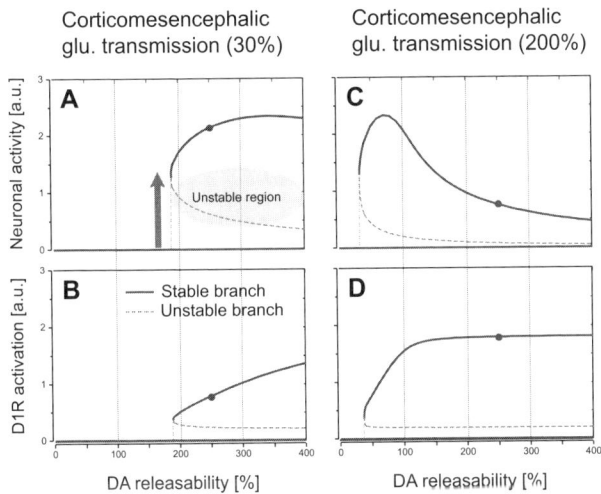

Figure 16.4 Bifurcation diagrams of PFC activity (A and C) and D1 receptor activation (B and D). When the glutamatergic transmission efficacy through the corticomesencephalic projection is low (A and B), a large unstable region appears, which separates high and low levels of PFC activity. Even a slight increase in DA releasability can cause a catastrophic jump in PFC activity (the arrow). Note that, even if the PFC is activated high (A), the D1 receptor activation is rather low (B). In contrast, the PFC activity is modest in C with high D1 receptor activation (D) when the glutamatergic transmission efficacy through the corticomesencephalic projection is high.

16.8. Concluding Remarks

This chapter argued how the PFC changes its activity through dopaminergic modulation. The model reviewed here predicts the following features on PFC activation: (i) Stability of the PFC circuit depends on the DA tone in the PFC; (ii) Instability occurs in the PFC circuit under hypodopaminergic conditions, yielding abnormal activation of the PFC. PFC activity shows a catastrophic jump by a slight increase in the DA releasability of the mesocortical dopaminergic system. This may account for seemingly inconsistent activation of the PFC that has been repeatedly observed in the brains of patients with schizophrenia; (iii) The PFC circuit is stable when the PFC is hyperdopaminergic. In this case, the prefronto-mesoprefrontal system works as a *regulator* of PFC activity, which is enabled by the negative feedback control of the prefronto-mesoprefrontal system; (iv) Under hyperdopaminergic conditions, the activity of the PFC is modest with high D1 receptor activation. This is due to effective GABAergic inhibition in the PFC circuit. If the GABAergic inhibition is weaker, as suggested in schizophrenia (Benes and Berretta, 2001; Reynolds et al., 2001; Lewis et al., 2005), this state would lose its stability, causing cognitive deficits along with unregulated hyperactivity of the PFC.

The hypothesis that the cortical dopaminergic tone is regulated by a feedback mechanism of the closed-loop cortico-mesocortical circuit is interesting because the DA release in the PFC is automatically regulated by the system and, therefore, the PFC can control the activity of itself through dopaminergic modulation. A regulator would be one of the fundamental functions of the prefronto-mesoprefrontal system. This regulator is, however, *imperfect* as argued above. Because of this, the DA tone in the PFC is not always successfully regulated, and hence PFC activity can fluctuate. This may be one of the reasons for the vulnerability of our cognitive system.

References

Abi-Dargham, A., Mawlawi, O., Lombardo, I., Gil, R., Martinez, D., Huang, Y., Hwang, D.-R., Keilp, J., Kochan, L., Heertum, R. V., Gorman, J. M. and Laruelle, M. (2002) Prefrontal dopamine D1 receptors and working memory in schizophrenia. J. Neurosci. 22, 3708–3719.

Andreasen, N., Rezai, K., Alliger, R., Swayze II, V., Flaum, M., Kirchner, P., Cohen, G. and O'Leary, D. S. (1992) Hypofrontality in neuroleptic-naive patients and in patients with chronic schizophrenia. Assessment with xenon 133 single-photon emission computed tomography and the Tower of London. Arch. Gen. Psychiatry. 49, 943–958.

Benes, F. M. and Berretta, S. (2001) GABAergic interneurons: implications for understanding schizophrenia and bipolar disorder. Neuropsychopharmacology. 25, 1–27.

Brozoski, T. J., Brown, R. M., Rosvold, H. E. and Goldman, P. S. (1979) Cognitive deficit caused by regional depletion of dopamine in prefrontal cortex of rhesus monkey. Science. 205, 929–932.

Brunel, N. and Wang, X.-J. (2001) Effects of neuromodulation in a cortical network model of object working memory dominated by recurrent inhibition. J. Comput. Neurosci. 11, 63–85.

Callicott, J. H., Bertolino, A., Mattay, V. S., Langheim, F. J., Duyn, J., Coppola, R., Goldberg, T. E. and Weinberger, D. R. (2000) Physiological dysfunction of the dorsolateral prefrontal cortex in schizophrenia revisited. Cereb. Cortex. 10, 1078–1092.

Callicott, J. H., Mattay, V. S., Verchinski, B. A., Marenco, S., Egan, M. F. and Weinberger, D. R. (2003) Complexity of prefrontal cortical dysfunction in schizophrenia: more than up or down. Am. J. Psychiatry. 160, 2209–2215.

Carter, C. S., Perlstein, P., Ganguli, R., Brar, J., Mintun, M. and Cohen, J. D. (1998) Functional hypofrontality and working memory dysfunction in schizophrenia. Am. J. Psychiatry. 155, 1285–1287.

Davis, K. L., Kahn, R. S., Ko, G. and Davidson, M. (1991) Dopamine in schizophrenia: a review and reconceptualization. Am. J. Psychiatry. 148, 1474–1486.

Durstewitz, D., Seamans, J. K. and Sejnowski, T. J. (2000) Dopamine-mediated stabilization of delay-period activity in a network model of prefrontal cortex. J. Neurophysiol. 83, 1733–1750.

Durstewitz, D. and Seamans, J. K. (2002) The computational role of dopamine D1 receptors in working memory. Neural Netw. 15, 561–572.

Elvevag, B. and Goldberg, T. E. (2000) Cognitive impairment in schizophrenia is the core of the disorder. Crit. Rev. Neurobiol. 14, 1–21.

Frankle, W. G., Laruelle, M. and Haber, S. N. (2006) Prefrontal cortical projections to the midbrain in primates: evidence for a sparse connection. Neuropsychopharmacology 31, 1627–1636.

Funahashi, S., Bruce, C. J. and Goldman-Rakic, P. S. (1989) Mnemonic coding of visual space in the monkey's dorsolateral prefrontal cortex. J. Neurophysiol. 61, 331–349.

Fuster, J. M. (1997) *The prefrontal cortex*, third edition. Lippincott-Raven. Philadelphia, New York.

Goldberg, T. E., David, A. and Gold, J. M. (2003) Neurocognitive deficits in schizophrenia. In: Hirsch SR and Weinberger DR (Eds.) *Schizophrenia*, 2e. Blackwell, Malden, MA, pp. 168–184.

Goldman-Rakic, P. S. (1987) Circuitry of primate prefrontal cortex and regulation of behavior by representational memory. *Handbook of physiology, Sec 1: The nervous system, Vol V: Higher functions of the brain, Part 1*. American Physiological Society, Bethesda, MD, 373–417.

Goldman-Rakic, P. S. (1994) Working memory dysfunction in schizophrenia. J. Neuropsychiatry Clin. Neurosci. 6, 348–357.

Grace, A. A. (1995) The tonic/phasic model of dopamine system regulation: its relevance for understanding how stimulant abuse can alter basal ganglia function. Drug Alcohol Depend. 37, 111–129.

Guo, N., Hwang, D. R., Lo, E. S., Huang, Y. Y., Laruelle, M. and Abi-Dargham, A. (2003) Dopamine depletion and in vivo binding of PET D1 receptor radioligands: implications for imaging studies in schizophrenia. Neuropsychopharmacology. 28, 1703–1711.

Kahn, R. S. and Davis, K. L. (2000) New Developments in Dopamine and Schizophrenia. *Psychopharmacology: The Fourth Generation of Progress*. http://www.acnp.org/G4/GN401000115/CH113.html.

Khalil, H. K. (2000) Nonlinear systems, third edition. Prentice-Hall: Upper Saddle River, NJ.

Kuperberg, G. and Heckers, S. (2000) Schizophrenia and cognitive function. Curr. Opin. Neurobiol. 10, 205–210.

Laruelle, M. (2000) Imaging synaptic neurotransmission with in vivo binding competition techniques: a critical review. J Cereb. Blood Flow Metab. 20, 423–451.

Lewis, D. A., Hashimoto, T. and Volk, D. W. (2005) Cortical inhibitory neurons and schizophrenia. Nat. Rev. Neurosci. 6, 312–324.

Lewis, R. (2004) Should cognitive deficit be a diagnostic criterion for schizophrenia? J. Psychiatry Neurosci. 29, 102–113.

Manoach, D. S. (2003) Prefrontal cortex dysfunction during working memory performance in schizophrenia: reconciling discrepant findings. Schizophr. Res. 60, 285–298.

Manoach, D. S., Gollub, R. L., Benson, E. S., Searl, M. M., Goff, D. C., Halpern, E., Saper, C. B. and Rauch, S. L. (2000) Schizophrenic subjects show aberrant fMRI activation of dorsolateral prefrontal cortex and basal ganglia during working memory performance. Biol. Psychiatry. 48, 99–109.

Manoach, D. S., Press, D. Z., Thangaraj, V., Searl, M. M., Goff, D. C., Halpern, E., Saper, C. B. and Warach, S. (1999) Schizophrenic subjects activate dorsolateral prefrontal cortex during a working memory task, as measured by fMRI. Biol. Psychiatry. 45, 1128–1137.

Meyer-Lindenberg, A., Miletich, R. S., Kohn, P. D., Esposito, G., Carson, R. E., Quarantelli, M., Weinberger, D. R. and Berman, K. F. (2002) Reduced prefrontal activity predicts exaggerated striatal dopaminergic function in schizophrenia. Nat. Neurosci. 5, 267–271.

Okubo, Y., Suhara, T., Suzuki, K., Kobayashi, K., Inoue, O., Terasaki, O., Someya, Y., Sassa, T., Sudo, Y., Matsushima, E., Iyo, M., Tateno, Y. and Toru, M. (1997) Decreased prefrontal dopamine D1 receptors in schizophrenia revealed by PET. Nature. 385, 634–636.

Paquet, M., Tremblay, M., Soghomonian, J.-J. and Smith, Y. (1997) AMPA and NMDA glutamate receptor subunits in midbrain dopaminergic neurons in the squirrel monkey: an immunohistochemical and in situ hybridization study. J. Neurosci. 17, 1377–1396.

Park, S. and Holzman, P. S. (1992) Schizophrenics show spatial working memory deficits. Arch. Gen. Psychiatry. 49, 975–982.

Paulman, R. G., Devous, M. D., Sr., Gregory, R. R., Herman, J. H., Jennings, L., Bonte, F. J., Nasrallah, H. A., Raese, J. D. (1990) Hypofrontality and cognitive impairment in schizophrenia: dynamic single-photon tomography and neuropsychological assessment of schizophrenic brain function. Biol. Psychiatry. 27, 377–399.

Ramsey, N. F., Koning, H. A., Welles, P., Cahn, W., van der Linden, J. A. and Kahn, R. S. (2002) Excessive recruitment of neural systems subserving logical reasoning in schizophrenia. Brain. 125, 1793–1807.

Reynolds, G. P., Zhang, Z. J. and Beasley, C. L. (2001) Neurochemical correlates of cortical GABAergic deficits in schizophrenia: selective losses of calcium binding protein immunoreactivity. Brain Res. Bull. 55, 579–584.

Sesack, S. R. and Carr, D. B. (2002) Selective prefrontal cortex inputs to dopamine cells: implications for schizophrenia. Physiol. Behav. 77, 513–517.

Sesack, S. R., Carr, D. B., Omelchenko, N. and Pinto, A. (2003) Anatomical substrates for glutamate-dopamine interactions: evidence for specificity of connections and extrasynaptic actions. Ann. N. Y. Acad. Sci. 1003, 36–52.

Tanaka, S. (2002a) Dopamine controls fundamental cognitive operations of multi-target spatial working memory. Neural Netw. 15, 573–582.

Tanaka, S. (2002b) Multi-directional representation of spatial working memory in a model prefrontal cortical circuit. Neurocomputing. 44–46, 1001–1008.

Tanaka, S. (2006) Dopaminergic control of working memory and its relevance to schizophrenia: a circuit dynamics perspective. Neuroscience. 139, 153–171.

Tzschentke, T. M. (2001) Pharmacology and behavioral pharmacology of the mesocortical dopamine system. Prog. Neurobiol. 63, 241–320.

Weinberger, D. R., Egan, M. F., Bertolino, A., Callicott, J. H., Mattay, V. S., Lipska, B. K., Berman, K. F. and Goldberg, T. E. (2001) Prefrontal neurons and the genetics of schizophrenia. Biol. Psychiatry. 50, 825–844.

Weinberger, D. R. and Laruelle, M. (2002) Neurochemical And Neuropharmacological Imaging In Schizophrenia. In: Kenneth L Davis, Dennis Charney, Joseph T Coyle, and Charles Nemeroff (Eds.), Neuropsychopharmacology: The Fifth Generation of Progress. Lippincott Williams & Wilkins, Philadelphia, PA, pp. 833–855.

Williams, S. M. and Goldman-Rakic, P. S. (1998) Widespread origin of the primate mesofrontal dopamine system. Cereb. Cortex. 8, 321–345.

Winterer, G. and Weinberger, D. R. (2004) Genes, dopamine and cortical signal-to-noise ratio in schizophrenia. Trends. Neurosci. 27, 683–690.

Yamashita, K. and Tanaka, S. (2003) Circuit properties of the cortico-mesocortical system. Neurocomputing. 52–54, 969–975.

Yamashita, K. and Tanaka, S. (2005) Parametric study of dopaminergic neuromodulatory effects in a reduced model of the prefrontal cortex. Neurocomputing. 65–66, 579–586.

Chapter 17

Dopamine and Norepinephrine Modulation of Cortical and Subcortical Dynamics During Visuomotor Learning

Hernán G. Rey, Sergio E. Lew, and B. Silvano Zanutto
Instituto de Ingeniería Biomedica, Facultad de Ingeniería, Universidad de Buenos Aires and
    Instituto de Biologia y Medicina Experimental—CONICET

17.1 Introduction

Understanding the mechanisms of neuromodulation exerted by catecholamines over cortical and subcortical neurons has been one of the major challenges in neuroscience that remains unsolved. Although there is compelling neurophysiological evidence indicating that both dopamine (DA) and norepinephrine (NE) systems are critically involved in the control of neuronal functions, there are still many discrepancies on how these modulations occur, perhaps as result of the different experimental approaches used to investigate the effects. Of particular interest, experiments performed in behaving animals have provided significant amount of information about how these modulatory systems may regulate complex neural networks such as during behavioral conditions in which both excitatory and inhibitory neurons are engaged. In this regard, computational models could provide a better understanding on the neuronal dynamics underlying these interactions by formalizing biologically plausible hypotheses and predictions to be tested in realistic experimental conditions. Two main approaches have been successfully implemented to tackle this issue. The first one is bottom-up, that is, brain functions are modeled from neurophysiological bases, simulating the properties of neurons and synapses as close as possible. The other approach is top-down, that is, the behavioral and brain functions are simulated as close as possible to animal behavior. Unfortunately, both approaches suffer from the lack of experimental data to link neurons and higher brain functions. Because of this, it seems obvious to choose a position in between both approaches to start the modeling work. Behavioral results constrain the modeling process, giving top-down cues in the development of a model that incorporates molecular, physiological, and behavioral evidence parsimoniously. In this chapter, we will present and discuss how neural network models that take into account realistic biological data could be used to explain the roles of DA and NE underlying learning processes involved in simple and complex tasks performances.

17.2 Neurophysiological and Neuropsychological Hypotheses

In behaving monkeys and rats, dopaminergic neurons in the ventral tegmental area (VTA) and substantia nigra pars compacta (SNc) fire action potentials when appetitive unconditioned stimuli (US), such as food or sweet liquids, are delivered. In classical and operant conditioning, DA neurons increase their activity when conditioned stimuli (CS) are presented and an inhibition can be observed if predicted rewards are omitted (Schultz, 2002; Pan et al., 2005). Time difference models (TD) provide a good fit of the DA neurons firing pattern during classical and operant conditioning (Schultz et al., 1997; Pan et al., 2005). If the focus of interest is to determine the impact of DA on frontal lobe and motor-related areas, the Rescorla-Wagner rule could be applied since it allows the association of previously paired CSs-US preserving blocking and overshadowing effects (Rescorla and Wagner, 1972).

Although phasic DA lasts a few hundred of milliseconds, postsynaptic effects of DA in the prefrontal cortex (PFC) and motor-related structures persist through longer periods of time (Garris and Wightman, 1994; Gonon, 1997). This fact could support the hypothesis of a gating mechanism exerted by DA over these structures (O'Reilly et al., 2002).

In the PFC, DA regulates the excitability of neuronal clusters. VTA stimulation decreases the spontaneous firing of PFC pyramidal neurons, mainly by exciting local interneurons (Lewis and O'Donnell, 2000; Tseng et al., 2006). However, due to the synergist action between NMDA and D1 receptors (Wang and O'Donnell, 2001; Tseng and O'Donnell, 2005), it can be postulated that initially inhibited PFC pyramidal neurons will fire strongly when afferent inputs release larger amounts of glutamate. The activated cluster will then inhibit other clusters (Durstewitz et al., 2000). Thus, a winner-take-all mechanism could be applied to the PFC output.

Taking into consideration that during VD learning, activity changes in the basal ganglia (BG) and the premotor cortex (PMC) occur at the same time course (Brasted and Wise, 2004), BG and PMC can be modeled as a single layer of responsive neurons. It has also been proposed that DA, by modulating the excitability of striatal neurons, would allow the BG to inhibit competing programs and to release the correct one (Mink, 1996). The effect of DA would be to apply a "brake" over all possible motor programs and to release the one whose activity surpasses a fixed threshold.

In addition to neuronal excitability, DA also affects synaptic efficacy in the PFC via long-term potentiation (LTP) and depression (LTD) (Otani et al., 2003). In Reynolds and Wickens (2002), the authors conclude that low levels of DA are required for LTD induction. They also showed that reward-related burst firing of DA neurons induces sufficient DA release to activate intracellular cascades that lead to LTP induction and reinforcement learning. Conversely, omission of reward at an expected time leads to a relative depression of DA release, to a level that is compatible with the induction of LTD following the conjunction of presynaptic and postsynaptic activity. For these reasons, previous computational models have used DA signal to adjust changes in synaptic weight (Schmajuk and Zanutto, 1997; O'Reilly et al., 2002; Gutnisky and Zanutto, 2004a, 2004b).

Visual and somatosensory cortical neurons are modulated by NE neurons from the locus coeruleus (LC) (Berridge and Waterhouse, 2003). LC neurons exhibit at least two distinguishable firing modes (Aston-Jones and Cohen, 2005). In the phasic mode, the LC bursts are closely coupled with generally highly accurate behavioral responses. In the tonic mode, LC activity is increased when performance is erratic. The increasing tonic firing reflects an increase in the environmental uncertainty. When reward probability decreases, an increase in the LC tonic activity facilitates alternative behaviors to obtain reward. The LC also receives inputs from the medial prefrontal cortex (mPFC) (Aston-Jones et al., 2005), a brain region involved in decision making. This suggests that neuronal activity in the mPFC may contribute to adjust the firing pattern of LC neurons, which in turn would serves to optimize the balance between exploitation and exploration of opportunities for reward.

In a different model exploring the roles of NE and DA on exploration and exploitation (McClure et al., 2005) using a target detection task (similar in nature to the VD task that will be used later), the authors explicitly include the tonic and phasic components of LC firing. They assume that the phasic mode can have a role in the exploitation of rules. Nevertheless, in Clayton et al. (2004) the authors came up with a new hypothesis, where the phasic activation of the LC is engaged in a facilitation of the behavioral response and it is actually driven by the decision process. As the focus of interest here is the decision process mechanism, the model presented in the next section considers only the tonic component, which has shown a defined correlation with performance of behaving monkeys (Usher et al., 1999). Tonic firing within 2–3 Hz are associated with good performance epochs, while frequencies >3 Hz are related with periods where erratic performance and distractibility are observed. This gives a hint about the function of the NE system in the regulation of exploratory behavior (Aston-Jones et al., 2005).

In the midbrain, NE afferents also impact DA neurons function. In acute brain slices, Paladini and Williams (2004) have observed an inhibitory effect of NE on DA neurons activity through activation of α_1 NE receptors. In Grenhoff et al. (1993) in vivo study in anesthetized rats, stimulation of the LC produced a long-lasting depression of DA cell activity in the VTA and SNc.

Experimental data from behaving monkeys performing DMTS have shown that the PFC is a key component for its learning (Freedman et al., 2002). For instance, behavioral analyses in animals with partial or total PFC damage indicate that the PFC is indeed involved in working memory as well as in complex learning processes (Petrides, 2000). Conversely, simple rules such as in response to avoiding aversive stimuli or learning to respond to visual cues in order to get appetitive reward, can be achieved by animals whose PFC is not developmentally mature (Schmajuk et al., 1997) or damaged (Petrides, 2000).

In behaving monkeys, short-term memory activity (STM) is observed in the last stages of the visual ventral pathway, that is, in V4 and inferotemporal cortices (ITC) (Fuster, 1989) and also in the medial PFC, where reward related stimuli are processed. Anatomical data also indicates that other systems including the VTA/SNc, BG-PMC could be involved in the regulation of STM formation (Cavada et al., 2000).

17.3 The Tasks Under Study

In the VD task, the subject has to associate two different cues (conditioned stimuli CS1 and CS2) with saccadic movements in opposite directions (R1 and R2). A stimulus is presented for 500 ms duration followed by a delay period of 1 s and then the subject is allowed to respond. If a correct response is obtained, the unconditioned stimulus would be delivered 2 s after the cue onset.

In the DMTS task, the subject has to release (R1 = Go) or hold (R2 = No-Go) a lever press depending on the matching/nonmatching of two previous sequentially presented stimuli: the sample and the comparison. The sample stimulus is presented for 500 ms duration followed by 1 s delay, and then the comparison stimulus is presented. If the sample and comparison stimuli match, the subject will have to release the lever to obtain reward, which is delivered 2.4 s after the onset of the sample stimulus. When the sample and comparison stimuli failed to match, the reward will be delivered only if the subject holds the lever press.

Although the VD task is a simple operant conditioning task which can be solved by generating direct stimulus–response (S-R) mappings, once the S-R mappings are acquired (and successfully exploited), the contingencies can be reversed (i.e., if the original associations are CS1-R1 and CS2-R2, now the reinforced ones are CS1-R2 and CS2-R1), and thus a new exploration process will be required to solve the task. In contrast, the DMTS task is a more complex. After the presentation of the sample, the degree of uncertainty for the proper response remains high. It is only after the appearance of the comparison that the contingency can be correctly solved. This is due to the linear nonseparability of the DMTS task, as will be shown in the next section.

17.4 A Computational Theory

In Fig. 17.1, a scheme of the proposed model is shown. The input to the model is composed by different stimuli from the environment (CSs and US). The first layer of the model generates short-term traces for those stimuli. Biologically, this is not the result of a single structure information processing. Actually, it is the result of the interaction between sensory cortices, associative cortices (mPFC, ITC) and subcortical structures (amygdala, hippocampus).

The second layer also involves the interaction of many structures. It represents mainly the lateral PFC, which conveys information from mPFC, ITC, cingulate and parietal cortex. It allows further filtering of task relevant stimuli (by leading to highly selective PFC clusters that encode the S-R mappings), which are also actively maintained in working memory. In addition, it is a key component for learning the DMTS task, since joint representations between sample and comparison stimuli will be generated in this layer. These clusters will then be associated with the proper response (in the third layer) according to the contingencies of the task. Thus, although medial and lateral PFC are engaged in working memory processes, they represent different kinds of information (O'Reilly et al., 2002). Moreover, this is in agreement with the approach of Atallah et al. (2004), as the second layer selects the initially most activated clusters in order to establish the mappings, while the remaining clusters are inhibited (winner-take-all rule).

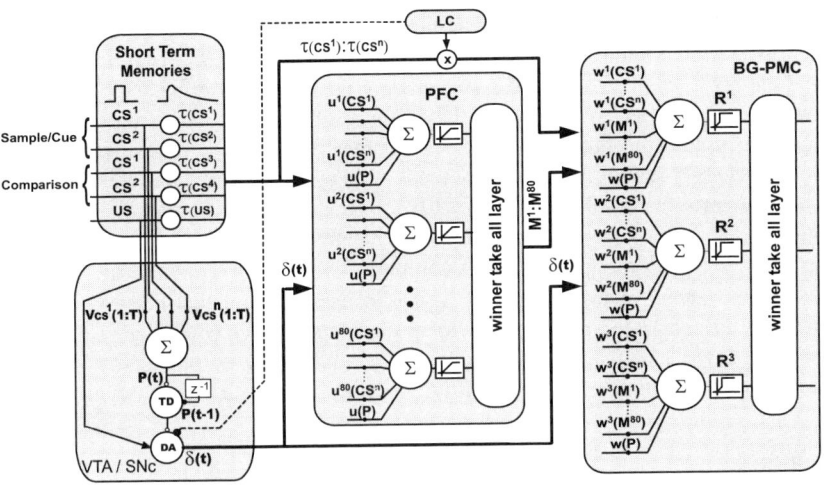

Figure 17.1 Schematic diagram of the computational model. In the PFC, 80 linear threshold neurons are used, 3 in the BG-PMC and a TD(λ_{TD}) model representing the VTA/SNc. The LC represents a modulation exerted by the Locus Coeruleus.

Three possible responses are included (even though the model still works well with more responses), but only two of them are associated with reward (the other one stands for any response nonrelated to the task). Responses are initially executed randomly, simulating a motivational state generated by deprivation. Reinforcement information reaches the VTA and LC mainly from frontal structures related to its processing, as the mPFC. However, the VTA and the LC process such information in a different way in order to produce different patterns of neural responses and/or learning mechanisms. Once the information about reinforcement reaches the VTA, DA is released, and those conditioned stimuli traces at that time active can be associated with reinforcement. As learning proceeds, the probability of being reinforced increases and the synaptic weights in VTA will represent the association between conditioned stimuli and reinforcement. Consequently, every time that a conditioned stimulus is presented, the VTA will fire strongly releasing DA over the PFC and BG-PMC structures. In order to generate this process the VTA/SNc is modeled by a TD model with eligibility traces like the one in Pan et al. (2005). Its output is a prediction error signal $\delta(t)$. The quantity W_t^δ is also introduced as a gating window for the DA postsynaptic effects. When DA firing is closed to baseline, $W_t^\delta = 0.5$. When DA bursts occur, if $\delta(t)$ surpasses a fixed threshold θ_{hebb}, $W_t^\delta = 1$ for the following T seconds. The duration of this window depends on the amplitude of $\delta(t)$, as shown by the experimental evidence (Garris et al., 1994; Gonon, 1997). If min (a,b) is defined as the minimum between a and b, then $T = min (5*\delta(t)+1, 3.5)$. When the predicted reward is omitted, DA firing goes below baseline. If δ(t) is below the threshold θ_antihebb, $W_t^\delta = 0$ for the following 1.5 sec.

The gating mechanism is also used in order to include DA effects on learning. When $W_t^\delta = 1$, hebbian learning is applied to PFC and BG-PMC neurons, and the synapses are potentiated. The opposite occurs when $W_t^\delta = 0$. No synaptic weight modifications take place when VTA/SNc firing is close to baseline. This mechanism allows the model to learn paradigms based on reinforcement and to build S-R mappings that, unless the task contingencies change, remain over time. After reaching criterion, the S-R mappings are represented in the synaptic weights of the model.

The LC processing of reinforcement is proposed as a mechanism that allows the model to react to environment or schedules changes in order to find better strategies (exploration). The tonic level of NE is computed as $1 - \alpha_{lc}*\tau_{US}$, where τ_{US} is the low-pass filtered reward history, and α_{lc} is a fixed constant parameter. This LC activation has a dual effect on the model behavior. First, it multiplies (modulates) the energy flow from layer 1 to the BG-PMC. Second, it inhibits the output of the VTA/SNc. These two actions increase the flexibility for updating learned S-R mappings.

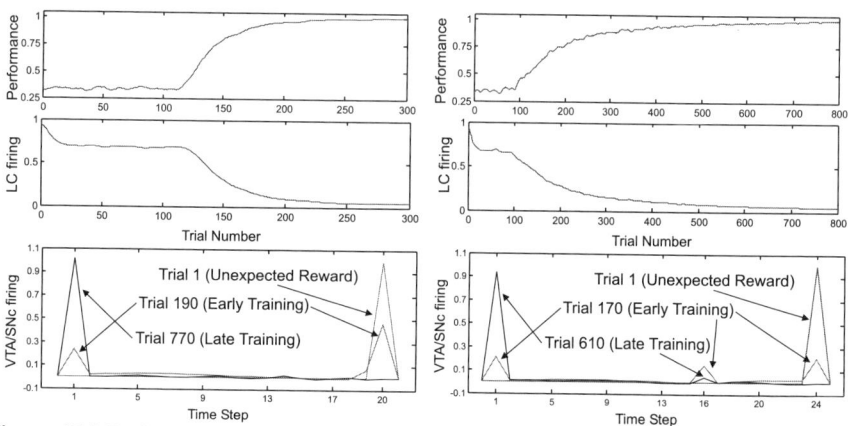

Figure 17.2 Performance, LC firing, and VTA/SNc firing for VD (*left*) and DMTS (*right*) tasks.

Despite the fact that in three-terms contingencies, as in the VD task, direct S-R mappings between input and output neurons are enough to learn the paradigms, major order contingencies require additional resources to be learned. A previous version of this model that does not include layer 2 can explain experiments such as the matching law, partial reinforcement extinction, blocking, learned helplessness, response selection, successive negative contrast effect, modulation of the avoidance response, transfer of control between conditioned stimuli and spontaneous recovery (Schmajuk et al., 1997; Gutnisky et al., 2004a, 2004b). However, when tested in the DMTS paradigm, its performance is close to chance level. Contrary to the VD task and those previously mentioned experiments, input patterns in the DMTS paradigm can be proved to be linearly nonseparable, that is, there is no plane in the input space that separates those input patterns associated with the Go response from those asso-ciated with the No-Go response. Under this condition, the DMTS paradigm cannot be learned with a unique layer of linear threshold neurons. To prove this characteris-tic of the input–response relation in the DMTS task, we proceed as in Roychowdhury et al. (1995), multiplying by –1 those input patterns that should generate a No-Go response to obtain reward. Then, the sum over all possible input patterns is a null vector, that is, there is a linear combination of input patterns that leads to a null vec-tor, which is a sufficient condition for linear nonseparability.

17.5 Simulation Results on Learning of Simple and Complex Tasks

The duration of each trial is 50 time steps (5 sec). Each experiment is conducted during 1,000 trials. The simulations are the result of the ensemble average of 100 independent experiments, unless otherwise stated.

17.5.1 PERFORMANCE, LC FIRING, AND VTA/SNc FIRING

During the VD task, 100% of simulations learned with more than 90% performance, reaching a 100% performance in 300 trials on average. In the DMTS paradigm, from the three possible responses, two responses are related to reward, that is, to release (R1) or hold (R2) the lever. Therefore, a total of four rules can be learned using stimuli CS1 and CS2. Eighty percent of simulations exceeded 90% performance, reaching a 90% performance in about 300 trials on average.

Figure 17.2 shows the average performance, LC firing, and VTA/SNc firing for the VD (*left*) and DMTS tasks (*right*). As expected, the NE level is large when the performance is near chance, and it decreases as the performance is improved. The VTA/SNc firing for both tasks is shown for a single run of the model. Three re-warded trials were chosen for each task in three different stages of the learning proc-ess. At the beginning of training, nonpredicted rewards elicit a phasic DA response at the time when reward is delivered. Early training DA responses show phasic activity at both cues and reward onset. In late training stages, DA neurons show activation at the CS's onset, while no activation is observed when the predicted reward is deliv-ered.

17.5.2 REVERSAL LEARNING

To force an exploration process, a reversal is applied when the model executes 30 correct consecutive responses for each of the two CSs (60 in total). At this stage, previous contingencies between CSs and responses are inverted (Pasupathy and Miller, 2005). At the top left of Fig. 17.3 it can be seen that due to previous learning, there is a time period of about ten trials of behavioral perseveration (the subject sticks to the old reinforced response). As the averaged received reward diminishes, tonic activity in the LC grows (not shown). Thus, there is an increase in the energy flow from the short-term memories layer to motor structures and an inhibition of the dopaminergic firing. This process increases the probability of executing responses randomly, restarting the exploration process.

When no LC modulation is used in the model, the VD task can still be properly acquired. However, it can be seen at the bottom left of Fig. 17.3 that the perseveration effect during a reversal procedure increases from ten trials with LC to 225 trials without it. This shows the importance of the LC inhibition over dopaminergic neurons. When it is absent, the DA neurons take a long time to decrease their firing, preserving the exploitation mechanisms exerted by DA. This effect is in agreement with the results from McClure et al. (2005) although different mechanisms are used to explain it (the latter authors do not include an inhibitory path from the LC to the VTA; actually, they have the opposite path, i.e., an inhibition from the VTA to the LC).

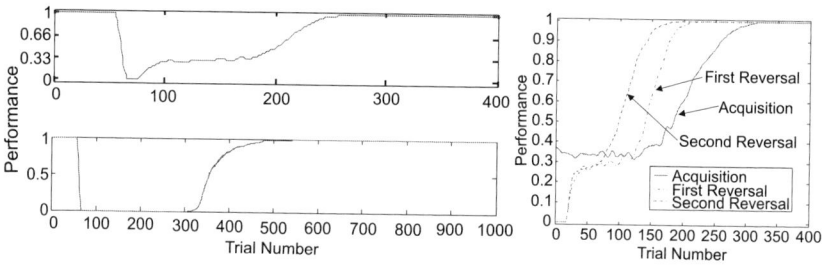

Figure 17.3 *Left*: A reversal procedure in a VD task was applied after 30 correct consecutive trials for each stimulus were executed. Average performance was computed in a range that starts 60 trials before reversal for a model with LC modulation (*top*) and without it (*bottom*). *Right*: For the VD task, two consecutive reversals were performed. The performance of the acquisition and the two reversals are shown.

Two consecutive reversal procedures were also performed in order to test catastrophic interference in the model, a phenomenon that takes place when learning of new associations disrupts the previously learned ones. On the right side of Fig. 17.3, average performance of the acquisition and the two consecutive reversal procedures are compared. As the time of the reversal is random on each run of the model, the reversal onsets were aligned in order to construct the curves. As it can be seen in the figure, reversals reach criterion much faster than the acquisition. If the interference is catastrophic, one will expect the performance of the second reversal to have the same dynamics as the one of the acquisition (as the contingencies are the same).

17.5.3 THE ROLE OF NE IN DMTS TASK

In contrast to the VD task, when the LC is not modeled, the DMTS task performance is highly deteriorated. Only 26% of the simulations reach criterion. As mentioned before, one of the differences between VD and DMTS tasks is that the latter is a linearly nonseparable problem. In the VD task the CS1 is mapped to R1. This can be done by using only the direct path from layer 1 to the BG-PMC layer. In the DMTS task, this is not possible. When CS1 appears as the sample, two possible rules might come out depending on the comparison, and each of them will be associated with a different response. Thus, the correct mapping cannot be solved with the direct path from layer 1 to the BG-PMC layer alone. However, this path is important in the exploration process. Hence, the hypothesis is that the LC allows the energy to flow through this path during the initial exploration process, but then it attenuates it in order to allow the development of the exploitation process.

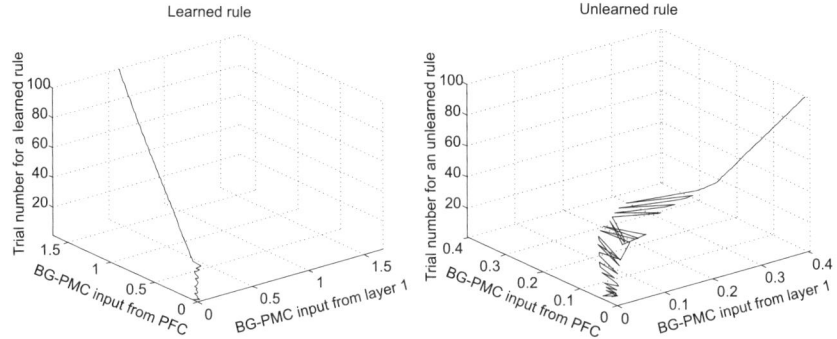

Figure 17.4 Trajectories for the inputs to the BG-PMC from PFC and layer 1 during a DMTS experiment. On the left side, a certain rule under analysis was correctly learned, while the one on the right side was not properly learned (mapped to the incorrect response). When a rule is learned, the steady-state input from PFC is larger than the one from layer 1. Whenever this relation is reversed, the rule will be unlearned.

The energy flowing to the BG-PMC layer from PFC and from layer 1 is analyzed in Fig. 17.4. By looking at the trajectories of this energy, which are conditioned to a certain rule, it can be seen that: (i) when a rule is learned, the steady-state energy from PFC is higher than the one from layer 1. (ii) when the steady-state energy from PFC is lower than the one from layer 1, the rule will not be learned as the direct path cannot solve the contingency. However, a rule might not be learned and still present the trajectory as on the left side of Fig. 17.4. In this case, a PFC neuron encodes two different rules and projects to BG-PMC with a synaptic weight large enough to execute a response (that will be right for one of the rules but wrong for the other one). Nevertheless, most of the unlearned rules when no LC is included in the model are caused by an excessive growth of the direct path during the learning phase.

To further clarify this, the analysis shown in Fig. 17.5 was performed. One hundred DMTS experiments with and without LC were run, which led to 400 rules in total for each case. The rules were divided in two groups, the learned and unlearned ones. For each group the average of the ratios between the steady-state energy from PFC to BG-PMC and the one from layer 1 to BG-PMC was computed. In the case of the learned rules with LC modulation, this ratio is high, indicating that the PFC is guiding the decision process. The value drops for the unlearned case, but it is still larger than 1. Here, there are unlearned rules due to both excessive energy through the direct path (in this case the proposed LC firing dynamics is not enough to compensate the direct path growth) and incorrect mapping from PFC to BG-PMC. When LC modulation is not modeled, the ratio drops considerably. The small value for the learned rules shows that although the learning was achieved successfully, it was by a narrow margin. In the case of the unlearned rules, the ratio is below 1, indicating that in most of these rules the failure to learn was due to an excessive growth of the direct path. Thus, the LC modulation attenuates the direct path as the reward probability grows during the exploration phase, increasing the ability to learn the correct mappings and letting the model get into the exploitation mode.

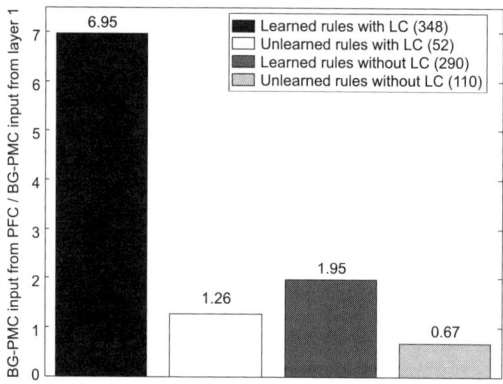

Figure 17.5 Average values of the ratio between the steady-state input to the BG-PMC from the PFC and the one from layer 1. Each group represents learned or unlearned rules with or without LC modulation. Between brackets is shown the total number of rules for each group.

17.6 Conclusions and Discussion

Our computational theory explains the roles of DA and NE and how these mono-aminergic systems interact with cortical and subcortical circuits during learning performances of simple (Visual Discrimination) and complex (Delayed Matching to Sample) tasks.

In cortical and subcortical structures, DA is involved in exploitation response by two means: neuron excitability and synaptic plasticity. In the PFC, as DA is released, interneurons inhibit pyramidal neurons. However, due to the NMDA-D1 synergism, the firing rate of those pyramidal neurons that receive a strong glutamatergic input differentiates from the rest of the population. In motor structures, DA allows reducing the search space by extinguishing those responses not associated with reward and increasing the probability of executing rewarded responses. In the model, DA is also involved in learning; synaptic strength is potentiated or depressed by Hebbian or anti-Hebbian law with a gating mechanism depending on the level of released DA. Those responses associated to reward or predictions of reward are exploited, and the synaptic paths connecting input structures, PFC and motor structures are reinforced.

While DA is correlated with response exploitation, experimental data suggest the role of NE in exploration. Due to the strong afferent input from the orbitofrontal cortex (which is part of the mPFC), the modulation of the LC tonic firing can be done by the reward history. Depending on the complexity of the task, NE actions can be found at different structures and task stages. In simple operant paradigms, its principal effect could be seen as a neuromodulation of dopaminergic excitability, allowing fast changes in S-R mapping as required during a reversal. In complex paradigms, as in DMTS, the principal action of NE would be to modulate the energy from visual and somatosensorial input neurons to motor-related structures. This mechanism allows fast learning of the subset of task relevant responses giving time to the PFC to learn those high order contingencies underlying in the task.

Other proposals can be found in the literature. In Atallah et al. (2004) the authors suggest that the BG-PMC are involved in reversal learning through the adjustment of its synaptic inputs from the PFC. To do so, they use the history of reinforcement and a DA-based modulation. We agree about the plausibility of this effect, but faster reversal learning could be mediated by an inhibition of VTA/SNc neurons from the tonic LC firing. However, as mentioned before, there is also a direct effect of LC neurons on the modulation of the energy flow of the direct input-output path. This is a key component for learning the DMTS task.

In Gisiger and Kerszberg (2006), the authors introduced a model with the ability to solve a DMTS task. Although they have a Hebbian learning mechanism, no neuromodulatory information is explicitly used. In fact, the model does not learn the task, given that a mature gating mechanism is included to guide the network. Another important fact is the ability to learn a delayed nonmatching to sample task. In the model analyzed in this chapter, this is the same as the DMTS task with the reward contingencies reversed (if sample and comparison are equal, the No-Go response should be rewarded). In Gisiger's model, a DMTS task should be learned first, and then a delayed pair association task should be used to associate each sample with the nonmatching stimulus. There is no experimental evidence that supports this kind of sequential learning to solve the delayed nonmatching to sample task.

Another model of exploration–exploitation was proposed in McClure et al. (2005). Although they simulate the DA effect on learning, they do not include its role in the excitability of PFC neurons (by the NMDA-D1 synergism). They also share the approach that the executed response is the first one that surpasses a certain threshold. However, they include a response conflict monitoring mechanism as an outcome from the interaction between the cingulate and LC neurons. Despite the possible existence of this mechanism, it was not necessary to include it in the model analyzed throughout this chapter to account for the presented results. Here, the exploration–exploitation trade-off was controlled with the tonic firing of the LC mediated by the reward history. Actually, as previously mentioned, an important aspect is the inhibition of DA neurons by the LC output. This mechanism is not included in the model by McClure and collaborators. In fact, they propose an inhibitory connection from the DA neurons to the LC, which is not discussed in detail in their paper. These interactions between DA and LC neurons are not mutually exclusive. An inhibition from LC to DA neurons will improve exploratory behavior, while the opposite interaction will strengthen exploitative behavior. On the other hand, the involvement of other neuromodulators in exploration (e.g., the role of serotonin and acetylcholine in PFC and BG respectively) cannot be ruled out.

The results on reversal learning in the VD task are qualitatively in agreement with their results. If only DA modulation is included (without the LC), there is an increase in the time to reach criterion after reversal without noticeably affecting the performance during the acquisition. However, the prediction here is that during the DMTS task, the LC has a key role (in addition to the DA modulation) for both acquisition and reversal learning. This hypothesis has not been tested before, either experimentally or theoretically.

The key assumptions of the model analyzed in this chapter are quite simple. The model relies on a layer computing short-term memory traces, a reward prediction neural cluster, the PFC layer, a winner-take-all mechanism, the layer with possible responses and a mechanism governed by NE release to balance the energy flow between input–output layers. All these assumptions are widely accepted in the field. Moreover, the reward prediction cluster has been implemented using the TD model and the Rescorla-Wagner rule leading to similar results. We believe that the overall structure of the neural network model, and not its particular implementation, is responsible for the results discussed throughout this chapter.

Together, dopamine and norepinephrine modulation mechanisms allow shifting between exploitation and exploration stages depending on the changes in the environment. Giving the results obtained in simulations with changing environments, the model provides a computational framework to the exploration–exploitation dilemma.

Acknowledgments

This research was funded by grants from ANPCYT: PICT 11-990, UBACYT I007, and CONICET: PIP 5876. We thank Diego Gutnisky and Prof. Anthony A. Wright for their helpful comments.

References

Aston-Jones, G. and Cohen, J.D. (2005) An integrative theory of locus coeruleusnorpine-phrine function: Adaptive gain and optimal performance. Annual Review of Neu-roscience, 28, 403–450.

Atallah, H.E., Frank, M.J. and O'Reilly, R.C. (2004) Hippocampus, cortex, and basal ganglia: insights from computational models of complementary learning systems. Neurobiology of Learning and Memory, 82(3), 253–267.

Berridge, C.W. and Waterhouse, B.D. (2003) The locus coeruleus-noradrenergic system: modulation of behavioral state and state-dependent cognitive processes. Brain Research Reviews, 42(1), 33–84.

Brasted, P.J. and Wise, S.P. (2004) Comparison of learning-related neuronal activity in the dorsal premotor cortex and striatum. European Journal of Neuroscience, 19(3), 721–740.

Cavada, C., Company, T., Tejedor, J., Cruz-Rizzolo, R.J. and Reinoso-Suarez, F. (2000) The anatomical connections of the macaque monkey orbitofrontal cortex. A review. Cerebral Cortex, 10(3), 220–242.

Clayton, E.C., Rajkowski, J., Cohen, J.D. and Aston-Jones, G. (2004) Phasic activation of monkey locus coeruleus by simple decisions in a forced-choice task. Journal of Neuroscience, 24(44), 9914–9920.

Durstewitz, D., Seamans, J.K. and Sejnowski, T.J. (2000) Dopamine-mediated stabilization of delay-period activity in a network model of prefrontal cortex. Journal of Neurophysiology, 83(3), 1733–1750.

Freedman, D.J., Riesenhuber, M., Poggio, T. and Miller, E.K. (2002) Visual categorization and the primate prefrontal cortex: neurophysiology and behavior. Journal of Neurophysiology, 88(2), 929–941.

Fuster, J.M. (1989) The prefrontal cortex: anatomy, physiology, and neuropsychology of the frontal lobe. Raven Press New York, NY.

Garris, P.A. and Wightman, R.M. (1994) Different kinetics govern dopaminergic transmission in the amygdala, prefrontal cortex, and striatum: an in vivo voltammetric study. Journal of Neuroscience, 14(1), 442–450.

Gisiger, T. and Kerszberg, M. (2006). A model for integrating elementary neural functions into delayed-response behavior. PLoS Computational Biology, 2(4), e25, 211–227.

Gonon, F. (1997) Prolonged and extrasynaptic excitatory action of dopamine mediated by D1 receptors in the rat striatum in vivo. Journal of Neuroscience, 17(15), 5972–5978.

Grenhoff, J., Nisell, M., Ferre, S., Aston-Jones, G. and Svensson, T.H. (1993) Noradrenergic modulation of midbrain dopamine cell firing elicited by stimulation of the locus coeruleus in the rat. Journal of Neural Transmission. General Section, 93(1), 11–25.

Gutnisky, D.A. and Zanutto, B.S. (2004a) Cooperation in the iterated prisoner's dilemma is learned by operant conditioning mechanisms. Artificial Life, 10(4), 433–461.

Gutnisky, D.A. and Zanutto, B.S. (2004b) Learning obstacle avoidance with an operant behavior model. Artificial Life, 10(1), 65–81.

Lewis, B.L. and O'Donnell, P. (2000) Ventral tegmental area afferents to the prefrontal cortex maintain membrane potential 'up' states in pyramidal neurons via D1 dopamine receptors. Cerebral Cortex, 10, 1168–1175.

McClure, S.M., Gilzenrat, M.S. and Cohen, J.D. (2005) An exploration-exploitation model based on norepinephrine and dopamine activity. Advances in Neural Information Processing Systems, 18, 867–874.

Mink, J.W. (1996) The basal ganglia: focused selection and inhibition of competing motor programs. Progress in Neurobiology, 50(4), 381–425.

O'Reilly, R.C., Noelle, D.C., Braver, T.S. and Cohen, J.D. (2002) Prefrontal cortex and dynamic categorization tasks: representational organization and neuromodulatory control. Cerebral Cortex, 12(3), 246–257.

Otani, S., Daniel, H., Roisin, M.P. and Crepel, F. (2003) Dopaminergic modulation of long-term synaptic plasticity in rat prefrontal neurons. Cerebral Cortex, 13(11), 1251–1256.

Paladini, C.A. and Williams, J.T. (2004) Noradrenergic inhibition of midbrain dopamine neurons. Journal of Neuroscience, 24(19), 4568–4575.

Pan, W.X., Schmidt, R. and Wickens, J.R. and Hyland, B.I. (2005) Dopamine cells respond to predicted events during classical conditioning: evidence for eligibility traces in the reward-learning network. Journal of Neuroscience, 25(26), 6235–6242.

Pasupathy, A. and Miller, E.K. (2005) Different time courses of learning-related activity in the prefrontal cortex and striatum. Nature, 433(7028), 873–876.

Petrides, M. (2000) Dissociable roles of mid-dorsolateral prefrontal and anterior inferotemporal cortex in visual working memory. Journal of Neuroscience, 20(19), 7496–7503.

Rescorla, R.A. and Wagner, A.R. (1972) A theory of Pavlovian conditioning: variations in the effectiveness of reinforcement and nonreinforcement. In: A. H. Black and W. F. Prokasy (Eds), Classical Conditioning II: Current Research and Theory. Appleton Century Crofts, New York, pp. 64–99.

Reynolds, J.N.J. and Wickens, J.R. (2002) Dopamine-dependent plasticity of corticostriatal synapses. Neural Networks, 15(4–6), 507–521.

Roychowdhury, V.P., Kai-Yeung Siu, and Kailath, T. (1995) Classification of linearly non-separable patterns by linear threshold elements. IEEE Transactions on Neural Networks, 6(2), 318–331.

Schmajuk, N.A. and Zanutto, B.S. (1997) Escape, avoidance, and imitation: a neural network approach. Adaptive Behavior, 6(1), 63–129.

Schultz, W., Dayan, P. and Montague, P.R (1997). A neural substrate of prediction and reward. Science, 275 (5306), 1593–1599.

Schultz, W. (2002) Getting formal with dopamine and reward. Neuron, 36(2), 241–263.

Tseng, K.Y., Mallet, N., Toreson, K., Le Moine, C., Gonon, F. and O'Donnell, P. (2006) Excitatory response of prefrontal cortical fast-spiking interneurons to ventral tegmental area stimulation in vivo. Synapse, 59, 412–417.

Tseng, K.Y. and O'Donnell, P. (2005) Post-pubertal emergence of prefrontal cortical up states induced by D1-NMDA co-activation. Cerebral Cortex, 15, 49–57.

Usher, M., Cohen, J.D., Servan-Schreiber, D., Rajkowski, J. and Aston-Jones, G. (1999) The role of locus coeruleus in the regulation of cognitive performance. Science, 283(5401), 549–554.

Wang, J. and O'Donnell, P. (2001) D(1) dopamine receptors potentiate NMDA-mediated excitability increase in layer V prefrontal cortical pyramidal neurons. Cerebral Cortex, 11(5), 452–462.

Chapter 18

Basal Ganglia – Cortex Interactions: *Regulation of Cortical Function by D1 Dopamine Receptors in the Striatum*

Heinz Steiner
Department of Cellular & Molecular Pharmacology, Rosalind Franklin University of Medicine
  & Science, The Chicago Medical School, North Chicago, IL 60064, USA

Abstract. This paper reviews recent findings of molecular imaging studies that investigated the role of striatal dopamine in the regulation of basal ganglia output and cortical function. These studies employed immediate-early genes such as *c-fos* and *zif 268* as functional markers to determine the effects of dopamine depletion and local dopamine receptor stimulation in the striatum on cortical function. The results indicate that the D1 receptor-regulated direct striatal output pathway provides widespread activation of the cortex. The various anatomical pathways that could mediate this basal ganglia-cortical regulation are discussed. It is concluded that likely several pathways act in concert, some signaling specific motor commands, others providing more general (and widespread) cortical activation, perhaps related to arousal and attentional states, that is necessary for normal motor functioning. All these basal ganglia-cortical activating mechanisms appear to be facilitated by striatal dopamine.

18.1 Introduction

Interactions between the basal ganglia and the cerebral cortex are critical for the generation of normal motivated behavior. Indeed, dysfunctional interactions between these two brain systems have been implicated in many neurological and psychiatric disorders, including Parkinson's disease, schizophrenia, addiction to psychostimulants, obsessive compulsive disorders and attention-deficit hyperactivity disorder (e.g., Albin et al., 1989; Carlsson and Carlsson, 1990; DeLong, 1990; Hyman and Nestler, 1996; Graybiel, 1997; Robbins et al., 1998; Graybiel et al., 2000; Solanto, 2002). Deficient dopamine function in the striatum – the main input station of the basal ganglia – is perceived at the core of many of these disorders.

ANATOMY OF CORTICO-BASAL GANGLIA-CORTICAL CIRCUITS

This section provides a highly simplified description of the functional organization of the basal ganglia as it pertains to the present review (for more details, see, e.g.,

Gerfen, 1992; Parent and Hazrati, 1995; Gerfen and Wilson, 1996). According to current models of basal ganglia function, interactions between the basal ganglia and the cortex are mediated by several parallel and interconnected anatomical circuits/ loops. These circuits arise throughout the cortex and project to the striatum (caudate-putamen, nucleus accumbens) and from there via the basal ganglia output nuclei and thalamus back to the cortex (Alexander et al., 1986; Albin et al., 1989; Alexander et al., 1990; Groenewegen et al., 1990; Joel and Weiner, 1994; Parent and Hazrati, 1995; Gerfen and Wilson, 1996; Middleton and Strick, 2002; Haber, 2003) (Fig. 18.1).

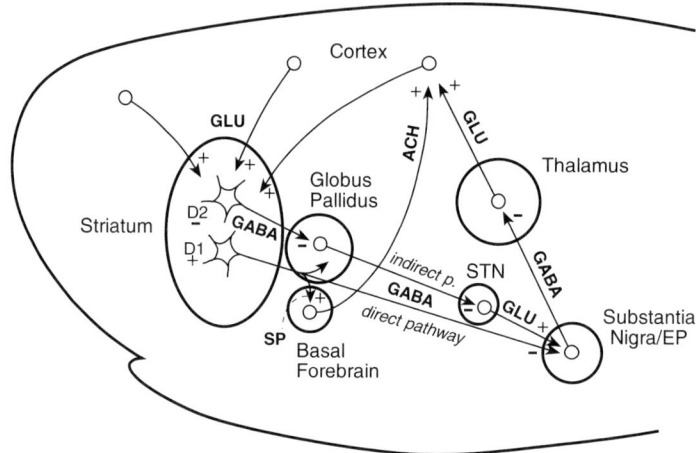

Figure 18.1 Simplified diagram of cortico-basal ganglia-cortical circuits. The striatum receives excitatory input from all parts of the cortex. Striatal output is conveyed to the basal ganglia output nuclei, substantia nigra and internal pallidum/entopeduncular nucleus (EP), by two pathways, the mostly D1 receptor-regulated direct (striatonigral) pathway and the indirect pathway, beginning with the mostly D2 receptor-regulated striatopallidal projection. Basal ganglia output is projected, via the thalamus, back to the cortex and also to the brainstem (not shown). Direct pathway neurons give off axon collaterals in the region of the globus pallidus and release substance P (SP). Substance P appears to facilitate activity in basal forebrain cholinergic (ACH) neurons that project to the cortex. Note that several other basal ganglia pathways are omitted from this scheme. STN, subthalamic nucleus.

Striatal information is conveyed to the basal ganglia output nuclei [substantia nigra pars reticulata and entopeduncular nucleus (rat)/internal pallidum (primates)] by two pathways: (i) the so-called "indirect" pathway that projects via globus pallidus (external pallidum in primates) and subthalamic nucleus to the output nuclei; and (ii) the "direct" or "striatonigral" pathway that sends axons directly to the output nuclei and also gives off axon collaterals in the region of the globus pallidus (Kawaguchi et al., 1990). While the two subtypes of striatal projection neurons use GABA as their main neurotransmitter, they differ in a number of receptors and neuropeptides they express (Gerfen and Wilson, 1996). For example, striatopallidal neurons, the initial segment of the indirect pathway, mostly contain the opioid peptide enkephalin, whereas striatonigral neurons co-release the peptides dynorphin and substance P (Fig. 18.1).

Basal ganglia output is projected back to the cortex via several thalamic nuclei, but also reaches the brainstem and tectum (Gerfen and Wilson, 1996). These pathways are discussed in more detail below (Section 18.3).

Dopamine in the striatum, released from terminals of midbrain dopamine neurons, regulates activity flow in cortico-basal ganglia-cortical circuits by gating cortical and other inputs to striatal projection neurons (Cepeda and Levine, 1998; Nicola et al., 2000). Notably, the effects of dopamine are opposite in the two subtypes of striatal projection neurons due to differential expression of dopamine receptors (Fig. 18.1). Striatopallidal neurons mostly contain D2 receptors, while striatonigral neurons predominantly express D1 receptors, and only a small proportion of these neurons contains both dopamine receptor subtypes at comparable levels (Gerfen et al., 1990; Le Moine et al., 1990; Curran and Watson, 1995; Le Moine and Bloch, 1995; Surmeier et al., 1996). [It should be noted, however, that the anatomy of ventral striatal circuits is somewhat more complex, as some neurons also feature other dopamine receptors (D3), neuropeptide combinations, and/or projection patterns (Curran and Watson, 1995; Le Moine and Bloch, 1995; Robertson and Jian, 1995; Le Moine and Bloch, 1996).]

FUNCTIONAL IMAGING IN CORTICO-BASAL GANGLIA-CORTICAL CIRCUITS USING IMMEDIATE-EARLY GENE MARKERS

Functional markers employed to study the effects of dopamine in striatal projection neurons and their downstream targets include immediate-early genes (IEGs) such as *c-fos* and *zif 268* (for reviews, see Harlan and Garcia, 1998; Steiner and Gerfen, 1998). IEGs are rapidly and transiently activated by a variety of stimuli, and their expression correlates to some degree with neuronal activity (Sagar et al., 1988; Sheng and Greenberg, 1990; Sharp et al., 1993; Chaudhuri, 1997).

Double-labeling studies demonstrated that selective D1 and mixed D1/D2 (apomorphine) receptor agonists, or psychostimulants such as cocaine and amphetamine, induce IEGs predominantly (but not exclusively) in striatonigral neurons (Robertson et al., 1990; Berretta et al., 1992; Cenci et al., 1992; Johansson et al., 1994; Gerfen et al., 1995; Kosofsky et al., 1995; Badiani et al., 1999; Uslaner et al., 2001). This response is mediated by striatal D1 receptors, as shown by D1 receptor antagonist studies (e.g., Graybiel et al., 1990; Young et al., 1991; Cole et al., 1992; Steiner and Gerfen, 1995) or targeted deletion of D1 receptors (Drago et al., 1996; Moratalla et al., 1996; Zhang et al., 2004), and is dependent on glutamate inputs (Hyman et al., 1996; Wang and McGinty, 1996). Concurrent stimulation of D2 receptors potentiates the D1 receptor-mediated IEG expression in striatonigral neurons (e.g., Paul et al., 1992; LaHoste et al., 1993; Gerfen et al., 1995; Le Moine et al., 1997), with the D2 receptor effect apparently being mediated by cholinergic interneurons (Wang and McGinty, 1996). In contrast, robust stimulation of D2 receptors inhibits gene expression in striatopallidal neurons (e.g., Gerfen et al., 1990; Le Moine et al., 1997; Pinna et al., 1997), whereas blockade of D2 receptors increases IEG expression in these neurons (Robertson and Fibiger, 1992; Robertson et al., 1992), responses that are also driven by glutamate inputs. [Note that because of the antagonistic control of IEG expression by glutamate (NMDA) vs. D2 receptors in striatopallidal neurons, drug treatments that are associated with pronounced corticostriatal input plus relatively weak D2 receptor stimulation can result in increased rather than decreased IEG expression in these neurons (Badiani et al., 1999; Uslaner et al., 2001; Ferguson and Robinson, 2004)].

Results of these gene regulation studies have been taken to indicate that stimulation of striatal D1 receptors facilitates glutamate input (Cepeda et al., 1993; Cepeda and Levine, 1998) and neuronal firing in the striatonigral (direct) pathway (You et al., 1994), whereas stimulation of D2 receptors inhibits glutamate-driven responses in striatopallidal neurons (Cepeda and Levine, 1998). According to the current basal ganglia models, stimulation of either dopamine receptor would result in reduced basal ganglia output and disinhibited thalamocortical activation (Albin et al., 1989; Chevalier and Deniau, 1990) (Fig. 18.1).

Consistent with such cortical activation (disinhibition) by striatal dopamine action, various treatments with direct and indirect dopamine receptor agonists, including apomorphine (see Fig. 18.3), cocaine and amphetamine, have been shown to produce increased IEG expression in the cortex (e.g., Graybiel et al., 1990; Paul et al., 1992; Dilts et al., 1993; Johansson et al., 1994; Steiner and Gerfen, 1994; Wang et al., 1995; LaHoste et al., 1996; Badiani et al., 1998; Yano and Steiner, 2005). However, all of these studies used systemic drug administration, thus precluding conclusions regarding possible contributions of extrastriatal dopamine receptors and, in the case of uptake blockers/psychostimulants, other neurotransmitter systems. Moreover, these studies showed that IEG induction in the cortex typically was much more widespread than predicted by the current basal ganglia models, which tend to emphasize premotor/motor cortical areas as targets of basal ganglia output.

The following sections summarize work that investigated the role of striatal dopamine receptors in cortical IEG induction, and by extension, cortical function (Section 18.2). Moreover, as these studies demonstrate that especially D1 receptor-regulated striatal output produces widespread and pronounced facilitation of gene induction throughout much of the cortex, the anatomical pathways that potentially mediate these effects are discussed (Section 18.3).

18.2 Striatal Dopamine Receptors Regulate Cortical Function

Early evidence for a role of extrinsic or corticocortical inputs, as opposed to local dopamine receptor stimulation, in cortical IEG induction after (systemic) dopamine agonist treatments was provided by regional mapping studies (LaHoste et al., 1996). These studies revealed a principal mismatch between the regional patterns of IEG induction and the distribution of dopamine receptor-expressing cells throughout the cortex. [It should be noted that, in contrast to 2-deoxyglucose and other metabolic markers, IEG mRNAs (and proteins) typically used in molecular imaging studies (e.g., *c-fos*, *zif 268*) are largely confined to the cell body, thus allowing unequivocal identification of the responding cells (Sharp et al., 1993)].

Other early studies used sensory deprivation to investigate the role of thalamic input to the cortex in cortical IEG expression. These studies showed that "basal" expression of certain IEGs in the visual and somatosensory cortex is dependent on sensory input (and thus likely reflects ongoing synaptic activity) (Chaudhuri and Cynader, 1993; Steiner and Gerfen, 1994; Kaczmarek and Chaudhuri, 1997; Staiger et al., 2000). Importantly, such studies also demonstrated that sensory deprivation (by whisker clipping) almost completely eliminated apomorphine-induced IEG expression in the affected somatosensory cortex (Steiner and Gerfen, 1994; Filipkowski et al., 2001) (see Fig. 18.4).

Together, these findings indicated a critical role for excitatory input from the thalamus in basal and dopamine agonist-induced IEG expression in the cortex.

CORTICAL IEG EXPRESSION IS REDUCED BY ACUTE DOPAMINE DEPLETION

Consistent with the notion that striatal dopamine action enhances cortical activation and gene expression, several studies demonstrated altered gene regulation in the cortex after striatal dopamine loss. For example, we recently investigated the effects of dopamine depletion (by 6-OHDA infusion into the midbrain) on basal IEG expression in the cortex (Steiner and Kitai, 2001). Our results showed that a unilateral 6-OHDA lesion that produced near-total dopamine loss in the ipsilateral striatum caused significantly reduced *zif 268* expression throughout large portions of the cortex (Fig. 18.2) (Steiner and Kitai, 2001). While this effect was considerably more robust in the ipsilateral cortex, several contralateral cortical regions also displayed decreased *zif 268* expression (Fig. 18.2). This reduction in cortical *zif 268* expression was time-dependent; the decrease was present 2 days after 6-OHDA administration, when lesion-induced behavioral deficits were maximal, but not 21 days after the lesion, when some behavioral recovery had occurred (Steiner and Kitai, 2001).

zif 268 expression in the cortex after dopamine depletion

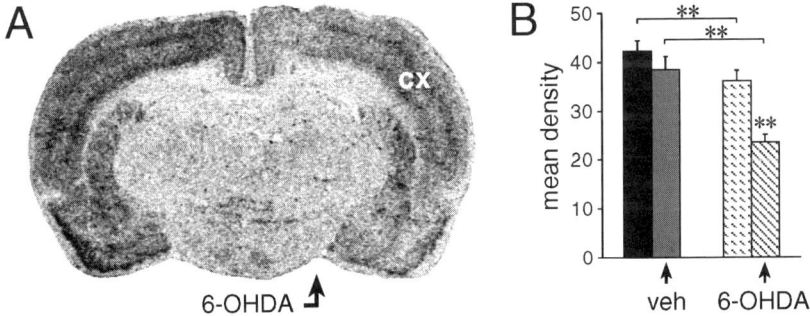

Figure 18.2 Effects of acute dopamine depletion on *zif 268* expression in the cortex (Steiner and Kitai, 2001). (A) Illustration of film autoradiogram shows *zif 268* expression in a coronal section from a rat killed 2 days after 6-OHDA infusion into the right midbrain. (B) Mean density values (mean ± SEM, arbitrary units) measured throughout the cortex (cx) on the noninfused side (left bar) and infused side (right bar) are given for animals that had received a vehicle (veh) or a 6-OHDA infusion 2 days before killing. The dopamine depletion produced a bilateral decrease in cortical *zif 268* expression, which was more robust on the side ipsilateral to the lesion. **$P<0.01$.

In order to assess more directly how dopamine depletion affects cortical function, we determined the effects of a unilateral 6-OHDA lesion on a sensory-evoked response – IEG induction in the somatosensory (barrel) cortex evoked by whisker stimulation. Brief stimulation of individual whiskers by a pulsating magnetic field in freely moving rats is sufficient to evoke IEG expression in the stimulated barrel column (Melzer and Steiner, 1997), an effect that most likely reflects excitatory input via the ventrobasal thalamus (cf. Melzer and Steiner, 1997). After unilateral dopamine depletion, this sensory-evoked IEG response in the barrel cortex was largely abolished (Steiner and Kitai, 2001). Again, this effect was time-dependent, as the deficit was present 2 days, but not 21 days, after the lesion (Steiner and Kitai, 2001). These findings suggested that acute dopamine depletion impaired cortical processes (causing sensorimotor deficits), and that subsequent compensatory neuronal changes in the basal ganglia (Schwarting and Huston, 1996) and/or the cortex (see below) enabled some "normalization" in sensorimotor functions (see Steiner and Kitai, 2001, for discussion).

The above 6-OHDA lesion likely also affected the mesocortical dopamine projections (norepinephrine neurons were protected by administration of desipramine) (Steiner and Kitai, 2001). However, as it was the case for IEG induction by selective dopamine receptor agonists (e.g., LaHoste et al., 1996), the widespread lesion effects in the cortex (Steiner and Kitai, 2001) did not match the rather limited distribution of dopamine terminal fields (Björklund and Lindvall, 1984; Berger et al., 1991) and dopamine receptors (Mansour et al., 1990; Bouthenet et al., 1991; Mengod et al., 1991; Gaspar et al., 1995; Defagot et al., 1997; Ciliax et al., 2000) in the rat cortex, and thus supported a role for extracortical (or corticocortical) inputs in these changes. Furthermore, the lesion-induced reduction in whisker stimulation-evoked IEG expression in the barrel cortex was directly correlated with changes in a marker for dopamine loss in the striatum (enkephalin expression), indicating a close relationship between striatal dopamine and barrel cortex function (Steiner and Kitai, 2001).

The above findings are consistent with results of several other studies on the effects of dopamine depletion in thalamocortical pathways and their cortical targets. First, dopamine lesions with 6-OHDA in rats or MPTP in monkeys were found to produce a decrease in metabolic activity in excitatory thalamocortical projection neurons (Rolland et al., 2007), as well as in several cortical regions (Orieux et al., 2002; Brownell et al., 2003; Oueslati et al., 2005; but see Pelled et al., 2002). For example, bilateral infusion of 6-OHDA into the dorsolateral striatum resulted in reduced COI mRNA levels (metabolic marker) in the sensorimotor cortex, and this decrease was correlated with lesion-induced akinesia (Oueslati et al., 2005). Second, widespread and in part bilateral changes in (basal) gene expression in the cortex were reported for neuropeptides, glutamic acid decarboxylase and glutamate receptors (Lindefors et al., 1989, 1990; Rodriguez-Puertas et al., 1999). Most of these effects were observed weeks after the dopamine lesion and may thus represent compensatory neuroadaptations. Third, dopamine depletion altered dopamine receptor agonist-induced gene expression in the cortex. Thus, after unilateral 6-OHDA lesions, selective D1 receptor agonists produced dramatically enhanced induction of IEGs and other markers (RGS2) in the cortex of both hemispheres (Steiner and Gerfen, 1996; Berke et al., 1998; Rodriguez-Puertas et al., 1999; Taymans et al., 2005), probably due to stimulation of supersensitive D1 receptors in the dopamine-depleted striatum.

In summary, various studies demonstrated that unilateral dopamine lesions targeting mesostriatal dopamine projections produce widespread, often bilateral changes in gene expression in the cortex.

STRIATAL D1 RECEPTORS FACILITATE CORTICAL IEG EXPRESSION

The above-reviewed studies employing systemic dopamine agonist treatments and/or dopamine depletion are highly suggestive of a role for striatal dopamine in cortical regulation. However, a contribution of corticocortical (or cortico-thalamo-cortical; Haber and McFarland, 2001) connections arising from cortical areas that do have dopamine receptors could not be ruled out. In order to directly address the role of striatal dopamine receptors, other studies have used intrastriatal administration of agents that only affected striatal dopamine transmission.

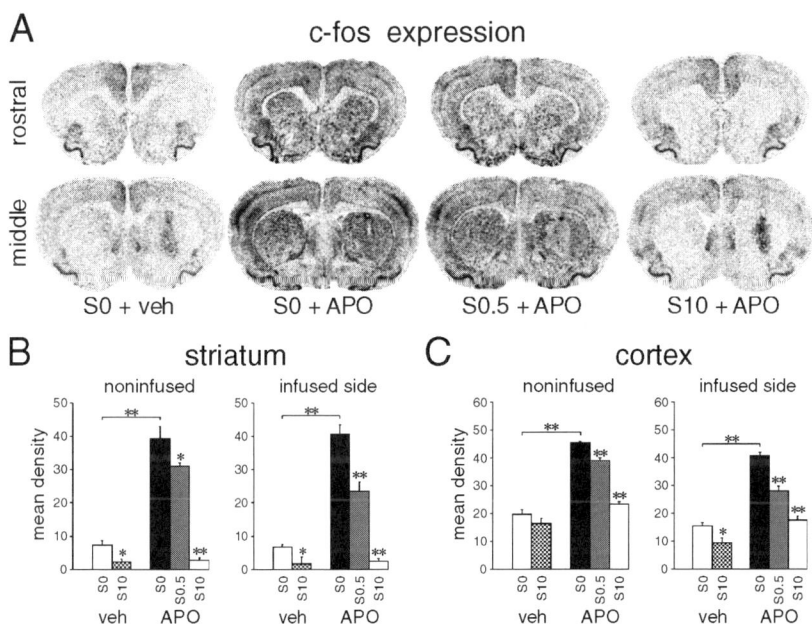

Figure 18.3 Effects of blocking D1 receptors in the striatum on *c-fos* expression in striatum and cortex (Steiner and Kitai, 2000). (A) Illustrations of film autoradiograms depict *c-fos* expression in coronal sections from the rostral and middle striatum in rats that received an intrastriatal infusion of vehicle (S0), 0.5 μg (S0.5), or 10 μg (S10) of the D1 receptor antagonist SCH-23390 into the right striatum (middle level) before an injection of vehicle (veh) or 3 mg/kg (s.c.) apomorphine (APO). (B and C) Mean density values (mean ± SEM) measured throughout the striatum (middle level) (B) or cortex (rostral level) (C) on the noninfused or infused side are given for the above treatment groups. Unilateral blockade of striatal D1 receptors produced bilateral inhibition of apomorphine-induced IEG expression in striatum and cortex, which was more robust on the infused side. * $P<0.05$, ** $P<0.01$, versus S0.

In an early study on the opioid regulation of striatal dopamine action, we used intrastriatal administration of a kappa receptor agonist to block cocaine-induced IEG expression in the striatum and observed that cortical IEG expression was attenuated as well (Steiner and Gerfen, 1995). This finding suggested that dopamine-regulated striatal output played a role in cortical gene regulation.

In a subsequent study, we investigated the effects of blocking D1 receptors in the striatum on IEG expression in the cortex induced by the D1/D2 receptor agonist apomorphine (Steiner and Kitai, 2000). Rats received an infusion of the D1 receptor antagonist SCH-23390 (via chronically implanted cannula) into the central striatum, followed by systemic administration of apomorphine. Apomorphine (3 mg/kg) produced pronounced IEG induction in the striatum and throughout most of the cortex (Fig. 18.3). Intrastriatal administration of SCH-23390 (0.5–10 µg) inhibited this IEG response in the cortex (and striatum) in a dose-dependent manner (Steiner and Kitai, 2000). Importantly, unilateral administration of the D1 receptor antagonist attenuated cortical (and striatal) gene induction in both hemispheres. Even a low dose (0.5 µg) with an apparently restricted local distribution in the central striatum affected cortical gene expression bilaterally in a "global" manner (Fig. 18.3).

It is important to note that, although widespread, these effects were regionally selective, affecting specific functional subdivisions of the cortex. Thus, IEG induction

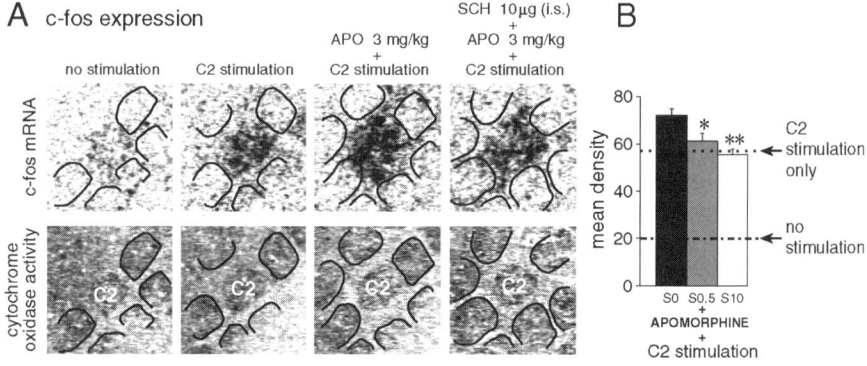

Figure 18.4 Effects of blocking D1 receptors in the striatum on whisker stimulation-evoked *c-fos* expression in the barrel cortex (Steiner and Kitai, 2000). (A) Illustrations of film autoradiograms show *c-fos* expression in barrel C2 in tangential sections through layer IV of the right barrel cortex (top). Barrels were identified by cytochrome oxidase staining in adjacent sections (bottom). Rats received an intrastriatal infusion of vehicle (S0), 0.5 µg (S0.5), or 10 µg (S10) of SCH-23390 into the right striatum before an injection of vehicle or 3 mg/kg apomorphine (APO), followed 10 min later by stimulation of left whisker C2 (15 min) in a pulsating magnetic field. Controls were placed in the stimulation chamber without magnetic field (no stimulation). All surrounding whiskers were clipped. (B) Mean density values (mean ± SEM) measured in barrel C2 for the above treatment groups are given. Blockade of D1 receptors in the ipsilateral striatum attenuated the apomorphine-induced facilitation of whisker stimulation-evoked *c-fos* expression in barrel C2. Note also that apomorphine did not enhance *c-fos* expression in the surrounding, sensory-deprived barrels. *$P<0.05$, **$P<0.01$, versus S0.

was inhibited in premotor, motor, sensory, and insular cortical regions, but not in the cingulate cortex (Steiner and Kitai, 2000). Notably, the same cortical regions were also affected by dopamine depletion (Steiner and Kitai, 2001).

In order to investigate the effects of apomorphine and striatal D1 receptor antagonism on a more direct measure of cortical function, sensory-evoked responses in the barrel cortex were assessed. Thus, rats received the above drug treatments in conjunction with whisker stimulation (Fig. 18.4). Results showed that systemic apomorphine facilitated whisker stimulation-evoked IEG expression in the stimulated cortical barrel, while infusion of SCH-23390 into the striatum ipsilateral to the stimulated cortex attenuated this effect in a dose-dependent manner (Steiner and Kitai, 2000). These findings indicate that stimulation of D1 receptors in the striatum regulates sensory-evoked responses in the cortex.

A more recent study confirmed the role of striatal D1 receptors in cortical IEG expression by using a complementary approach (Blandini et al., 2003). The effects of intrastriatal administration of selective D1 or D2 receptor agonists in combination with systemic administration of D1 or D2 receptor antagonists were assessed. Intrastriatal infusion of a D1, but not a D2, receptor agonist induced widespread cortical IEG expression, an effect that was blocked by a D1 receptor antagonist (Blandini et al., 2003).

Together, these findings demonstrate that dopamine, by stimulating D1 receptor-regulated striatal output, facilitates cortical function in a rather global manner.

18.3 Connections: From Striatal D1 Receptors to the Cortex

This section provides an overview of the different anatomical pathways that are thought to convey D1 receptor-regulated striatal output to the cortex (Fig. 18.5). A wealth of studies indicates that several of these pathways act in concert to regulate cortical function, some in a more specific manner (role in premotor, motor functions), others likely in a more general fashion (e.g., setting of activity levels). These include the well-established thalamocortical pathways, but also a recently discovered, yet controversial connection via the basal forebrain. The former have often been reviewed before and will thus only be summarized here, while the latter will be described in more detail.

VIA "MOTOR" NUCLEI OF THE THALAMUS

The D1 receptor-regulated direct (striatonigral) pathway projects to the entopeduncular nucleus (rat) or internal segment of the globus pallidus (primates) and extends to the substantia nigra pars reticulata (both), in part by axon collaterals (Fig. 18.1). These output nuclei are connected with the cortex via several thalamic nuclei (for reviews, see e.g., Parent and Hazrati, 1995; Gerfen and Wilson, 1996; Middleton and Strick, 2000; Haber and McFarland, 2001). Pallido-/nigrothalamic neurons are inhibitory (GABA), whereas thalamocortical neurons are excitatory (glutamate). The pathways arising in the entopeduncular nucleus and internal pallidum project to the frontal lobe (premotor, supplementary motor, motor areas) mostly via the ventral lateral and parts of the ventral anterior thalamic nuclei (Fig. 18.5). In primates, connections from the substantia nigra pars reticulata project mostly to other parts of the ventral anterior nucleus and from there to several cortical regions in the frontal lobe and to some sensory cortices (Middleton and Strick, 1996, 1997, 2000). In rats, the pars reticulata projects predominantly to the ventromedial thalamic nucleus, and to the superior colliculus and the pedunculopontine nucleus (Parent and Hazrati, 1995; Gerfen and Wilson, 1996). The ventromedial nucleus in turn gives rise to widespread projections to superficial layers of most cortical regions, including the sensorimotor cortex (Herkenham, 1979, 1980; Arbuthnott et al., 1990). These connections to the cortex via the thalamic "motor" nuclei are the most prominent.

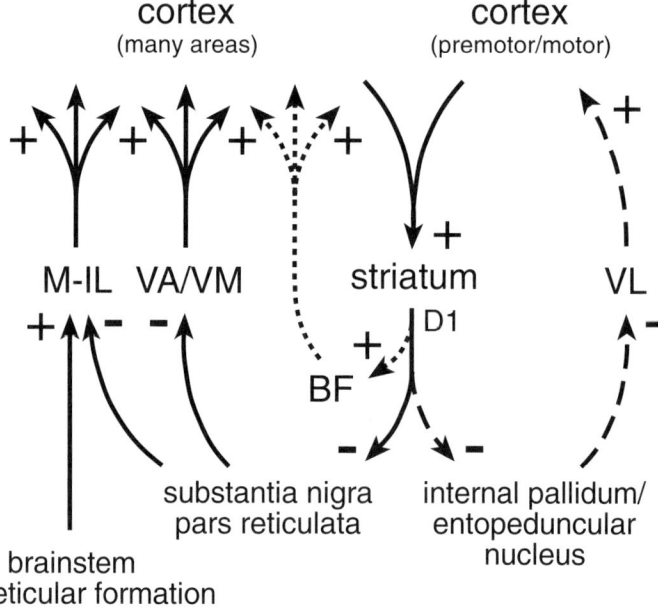

Figure 18.5 Simplified diagram illustrating the anatomical connections from the D1 receptor-regulated direct striatal output pathway to the cortex. The well-established pathways through the thalamus and the proposed connection via the basal forebrain (BF) are shown. M-IL, midline–intralaminar complex; VA, ventral anterior nucleus; VM, ventromedial nucleus; VL, ventrolateral nucleus.

VIA "NONSPECIFIC" THALAMIC NUCLEI

Basal ganglia output nuclei also project to midline and intralaminar thalamic nuclei (Fig. 18.5), either directly or indirectly via the superior colliculus or pedunculopontine nucleus (Groenewegen and Berendse, 1994; Parent and Hazrati, 1995; Smith et al., 2004; McHaffie et al., 2005). The midline–intralaminar thalamic complex is known to have widespread cortical connections (Groenewegen and Berendse, 1994; Parent and Hazrati, 1995; Gerfen and Wilson, 1996) (in addition to prominent projections back to the striatum; Smith et al., 2004). These thalamocortical projections are organized in a highly topographical manner, connecting individual thalamic subnuclei with specific cortical targets (Groenewegen and Berendse, 1994; Smith et al., 2004). The midline–intralaminar complex is one of the major components of the "nonspecific" thalamocortical system that, together with the reticular formation, is associated with the regulation of arousal, attention, and consciousness (Groenewegen and Berendse, 1994).

PROBABLY NOT VIA THE MEDIODORSAL THALAMIC NUCLEUS?

The mediodorsal thalamic nucleus is reciprocally connected with the prefrontal cortex (which indeed provided one of the anatomical definitions of the prefrontal cortex; e.g., Groenewegen et al., 1990; Uylings et al., 2003). This thalamic nucleus receives little input from the basal ganglia output nuclei, with the exception of some projections from the substantia nigra pars reticulata to its most lateral part (pars paralaminaris) (Groenewegen et al., 1990; Parent and Hazrati, 1995; Gerfen and Wilson, 1996). On the other hand, the mediodorsal nucleus is considered a node in the "limbic" cortico-basal ganglia-cortical circuit from the prefrontal cortex via nucleus accumbens, ventral pallidum, and mediodorsal nucleus back to the prefrontal cortex (Alexander et al., 1990; Groenewegen et al., 1990). It is currently unclear whether this circuit participates in the cortical regulation discussed here. Notably, the parts of the prefrontal cortex that were routinely assessed in our studies (anterior cingulate cortex) showed minimal or no effects of the dopamine manipulations employed (see above; Steiner and Kitai, 2000, 2001).

VIA BASAL FOREBRAIN CHOLINERGIC NEURONS?

As mentioned above, D1 receptor-regulated direct pathway neurons give off axon collaterals as their axon travels through the area of the globus pallidus, in rats and primates (Kawaguchi et al., 1990; Parent and Hazrati, 1995). The function of these axon collaterals is presently unknown. Tract tracing studies indicated that cholinergic neurons in the ventral pallidal complex ("basal forebrain," basal nucleus of Meynert) are among the targets of projections from the dorsal (and ventral) striatum (Grove et al., 1986; Henderson, 1997), and at least some of these connections arise from direct pathway axon collaterals (Henderson, 1997) (Fig. 18.5). Nauta and colleagues thus suggested that these cholinergic neurons are an integral part of the basal ganglia circuitry (Grove et al., 1986) and could provide an extrathalamic feedback to the cortex. The following paragraphs summarize further findings that support such a role.

Substance P terminals form synapses with cholinergic neurons. Direct pathway neurons contain and release substance P (in addition to GABA) (Gerfen and Wilson, 1996). Ultrastructural studies showed substance P-containing terminals in synaptic contact with cell bodies and proximal dendrites of basal forebrain cholinergic neurons in the ventral pallidum and ventromedial globus pallidus/basal nucleus of Meynert (Ch4) (Bolam et al., 1986; Csillik et al., 1998). Based on cell morphology, such inputs were found to target cholinergic neurons that project to the cortex (Ingham et al., 1985). Although the source of these substance P inputs was not determined in these studies, it was suggested that some derived from the dorsal striatum, others probably from the nucleus accumbens (Bolam et al., 1986).

Substance P facilitates acetylcholine release. Substance P-stimulated acetylcholine release is perhaps best-established for cholinergic neurons in the striatum (Arenas et al., 1991; Petitet et al., 1991; Anderson et al., 1993), but there is good evidence for a similar mechanism in basal forebrain cholinergic neurons. In the basal forebrain, substance P (NK-1) receptors are selectively expressed by cholinergic neurons (Gerfen, 1991). Stimulation of these receptors excites cholinergic neurons (Nakajima et al., 1991; Takano et al., 1995). Consistent with these findings, substance P infused into the basal forebrain enhanced acetylcholine release in the cortex (e.g., Nakajima et al., 1991; De Souza Silva et al., 2000).

Dopamine (D1) receptor stimulation promotes substance P release from direct pathway terminals as well as acetylcholine release in the cortex. Treatment with apomorphine produces substance P release at terminal sites of direct pathway neurons (substantia nigra) (Orosz and Bennett, 1990). Moreover, intrastriatal administration of a D1 receptor agonist stimulated local acetylcholine release, an effect that was mediated by NK-1 receptors and thus presumably by substance P released from axon collaterals of direct pathway neurons (Anderson et al., 1994). Other studies showed that dopamine regulates basal forebrain cholinergic neurons and cortical acetylcholine release. For example, D1 receptor agonist treatment induced *c-fos* in these cholinergic neurons (Robertson and Staines, 1994). Furthermore, systemic treatments with the mixed D1/D2 receptor agonist apomorphine, selective D1 (but not D2) receptor agonists or amphetamine increased cortical acetylcholine release (Day and Fibiger, 1992, 1993; Day et al., 1994; Arnold et al., 2000). The amphetamine-induced release was eliminated by dopamine depletion (6-OHDA) but not by norepinephrine depletion (Day et al., 1994).

Role of the nucleus accumbens? The above reviewed studies that used intrastriatal drug administration implicated direct pathway neurons arising in the dorsal striatum (caudate-putamen) (Fig. 18.3). However, a subtype of output neurons of the nucleus accumbens also releases substance P. Moreover, some nucleus accumbens neurons project to the cholinergic cell groups in the basal forebrain (Zaborszky et al., 1991; Zaborszky and Cullinan, 1992), and are thought to provide a GABAergic regulation (inhibition) of corticopetal cholinergic projections (Moore et al., 1995). There is evidence for a role of nucleus accumbens dopamine receptors (mostly D2) in regulating cortical acetylcholine release (Moore et al., 1999). However, intra-accumbens administration of amphetamine failed to affect cortical acetylcholine release (Arnold et al., 2000), nor did intra-accumbens infusion of D1 or D2 receptor antagonists inhibit systemic amphetamine-induced cortical acetylcholine efflux (Arnold et al., 2000). Therefore, it appears unlikely that output from the nucleus accumbens plays a major role in the cortical regulation discussed in the present chapter.

Summary. Together, these findings suggest that stimulation of (dorsal) striatal D1 receptors facilitates activity in direct pathway neurons, and that substance P released in the basal forebrain from direct pathway axon collaterals could – via NK-1 receptors – facilitate activity in cholinergic cortical afferents. Our finding of increased sensory-evoked responses in the barrel cortex after striatal D1 receptor stimulation (see above) is consistent with enhanced cholinergic cortical input, which facilitates sensory-evoked responses (e.g., Donoghue and Carroll, 1987; Howard and Simons, 1994; see Rasmusson, 2000, for review). Although such a basal ganglia-cortical pathway is supported by anatomical evidence on several levels, as well as by some functional findings (see above), it is currently unclear how prominent this pathway is. Future studies will have to determine the functional significance of this cortical afferent system.

INTERHEMISPHERIC CONNECTIONS

A consistent observation after unilateral manipulations of the dopamine transmission (dopamine depletion, intrastriatal drug infusion) is the finding of *bilateral* changes in gene regulation in the cortex (and striatum). In studies showing decreased IEG expression in the cortex (after a 6-OHDA lesion or intrastriatal D1 receptor antagonism), the contralateral cortical effects were often less robust (Steiner and Kitai, 2000, 2001). However, in studies that found increased cortical gene induction (by a D1 receptor agonist presumably acting on supersensitive D1 receptors ipsilateral to a 6-OHDA lesion), the bilateral cortical activation was fairly symmetrical (Steiner and Gerfen, 1996; Berke et al., 1998; Taymans et al., 2005).

   These bilateral effects are consistent with anatomical findings in several species (if not reflected in the current models of basal ganglia function). Studies show that most of the above transthalamic pathways have an interhemispheric component, mostly in their thalamic afferent segment (Gerfen et al., 1982; Krout et al., 2001; Krout et al., 2002; reviewed in Parent and Hazrati, 1995). These crossed projections, together with corticocortical connections, could conceivably mediate the observed interhemispheric effects. Although these contralateral thalamic afferents are typically only a fraction of the ipsilateral projections, there is some evidence indicating that they play a role in compensatory mechanisms after unilateral brain lesions (see Parent and Hazrati, 1995, for discussion).

18.4 Functional Significance

The studies reviewed in this chapter indicate that stimulation of D1 receptors in the striatum provides widespread regulation of cortical function, most likely via facilitation of striatal output in the direct pathway. Facilitation of cortical function by striatal D1 receptors is consistent with current models of basal ganglia organization (Fig. 18.1), which propose disinhibition of thalamocortical activation as a consequence of increased activity in the direct pathway. However, the present findings demonstrate that, in the rat, this cortical regulation is much more widespread than that proposed by these models (which indicate premotor/motor cortical targets of basal ganglia-thalamocortical projections). Indeed, this regulation is readily apparent in many major functional subdivisions of the cortex (motor, sensory, insular), perhaps with the exception of the prefrontal cortex. As discussed here, more than one of the anatomical pathways that connect the basal ganglia with the cortex could potentially mediate these widespread cortical effects, including some of the main "motor" tha-lamocortical pathways (via ventral anterior nucleus in the primate, or ventromedial nucleus in the rat), and the "nonspecific" thalamocortical projections. Although there are species differences in these pathways (Parent and Hazrati, 1995), it is important to note that basal ganglia-thalamocortical projections that target areas outside the frontal lobe also exist in primates (Middleton and Strick, 1996, 2000). Moreover, it is argued here that connections via the basal forebrain may also play a role in a basal ganglia-cortical regulation.

Most likely several of these pathways act in concert. Indeed, such dual (or multi-ple) basal ganglia-cortical feedforward activation may reflect a general organizing principle well-known for other brain systems (e.g., specific ascending sensory path-ways vs. sensory input to the ascending reticular activating system). In addition to "specific" (premotor/motor) basal ganglia-cortical input, a more general cortical activation may be necessary to provide certain activity states (arousal, conscious-ness) as a prerequisite for specific cortical functions. For one, basal ganglia-cortical activation via the midline–intralaminar complex may contribute to such states. Con-versely, cortical activation through the basal forebrain cholinergic system may play a role in attention. It is known that frontostriatal circuits are engaged in attentional performances (e.g., Robbins et al., 1998; Berger and Posner, 2000; Robbins, 2000), as are the cholinergic pathways from the basal forebrain to the cortex (e.g., Sarter and Bruno, 1999; Sarter et al., 2001, for reviews).

What can changes in IEG responses indicate? IEG expression can reflect changes in neuronal activity within a several-minutes timeframe at best (Melzer and Steiner, 1997). Such IEG signals may thus not be able to represent "specific" motor commands (e.g., related to action selection or switching; Mink, 1996; Redgrave et al., 1999) presumably in "motor" pathways. However, these signals could indicate changes in general activation (e.g., arousal) or attentional states that span minutes and more.

18.5 Conclusions

Anatomical findings indicate that striatal output in the direct pathway provides dif-ferent cortical activation by connecting to functionally different basal ganglia output systems. The above reviewed studies suggest that dopamine in the striatum, by facili-tating D1 receptor-modulated output in the direct pathway, not only affects motor commands, but also provides a powerful regulation of more general activity states in the cortex without which the cortex cannot function.

Acknowledgments

This work was supported by USPHS grant DA11261.

References

Albin, R.L., Young, A.B. and Penney, J.B. (1989) The functional anatomy of basal ganglia disorders. Trends Neurosci. 12, 366–375.

Alexander, G.E., Crutcher, M.D. and DeLong, M.R. (1990) Basal ganglia-thalamocortical circuits: parallel substrates for motor, oculomotor, "prefrontal" and "limbic" functions. Prog. Brain Res. 85, 119–146.

Alexander, G.E., DeLong, M.R. and Strick, P.L. (1986) Parallel organization of functionally segregated circuits linking basal ganglia and cortex. Annu. Rev. Neurosci. 9, 357–381.

Anderson, J.J., Chase, T.N. and Engber, T.M. (1993) Substance P increases release of acetylcholine in the dorsal striatum of freely moving rats. Brain Res. 623, 189–194.

Anderson, J.J., Kuo, S., Chase, T.N. and Engber, T.M. (1994) Dopamine D1 receptor-stimulated release of acetylcholine in rat striatum is mediated indirectly by activation of striatal neurokinin1 receptors. J. Pharmacol. Exp. Ther. 269, 1144–1151.

Arbuthnott, G.W., MacLeod, N.K., Maxwell, D.J. and Wright, A.K. (1990) Distribution and synaptic contacts of the cortical terminals arising from neurons in the rat ventromedial thalamic nucleus. Neuroscience 38, 47–60.

Arenas, E., Alberch, J., Perez-Navarro, E., Solsona, C. and Marsal, J. (1991) Neurokinin receptors differentially mediate endogenous acetylcholine release evoked by tachykinins in the neostriatum. J. Neurosci. 11, 2332–2338.

Arnold, H.M., Nelson, C.L., Neigh, G.N., Sarter, M. and Bruno, J.P. (2000) Systemic and intra-accumbens administration of amphetamine differentially affects cortical acetylcholine release. Neuroscience 96, 675–685.

Badiani, A., Oates, M.M., Day, H.E., Watson, S.J., Akil, H. and Robinson, T.E. (1998) Amphetamine-induced behavior, dopamine release, and c-fos mRNA expression: modulation by environmental novelty. J. Neurosci. 18, 10579–10593.

Badiani, A., Oates, M.M., Day, H.E., Watson, S.J., Akil, H. and Robinson, T.E. (1999) Environmental modulation of amphetamine-induced c-fos expression in D1 versus D2 striatal neurons. Behav. Brain Res. 103, 203–209.

Berger, A. and Posner, M.I. (2000) Pathologies of brain attentional networks. Neurosci. Biobehav. Rev. 24, 3–5.

Berger, B., Gaspar, P and Verney, C. (1991) Dopaminergic innervation of the cerebral cortex: unexpected differences between rodents and primates. Trends Neurosci. 14, 21 27.

Berke, J.D., Paletzki, R.F., Aronson, G.J., Hyman, S.E. and Gerfen, C.R. (1998) A complex program of striatal gene expression induced by dopaminergic stimulation. J. Neurosci. 18, 5301–5310.

Berretta, S., Robertson, H.A. and Graybiel, A.M. (1992) Dopamine and glutamate agonists stimulate neuron-specific expression of Fos-like protein in the striatum. J. Neurophysiol. 68, 767–777.

Björklund, A. and Lindvall, O. (1984) Dopamine containing systems in the CNS. In: A. Björklund and T. Hökfelt (Eds.), Handbook of Chemical Neuroanatomy, Vol 2, Classical Transmitters in the CNS, Part 1. Elsevier, London, pp. 55–122.

Blandini, F., Fancellu, R., Orzi, F., Conti, G., Greco, R., Tassorelli, C. and Nappi, G. (2003) Selective stimulation of striatal dopamine receptors of the D1- or D2-class causes opposite changes of fos expression in the rat cerebral cortex. Eur. J. Neurosci. 17, 763–770.

Bolam, J.P., Ingham, C.A., Izzo, P.N., Levey, A.I., Rye, D.B., Smith, A.D. and Wainer, B.H. (1986) Substance P-containing terminals in synaptic contact with cholinergic neurons in the neostriatum and basal forebrain: a double immunocytochemical study in the rat. Brain Res. 397, 279–289.

Bouthenet, M.-L., Souil, E., Martres, M.-P., Sokoloff, P., Giros, B. and Schwartz, J.-C. (1991) Localization of dopamine D3 receptor mRNA in the rat brain using in situ hybridization histochemistry: comparison with dopamine D2 receptor mRNA. Brain Res. 564, 203–219.

Brownell, A.L., Canales, K., Chen, Y.I., Jenkins, B.G., Owen, C., Livni, E., Yu, M., Cicchetti, F., Sanchez-Pernaute, R. and Isacson, O. (2003) Mapping of brain function after MPTP-induced neurotoxicity in a primate Parkinson's disease model. Neuroimage 20, 1064–1075.

Carlsson, M. and Carlsson, A. (1990) Interactions between glutamatergic and monoaminergic systems within the basal ganglia – implications for schizophrenia and Parkinson's disease. Trends Neurosci. 13, 272–276.

Cenci, M.A., Campbell, K., Wictorin, K. and Björklund, A. (1992) Striatal *c-fos* induction by cocaine or apomorphine occurs preferentially in output neurons projecting to the substantia nigra in the rat. Eur. J. Neurosci. 4, 376–380.

Cepeda, C., Buchwald, N.A. and Levine, M.S. (1993) Neuromodulatory actions of dopamine in the neostriatum are dependent upon the excitatory amino acid receptor subtypes activated. Proc. Natl. Acad. Sci. USA 90, 9576–9580.

Cepeda, C. and Levine, M.S. (1998) Dopamine and N-methyl-D-aspartate receptor interactions in the neostriatum. Dev. Neurosci. 20, 1–18.

Chaudhuri, A. (1997) Neural activity mapping with inducible transcription factors. Neuroreport 8, v–ix.

Chaudhuri, A. and Cynader, M.S. (1993) Activity-dependent expression of the transcription factor Zif268 reveals ocular dominance columns in monkey visual cortex. Brain Res. 605, 349–353.

Chevalier, G. and Deniau, J.M. (1990) Disinhibition as a basic process in the expression of striatal functions. Trends Neurosci. 13, 277–280.

Ciliax, B.J., Nash, N., Heilman, C., Sunahara, R., Hartney, A., Tiberi, M., Rye, D.B., Caron, M.G., Niznik, H.B. and Levey, A.I. (2000) Dopamine D5 receptor immunolocalization in rat and monkey brain. Synapse 37, 125–145.

Cole, A.J., Bhat, R.V., Patt, C., Worley, P.F. and Baraban, J.M. (1992) D1 dopamine receptor activation of multiple transcription factor genes in rat striatum. J. Neurochem. 58, 1420–1426.

Csillik, B., Rakic, P. and Knyihar-Csillik, E. (1998) Peptidergic innervation and the nicotinic acetylcholine receptor in the primate basal nucleus. Eur. J. Neurosci. 10, 573–585.

Curran, E.J. and Watson, S.J. (1995) Dopamine receptor mRNA expression patterns by opioid peptide cells in the nucleus accumbens of the rat: a double in situ hybridization study. J. Comp. Neurol. 361, 57–76.

Day, J. and Fibiger, H.C. (1992) Dopaminergic regulation of cortical acetylcholine release. Synapse 12, 281–286.

Day, J. and Fibiger, H.C. (1993) Dopaminergic regulation of cortical acetylcholine release: effects of dopamine receptor agonists. Neuroscience 54, 643–648.

Day, J.C., Tham, C.S. and Fibiger, H.C. (1994) Dopamine depletion attenuates amphetamine-induced increases of cortical acetylcholine release. Eur. J. Pharmacol. 263, 285–292.

De Souza Silva, M.A., Hasenohrl, R.U., Tomaz, C., Schwarting, R.K.W. and Huston, J.P. (2000) Differential modulation of frontal cortex acetylcholine by injection of substance P into the nucleus basalis magnocellularis region in the freely-moving vs. the anesthetized preparation. Synapse 38, 243–253.

Defagot, M.C., Malchiodi, E.L., Villar, M.J. and Antonelli, M.C. (1997) Distribution of D4 dopamine receptor in rat brain with sequence-specific antibodies. Mol. Brain Res. 45, 1–12.

DeLong, M.R. (1990) Primate models of movement disorders of basal ganglia origin. Trends Neurosci. 13, 281–285.

Dilts, R.P.J., Helton, T.E. and McGinty, J.F. (1993) Selective induction of Fos and FRA immunoreactivity within the mesolimbic and mesostriatal dopamine terminal fields. Synapse 13, 251–263.

Donoghue, J.P. and Carroll, K.L. (1987) Cholinergic modulation of sensory responses in rat primary somatic sensory cortex. Brain Res. 408, 367–371.

Drago, J., Gerfen, C.R., Westphal, H. and Steiner, H. (1996) D1 dopamine receptor-deficient mouse: Cocaine-induced regulation of immediate-early gene and substance P expression in the striatum. Neuroscience 74, 813–823.

Ferguson, S.M. and Robinson, T.E. (2004) Amphetamine-evoked gene expression in striato-pallidal neurons: regulation by corticostriatal afferents and the ERK/MAPK signaling cascade. J. Neurochem. 91, 337–348.

Filipkowski, R.K., Rydz, M. and Kaczmarek, L. (2001) Expression of c-fos, Fos B, Jun B, and Zif268 transcription factor proteins in rat barrel cortex following apomorphine-evoked whisking behavior. Neuroscience 106, 679–688.

Gaspar, P., Bloch, B. and Le Moine, C. (1995) D1 and D2 receptor gene expression in the rat frontal cortex: cellular localization in different classes of efferent neurons. Eur. J. Neurosci. 7, 1050–1063.

Gerfen, C.R. (1991) Substance P (neurokinin-1) receptor mRNA is selectively expressed in cholinergic neurons in the striatum and basal forebrain. Brain Res. 556, 165–170.

Gerfen, C.R. (1992) The neostriatal mosaic: multiple levels of compartmental organization. Trends Neurosci. 15, 133–139.

Gerfen, C.R., Engber, T.M., Mahan, L.C., Susel, Z., Chase, T.N., Monsma, F.J., Jr. and Sibley, D.R. (1990) D1 and D2 dopamine receptor-regulated gene expression of striatonigral and striatopallidal neurons. Science 250, 1429–1432.

Gerfen, C.R., Keefe, K.A. and Gauda, E.B. (1995) D1 and D2 dopamine receptor function in the striatum: coactivation of D1- and D2-dopamine receptors on separate populations of neurons results in potentiated immediate-early gene response in D1-containing neurons. J. Neurosci. 15, 8167–8176.

Gerfen, C.R., Staines, W.A., Arbuthnott, G.W. and Fibiger, H.C. (1982) Crossed connections of the substantia nigra in the rat. J. Comp. Neurol. 207, 283–303.

Gerfen, C.R. and Wilson, C.J. (1996) The basal ganglia. In: L.W. Swanson, A. Björklund and T. Hökfelt (Eds.), Handbook of Chemical Neuroanatomy. Elsevier, Amsterdam, pp. 371–468.

Graybiel, A.M. (1997) The basal ganglia and cognitive pattern generators. Schizophr. Bull. 23, 459–469.

Graybiel, A.M., Canales, J.J. and Capper-Loup, C. (2000) Levodopa-induced dyskinesias and dopamine-dependent stereotypies: a new hypothesis. Trends Neurosci. 23, S71–S77.

Graybiel, A.M., Moratalla, R. and Robertson, H.A. (1990) Amphetamine and cocaine induce drug-specific activation of the c-fos gene in striosome-matrix compartments and limbic subdivisions of the striatum. Proc. Natl. Acad. Sci. USA 87, 6912–6916.

Groenewegen, H.J. and Berendse, H.W. (1994) The specificity of the 'nonspecific' midline and intralaminar thalamic nuclei. Trends Neurosci. 17, 52–57.

Groenewegen, H.J., Berendse, H.W., Wolters, J.G. and Lohman, A.H. (1990) The anatomical relationship of the prefrontal cortex with the striatopallidal system, the thalamus and the amygdala: evidence for a parallel organization. Prog. Brain Res. 85, 95–116.

Grove, E.A., Domesick, V.B. and Nauta, W.J. (1986) Light microscopic evidence of striatal input to intrapallidal neurons of cholinergic cell group Ch4 in the rat: a study employing the anterograde tracer Phaseolus vulgaris leucoagglutinin (PHA-L). Brain Res. 367, 379–384.

Haber, S. and McFarland, N.R. (2001) The place of the thalamus in frontal cortical-basal ganglia circuits. Neuroscientist 7, 315–324.

Haber, S.N. (2003) The primate basal ganglia: parallel and integrative networks. J. Chem. Neuroanat. 26, 317–330.

Harlan, R.E. and Garcia, M.M. (1998) Drugs of abuse and immediate-early genes in the forebrain. Mol. Neurobiol. 16, 221–267.

Henderson, Z. (1997) The projection from the striatum to the nucleus basalis in the rat: an electron microscopic study. Neuroscience 78, 943–955.

Herkenham, M. (1979) The afferent and efferent connections of the ventromedial thalamic nucleus in the rat. J. Comp. Neurol. 183, 487–517.

Herkenham, M. (1980) Laminar organization of thalamic projections to the rat neocortex. Science 207, 532–535.

Howard, M.A. and Simons, D.J. (1994) Physiologic effects of nucleus basalis magnocellularis stimulation on rat barrel cortex neurons. Exp. Brain Res. 102, 21–33.

Hyman, S.E., Cole, R.L., Schwarzschild, M., Cole, D., Hope, B. and Konradi, C. (1996) Molecular mechanisms of striatal gene regulation: a critical role for glutamate in dopamine-mediated gene induction. In: K.M. Merchant (Ed.), *Pharmacological Regulation of Gene Expression in the CNS.* CRC, Boca Raton, FL, pp. 115–131.

Hyman, S.E. and Nestler, E.J. (1996) Initiation and adaptation: a paradigm for understanding psychotropic drug action. Am. J. Psychiatry 153, 151–162.

Ingham, C.A., Bolam, J.P., Wainer, B.H. and Smith, A.D. (1985) A correlated light and electron microscopic study of identified cholinergic basal forebrain neurons that project to the cortex in the rat. J. Comp. Neurol. 239, 176–192.

Joel, D. and Weiner, I. (1994) The organization of the basal ganglia-thalamocortical circuits: open interconnected rather than closed segregated. Neuroscience 63, 363–379.

Johansson, B., Lindström, K. and Fredholm, B.B. (1994) Differences in the regional and cellular localization of *c-fos* messenger RNA induced by amphetamine, cocaine and caffeine in the rat. Neuroscience 59, 837–849.

Kaczmarek, L. and Chaudhuri, A. (1997) Sensory regulation of immediate-early gene expression in mammalian visual cortex: implications for functional mapping and neural plasticity. Brain Res. Rev. 23, 237–256.

Kawaguchi, Y., Wilson, C.J. and Emson, P.C. (1990) Projection subtypes of rat neostriatal matrix cells revealed by intracellular injection of biocytin. J. Neurosci. 10, 3421–3438.

Kosofsky, B.E., Genova, L.M. and Hyman, S.E. (1995) Substance P phenotype defines specificity of *c-fos* induction by cocaine in developing rat striatum. J. Comp. Neurol. 351, 41–50.

Krout, K.E., Belzer, R.E. and Loewy, A.D. (2002) Brainstem projections to midline and intralaminar thalamic nuclei of the rat. J. Comp. Neurol. 448, 53–101.

Krout, K.E., Loewy, A.D., Westby, G.W. and Redgrave, P. (2001) Superior colliculus projections to midline and intralaminar thalamic nuclei of the rat. J. Comp. Neurol. 431, 198–216.

LaHoste, G.J., Ruskin, D.N. and Marshall, J.F. (1996) Cerebrocortical Fos expression following dopaminergic stimulation: D1/D2 synergism and its breakdown. Brain Res. 728, 97–104.

LaHoste, G.J., Yu, J. and Marshall, J.F. (1993) Striatal Fos expression is indicative of dopamine D1/D2 synergism and receptor supersensitivity. Proc. Natl. Acad. Sci. USA 90, 7451–7455.

Le Moine, C. and Bloch, B. (1995) D1 and D2 dopamine receptor gene expression in the rat striatum: sensitive cRNA probes demonstrate prominent segregation of D1 and D2 mRNAs in distinct neuronal populations of the dorsal and ventral striatum. J. Comp. Neurol. 355, 418–426.

Le Moine, C. and Bloch, B. (1996) Expression of the D3 dopamine receptor in peptidergic neurons of the nucleus accumbens: comparison with the D1 and D2 dopamine receptors. Neuroscience 73, 131–143.

Le Moine, C., Normand, E., Guitteny, A.F., Fouque, B., Teoule, R. and Bloch, B. (1990) Dopamine receptor gene expression by enkephalin neurons in rat forebrain. Proc. Natl. Acad. Sci. USA 87, 230–234.

Le Moine, C., Svenningsson, P., Fredholm, B.B. and Bloch, B. (1997) Dopamine-adenosine interactions in the striatum and the globus pallidus: inhibition of striatopallidal neurons through either D2 or A2A receptors enhances D1 receptor-mediated effects on *c-fos* expression. J. Neurosci. 17, 8038–8048.

Lindefors, N., Brene, S., Herrera-Marschitz, M. and Persson, H. (1989) Region specific regulation of glutamic acid decarboxylase mRNA expression by dopamine neurons in rat brain. Exp. Brain Res. 77, 611–620.

Lindefors, N., Brene, S., Herrera-Marschitz, M. and Persson, H. (1990) Neuropeptide gene expression in brain is differentially regulated by midbrain dopamine neurons. Exp. Brain Res. 80, 489–500.

Mansour, A., Meador-Woodruff, J.H., Bunzow, J.R., Civelli, O., Akil, H. and Watson, S.J. (1990) Localization of dopamine D2 receptor mRNA and D1 and D2 receptor binding in the rat brain and pituitary: an *in situ* hybridization-receptor autoradiographic analysis. J. Neurosci. 10, 2587–2600.

McHaffie, J.G., Stanford, T.R., Stein, B.E., Coizet, V. and Redgrave, P. (2005) Subcortical loops through the basal ganglia. Trends Neurosci. 28, 401–407.

Melzer, P. and Steiner, H. (1997) Stimulus-dependent expression of immediate-early genes in rat somatosensory cortex. J. Comp. Neurol. 380, 145–153.

Mengod, G., Vilaro, M.T., Niznik, H.B., Sunahara, R.K., Seeman, P., O'Dowd, B.F. and Palacios, J.M. (1991) Visualization of a dopamine D_1 receptor mRNA in human and rat brain. Mol. Brain Res. 10, 185–191.

Middleton, F.A. and Strick, P.L. (1996) The temporal lobe is a target of output from the basal ganglia. Proc. Natl. Acad. Sci. USA 93, 8683–8687.

Middleton, F.A. and Strick, P.L. (1997) New concepts about the organization of basal ganglia output. Adv. Neurol. 74, 57–68.

Middleton, F.A. and Strick, P.L. (2000) Basal ganglia and cerebellar loops: motor and cognitive circuits. Brain Res. Rev. 31, 236–250.

Middleton, F.A. and Strick, P.L. (2002) Basal-ganglia 'projections' to the prefrontal cortex of the primate. Cereb. Cortex 12, 926–935.

Mink, J.W. (1996) The basal ganglia: focused selection and inhibition of competing motor programs. Prog. Neurobiol. 50, 381–425.

Moore, H., Fadel, J., Sarter, M. and Bruno, J.P. (1999) Role of accumbens and cortical dopamine receptors in the regulation of cortical acetylcholine release. Neuroscience 88, 811–822.

Moore, H., Sarter, M. and Bruno, J.P. (1995) Bidirectional modulation of cortical acetylcholine efflux by infusion of benzodiazepine receptor ligands into the basal forebrain. Neurosci. Lett. 189, 31–34.

Moratalla, R., Xu, M., Tonegawa, S. and Graybiel, A.M. (1996) Cellular responses to psychomotor stimulant and neuroleptic drugs are abnormal in mice lacking the D1 dopamine receptor. Proc. Natl. Acad. Sci. USA 93, 14928–14933.

Nakajima, Y., Stanfield, P.R., Yamaguchi, K. and Nakajima, S. (1991) Substance P excites cultured cholinergic neurons in the basal forebrain. Adv. Exp. Med. Biol. 295, 157–167.

Nicola, S.M., Surmeier, J. and Malenka, R.C. (2000) Dopaminergic modulation of neuronal excitability in the striatum and nucleus accumbens. Annu. Rev. Neurosci. 23, 185–215.

Orieux, G., Francois, C., Feger, J. and Hirsch, E.C. (2002) Consequences of dopaminergic denervation on the metabolic activity of the cortical neurons projecting to the subthalamic nucleus in the rat. J. Neurosci. 22, 8762–8770.

Orosz, D. and Bennett, J.P. (1990) Baseline and apomorphine-induced extracellular levels of nigral substance P are increased in an animal model of Parkinson's disease. Eur. J. Pharmacol. 182, 509–514.

Oueslati, A., Breysse, N., Amalric, M., Kerkerian-Le Goff, L. and Salin, P. (2005) Dysfunction of the cortico-basal ganglia-cortical loop in a rat model of early parkinsonism is reversed by metabotropic glutamate receptor 5 antagonism. Eur. J. Neurosci. 22, 2765–2774.

Parent, A. and Hazrati, L.N. (1995) Functional anatomy of the basal ganglia. I. The cortico-basal ganglia-thalamo-cortical loop. Brain Res. Rev. 20, 91–127.

Paul, M.L., Graybiel, A.M., David, J.-C. and Robertson, H.A. (1992) D1-like and D2-like dopamine receptors synergistically activate rotation and *c-fos* expression in the dopamine-depleted striatum in a rat model of Parkinson's disease. J. Neurosci. 12, 3729–3742.

Pelled, G., Bergman, H. and Goelman, G. (2002) Bilateral overactivation of the sensorimotor cortex in the unilateral rodent model of Parkinson's disease - a functional magnetic resonance imaging study. Eur. J. Neurosci. 15, 389–394.

Petitet, F., Glowinski, J. and Beaujouan, J.C. (1991) Evoked release of acetylcholine in the rat striatum by stimulation of tachykinin NK-1 receptors. Eur. J. Pharmacol. 192, 203–204.

Pinna, A., Wardas, J., Cristalli, G. and Morelli, M. (1997) Adenosine A2A receptor agonists increase Fos-like immunoreactivity in mesolimbic areas. Brain Res. 759, 41–49.

Rasmusson, D.D. (2000) The role of acetylcholine in cortical synaptic plasticity. Behav. Brain Res. 115, 205–218.

Redgrave, P., Prescott, T.J. and Gurney, K. (1999) The basal ganglia: a vertebrate solution to the selection problem? Neuroscience 89, 1009–1023.

Robbins, T.W. (2000) Chemical neuromodulation of frontal-executive functions in humans and other animals. Exp. Brain Res. 133, 130–138.

Robbins, T.W., Granon, S., Muir, J.L., Durantou, F., Harrison, A. and Everitt, B.J. (1998) Neural systems underlying arousal and attention. Implications for drug abuse. Ann. N. Y. Acad. Sci. 846, 222–237.

Robertson, G.S. and Fibiger, H.C. (1992) Neuroleptics increase c-fos expression in the forebrain: contrasting effects of haloperidol and clozapine. Neuroscience 46, 315–328.

Robertson, G.S. and Jian, M. (1995) D1 and D2 dopamine receptors differentially increase Fos-like immunoreactivity in accumbal projections to the ventral pallidum and midbrain. Neuroscience 64, 1019–1034.

Robertson, G.S. and Staines, W.A. (1994) D1 dopamine receptor agonist-induced Fos-like immunoreactivity occurs in basal forebrain and mesopontine tegmentum cholinergic neurons and striatal neurons immunoreactive for neuropeptide Y. Neuroscience 59, 375–387.

Robertson, G.S., Vincent, S.R. and Fibiger, H.C. (1990) Striatonigral projection neurons contain D1 dopamine receptor-activated c-fos. Brain Res. 523, 288–290.

Robertson, G.S., Vincent, S.R. and Fibiger, H.C. (1992) D1 and D2 dopamine receptors differentially regulate c-fos expression in striatonigral and striatopallidal neurons. Neuroscience 49, 285–296.

Rodriguez-Puertas, R., Herrera-Marschitz, M., Koistinaho, J. and Hokfelt, T. (1999) Dopamine D1 receptor modulation of glutamate receptor messenger RNA levels in the neocortex and neostriatum of unilaterally 6-hydroxydopamine-lesioned rats. Neuroscience 89, 781–797.

Rolland, A.S., Herrero, M.T., Garcia-Martinez, V., Ruberg, M., Hirsch, E.C. and Francois, C. (2007) Metabolic activity of cerebellar and basal ganglia-thalamic neurons is reduced in parkinsonism. Brain 130, 265–275.

Sagar, S.M., Sharp, F.R. and Curran, T. (1988) Expression of c-fos protein in brain: metabolic mapping at the cellular level. Science 240, 1328–1331.

Sarter, M. and Bruno, J.P. (1999) Abnormal regulation of corticopetal cholinergic neurons and impaired information processing in neuropsychiatric disorders. Trends Neurosci. 22, 67–74.

Sarter, M., Givens, B. and Bruno, J.P. (2001) The cognitive neuroscience of sustained attention: where top-down meets bottom-up. Brain Res. Rev. 35, 146–160.

Schwarting, R.K.W. and Huston, J.P. (1996) Unilateral 6-hydroxydopamine lesions of mesostriatal dopamine neurons and their physiological squeal. Prog. Neurobiol. 49, 215–266.

Sharp, F.R., Sagar, S.M. and Swanson, R.A. (1993) Metabolic mapping with cellular resolution: c-fos vs. 2-deoxyglucose. Crit. Rev. Neurobiol. 7, 205–228.

Sheng, M. and Greenberg, M.E. (1990) The regulation and function of c-fos and other immediate early genes in the nervous system. Neuron 4, 477–485.

Smith, Y., Raju, D.V., Pare, J.F. and Sidibe, M. (2004) The thalamostriatal system: a highly specific network of the basal ganglia circuitry. Trends Neurosci. 27, 520–527.

Solanto, M.V. (2002) Dopamine dysfunction in AD/HD: integrating clinical and basic neuroscience research. Behav. Brain Res. 130, 65–71.

Staiger, J.F., Bisler, S., Schleicher, A., Gass, P., Stehle, J.H. and Zilles, K. (2000) Exploration of a novel environment leads to the expression of inducible transcription factors in barrel-related columns. Neuroscience 99, 7–16.

Steiner, H. and Gerfen, C.R. (1994) Tactile sensory input regulates basal and apomorphine-induced immediate-early gene expression in rat barrel cortex. J. Comp. Neurol. 344, 297–304.

Steiner, H. and Gerfen, C.R. (1995) Dynorphin opioid inhibition of cocaine-induced, D1 dopamine receptor-mediated immediate-early gene expression in the striatum. J. Comp. Neurol. 353, 200–212.

Steiner, H. and Gerfen, C.R. (1996) Dynorphin regulates D1 dopamine receptor-mediated responses in the striatum: relative contributions of pre- and postsynaptic mechanisms in dorsal and ventral striatum demonstrated by altered immediate-early gene induction. J. Comp. Neurol. 376, 530–541.

Steiner, H. and Gerfen, C.R. (1998) Role of dynorphin and enkephalin in the regulation of striatal output pathways and behavior. Exp. Brain Res. 123, 60–76.

Steiner, H. and Kitai, S.T. (2000) Regulation of rat cortex function by D1 dopamine receptors in the striatum. J. Neurosci. 20, 5449–5460.

Steiner, H. and Kitai, S.T. (2001) Unilateral striatal dopamine depletion: time-dependent effects on cortical function and behavioural correlates. Eur. J. Neurosci. 14, 1390–1404.

Surmeier, D.J., Song, W.-J. and Yan, Z. (1996) Coordinated expression of dopamine receptors in neostriatal medium spiny neurons. J. Neurosci. 16, 6579–6591.

Takano, K., Stanfield, P.R., Nakajima, S. and Nakajima, Y. (1995) Protein kinase C-mediated inhibition of an inward rectifier potassium channel by substance P in nucleus basalis neurons. Neuron 14, 999–1008.

Taymans, J.M., Kia, H.K., Groenewegen, H.J., Leysen, J.E. and Langlois, X. (2005) Bilateral control of brain activity by dopamine D1 receptors: evidence from induction patterns of regulator of G protein signaling 2 and c-fos mRNA in D1-challenged hemiparkinsonian rats. Neuroscience 134, 643–656.

Uslaner, J., Badiani, A., Norton, C.S., Day, H.E., Watson, S.J., Akil, H. and Robinson, T.E. (2001) Amphetamine and cocaine induce different patterns of c-fos mRNA expression in the striatum and subthalamic nucleus depending on environmental context. Eur. J. Neurosci. 13, 1977–1983.

Uylings, H.B., Groenewegen, H.J. and Kolb, B. (2003) Do rats have a prefrontal cortex? Behav. Brain Res. 146, 3–17.

Wang, J.Q. and McGinty, J.F. (1996) Glutamatergic and cholinergic regulation of immediate-early gene and neuropeptide gene expression in the striatum. In: K.M. Merchant (Ed.), *Pharmacological Regulation of Gene Expression in the CNS*. CRC, Boca Raton, FL, pp. 81–113.

Wang, J.Q., Smith, A.J.W. and McGinty, J.F. (1995) A single injection of amphetamine or methamphetamine induces dynamic alterations in c-fos, zif/268 and preprodynorphin messenger RNA expression in rat forebrain. Neuroscience 68, 83–95.

Yano, M. and Steiner, H. (2005) Methylphenidate (Ritalin) induces Homer 1a and *zif 268* expression in specific corticostriatal circuits. Neuroscience 132, 855–865.

You, Z.-B., Herrera-Marschitz, M., Nylander, I., Goiny, M., O'Connor, W.T., Ungerstedt, U. and Terenius, L. (1994) The striatonigral dynorphin pathway of the rat studied with *in vivo* microdialysis - II. Effects of dopamine D1 and D2 receptor agonists. Neuroscience 63, 427–434.

Young, S.T., Porrino, L.J. and Iadarola, M.J. (1991) Cocaine induces striatal c-fos-immunoreactive proteins via dopaminergic D_1 receptors. Proc. Natl. Acad. Sci. USA 88, 1291–1295.

Zaborszky, L. and Cullinan, W.E. (1992) Projections from the nucleus accumbens to cholinergic neurons of the ventral pallidum: a correlated light and electron microscopic double-immunolabeling study in rat. Brain Res. 570, 92–101.

Zaborszky, L., Cullinan, W.E. and Braun, A. (1991) Afferents to basal forebrain cholinergic projection neurons: an update. Adv. Exp. Med. Biol. 295, 43–100.

Zhang, L., Lou, D., Jiao, H., Zhang, D., Wang, X., Xia, Y., Zhang, J. and Xu, M. (2004) Cocaine-induced intracellular signaling and gene expression are oppositely regulated by the dopamine D1 and D3 receptors. J. Neurosci. 24, 3344–3354.

Chapter 19

Interplay Between Dopamine and Acetylcholine in the Modulation of Attention

Marco Atzori[1] and Rodrigo Paz[2]

[1]School for Behavioral & Brain Sciences, University of Texas at Dallas, Richardson, TX 75080, USA; [2]Departamento de Psiquiatría y Neurociencias Universidad Diego Portales, Instituto Psiquiátrico Dr. José Horwitz Barak, Santiago de Chile, Chile

Abstract. Attention involves several functions such as *alertness, shift, stabilization,* and *distractor suppression. Alertness* is a global mental state characterized by increased motivation and lowered thresholds for encoding new information. Attention *shift* allows either transiting from a passive inattentive to an active focused state, or refocusing perceptual resources from a previously targeted perceptual object to a more salient one. During attention stabilization perceptual resources are kept concentrated onto a particular target in a manner that the neural activity evoked by selected stimuli is momentarily enhanced. Suppression is an active process by which neural activity evoked by task irrelevant stimuli is diminished. These functions are impaired in several neuropsychiatric conditions. We review clinical and neurophysiological data in humans and laboratory animals suggesting that acetylcholine and dopamine interact in the neocortex to produce purposeful attention.

It is proposed that a more satisfactory theory of attention needs to integrate both tonic and phasic effects produced by acetylcholine and dopamine. In the model here proposed, nicotinic receptors are thought to play a pivotal role in the enhancement of neural activity evoked by task relevant stimuli. Muscarinic receptors are proposed to be involved in alertness, and dopaminergic receptors in the temporary representation of intermediate goals. A combination of signals triggered by muscarinic and dopaminergic receptor coactivation may facilitate the suppression of neural activity evoked by task irrelevant stimuli. A better understanding of the interplay between dopamine and acetylcholine in attention modulation may help to develop better psychopharmacological interventions for neuropsychiatric conditions in which attention is impaired.

19.1 Introduction

Attention is a psychological term that can be defined by the temporary allocation of perceptual and representational resources onto those stimuli that are relevant for the execution of a particular task. All vertebrates – from anurans, urodels, and avians, to human primates – have the capability of executing goal-directed behaviors that vary greatly depending on their motivational state and the environmental conditions (De Lanerolle and Millam, 1980; Ewert et al., 1999, 2001). Interactions between motivation and environment are essential determinants of attention, situating it at the intersection between top-down and bottom-up processes. In fact, perceptual resources can be focused onto a particular percept when the execution of the ongoing task demands attention to the selected item (top-down process), or when some physical property of the object is salient enough to capture perceptual resources independently of the task bottom-up process). An operative definition of attention requires the analyses of several mental functions, which for the purpose of this discussion will be grouped under:

– *Alertness*: early attentive phase in which sensory thresholds for encoding new information are lowered in a manner facilitating the production of more adaptive responses.
– *Attentive shift*: sudden change from idleness to a purposeful activity state, or from executing one motor plan to executing another, following the appearance of new stimuli in the environment.
– *Stabilization*: focusing mental resources on a specific object or mental representation for the whole duration of the task.
– *Distractor Suppression*: active depression of neural activity evoked by irrelevant stimuli.

The rise of attention results probably from the complex interaction between (i) limbic cortices evaluating of the emotional value of perceived stimuli, (ii) prefrontal cortex (PFC) allowing the initiation and coordination of purposeful motor activity, (iii) sensory cortices collecting and furthering relevant perceptual data and filtering of irrelevant information, and (iv) parietal, temporal, and orbitofrontal cortices, supplying the substrate for storage and retrieval of associations between perceptual representations and procedures necessary to obtain them.

Given the extent of cortical areas recruited during energetically highly demanding goal-oriented activity, the existence of a large number of subcortical regulatory mechanisms for the timely initiation and coordination of the activity in the many structures involved in attention is not surprising. A wealth of experimental data has underscored the role of monoamines and other neuromodulators in determining the transitions between different attentive states. Many laboratories have highlighted the importance of *dopaminergic* and *cholinergic* systems in attention regulation. These studies typically address separately the role of each system in attention modulation. In this chapter we review recent studies suggesting that dopaminergic and the cholinergic systems rather than working separately, are exquisitely integrated.

19.2 Corticopetal Dopaminergic and Cholinergic Systems

19.2.1 PROJECTIONS

Prefrontal and limbic cortices have bidirectional excitatory connections with the nucleus accumbens (NA) and the major corticopetal cholinergic and dopaminergic area, namely; the basal forebrain (BF, including the nucleus basalis of Meynert, the diagonal band of Broca, and the septal nuclei) and the ventral tegmental area (VTA).

In addition to back projections to the PFC and limbic cortices (LC), the BF projects massively to the rest of the neocortex and hippocampus, including sensory, motor, and associative cortices, as well as the nucleus accumbens (NA) (Fig. 19.1).

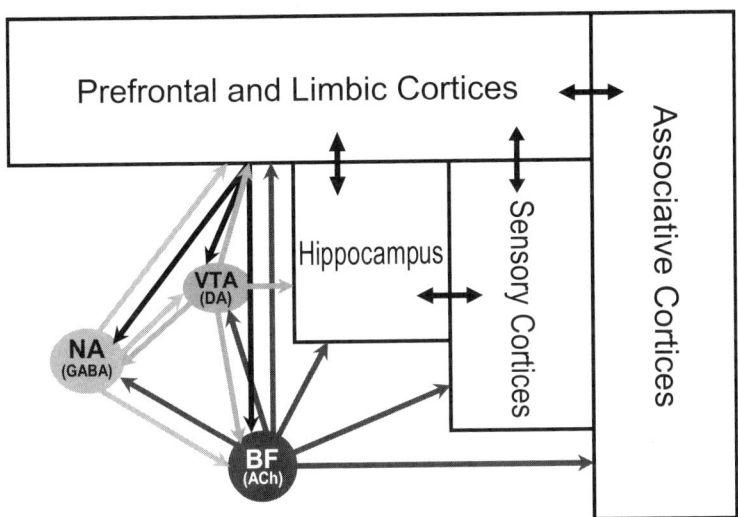

Figure 19.1 Scheme of the putative brain areas involved in attention. The modulation of attention presumably involves the interaction of sensory and "control" cortices. The activity of these extended brain regions is modulated by few subcortical areas including the dopaminergic VTA, the basal forebrain (Basalis of Meynert, Medial Septum and Diagonal Band of Broca, BF) and the nucleus accumbens (NA) Other subcortical areas likely involved in attention modulation such as thalamus, amygdala, locus coeruleus, tuberomammillary hypothalamic and pedunculopontine nucleus have been deliberately omitted in the attempt to simplify the circuit accounting for the basic features of attention.

19.2.2 RECEPTORS

Dopamine (DA) receptors are grouped into two families; low affinity D1-like receptors (D1- and D5Rs), predominantly localized in prefrontal and limbic cortices, and high affinity D2-like receptors (D2-, D3-, and D4Rs), which are more abundantly although not exclusively localized in the striatum (Sealfon and Olanow, 2000; Seamans et al., 2001a, b).

Acetylcholine (ACh) activates two main receptor types: fast, ionotropic nicotinic and slow, metabotropic, G-protein coupled receptors. Similar to those localized at the neuromuscular junction, cortical nicotinic receptors are mostly heterotetramers formed by $\alpha 4$ and $\beta 2$ subunits, or homomers constituted by $\alpha 7$ subunits (Alkondon and Albuquerque, 2004). These receptors exhibit a very low affinity for ACh [tens to several hundreds μM (Giniatullin et al., 2005)], and are only activated by phasic ACh release.

Five classes of cortical muscarinic receptors have been described. The M1-, M2-, and M4-subtypes are most abundantly expressed in the cortex (Mash and Potter, 1986; Ferrari-Dileo et al., 1994), and exhibit higher affinity for ACh than nicotinic receptors. Among cortical muscarinic receptors, a gradient of affinity for ACh has been identified (50–100 nM for M1Rs, in the low 10 nM for M2Rs and M4Rs (Mash and Potter, 1986; Ishii and Kurachi, 2006). This feature suggests that at low tonic ACh concentrations only M2 receptors would be activated, whereas M1 receptors become activated only by high tonic to low phasic ACh release. By contrast, nicotinic receptors to activate only during transient phasic ACh release

19.3 Disruption of Attention in Neuropsychiatric Disorders

Attention deficits are present in multiple neuropsychiatric disorders such as Alzheimer disease, schizophrenia, attention-deficit disorder, autism, Lewy's bodies disease, and drug dependence. Clinical and pharmacological findings suggest that abnormalities in cholinergic and/or dopaminergic function may play a role in the pathogenesis of these disorders. Dissecting the specific contribution of each neurotransmitter system in these conditions is complicated. The classification that follows is by no means an attempt to systematically review this topic, nor to attribute the causes of these disorders to the failure of one or another transmitter system, but rather to highlight potential links between deficits in these neurotransmitter systems and some abnormalities in attention modulation observed in these conditions.

19.3.1 INVOLVEMENT OF THE DOPAMINERGIC SYSTEM IN ATTENTION

Attention Deficit Disorder

Subjects falling under the umbrella of attention-deficit disorder (ADD) with and without hyperactivity display the peculiar tendency to quickly shift attention from one target to another in a manner that makes impossible to maintain focus during the time necessary to finish the required task (Riccio and Reynolds, 2001). In some cases ADD symptoms disappear after puberty, coinciding with the end of the maturation of the cortical dopaminergic system (Noisin and Thomas, 1988). Among the most common FDA-approved drugs prescribed for ADD are the so-called "stimulant" methylphenidate and amphetamine, which modulate dopaminergic function by decreasing DA reuptake in synaptic terminals.

What mechanism underlies the stabilization of attention operated by stimulant treatments used against ADD symptoms? Direct or indirect DA agonists may exert a therapeutic effect in ADD by modulating PFC glutamatergic activity. Consistent with this idea, animal research suggest that DA modulates PFC glutamate release (Gao et al., 2001). Another mechanism by which stimulant drugs may lead to attention stabilization in ADD people could be throughout the excitatory effect of both D1 and D2 agonists on different subtypes of PFC GABAergic interneurons (Tseng and O'Donnell, 2006).

Drug Dependence

The core of drug dependence comprises a pathological interest toward environment stimuli and circumstances previously associated with drug intake at expenses of directing attentive resources to other activities that are relevant for social adaptation and survival (job, family, etc.) (Johnson et al., 2005). Most drugs of abuse directly or indirectly increase dopaminergic function, either as postsynaptic DA receptor agonists (Di Chiara, 1999), or by preventing DA reuptake in the synaptic terminal. The pioneer work of Schultz (reviewed in Schultz, 2002) suggests that DA is released from the VTA following unexpected reward, which may explain the addictive aspects of pathological drug intake. The existence of mesolimbic projections from the VTA to the hippocampus, and the crucial role played by the hippocampus in memory, where DA release is necessary for inducing long-lasting synaptic changes, points to the hippocampus as one of the main temporary recipients for reward-associated information (Lisman and Grace, 2005).

The recent discovery that the hippocampus continuously plays backward contextual sensory information (Foster and Wilson, 2006) might be crucial for understanding the hippocampal DA-ACh interplay that initiates permanent storage of reward-related associations (conditional and unconditional stimuli). During the learning procedure (consisting in the association between contextual information – or conditional stimulus – and reward), the pattern of VTA activation would progressively shift from being triggered by reward to being triggered by the associative stimulus itself. At this stage, receiving the expected reward no longer increases DA release, and failure in receiving reward decreases VTA tonic firing (Schultz, 1998). These features could explain the need of the drug addict to constantly increase drug doses, and is likely to be central for any working hypothesis on motivation based attention.

Autism and Obsessive Compulsive Disorder

Attention hyperstabilization may explain the emergence of the repetitive movements characterizing autism as well as obsessive compulsive disorder (OCD). Among the central clinical features of autism it is conventionally accepted a "restricted, repetitive and stereotypical patterns of behavior, interests, and activities," associated, more specifically with "encompassing preoccupation with one or more stereotyped and restricted patterns of interest that is abnormal either in intensity or focus" and "apparently inflexible adherence to specific, nonfunctional routines or rituals; stereo-typed and repetitive motor mannerisms, such as hand or finger flapping, or complex whole body movements; persistent preoccupation with parts of objects" (Eilam et al., 2006).

Autistic symptoms are at least partially alleviated by administration of the D2 antagonists (Goforth and Rao, 2003). In autistic patients, DA agonists and antagonist have been found to act selectively on a different range of autistic symptoms, suggesting a complex pharmacological anatomy of this disorder (Dollfus et al., 1992). Besides a hyperactive direct striatal pathway, the involvement of prefrontal cortex in repetitive behavior is suggested by the fact that D1 agonist exposure induces repetitive behavior, which is alleviated in OCD patients by D1 antagonists (Carlsson, 2001).

Schizophrenia

Attentive impairments are frequently observed in patients with schizophrenia Higashima et al., 2005). These deficits usually include exaggerated shifting attentive focus, and inability to suppress distractors (Green et al., 2004). Besides being released after rewarding stimuli, mesolimbic DA is released following the presentation of salient stimuli (Ungless, 2004). PFC neurons have the unique capability of holding an increased firing rate during the execution of tasks requiring to maintain the representation of an object in the absence of the object itself (Fuster, 1973; Constantinidis et al., 2001). This increased firing rate subsides after task completion. These features suggest the existence of neuronal ensembles within the PFC whose simultaneous firing represents either the object, the task itself, or other information necesary to the task execution.

The existence of simultaneously active neuronal ensembles has been assimilated with a peculiar property of cortical – as well as striatal cells – that is the capability to present bi-stable membrane potentials denominated up- and down-states (Plenz and Kitai, 1999; Tseng and O'Donnell, 2005). This cellular property appears as a likely candidate to explain the capacity of simultaneous firing within PFC neuronal assemblies (Tseng and O'Donnell, 2005). DA may play a crucial role in the stabilization of these ensembles by preventing PFC and striatal neurons to switch from a down-state to an up-state and viceversa (Durstewitz et al., 1999; Brunel and Wang, 2001; Durstewitz and Seamans, 2002). In the mathematical terminology used in the linear dynamics description, this effect would correspond to a deepening and widening of the basins of attraction of the up- and down-states. In this manner, the presence of DA would stabilize temporary representations of mental objects – and possibly motor strategies – in the prefrontal cortex.

It has been proposed that a prolonged alteration in dopaminergic cortical function may contribute to destabilize the representation of objects and tasks in schizophrenia, preventing proper associations between objects and motor outputs. Seamans and coauthors hypothesized a distinct physiological role for D1-like (D1R) and D2-like (D2R) PFC DA receptors (Seamans and Yang, 2004). In this model the activation of high affinity D2Rs would promote ensemble synchronization in a subthreshold fashion. Conversely, only the activation of low-affinity D1Rs by phasic DA release may allow pushing above this threshold the activity of one selected ensemble in a manner that this ensemble and no others would become able to produce a motor output. A corollary of this hypothesis is that in case of decreased DA release or impaired D1R-function, PFC neuronal ensembles would keep on activating synchronously but none of them would be able to "stand out" from the PFC background synchronized activity, preventing the emergence of a clear-cut motor output. In this case – possibly corresponding to schizophrenic psychosis – symptoms would be alleviated by decreasing D2R activation (see chapter by Seamans in this book). PFC dopaminergic hypofunction seems to be associated with decrease in sensory gating, which is indeed alleviated by intake of D2 antagonists (Takahashi et al., 2006). Consistent with this interpretation, sensory gating deficits (defined as the physiologic decrease of a response preceded by a similar one in the 40–200 ms range), a hallmark of schizophrenia, can be induced by 6-OHDA-lesions of cortical dopaminergic fibers (Bubser and Koch, 1994; Koch and Bubser, 1994).

19.3.2 INVOLVEMENT OF THE CHOLINERGIC SYSTEM IN ATTENTION

Alzheimer Disease

Patients with Alzheimer disease (AD) display extensive deficits in attention. For instance, AD patients perform far from normal in a visual selective test (Foldi et al., 2005). These deficits can be ameliorated by treatment with AchE inhibitors (Behl et al., 2007), while the use of nicotinic agonists has been suggested to prevent the occurrence of AD symptoms (Arroyo et al., 2002). The relationship between cholinergic dysfunction and AD is not surprising since one of the earliest finding associated with AD symptomatology is the selective destruction of the magnocellular component of Nucleus Basalis and Substantia Innominata (Mesulam, 2004), and lesions of NB impair attention (Voytko et al., 1994).

Lewy's Body Disease and Attention

Failure in tests of attention and executive functions are considered a central feature in patients affected by dementia with Lewy bodies (Crowell et al., 2007). This subtype of dementia has been associated with a severe cholinergic deficit (McKeith and Dickson, 2005). Abnormal prepulse inhibition, a clinical indicator of sensory gating deficits, has also been reported in these patients (Perriol et al., 2005). Attention deficits in AD and DLB patients may be caused by a cholinergic hypofunction.

Attention Deficit Disorder

To our knowledge, no cholinergic drugs have been clinically tested for ADD patients. Yet, anecdotic reports suggest a high incidence of nicotine consume among ADD subjects, suggesting that nicotine self-administration might alleviate ADD symptoms, perhaps through desensitization of a hyperactive or over-enhanced perceptual pathway.

Obsessive Compulsive Disorder

A hyperactive cholinergic system has been hypothesized to be associated with OCD (Carlsson, 2001). The ACh-esterase inhibitor pyridostigmine exacerbates OCD symptoms, while clomipramine, which is regarded as a more effective drug against OCD symptoms than pure serotonin reuptake inhibitors, has anticholinergic properties.

Autism

One of the most highly reproduced molecular finding on autism is a downregulation in nicotinic receptor expression (Blatt et al., 2001; Martin-Ruiz et al., 2004). The abundance of nicotinic receptors along the sensory pathway including thalamocortical relays and sensory cortices (Lavine et al., 1997; Penschuck et al., 2002; Metherate and Hsieh, 2003) suggest that nicotinic receptors in the cortex may also play a crucial role in the enhancement of external inputs versus local cortico-cortical activity

(Kimura, 2000). Thus, a downregulation of the nicotinic receptor function could explain the autistic trait of neglecting external sensory information at expenses of excessive processing of internal signals.

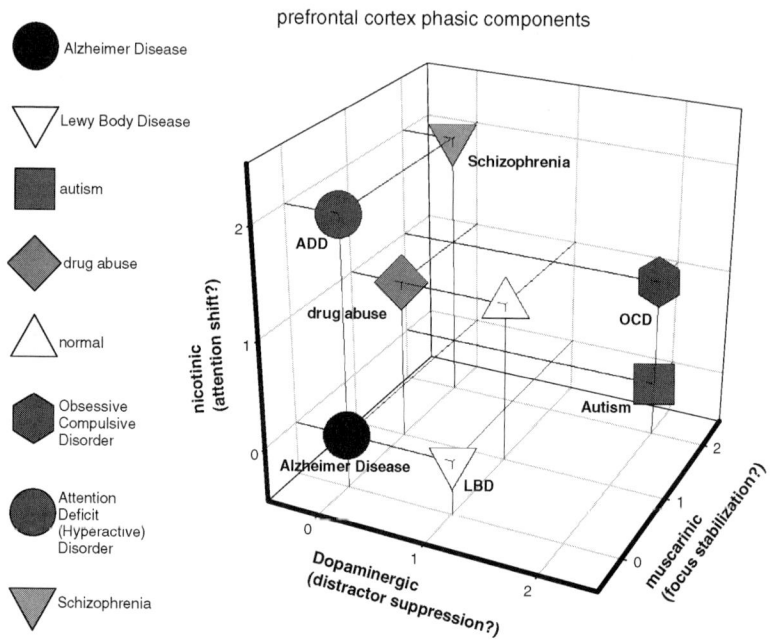

Figure 19.2 Proposed roles for DA and ACh in attention modulation under normal and pathological conditions.

Schizophrenia

The involvement of ACh in the etiology of schizophrenia has been proposed by many authors including Tandon (1999) and (Sarter et al., 2006). An excessive NA stimulation by DA may increase ACh release from the BF (Zhang et al., 2004), which may in turn increase general cortical excitability by blocking K^+ channels (Krnjevic, 1993) in pyramidal cells with detrimental effects for sensory processing and working memory.

We propose different functions for tonic versus phasic DA and ACh levels during different aspects of attention (Grace, 1991). We represented the possible relationship between brain disease and alteration of the phasic actions of dopaminergic, muscarinic, and nicotinic receptors in the prefrontal cortex (Fig. 19.2). A more accurate representation of the relationship neurotransmitter-disease would need the inclusion of many other variables: kinetic (tonic vs. phasic), spatial (ventral striatum, other cortical areas, etc.), and cell types (pyramidal cells vs. fast-spiking and non fast-spiking

GABAergic neurons) that exceed the limits of the present discussion (see review by Seamans and Yang, 2004).

19.4 Dopamine Modulation of Cortical Release of Acetylcholine

Sarter and colleagues proposed (Sarter et al., 2005, 2006; Kozak et al., 2006) that excitatory projections coming from PFC induce cortical ACh release from the BF (representing the "attentional effort"). This effect would be achieved throughout multiple anatomical pathways: a direct excitatory pathway from the PFC to the BF, a second pathway through the NA, and a further pathway involving feedback excitation of dopaminergic neurons within the VTA. Attention-driven increases in PFC ACh release are supported by several studies (Passetti et al., 2000; Arnold et al., 2002; McGaughy et al., 2002; Robbins and Everitt, 2002; Kozak et al., 2006) and is further corroborated by the finding that amphetamine-induced increase in ACh release requires the integrity of the mesocortical dopaminergic pathway (Day et al., 1994; Nelson et al., 2000). PFC activity is able to increase ACh release from the BF through DA-independent and DA-dependent pathways. The dopaminergic circuit connecting PFC with the BF would be represented by the VTA, which stimulated by the PFC may, in turn stimulate DA release in the NA, increasing ACh release in cortical areas other than the prefrontal cortex, like parietal and other sensory cortices. ACh release to nonprefrontal areas would eventually subserve multiple purposes, like increasing general excitability (Cruikshank and Weinberger, 2001), slow waves-desynchronized EEG), sensory receptive field sharpening (Metherate and Weinberger, 1989; Metherate and Ashe, 1991), and promotion of beta and gamma oscillations (Buhl et al., 1998; Shimono et al., 2000).

19.5 Do Dopamine and Acetylcholine Encode Different Aspects of Attention?

In the previous section, we have tried to integrate some of the clinical aspects of attentive dysfunction observed in different neuropsychiatric conditions into a model of attention that incorporates the main functions recognized so far for DA and ACh neurotransmitter systems. Far from explaining the etiology and the detailed mechanisms involved in these conditions, we observe that some of their important features could be summarized in a description representing the temporal development of sustained attention, and the corresponding transients in ACh and DA (Fig. 19.3).

DA release within the cortex (neocortex and hippocampus) would have two different functions depending on whether DA release follows either reward or, on the other hand, novel salient stimuli. In the first case DA would use the hippocampus (and other medial temporal lobe structures as well as orbitofrontal structures) for promoting the storage of the association between contextual information or conditioning stimuli and the actual reward, by rehearsing backward a positively reinforced representation of the circumstances and modality of the reward (Foster and Wilson, 2006).

In the second case it would temporarily stabilize mental representations of previously rewarded objects or behaviors before and during the executive (mental and/or motor) phase leading to pursuing reward. It is important to note that this phase, which in animal models hardly lasts more than a few seconds, in real situations can encompass much longer time spans, interrupted possibly by nonrelated mental and motor activities.

ACETYLCHOLINE-DOPAMINE INTERPLAY IN MOTIVATION-BASED ATTENTION

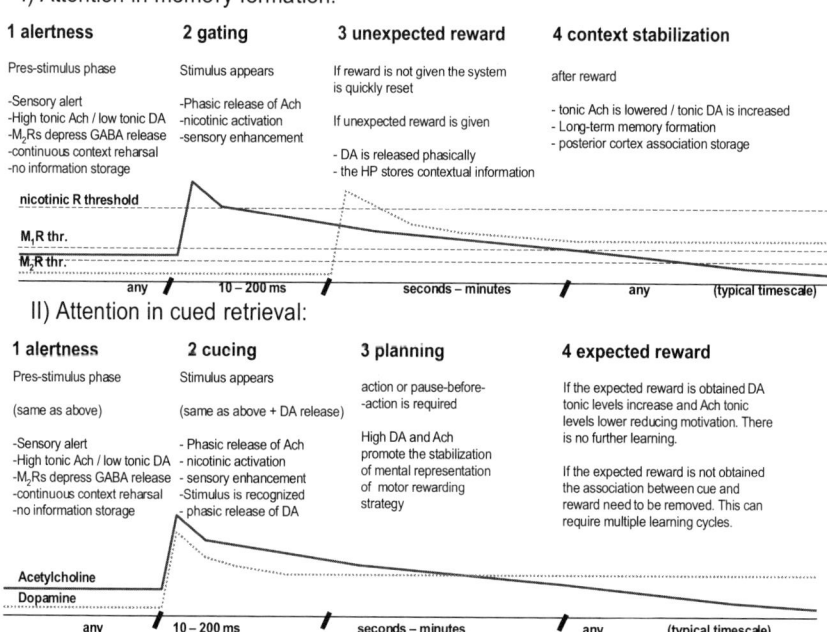

Figure 19.3 Involvement of the dopaminergic and cholinergic system in attention. The sketch illustrates the proposed timeframe for the dopaminergic and cholinergic interaction during different phases of attention. (I) Attention in memory formation: the rewarding (emotional) value of the stimulus is not known in advance. The combined action of ACh and DA, released at different times, allows the formation of memories by involving the hippocampus and other medial temporal lobe structures as well as the orbitofrontal cortex. See legend inside the figure. (II) Attention in cue retrieval: the emotional value of the perceived stimulus is already known and is retrieved from a past experience. High PFC DA release is critical for sustaining goal plus motor plan representations until the motor task is completed. Note the inversion of the levels of ACh and DA in the early vs the late phase of the attentive episode (high ACh and low DA initially, and vice-versa at the end)   .

ACh release would have similar functions in the two cases described before. In this operative definition, attention is gated by motivation. Motivation would increase the general alertness state by increasing the tonic release of ACh. It is long known that ACh release stops slow cortical waves associated with slow wave sleep or quiet wake (idleness) desynchronizing cortical EEG (Cruikshank and Weinberger, 2001). This could be due to activation of high affinity M2Rs depressing massively the release of the inhibitory transmitter GABA (Salgado H., personal communication). Sensory stimuli could trigger rapid and large transients in ACh release triggering a similarly rapid activation and subsequent inactivation of the nicotinic receptors available along the corresponding perceptual pathway. The behavioral effect of the quickly-subsiding increase in ACh concentration would then differ in case of reward-related than novelty-related situations.

In the first case activation of M1Rs would enhance attentive focusing needed to obtain a reward, until either the reward is obtained, or the possibility of obtaining it subsides. If reward is detected the combined action of DA and ACh contribute to the storage of relevant associations using medial temporal lobe and orbitofrontal structures. If reward is not detected the system is reset on the high alert state.

In the second scenario, in which the stimulus is recognized as signaling a potential reward (it can be a novel salient stimulus or a stimulus previously associated with reward), ACh would suppress background activity through M1R-mediated block of glutamate release (Cox et al., 1994; Atzori et al., 2005) while DA temporarily stabilizes the representation of objects and motor strategies as described above. The decrease in motivation following the fulfillment of the biological need would result in decreased vigilance induced by lowering tonic ACh levels; while an increase in tonic residual DA due to the rewarding activity would decrease motivation, reversing the relationship between DA and ACh. In other words, the tonic levels of ACh and DA would be, respectively, high and low at the rise of the attentive phase, and would switch after obtaining the reward.

Acknowledgments

This work was in part supported by grant: NIDCD 1R01-DC005986-01A1 and NARSAD foundation/Sidney Baer Trust (M.A.).

References

Alkondon, M. and Albuquerque, E.X. (2004) The nicotinic acetylcholine receptor subtypes and their function in the hippocampus and cerebral cortex. Prog Brain Res. 145, 109–120.

Arnold, H.M., Burk, J.A., Hodgson, E.M., Sarter, M., and Bruno, J.P. (2002) Differential cortical acetylcholine release in rats performing a sustained attention task versus behavioral control tasks that do not explicitly tax attention. Neuroscience. 114, 451–460.

Arroyo, G., Aldea, M., Fuentealba, J., and Garcia, A.G. (2002) [Nicotinic Receptor, galantamine and Alzheimer disease]. Rev Neurol. 34, 1057–1065.

Atzori, M., Kanold, P.O., Pineda, J.C., Flores-Hernandez, J., and Paz, R.D. (2005) Dopamine prevents muscarinic-induced decrease of glutamate release in the auditory cortex. Neuroscience. 134, 1153–1165.

Behl, P., Bocti, C., Swartz, R.H., Gao, F., Sahlas, D.J., Lanctot, K.L., Streiner, D.L., and Black, S.E. (2007) Strategic subcortical hyperintensities in cholinergic pathways and executive function decline in treated Alzheimer patients. Arch Neurol. 64, 266–272.

Blatt, G.J., Fitzgerald, C.M., Guptill, J.T., Booker, A.B., Kemper, T.L., and Bauman, M.L. (2001) Density and distribution of hippocampal neurotransmitter receptors in autism: an autoradiographic study. J Autism Dev Disord. 31, 537–543.

Brunel, N. and Wang, X.J. (2001) Effects of neuromodulation in a cortical network model of object working memory dominated by recurrent inhibition. J Comput Neurosci. 11, 63–85.

Bubser, M. and Koch, M. (1994) Prepulse inhibition of the acoustic startle response of rats is reduced by 6-hydroxydopamine lesions of the medial prefrontal cortex. Psychopharmacology (Berl). 113, 487–492.

Buhl, E.H., Tamas, G., and Fisahn, A. (1998) Cholinergic activation and tonic excitation induce persistent gamma oscillations in mouse somatosensory cortex in vitro. J Physiol. 513 (Pt 1), 117–126.

Carlsson, M.L. (2001) On the role of prefrontal cortex glutamate for the antithetical phenomenology of obsessive compulsive disorder and attention deficit hyperactivity disorder. Prog Neuropsychopharmacol Biol Psychiatry. 25, 5–26.

Constantinidis, C., Franowicz, M.N., and Goldman-Rakic, P.S. (2001) The sensory nature of mnemonic representation in the primate prefrontal cortex. Nat Neurosci. 4, 311–316.

Cox, C.L., Metherate, R., and Ashe, J.H. (1994) Modulation of cellular excitability in neocortex: muscarinic receptor and second messenger-mediated actions of acetylcholine. Synapse. 16, 123–136.

Crowell, T.A., Luis, C.A., Cox, D.E., and Mullan, M. (2007) Neuropsychological comparison of Alzheimer's disease and dementia with lewy bodies. Dement Geriatr Cogn Disord. 23, 120–125.

Cruikshank, S.J. and Weinberger, N.M. (2001) In vivo Hebbian and basal forebrain stimulation treatment in morphologically identified auditory cortical cells. Brain Res. 891, 78–93.

Day, J.C., Tham, C.S., and Fibiger, H.C. (1994) Dopamine depletion attenuates amphetamine-induced increases of cortical acetylcholine release. Eur J Pharmacol. 263, 285–292.

de Lanerolle, N.C. and Millam, J.R. (1980) Dopamine, chick behavior, and states of attention. J Comp Physiol Psychol. 94, 346–352.

Di Chiara, G. (1999) Drug addiction as dopamine-dependent associative learning disorder. Eur J Pharmacol. 375, 13–30.

Dollfus, S., Petit, M., Menard, J.F., and Lesieur, P. (1992) Amisulpride versus bromocriptine in infantile autism: a controlled crossover comparative study of two drugs with opposite effects on dopaminergic function. J Autism Dev Disord. 22, 47–60.

Durstewitz, D. and Seamans, J.K. (2002) The computational role of dopamine D1 receptors in working memory. Neural Netw. 15, 561–572.

Durstewitz, D., Kelc, M., and Gunturkun, O. (1999) A neurocomputational theory of the dopaminergic modulation of working memory functions. J Neurosci. 19, 2807–2822.

Eilam, D., Zor, R., Szechtman, H., and Hermesh, H. (2006) Rituals, stereotypy and compulsive behavior in animals and humans. Neurosci Biobehav Rev. 30, 456–471.

Ewert, J.P., Buxbaum-Conradi, H., Glagow, M., Rottgen, A., Schurg-Pfeiffer, E., and Schwippert, W.W. (1999) Forebrain and midbrain structures involved in prey-catching behaviour of toads: stimulus-response mediating circuits and their modulating loops. Eur J Morphol. 37, 172–176.

Ewert, J.P., Buxbaum-Conradi, H., Dreisvogt, F., Glagow, M., Merkel-Harff, C., Rottgen, A., Schurg-Pfeiffer, E., and Schwippert, W.W. (2001) Neural modulation of visuomotor functions underlying prey-catching behaviour in anurans: perception, attention, motor performance, learning. Comp Biochem Physiol A Mol Integr Physiol. 128, 417–461.

Ferrari-Dileo, G., Waelbroeck, M., Mash, D.C., and Flynn, D.D. (1994) Selective labeling and localization of the M4 (m4) muscarinic receptor subtype. Mol Pharmacol. 46, 1028–1035.

Foldi, N.S., White, R.E., and Schaefer, L.A. (2005) Detecting effects of donepezil on visual selective attention using signal detection parameters in Alzheimer's disease. Int J Geriatr Psychiatry. 20, 485–488.

Foster, D.J. and Wilson, M.A. (2006) Reverse replay of behavioural sequences in hippocampal place cells during the awake state. Nature. 440, 680–683.

Fuster, J.M. (1973) Unit activity in prefrontal cortex during delayed-response performance: neuronal correlates of transient memory. J Neurophysiol. 36, 61–78.

Gao, W.J., Krimer, L.S., and Goldman-Rakic, P.S. (2001) Presynaptic regulation of recurrent excitation by D1 receptors in prefrontal circuits. Proc Natl Acad Sci USA. 98, 295–300.

Giniatullin, R., Nistri, A., and Yakel, J.L. (2005) Desensitization of nicotinic ACh receptors: shaping cholinergic signaling. Trends Neurosci. 28, 371–378.

Goforth, H.W. and Rao, M.S. (2003) Improvement in behaviour and attention in an autistic patient treated with ziprasidone. Aust N Z J Psychiatry. 37, 775–776.

Grace, A.A. (1991) Phasic versus tonic dopamine release and the modulation of dopamine system responsivity: a hypothesis for the etiology of schizophrenia. Neuroscience. 41, 1–24.

Green, M.F., Nuechterlein, K.H., Gold, J.M., Barch, D.M., Cohen, J., Essock, S., Fenton, W.S., Frese, F., Goldberg, T.E., Heaton, R.K., Keefe, R.S., Kern, R.S., Kraemer, H., Stover, E., Weinberger, D.R., Zalcman, S., and Marder, S.R. (2004) Approaching a consensus cognitive battery for clinical trials in schizophrenia: the NIMH-MATRICS conference to select cognitive domains and test criteria. Biol Psychiatry. 56, 301–307.

Higashima, M., Nagasawa, T., Oka, T., Tsukada, T., Okamoto, T., Komai, Y., Kawasaki, Y., and Koshino, Y. (2005) Neuropsychological correlates of an attention-related negative component elicited in an auditory oddball paradigm in schizophrenia. Neuropsychobiology. 51, 177–182.

Ishii, M. and Kurachi, Y. (2006) Muscarinic acetylcholine receptors. Curr Pharm Des. 12, 3573–3581.

Johnson, B.A., Roache, J.D., Ait-Daoud, N., Wallace, C., Wells, L.T., and Wang, Y. (2005) Effects of isradipine on methamphetamine-induced changes in attentional and perceptual-motor skills of cognition. Psychopharmacology (Berl). 178, 296–302.

Kimura, F. (2000) Cholinergic modulation of cortical function: a hypothetical role in shifting the dynamics in cortical network. Neurosci Res. 38, 19–26.

Koch, M. and Bubser, M. (1994) Deficient sensorimotor gating after 6-hydroxydopamine lesion of the rat medial prefrontal cortex is reversed by haloperidol. Eur J Neurosci. 6, 1837–1845.

Kozak, R., Bruno, J.P., and Sarter, M. (2006) Augmented prefrontal acetylcholine release during challenged attentional performance. Cereb Cortex. 16, 9–17.

Krnjevic, K. (1993) Central cholinergic mechanisms and function. Prog Brain Res. 98, 285–292.

Lavine, N., Reuben, M., and Clarke, P.B. (1997) A population of nicotinic receptors is associated with thalamocortical afferents in the adult rat: laminal and areal analysis. J Comp Neurol. 380, 175–190.

Lisman, J.E. and Grace, A.A. (2005) The hippocampal-VTA loop: controlling the entry of information into long-term memory. Neuron. 46, 703–713.

Martin-Ruiz, C.M., Lee, M., Perry, R.H., Baumann, M., Court, J.A., and Perry, E.K. (2004) Molecular analysis of nicotinic receptor expression in autism. Brain Res Mol Brain Res. 123, 81–90.

Mash, D.C. and Potter, L.T. (1986) Autoradiographic localization of M1 and M2 muscarine receptors in the rat brain. Neuroscience. 19, 551–564.

McGaughy, J., Dalley, J.W., Morrison, C.H., Everitt, B.J., and Robbins, T.W. (2002) Selective behavioral and neurochemical effects of cholinergic lesions produced by intrabasalis infusions of 192 IgG-saporin on attentional performance in a five-choice serial reaction time task. J Neurosci. 22, 1905–1913.

McKeith, I.G., Dickson, D.W., Lowe, J., Emre, M., O'Brien, J.T., Feldman, H., Cummings, J., Duda, J.E., Lippa, C., Perry, E.K., Aarsland, D., Arai, H., Ballard, C.G., Boeve, B., Burn, D.J., Costa, D., Del Ser, T., Dubois, B., Galasko, D., Gauthier, S., Goetz, C.G., Gomez-Tortosa, E., Halliday, G., Hansen, L.A., Hardy, J., Iwatsubo, T., Kalaria, R.N., Kaufer, D.,

Kenny, R.A., Korczyn, A., Kosaka, K., Lee, V.M., Lees, A., Litvan, I., Londos, E., Lopez, O.L., Minoshima, S., Mizuno, Y., Molina, J.A., Mukaetova-Ladinska, E.B., Pasquier, F., Perry, R.H., Schulz, J.B., Trojanowski, J.Q., and Yamada, M. (2005) Diagnosis and management of dementia with Lewy bodies: third report of the DLB Consortium. Neurology. 65, 1863–1872.

Mesulam, M. (2004) The cholinergic lesion of Alzheimer's disease: pivotal factor or side show? Learn Mem. 11, 43–49.

Metherate, R. and Ashe, J.H. (1991) Basal forebrain stimulation modifies auditory cortex responsiveness by an action at muscarinic receptors. Brain Res. 559, 163–167.

Metherate, R. and Hsieh, C.Y. (2003) Regulation of glutamate synapses by nicotinic acetyl-choline receptors in auditory cortex. Neurobiol Learn Mem. 80, 285–290.

Metherate, R. and Weinberger, N.M. (1989) Acetylcholine produces stimulus-specific receptive field alterations in cat auditory cortex. Brain Res. 480, 372–377.

Nelson, C.L., Sarter, M., and Bruno, J.P. (2000) Repeated pretreatment with amphetamine sensitizes increases in cortical acetylcholine release. Psychopharmacology (Berl). 151, 406–415.

Noisin, E.L. and Thomas, W.E. (1988) Ontogeny of dopaminergic function in the rat midbrain tegmentum, corpus striatum and frontal cortex. Brain Res. 469, 241–252.

Passetti, F., Dalley, J.W., O'Connell, M.T., Everitt, B.J., and Robbins, T.W. (2000) Increased acetylcholine release in the rat medial prefrontal cortex during performance of a visual attentional task. Eur J Neurosci. 12, 3051–3058.

Penschuck, S., Chen-Bee, C.H., Prakash, N., and Frostig, R.D. (2002) In vivo modulation of a cortical functional sensory representation shortly after topical cholinergic agent application. J Comp Neurol. 452, 38–50.

Perriol, M.P., Dujardin, K., Derambure, P., Marcq, A., Bourriez, J.L., Laureau, E., Pasquier, F., Defebvre, L., and Destee, A. (2005) Disturbance of sensory filtering in dementia with Lewy bodies: comparison with Parkinson's disease dementia and Alzheimer's disease. J Neurol Neurosurg Psychiatry. 76, 106–108.

Plenz, D. and Kital, S.T. (1999) A basal ganglia pacemaker formed by the subthalamic nucleus and external globus pallidus. Nature. 400, 677–682.

Riccio, C.A. and Reynolds, C.R. (2001) Continuous performance tests are sensitive to ADHD in adults but lack specificity. A review and critique for differential diagnosis. Ann N Y Acad Sci. 931, 113–139.

Robbins, T.W. and Everitt, B.J. (2002) Limbic-striatal memory systems and drug addiction. Neurobiol Learn Mem. 78, 625–636.

Sarter, M., Nelson, C.L., and Bruno, J.P. (2005) Cortical cholinergic transmission and cortical information processing in schizophrenia. Schizophr Bull. 31, 117–138.

Sarter, M., Bruno, J.P., Parikh, V., Martinez, V., Kozak, R., and Richards, J.B. (2006) Forebrain dopaminergic-cholinergic interactions, attentional effort, psychostimulant addiction and schizophrenia. Exs. 98, 65–86.

Schultz, W. (1998) Predictive reward signal of dopamine neurons. J Neurophysiol. 80, 1–27.

Schultz, W. (2002) Getting formal with dopamine and reward. Neuron. 36, 241–263.

Sealfon, S.C. and Olanow, C.W. (2000) Dopamine receptors: from structure to behavior. Trends Neurosci. 23, S34–S40.

Seamans, J.K., Durstewitz, D., Christie, B.R., Stevens, C.F., and Sejnowski, T.J. (2001a) Dopamine D1/D5 receptor modulation of excitatory synaptic inputs to layer V prefrontal cortex neurons. Proc Natl Acad Sci USA. 98, 301–306.

Seamans, J.K., Gorelova, N., Durstewitz, D., and Yang, C.R. (2001b) Bidirectional dopamine modulation of GABAergic inhibition in prefrontal cortical pyramidal neurons. J Neurosci. 21, 3628–3638.

Seamans, J.K. and Yang, C.R. (2004) The principal features and mechanisms of dopamine modulation in the prefrontal cortex. Prog Neurobiol. 74, 1–58.

Shimono, K., Brucher, F., Granger, R., Lynch, G., and Taketani, M. (2000) Origins and distribution of cholinergically induced beta rhythms in hippocampal slices. J Neurosci. 20, 8462–8473.

Takahashi, H., Higuchi, M., and Suhara, T. (2006) The role of extrastriatal dopamine D2 receptors in schizophrenia. Biol Psychiatry. 59, 919–928.

Tandon, R. (1999) Cholinergic aspects of schizophrenia. Br J Psychiatry Suppl, 7–11.

Tseng, K.Y. and O'Donnell, P. (2005) Post-pubertal emergence of prefrontal cortical up states induced by D1-NMDA co-activation. Cereb Cortex. 15, 49–57.

Tseng, K.Y. and O'Donnell, P. (2006) Dopamine Modulation of Prefrontal Cortical Interneurons Changes during Adolescence. Cereb Cortex. Jul 3; [Epub ahead of print]

Ungless, M.A. (2004) Dopamine: the salient issue. Trends Neurosci. 27, 702–706.

Voytko, M.L., Olton, D.S., Richardson, R.T., Gorman, L.K., Tobin, J.R., and Price, D.L. (1994) Basal forebrain lesions in monkeys disrupt attention but not learning and memory. J Neurosci. 14, 167–186.

Zhang, L., Zhou, F.M., and Dani, J.A. (2004) Cholinergic drugs for Alzheimer's disease enhance in vitro dopamine release. Mol Pharmacol. 66, 538–544.

Chapter 20

Monoamine-Based Treatments in Schizophrenia: *Time to Change the Paradigm?*

Dr. Rodrigo D. Paz[1] and Dr. Kuei Y. Tseng[2]
[1]Department of Psychiatry & Neuroscience, Universidad Diego Portales, Santiago de Chile, Chile; [2]Cellular & Molecular Pharmacology, Rosalind Franklin University of Medicine & Science, The Chicago Medical School, North Chicago, IL 60064, USA

20.1 Introduction

Functional deficit in the prefrontal cortex (PFC), also known as hypofrontality, is one of the major manifestations in schizophrenia. This is characterized by a failure of the PFC to support behavior. Although the neuronal mechanisms underlying hyporontality are still under debate, most of the neuroimaging findings have been consistent with a hypofunctional PFC state in response to working memory tasks. These cognitive impairments are typically evidenced by lack of PFC activation (for review see Manoach, 2003). However, there are also studies showing either no changes or increased activation of the dorsolateral PFC (Manoach et al., 1999, 2000; Callicott et al., 2000, 2003a, b; Honey et al., 2002). Perhaps these apparently contradictory findings could simply reflect the different strategy used to examine PFC activation and the nonlinear relationship between PFC activation and task demand (Braver et al., 1997; Goldberg et al., 1998; Callicott et al., 1999; Manoach, 2003). PFC response increases parametrically with working memory load until the demand exceeds the functional capacity, during which activation decrease. Despite the fact that patients suffering schizophrenia usually exhibit a relative lower task performance than controls, a similar nonlinear relationship between PFC activation and task demand may also exist in schizophrenia, but shifted toward the left of the curve. A relative decrease or increase of PFC activation could be obtained in schizophrenia subjects depending on the loads required to perform the task and the actual cortical functional level.

Since the early 1950s to our days, pharmacologically oriented treatments in schizophrenia have been monopolized by monoamine modulating drugs. Beginning with the first generation of relatively pure dopamine (DA) antagonists followed by a second generation of antipsychotic drugs with dual DA and serotonin (5HT) antagonist effects, and more recently by partial DA agonists and acetylcholine (Ach) receptor modulating drugs, modern pharmacological interventions in schizophrenia have been fueled with the idea of changing the curse of this devastating disorder by modifying brain monoamine function. However, despite the effectiveness of treating psychotic symptoms, these monoamine-based drugs have very limited effect in restoring the cognitive and emotional disabilities associated to schizophrenia.

In this chapter we will discuss and provide insights on how and why drugs that target the central monoamine system are not necessary effective to prevent cognitive and emotional impairments in schizophrenia. We will summarize evidence indicating that this failure in treating cognitive deficits could be caused by at least two factors. First, most of the cognitive and emotional deteriorations in schizophrenia appear during early adolescence, years before the emergence of psychosis. Secondly, DA antagonists may interfere with several neuronal signaling pathways critically involved in maintaining normal cognitive functions. We hypothesize that a new generation of glutamate modulating drugs could be used to prevent cognitive deteriorations related to schizophrenia in high risk adolescents.

20.2 The Good and the Bad of Antipsychotic Drugs

There is not question about how the discovery of clozapine in the 1960s changed and improved schizophrenia treatment ever since. Clozapine does not induce extrapyramidal side effects and has demonstrated to effectively reverse refractory psychotic symptoms (Kane et al., 1988). Unfortunately, several life-threatening side effects were also observed, particularly in association with chronic drug exposure (Cohen et al., 1990; Bustillo et al., 1996; Henderson et al., 2000; Lamberti et al., 2006). However, none of the so-called second-generation antipsychotic drugs developed during the last decades have achieved comparable efficacy to clozapine (Breier et al., 1999; Conley et al., 1999; Azorin et al., 2001; Chakos et al., 2001; Davis et al., 2003; Lewis et al., 2005; McEvoy et al., 2006; Shaw et al., 2006). The incidence of extrapyramidal side effects without inducing agranulocytosis was signiicantly reduced in patients treated with second-generation drugs (Freedman, 2005; Gardner et al., 2005); yet increased weight gain, hypercholesterolemia, diabetes, and hyperprolactinemia remain the major side effects (Kapur et al., 2002; Volavka et al., 2004; Lieberman et al., 2005). More importantly, both first and second-generation antipsychotic drugs exert a very limited therapeutic effect on cognitive and emotional impairments associated with schizophrenia (Carpenter and Gold, 2002; Mishara and Goldberg, 2004; Gardner et al., 2005; Rosenheck, 2006; Keefe et al., 2006b). This has become one of the major challenges today since the degree of cognitive deficits is one of the major factors that contribute to the unsatisfactory long-term prognosis in schizophrenia patients as assessed in terms of employment rates and independent living status (Hegarty et al., 1994; Wiersma et al., 2000; Malla and Payne, 2005).

It has been thought that the progressive gray and white matter reduction observed in subjects with first psychotic episode may be due to neurotoxic events triggered by psychosis itself (Lieberman, 1999; Lieberman et al., 2001a). However, this seems unlikely since no evidences of such association were found in schizophrenia (Harrison, 1999; Damadzic et al., 2001; Benes et al., 2003, 2006). On the other hand, it has been recently proposed that part of the cognitive deficits observed in schizophrenia could reflect the anatomical and molecular alterations induced by prolonged antipsychotic exposure (Lieberman, 1999; Lieberman et al., 2001b, 2005), in particular those antipsychotic drugs with anti-dopaminergic (DA) effects. Because DA neurotransmission facilitates signaling pathways critical for the formation of new dendritic spines during learning and memory related events and synaptic plasticity (Jay, 2003; Lisman and Grace, 2005), disruption of these neurotrophic effects by DA receptors blockade could explain some of the cellular and synaptic alterations observed after chronic antipsychotic treatment. In fact, a significant decrease of the neuroprotective enzyme manganese-superoxide dismutase was found after 6 month of olanzapine (Pillai et al., 2006). Similarly, two years exposure to olanzapine reduced cortical gray and white matter in nonhuman primates (Dorph-Petersen et al., 2005). Thus, these findings indicated that chronic treatments with second-generation antipsychotic drugs might induce PFC anatomical as well as functional abnormalities. New therapeutic strategies with better neuroprotective and neurocognitive profiles need to be examined, particularly in first psychotic episode patients.

20.3 The Glutamatergic Hypothesis of Schizophrenia Revised

The glutamatergic hypothesis of schizophrenia was originally based upon clinical observations of chronic abusers of phencyclidine (PCP). Thought disorder, emotional blunting, working memory disturbances, and auditory hallucinations resembling the symptoms presented by patients with schizophrenia were found in these subjects addicted to PCP (Luby et al., 1959; Javitt and Zukin, 1991). The observation that acute administration of sub-anesthetic doses of ketamine (which like PCP blocks NMDA receptors) induces similar psychopathological effects in healthy volunteers (Krystal et al., 1994; Malhotra et al., 1996; Adler et al., 1999) supported the hypothesis that hypofunctional NMDA receptors play a critical role in the pathophysiology of schizophrenia. Consequently, anti-DA drugs combined with agents that enhance NMDA function might ameliorate the residual cognitive, emotional and psychotic symptoms in schizophrenia (Goff et al., 1999; Goff and Coyle, 2001). Unfortunately, a recent multicentric study failed to provide evidence in favor of this therapeutic intervention (Carpenter, 2004).

A reformulation of the glutamatergic hypothesis of schizophrenia has emerged. According to this new paradigm, hyperactive glutamatergic neurons in several brain regions including the PFC may underlie the psychotic, cognitive, and emotional manifestations of schizophrenia (Krystal et al., 2003; Moghaddam, 2003). For insance, systemic administration of psychotomimetic doses of NMDA antagonists increased glutamate levels (Moghaddam and Adams, 1998) as well as pyramidal neurons firing in the PFC (Jackson et al., 2004), probably as result of increased tonic excitatory inputs from the hippocampus (Jodo et al., 2005). Interestingly, GABAergic interneurons in the hippocampus are particularly sensitive to psychotomimetic doses of NMDA antagonists (Grunze et al., 1996), an effect that would favor the excitation of the hippocampal → PFC glutamatergic pathway. Furthermore, PCP-induced working memory impairments are correlated with increased glutamate levels in the PFC (Adams and Moghaddam, 1998) and psychotic symptoms and cognitive abnormalities elicited by acute administration of ketamine in healthy volunteers concur with an enhancement of PFC metabolic activity (Breier et al., 1997; Holcomb et al.,

2001, 2005). Finally, proton magnetic resonance spectroscopy studies showing that glutamine levels, an indicator of synaptic glutamate recycling activity (Rothman et al., 1999), are increased in the medial PFC of healthy volunteers acutely exposed to sub-anesthetic doses of ketamine (Rowland et al., 2005). Similar PFC glutamine increase was observed in patients with schizophrenia with first psychotic episode that never received antipsychotic drugs (Bartha et al., 1997; Theberge et al., 2002) and in adolescents at-risk of developing schizophrenia (Tibbo et al., 2004).

Perhaps one of the most challenging questions is how a primary dysregulation of glutamatergic neurotransmission may explain the fact that most cases of schizophrenia emerge after puberty. Electrophysiological recordings of PFC pyramidal neurons at different maturational stages have revealed that DA-glutamate interactions in the PFC mature after puberty (Tseng and O'Donnell, 2004, 2005). It can be speculated that cortical disruption of NMDA functions may contribute to establish the pathophysiological changes observed in schizophrenia by altering the acquisition of mature PFC responses to DA. This is consistent with previous studies showing a delayed acquisition of adult levels of DA receptors (Leslie et al., 1991; Tarazi et al., 1999) and NMDA receptor subunits (Williams et al., 1993; Monyer et al., 1994) around puberty. Furthermore, the expression of BDNF mRNA, a neurotrophic factor that is tightly dependent on synaptic activation of NMDA receptors (Hardingham et al., 2002), reaches its maximum level in the PFC of late adolescence and young adulthood (Webster et al., 2002). Thus, subtle imbalances in glutamatergic excitability determined by genetic and environmental factors may drive some of the mild cognitive, emotional, and motor abnormalities observed in prepsychotic children and adolescents (Weinberger, 1987; Lewis and Murray, 1987; Lewis and Levitt, 2002). These changes may remain relatively silent until the acquisition of mature cognitive abilities dependent on D1-NMDA interactions that emerge around puberty (Tseng and O'Donnell 2005). A developmental disruption of either the glutamatergic or the DA system may therefore lead to the cognitive and emotional manifestations of prodromal schizophrenia (Weiser et al., 2001; Reichenberg et al., 2005; Lencz et al., 2006). Without treatment, postpubertal hyperglutamatergic states may escalate until full psychotic episodes emerge during late adolescence and early adulthood.

20.4 Mechanisms Underlying the Hyperreactive Prefrontal Cortex and its Relevance for Cognitive Deficits in Schizophrenia

Several DA-dependent and DA-independent factors could contribute to elicit the hyperactive NMDA state in schizophrenia. At the cellular level, an increased expression of calcyon (Koh et al., 2003), a D1 receptor interacting protein that allows calcium release from internal stores (Bergson et al., 2003), may potentiates NMDA function in a manner independent from D2 receptor activation. Similarly, reduction of calcineurin levels (Gerber et al., 2003; Eastwood et al., 2005), a phosphatase that normally downregulates NMDA receptors (Rycroft and Gibb, 2004; Smith et al., 2006), may also enhance NMDA function and promotes working memory deficits in response to low doses of the NMDA antagonist MK801 (Miyakawa et al., 2003). Finally, a recent study has identified a genetic defect affecting the expression of the phosphodiesterase 4B gene (PDE4B) in schizophrenia and mood disorders. This gene product plays a critical role in maintaining the normal D1-dependent protein kinase A signal transduction pathway. More importantly, it was found that the PDE4B enzyme interacts with the disrupted-in-schizophrenia 1 (DISC 1) gene in a manner that PDE4B can be released by DISC 1 when the dendritic levels of cyclic adenosine monophosphate (cAMP) are elevated (Millar et al., 2005). Thus, abnormalities in the expression or activity of different gene products such as calcyon, calcineurin, PDE4B, and DISC 1 may lead to hyperexcitable signal transduction pathways dependent on D1-NMDA receptor activation.

Disruption of GABAergic interneuron function may also contribute to initiate and sustain a hyperglutamatergic state in schizophrenia. PFC GABAergic interneurons play an important role in determining the responses of pyramidal neurons to glutamatergic and DA inputs (Tseng and O'Donnell, 2004; Tseng et al., 2006a) and functional disruption of this selective neuronal population could ultimately lead to the altered cognitive performances observed in schizophrenia (reviewed by Lewis et al., 2005). Because DA-dependent attenuation of PFC NMDA responses involves activation of local GABAergic interneurons (Tseng and O'Donnell, 2004), a reduction of this inhibitory tone may also contribute to engage the hyperglutamatergic state in the PFC. Although an overall enhancement of PFC activity may appear at odds with the traditional concept of hypofrontality (Manoach, 2003), recent studies conducted in a developmental animal model that exhibits cortical deficits resembling those observed in schizophrenia indicate that the hypofrontality could be associated with an hyperreactive PFC. The characteristic postpubertal emergence of PFC glutamatergic hyperactivity (O'Donnell et al., 2002) and hyperreactive PFC metabolic response to mesocortical stimulation (Tseng et al., 2006b) are correlated with selective downregulation of PFC glutamate decarboxylase-67 mRNA, a neuronal marker for GABA interneurons (Lipska et al., 2003). Furthermore, a postpubertal disruption of D2 receptor function that normally regulate PFC excitatory responses through several pre- and postsynaptic mechanisms (Tseng and O'Donnell 2004) combined with an abnormal response to D1 activation may also serve to increase pyramidal neurons excitability to NMDA (Tseng et al., 2007) and yield the concurrent hyperreactive and hypofunctional state in the PFC. Finally, it is also possible that other cortically released monoamines such as norepinephrine may further potentiate this abnormal enhancement of NMDA function, particularly the cortical system. Stress-dependent increases in norepinephrine release may contribute to sustain the hyperactive NMDA state through activation of $\alpha 1$ adrenergic receptors and up-regulate postsynaptic protein kinase C signaling pathways in the PFC (Arnsten, 2004) and thus, lead to the hyperreactive response to stress observed in schizophrenia.

We hypothesized that cortical deficits in schizophrenia (i.e., hypofrontality) could be compounded by a hyperglutamatergic PFC state elicited by a reduction of PFC GABAergic function concurrent with a disruption of D2 modulation and an abnormal potentiation of D1 and NMDA responses. These two pathophysiological conditions are not mutually exclusive and may not be restricted to the PFC. Markers of increased glutamatergic activity dependent on NMDA receptors have been shown to coexist with those indicative of decreased GABAergic function such as of GAD-67, GAD-65 and GAT1 in the cerebellar cortex of patients with schizophrenia (Paz et al., 2006).

20.5 Changing the Paradigm Towards a Non-Monoamine-Based Treatment

It has been proposed that anti-DA treatments during the early stages of first-episode psychoses may be effective to prevent cognitive and emotional decline in schizophrenia (Wyatt, 1991; Lieberman et al., 1999). However, recent studies indicate that this strategy does not truly improve the cognitive and emotional outcomes (Ho et al., 2003; Rund et al., 2004; Marshall et al., 2005; Perkins et al., 2005), perhaps because cognitive deficits in schizophrenia occur during early adolescence, before the emergence of psychotic episodes (Perkins et al., 2005). In fact, several retrospective studies have found that cognitive and emotional deterioration in a subgroup of schizophrenia patients become evident when postpubertal social and cognitive performances are compared with those exhibited in prepubertal ages (Cosway et al., 2000; Fuller et al., 2002; Rabinowitz et al., 2002; Ang and Tan, 2004; Reichenberg et al., 2005). Although limited by their retrospective nature, these studies are consistent with the idea that profound changes occurring in the adolescent brain make this neural development period vulnerable. Accelerated pruning of redundant connections in the PFC occurs during this developmental stage (Huttenlocher, 1979; Zecevic et al., 1989) and the myelination of axons projecting from the PFC is particularly intense during adolescence and young adulthood (Jernigan et al., 1991; Sowell et al., 2003; Paus et al., 1999; Giedd et al., 1999; Gogtay et al., 2004). Since these events are probably dependent on plastic changes affecting glutamatergic (Williams et al., 1993; Monyer et al., 1994) and DA systems (Leslie et al., 1991; Tarazi et al., 1999), it is not surprising that PFC pyramidal neurons acquire their mature response to D1+NMDA co-activation at early postpubertal stages of neural development (Tseng and O'Donnell, 2005). A similar delayed acquisition of mature responses to DA has been recently observed in PFC interneurons (Tseng and O'Donnell, 2006). All these interactions will ultimately determine the normal balance of excitation and inhibition that is needed for optimal cognitive performances including those dependent on PFC function during adulthood. Thus, several abnormalities at cellular and network levels compromising both glutamatergic and GABAergic systems may produce dramatic effects during critical periods of neural development that could potentially lead to permanent loss of synaptic connections resulting in cognitive and emotional alterations in adulthood. Consequently, acute exposure to low doses of NMDA antagonists induces psychotic symptoms and cognitive dysfunctions in pubertal/postpubertal rather than prepubertal ages (Olney and Farber, 1995).

A more direct modulation of glutamatergic neurotransmission within the PFC and other brain regions by drugs acting on metabotropic and NMDA glutamate receptors may allow a better control of persistent hyperglutamatergic states that are partially independent from D2 receptor blockade. In this regard, it is appealing that clozapine binds D1 receptors with a higher affinity than D2 receptors (Tauscher et al., 2004; Chou et al., 2006). Clozapine may exert a more direct interaction with NMDA receptors than other antipsychotic drugs and prevent the activation of hyperactive NMDA receptors in the PFC of animals sub-chronically exposed to low doses of PCP (Ninan et al., 2003). Thus, the therapeutic effects of clozapine in patients with refractory psychosis might be achieved by decreasing NMDA function through preventing activation of D1 receptors in the PFC.

A narrow window of NMDA activation mediated by D1 receptors and α adrenergic receptors is needed to maintain appropriate working memory performance. An excessive stimulation of D1 receptors leads to cognitive dysfunction (Zahrt et al., 1997; Cai and Arnsten, 1997; Arnsten and Goldman-Rakic, 1998) whereas insufficient recruitment of these receptors is associated with working memory deficits (Sawaguchi and Goldman-Rakic, 1994; Castner et al., 2000). Interestingly, a similar inverted U-shape response has been reported for noradrenergic manipulations (Arnsten and Li, 2005). Therefore, drugs with anti-metabotropic glutamate receptors (i.e., mGluR5) and partial NMDA agonist properties (e.g., acamprosate) might achieve a more physiological modulation of NMDA function allowing better outcomes in schizophrenia patients with severe cognitive dysfunction. Along with this hypothesis, transgenic mice overexpressing D2 receptors in the striatum exhibit severe deficits in working memory-dependent tasks in association with increased D1 receptor responses in the PFC (Kellendonk et al., 2006). Decreasing the expression of D2 receptors in the striatum failed to restore the increased PFC response. This suggests that blocking D2 receptors in the striatum does not prevent working memory impairments secondary to neurodevelopmentally induced abnormalities of D1-NMDA interactions within the PFC. Consequently, direct modulation of hyperactive NMDA receptors may allow targeting microcircuits within the PFC-subcortical network that are not accessible to conventional D2-based treatments.

20.6 Conclusion

Overall, these studies emphasize the vulnerability of the peri-pubertal brain to environmentally and genetically driven events that could potentially disrupt the balance of pro-excitatory and pro-inhibitory agents. These combined factors are particularly relevant when psychopharmacological treatments are being planned in at-risk adolescents exhibiting cognitive and emotional deterioration. The development of drugs that reduce the hyperglutamatergic state by virtue of normalizing the abnormal NMDA function without suppressing the normal glutamatergic transmission may be useful in chronic schizophrenia patients with persistent and residual psychotic symptoms as well as cognitive and emotional deficits.

References

Adams, B. and Moghaddam, B. (1998) Corticolimbic dopamine neurotransmission is temporally dissociated from the cognitive and locomotor effects of phencyclidine. J Neurosci. 18, 5545–5554.

Adler, C.M., Malhotra, A.K., Elman, I., Goldberg, T., Egan, M., Pickar, D. and Breier, A. (1999) Comparison of ketamine-induced thought disorder in healthy volunteers and thought disorder in schizophrenia. Am J Psychiatry. 156, 1646–1649.

Ang, Y.G. and Tan, H.Y. (2004) Academic deterioration prior to first episode schizophrenia in young Singaporean males. Psychiatry Res. 121, 303–307.

Arnsten, A.F. (2004) Adrenergic targets for the treatment of cognitive deficits in schizophrenia. Psychopharmacology (Berl). 174, 25–31.

Arnsten, A.F. and Goldman-Rakic, P.S. (1998) Noise stress impairs prefrontal cortical cognitive function in monkeys: evidence for a hyperdopaminergic mechanism. Arch Gen Psychiatry. 55, 362–368.

Arnsten, A.F. and Li, B.M. (2005) Neurobiology of executive functions: catecholamine influences on prefrontal cortical functions. Biol Psychiatry. 57, 1377–1384.

Azorin, J.M., Spiegel, R., Remington, G., Vanelle, J.M., Pere, J.J., Giguere, M. and Bourdeix, I. (2001) A double-blind comparative study of clozapine and risperidone in the management of severe chronic schizophrenia. Am J Psychiatry. 158, 1305–1313.

Bartha, R., Williamson, P.C., Drost, D.J., Malla, A., Carr, T.J., Cortese, L., Canaran, G., Rylett, R.J. and Neufeld, R.W. (1997) Measurement of glutamate and glutamine in the medial prefrontal cortex of never-treated schizophrenic patients and healthy controls by proton magnetic resonance spectroscopy. Arch Gen Psychiatry. 54, 959–965.

Benes, F.M., Walsh, J., Bhattacharyya, S., Sheth, A. and Berretta, S. (2003) DNA fragmentation decreased in schizophrenia but not bipolar disorder. Arch Gen Psychiatry. 60, 359–364.

Benes, F.M., Matzilevich, D., Burke, R.E. and Walsh, J. (2006) The expression of proapoptosis genes is increased in bipolar disorder, but not in schizophrenia. Mol Psychiatry. 11, 241–251.

Bergson, C., Levenson, R., Goldman-Rakic, P.S. and Lidow, M.S. (2003) Dopamine receptor-interacting proteins: the Ca(2+) connection in dopamine signaling. Trends Pharmacol Sci. 24, 486–492.

Braver, T.S., Cohen, J.D., Nystrom, L.E., Jonides, J., Smith, E.E. and Noll, D.C. (1997) A parametric study of prefrontal cortex involvement in human working memory. Neuroimage. 5, 49–62.

Breier, A., Malhotra, A.K., Pinals, D.A., Weisenfeld, N.I. and Pickar, D. (1997) Association of ketamine-induced psychosis with focal activation of the prefrontal cortex in healthy volunteers. Am J Psychiatry. 154, 805–811.

Breier, A.F., Malhotra, A.K., Su, T.P., Pinals, D.A., Elman, I., Adler, C.M., Lafargue, R.T., Clifton, A. and Pickar, D. (1999) Clozapine and risperidone in chronic schizophrenia: effects on symptoms, parkinsonian side effects, and neuroendocrine response. Am J Psychiatry. 156, 294–298.

Bustillo, J.R., Buchanan, R.W., Irish, D. and Breier, A. (1996) Differential effect of clozapine on weight: a controlled study. Am J Psychiatry. 153, 817–819.

Cai, J.X. and Arnsten, A.F. (1997) Dose-dependent effects of the dopamine D1 receptor agonists A77636 or SKF81297 on spatial working memory in aged monkeys. J Pharmacol Exp Ther. 283, 183–189.

Callicott, J.H., Mattay, V.S., Bertolino, A., Finn, K., Coppola, R., Frank, J.A., Goldberg, T.E. and Weinberger, D.R. (1999) Physiological characteristics of capacity constraints in working memory as revealed by functional MRI. Cereb Cortex. 9, 20–26.

Callicott, J.H., Bertolino, A., Mattay, V.S., Langheim, F.J., Duyn, J., Coppola, R., Goldberg, T.E. and Weinberger, D.R. (2000) Physiological dysfunction of the dorsolateral prefrontal cortex in schizophrenia revisited. Cereb Cortex. 10, 1078–1092.

Callicott, J.H., Egan, M.F., Mattay, V.S., Bertolino, A., Bone, A.D., Verchinksi, B. and Weinberger, D.R. (2003a) Abnormal fMRI response of the dorsolateral prefrontal cortex in cognitively intact siblings of patients with schizophrenia. Am J Psychiatry. 160, 709–719.

Callicott, J.H., Mattay, V.S., Verchinski, B.A., Marenco, S., Egan, M.F. and Weinberger, D.R. (2003b) Complexity of prefrontal cortical dysfunction in schizophrenia: more than up or down. Am J Psychiatry. 160, 2209–2215.

Carpenter, W.T. (2004) Carpenter WT. Is glutamatergic therapy efficacious in schizophrenia? ACNP, San Juan, Puerto Rico.

Carpenter, W.T. and Gold, J.M. (2002) Another view of therapy for cognition in schizophrenia. Biol Psychiatry. 51, 969–971.

Castner, S.A., Williams, G.V. and Goldman-Rakic, P.S. (2000) Reversal of antipsychotic-induced working memory deficits by short-term dopamine D1 receptor stimulation. Science. 287, 2020–2022.

Chakos, M., Lieberman, J., Hoffman, E., Bradford, D. and Sheitman, B. (2001) Effectiveness of second-generation antipsychotics in patients with treatment-resistant schizophrenia: a review and meta-analysis of randomized trials. Am J Psychiatry. 158, 518–526.

Chou, Y.H., Halldin, C. and Farde, L. (2006) Clozapine binds preferentially to cortical D1-like dopamine receptors in the primate brain: a PET study. Psychopharmacology (Berl). 185, 29–35.

Cohen, S., Chiles, J. and MacNaughton, A. (1990) Weight gain associated with clozapine. Am J Psychiatry. 147, 503–504.

Conley, R.R., Tamminga, C.A., Kelly, D.L. and Richardson, C.M. (1999) Treatment-resistant schizophrenic patients respond to clozapine after olanzapine non-response. Biol Psychiatry. 46, 73–77.

Cosway, R., Byrne, M., Clafferty, R., Hodges, A., Grant, E., Abukmeil, S.S., Lawrie, S.M., Miller, P. and Johnstone, E.C. (2000) Neuropsychological change in young people at high risk for schizophrenia: results from the first two neuropsychological assessments of the Edinburgh High Risk Study. Psychol Med. 30, 1111–1121.

Damadzic, R., Bigelow, L.B., Krimer, L.S., Goldenson, D.A., Saunders, R.C., Kleinman, J.E. and Herman, M.M. (2001) A quantitative immunohistochemical study of astrocytes in the entorhinal cortex in schizophrenia, bipolar disorder and major depression: absence of significant astrocytosis. Brain Res Bull. 55, 611–618.

Davis, J.M., Chen, N. and Glick, I.D. (2003) A meta-analysis of the efficacy of second-generation antipsychotics. Arch Gen Psychiatry. 60, 553–564.

Dorph-Petersen, K.A., Pierri, J.N., Perel, J.M., Sun, Z., Sampson, A.R. and Lewis, D.A. (2005) The influence of chronic exposure to antipsychotic medications on brain size before and after tissue fixation: a comparison of haloperidol and olanzapine in macaque monkeys. Neuropsychopharmacology. 30, 1649–1661.

Eastwood, S.L., Burnet, P.W. and Harrison, P.J. (2005) Decreased hippocampal expression of the susceptibility gene PPP3CC and other calcineurin subunits in schizophrenia. Biol Psychiatry. 57, 702–710.

Freedman, R. (2005) The choice of antipsychotic drugs for schizophrenia. N Engl J Med. 353, 1286–1288.

Fuller, R., Nopoulos, P., Arndt, S., O'Leary, D., Ho, B.C. and Andreasen, N.C. (2002) Longitudinal assessment of premorbid cognitive functioning in patients with schizophrenia through examination of standardized scholastic test performance. Am J Psychiatry. 159, 1183–1189.

Gardner, D.M., Baldessarini, R.J. and Waraich, P. (2005) Modern antipsychotic drugs: a critical overview. CMAJ. 172, 1703–1711.

Gerber, D.J., Hall, D., Miyakawa, T., Demars, S., Gogos, J.A., Karayiorgou, M. and Tonegawa, S. (2003) Evidence for association of schizophrenia with genetic variation in the 8p21.3 gene, PPP3CC, encoding the calcineurin gamma subunit. Proc Natl Acad Sci USA. 100, 8993–8998.

Giedd, J.N., Blumenthal, J., Jeffries, N.O., Castellanos, F.X., Liu, H., Zijdenbos, A., Paus, T., Evans, A.C. and Rapoport, J.L. (1999) Brain development during childhood and adolescence: a longitudinal MRI study. Nat Neurosci. 2, 861–863.

Goff, D.C. and Coyle, J.T. (2001) The emerging role of glutamate in the pathophysiology and treatment of schizophrenia. Am J Psychiatry. 158, 1367–1377.

Goff, D.C., Tsai, G., Levitt, J., Amico, E., Manoach, D., Schoenfeld, D.A., Hayden, D.L., McCarley, R. and Coyle, J.T. (1999) A placebo-controlled trial of D-cycloserine added to conventional neuroleptics in patients with schizophrenia. Arch Gen Psychiatry. 56, 21–27.

Gogtay, N., Giedd, J.N., Lusk, L., Hayashi, K.M., Greenstein, D., Vaituzis, A.C., Nugent, T.F., 3rd, Herman, D.H., Clasen, L.S., Toga, A.W., Rapoport, J.L. and Thompson, P.M. (2004) Dynamic mapping of human cortical development during childhood through early adulthood. Proc Natl Acad Sci USA. 101, 8174–8179.

Goldberg, T.E., Berman, K.F., Fleming, K., Ostrem, J., Van Horn, J.D., Esposito, G., Mattay, V.S., Gold, J.M. and Weinberger, D.R. (1998) Uncoupling cognitive workload and pre-frontal cortical physiology: a PET rCBF study. Neuroimage. 7, 296–303.

Grunze, H.C., Rainnie, D.G., Hasselmo, M.E., Barkai, E., Hearn, E.F., McCarley, R.W. and Greene, R.W. (1996) NMDA-dependent modulation of CA1 local circuit inhibition. J Neurosci. 16, 2034–2043.

Hardingham, G.E., Fukunaga, Y. and Bading, H. (2002) Extrasynaptic NMDARs oppose synaptic NMDARs by triggering CREB shut-off and cell death pathways. Nat Neurosci. 5, 405–414.

Harrison, P.J. (1999) The neuropathology of schizophrenia. A critical review of the data and their interpretation. Brain. 122 (Pt 4), 593–624.

Hegarty, J.D., Baldessarini, R.J., Tohen, M., Waternaux, C. and Oepen, G. (1994) One hun-dred years of schizophrenia: a meta-analysis of the outcome literature. Am J Psychiatry. 151, 1409–1416.

Henderson, D.C., Cagliero, E., Gray, C., Nasrallah, R.A., Hayden, D.L., Schoenfeld, D.A. and Goff, D.C. (2000) Clozapine, diabetes mellitus, weight gain, and lipid abnormalities: a five-year naturalistic study. Am J Psychiatry. 157, 975–981.

Ho, B.C., Alicata, D., Ward, J., Moser, D.J., O'Leary, D.S., Arndt, S. and Andreasen, N.C. (2003) Untreated initial psychosis: relation to cognitive deficits and brain morphology in first-episode schizophrenia. Am J Psychiatry. 160, 142–148.

Holcomb, H.H., Lahti, A.C., Medoff, D.R., Weiler, M. and Tamminga, C.A. (2001) Sequential regional cerebral blood flow brain scans using PET with H2(15)O demonstrate ketamine actions in CNS dynamically. Neuropsychopharmacology. 25, 165–172.

Holcomb, H.H., Lahti, A.C., Medoff, D.R., Cullen, T. and Tamminga, C.A. (2005) Effects of noncompetitive NMDA receptor blockade on anterior cingulate cerebral blood flow in volunteers with schizophrenia. Neuropsychopharmacology. 30, 2275–2282.

Honey, G.D., Bullmore, E.T. and Sharma, T. (2002) De-coupling of cognitive performance and cerebral functional response during working memory in schizophrenia. Schizophr Res. 53, 45–56.

Huttenlocher, P.R. (1979) Synaptic density in human frontal cortex - developmental changes and effects of aging. Brain Res. 163, 195–205.

Jackson, M.E., Homayoun, H. and Moghaddam, B. (2004) NMDA receptor hypofunction produces concomitant firing rate potentiation and burst activity reduction in the prefrontal cortex. Proc Natl Acad Sci USA. 101, 8467–8472.

Javitt, D.C. and Zukin, S.R. (1991) Recent advances in the phencyclidine model of schizo-phrenia. Am J Psychiatry. 148, 1301–1308.

Jay, T.M. (2003) Dopamine: a potential substrate for synaptic plasticity and memory mecha-nisms. Prog Neurobiol. 69, 375–390.

Jernigan, T.L., Trauner, D.A., Hesselink, J.R. and Tallal, P.A. (1991) Maturation of human cerebrum observed in vivo during adolescence. Brain. 114 (Pt 5), 2037–2049.

Jodo, E., Suzuki, Y., Katayama, T., Hoshino, K.Y., Takeuchi, S., Niwa, S. and Kayama, Y. (2005) Activation of medial prefrontal cortex by phencyclidine is mediated via a hippo-campo-prefrontal pathway. Cereb Cortex. 15, 663–669.

Kapur, S., Langlois, X., Vinken, P., Megens, A.A., De Coster, R. and Andrews, J.S. (2002) The differential effects of atypical antipsychotics on prolactin elevation are explained by their differential blood-brain disposition: a pharmacological analysis in rats. J Pharmacol Exp Ther. 302, 1129–1134.

Kane J, Honigfeld G, Singer J, Meltzer H. (1988) Clozapine for the treatment-resistant schizophrenic. Adouble-blind comparison with chlorpromazine. Arch Gen Psychiatry 45(9): 789–96.

Keefe, R.S., Seidman, L.J., Christensen, B.K., Hamer, R.M., Sharma, T., Sitskoorn, M.M., Rock, S.L., Woolson, S., Tohen, M., Tollefson, G.D., Sanger, T.M. and Lieberman, J.A. (2006a) Long-term neurocognitive effects of olanzapine or low-dose haloperidol in first-episode psychosis. Biol Psychiatry. 59, 97–105.

Keefe, R.S., Young, C.A., Rock, S.L., Purdon, S.E., Gold, J.M. and Breier, A. (2006b) One-year double-blind study of the neurocognitive efficacy of olanzapine, risperidone, and haloperidol in schizophrenia. Schizophr Res. 81, 1–15.

Kellendonk, C., Simpson, E.H., Polan, H.J., Malleret, G., Vronskaya, S., Winiger, V., Moore, H. and Kandel, E.R. (2006) Transient and selective overexpression of dopamine D2 receptors in the striatum causes persistent abnormalities in prefrontal cortex functioning. Neuron. 49, 603–615.

Koh, P.O., Bergson, C., Undie, A.S., Goldman-Rakic, P.S. and Lidow, M.S. (2003) Up-regulation of the D1 dopamine receptor-interacting protein, calcyon, in patients with schizophrenia. Arch Gen Psychiatry. 60, 311–319.

Krystal, J.H., Karper, L.P., Seibyl, J.P., Freeman, G.K., Delaney, R., Bremner, J.D., Heninger, G.R., Bowers, M.B., Jr. and Charney, D.S. (1994) Subanesthetic effects of the noncompetitive NMDA antagonist, ketamine, in humans. Psychotomimetic, perceptual, cognitive, and neuroendocrine responses. Arch Gen Psychiatry. 51, 199–214.

Krystal, J.H., D'Souza, D.C., Mathalon, D., Perry, E., Belger, A. and Hoffman, R. (2003) NMDA receptor antagonist effects, cortical glutamatergic function, and schizophrenia: toward a paradigm shift in medication development. Psychopharmacology (Berl). 169, 215–233.

Lamberti, J.S., Olson, D., Crilly, J.F., Olivares, T., Williams, G.C., Tu, X., Tang, W., Wiener, K., Dvorin, S. and Dietz, M.B. (2006) Prevalence of the metabolic syndrome among patients receiving clozapine. Am J Psychiatry. 163, 1273–1276.

Lencz, T., Smith, C.W., McLaughlin, D., Auther, A., Nakayama, E., Hovey, L. and Cornblatt, B.A. (2006) Generalized and specific neurocognitive deficits in prodromal schizophrenia. Biol Psychiatry. 59, 863–871.

Leslie, C.A., Robertson, M.W., Cutler, A.J. and Bennett, Jr., J.P., (1991) Postnatal development of D1 dopamine receptors in the medial prefrontal cortex, striatum and nucleus accumbens of normal and neonatal 6-hydroxydopamine treated rats: a quantitative autoradiographic analysis. Brain Res Dev Brain Res. 62, 109–114.

Lewis, D.A. and Levitt, P. (2002) Schizophrenia as a disorder of neurodevelopment. Annu Rev Neurosci. 25, 409–432.

Lewis, S.W. and Murray, R.M. (1987) Obstetric complications, neurodevelopmental deviance, and risk of schizophrenia. J Psychiatr Res. 21, 413–421.

Lewis, D.A., Hashimoto, T. and Volk, D.W. (2005) Cortical inhibitory neurons and schizophrenia. Nat Rev Neurosci. 6, 312–324.

Lieberman, J.A. (1999) Is schizophrenia a neurodegenerative disorder? A clinical and neurobiological perspective. Biol Psychiatry. 46, 729–739.

Lieberman, J., Chakos, M., Wu, H., Alvir, J., Hoffman, E., Robinson, D. and Bilder, R. (2001a) Longitudinal study of brain morphology in first episode schizophrenia. Biol Psychiatry. 49, 487–499.

Lieberman, J.A., Perkins, D., Belger, A., Chakos, M., Jarskog, F., Boteva, K. and Gilmore, J. (2001b) The early stages of schizophrenia: speculations on pathogenesis, pathophysiology, and therapeutic approaches. Biol Psychiatry. 50, 884–897.

Lieberman, J.A., Stroup, T.S., McEvoy, J.P., Swartz, M.S., Rosenheck, R.A., Perkins, D.O., Keefe, R.S., Davis, S.M., Davis, C.E., Lebowitz, B.D., Severe, J. and Hsiao, J.K. (2005) Effectiveness of antipsychotic drugs in patients with chronic schizophrenia. N Engl J Med. 353, 1209–1223.

Lipska, B.K., Lerman, D.N., Khaing, Z.Z., Weickert, C.S. and Weinberger, D.R. (2003) Gene expression in dopamine and GABA systems in an animal model of schizophrenia: effects of antipsychotic drugs. Eur J Neurosci. 18, 391–402.

Lisman, J.E. and Grace, A.A. (2005) The hippocampal-VTA loop: controlling the entry of information into long-term memory. Neuron. 46, 703–713.

Luby, E.D., Cohen, B.D., Rosenbaum, G., Gottlieb, J.S. and Kelley, R. (1959) Study of a new schizophrenomimetic drug; sernyl. AMA Arch Neurol Psychiatry. 81, 363–369.

Malhotra, A.K., Pinals, D.A., Weingartner, H., Sirocco, K., Missar, C.D., Pickar, D. and Breier, A. (1996) NMDA receptor function and human cognition: the effects of ketamine in healthy volunteers. Neuropsychopharmacology. 14, 301–307.

Malla, A. and Payne, J. (2005) First-episode psychosis: psychopathology, quality of life, and functional outcome. Schizophr Bull. 31, 650–671.

Manoach, D.S. (2003) Prefrontal cortex dysfunction during working memory performance in schizophrenia: reconciling discrepant findings. Schizophr Res. 60, 285–298.

Manoach, D.S., Press, D.Z., Thangaraj, V., Searl, M.M., Goff, D.C., Halpern, E., Saper, C.B. and Warach, S. (1999) Schizophrenic subjects activate dorsolateral prefrontal cortex during a working memory task, as measured by fMRI. Biol Psychiatry. 45, 1128–1137.

Manoach, D.S., Gollub, R.L., Benson, E.S., Searl, M.M., Goff, D.C., Halpern, E., Saper, C.B. and Rauch, S.L. (2000) Schizophrenic subjects show aberrant fMRI activation of dorsolateral prefrontal cortex and basal ganglia during working memory performance. Biol Psychiatry. 48, 99–109.

Marshall, M., Lewis, S., Lockwood, A., Drake, R., Jones, P. and Croudace, T. (2005) Association between duration of untreated psychosis and outcome in cohorts of first-episode patients: a systematic review. Arch Gen Psychiatry. 62, 975–983.

McEvoy, J.P., Lieberman, J.A., Stroup, T.S., Davis, S.M., Meltzer, H.Y., Rosenheck, R.A., Swartz, M.S., Perkins, D.O., Keefe, R.S., Davis, C.E., Severe, J. and Hsiao, J.K. (2006) Effectiveness of clozapine versus olanzapine, quetiapine, and risperidone in patients with chronic schizophrenia who did not respond to prior atypical antipsychotic treatment. Am J Psychiatry. 163, 600–610.

Millar, J.K., Pickard, B.S., Mackie, S., James, R., Christie, S., Buchanan, S.R., Malloy, M.P., Chubb, J.E., Huston, E., Baillie, G.S., Thomson, P.A., Hill, E.V., Brandon, N.J., Rain, J.C., Camargo, L.M., Whiting, P.J., Houslay, M.D., Blackwood, D.H., Muir, W.J. and Porteous, D.J. (2005) DISC1 and PDE4B are interacting genetic factors in schizophrenia that regulate cAMP signaling. Science. 310, 1187–1191.

Mishara, A.L. and Goldberg, T.E. (2004) A meta-analysis and critical review of the effects of conventional neuroleptic treatment on cognition in schizophrenia: opening a closed book. Biol Psychiatry. 55, 1013–1022.

Miyakawa, T., Leiter, L.M., Gerber, D.J., Gainetdinov, R.R., Sotnikova, T.D., Zeng, H., Caron, M.G. and Tonegawa, S. (2003) Conditional calcineurin knockout mice exhibit multiple abnormal behaviors related to schizophrenia. Proc Natl Acad Sci USA. 100, 8987–8992.

Moghaddam, B. (2003) Bringing order to the glutamate chaos in schizophrenia. Neuron. 40, 881–884.

Moghaddam, B. and Adams, B.W. (1998) Reversal of phencyclidine effects by a group II metabotropic glutamate receptor agonist in rats. Science. 281, 1349–1352.

Monyer, H., Burnashev, N., Laurie, D.J., Sakmann, B. and Seeburg, P.H. (1994) Developmental and regional expression in the rat brain and functional properties of four NMDA receptors. Neuron. 12, 529–540.

Ninan, I., Jardemark, K.E. and Wang, R.Y. (2003) Olanzapine and clozapine but not haloperidol reverse subchronic phencyclidine-induced functional hyperactivity of N-methyl-D-aspartate receptors in pyramidal cells of the rat medial prefrontal cortex. Neuropharmacology. 44, 462–472.

O'Donnell, P., Lewis, B.L., Weinberger, D.R. and Lipska, B.K. (2002) Neonatal hippocampal damage alters electrophysiological properties of prefrontal cortical neurons in adult rats. Cereb Cortex. 12, 975–982.

Olney, J.W. and Farber, N.B. (1995) Glutamate receptor dysfunction and schizophrenia. Arch Gen Psychiatry. 52, 998–1007.

Paus, T., Zijdenbos, A., Worsley, K., Collins, D.L., Blumenthal, J., Giedd, J.N., Rapoport, J.L. and Evans, A.C. (1999) Structural maturation of neural pathways in children and adolescents: in vivo study. Science. 283, 1908–1911.

Paz, R.D., Andreasen, N.C., Daoud, S.Z., Conley, R., Roberts, R., Bustillo, J. and Perrone-Bizzozero, N.I. (2006) Increased expression of activity-dependent genes in cerebellar glutamatergic neurons of patients with schizophrenia. Am J Psychiatry. 163, 1829–1831.

Perkins, D.O., Gu, H., Boteva, K. and Lieberman, J.A. (2005) Relationship between duration of untreated psychosis and outcome in first-episode schizophrenia: a critical review and meta-analysis. Am J Psychiatry. 162, 1785–1804.

Pillai, A., Terry, A.V., Jr. and Mahadik, S.P. (2006) Differential effects of long-term treatment with typical and atypical antipsychotics on NGF and BDNF levels in rat striatum and hippocampus. Schizophr Res. 82, 95–106.

Rabinowitz, J., De Smedt, G., Harvey, P.D. and Davidson, M. (2002) Relationship between premorbid functioning and symptom severity as assessed at first episode of psychosis. Am J Psychiatry. 159, 2021–2026.

Reichenberg, A., Weiser, M., Rapp, M.A., Rabinowitz, J., Caspi, A., Schmeidler, J., Knobler, H.Y., Lubin, G., Nahon, D., Harvey, P.D. and Davidson, M. (2005) Elaboration on premorbid intellectual performance in schizophrenia: premorbid intellectual decline and risk for schizophrenia. Arch Gen Psychiatry. 62, 1297–1304.

Rosenheck, R. (2006) Integration of mental health care and supported employment. Am J Psychiatry. 163, 940; author reply 940.

Rothman, D.L., Sibson, N.R., Hyder, F., Shen, J., Behar, K.L. and Shulman, R.G. (1999) In vivo nuclear magnetic resonance spectroscopy studies of the relationship between the glutamate-glutamine neurotransmitter cycle and functional neuroenergetics. Philos Trans R Soc Lond B Biol Sci. 354, 1165–1177.

Rowland, L.M., Bustillo, J.R., Mullins, P.G., Jung, R.E., Lenroot, R., Landgraf, E., Barrow, R., Yeo, R., Lauriello, J. and Brooks, W.M. (2005) Effects of ketamine on anterior cingulate glutamate metabolism in healthy humans: a 4-T proton MRS study. Am J Psychiatry. 162, 394–396.

Rund, B.R., Melle, I., Friis, S., Larsen, T.K., Midboe, L.J., Opjordsmoen, S., Simonsen, E., Vaglum, P. and McGlashan, T. (2004) Neurocognitive dysfunction in first-episode psychosis: correlates with symptoms, premorbid adjustment, and duration of untreated psychosis. Am J Psychiatry. 161, 466–472.

Rycroft, B.K. and Gibb, A.J. (2004) Inhibitory interactions of calcineurin (phosphatase 2B) and calmodulin on rat hippocampal NMDA receptors. Neuropharmacology. 47, 505–514.

Sawaguchi, T. and Goldman-Rakic, P.S. (1994) The role of D1-dopamine receptor in working memory: local injections of dopamine antagonists into the prefrontal cortex of rhesus monkeys performing an oculomotor delayed-response task. J Neurophysiol. 71, 515–528.

Shaw, P., Sporn, A., Gogtay, N., Overman, G.P., Greenstein, D., Gochman, P., Tossell, J.W., Lenane, M. and Rapoport, J.L. (2006) Childhood-onset schizophrenia: a double-blind, randomized clozapine-olanzapine comparison. Arch Gen Psychiatry. 63, 721–730.

Smith, K.E., Gibson, E.S. and Dell'Acqua, M.L. (2006) cAMP-dependent protein kinase postsynaptic localization regulated by NMDA receptor activation through translocation of an A-kinase anchoring protein scaffold protein. J Neurosci. 26, 2391–2402.

Sowell, E.R., Peterson, B.S., Thompson, P.M., Welcome, S.E., Henkenius, A.L. and Toga, A.W. (2003) Mapping cortical change across the human life span. Nat Neurosci. 6, 309–315.

Tarazi, F.I., Tomasini, E.C. and Baldessarini, R.J. (1999) Postnatal development of dopamine D1-like receptors in rat cortical and striatolimbic brain regions: an autoradiographic study. Dev Neurosci. 21, 43–49.

Tauscher, J., Hussain, T., Agid, O., Verhoeff, N.P., Wilson, A.A., Houle, S., Remington, G., Zipursky, R.B. and Kapur, S. (2004) Equivalent occupancy of dopamine D1 and D2 receptors with clozapine: differentiation from other atypical antipsychotics. Am J Psychiatry. 161, 1620–1625.

Theberge, J., Bartha, R., Drost, D.J., Menon, R.S., Malla, A., Takhar, J., Neufeld, R.W., Rogers, J., Pavlosky, W., Schaefer, B., Densmore, M., Al-Semaan, Y. and Williamson, P.C. (2002) Glutamate and glutamine measured with 4.0 T proton MRS in never-treated patients with schizophrenia and healthy volunteers. Am J Psychiatry. 159, 1944–1946.

Tibbo, P., Hanstock, C., Valiakalayil, A. and Allen, P. (2004) 3-T proton MRS investigation of glutamate and glutamine in adolescents at high genetic risk for schizophrenia. Am J Psychiatry. 161, 1116–1118.

Tseng, K.Y. and O'Donnell, P. (2004) Dopamine-glutamate interactions controlling prefrontal cortical pyramidal cell excitability involve multiple signaling mechanisms. J Neurosci. 24, 5131–5139.

Tseng, K.Y. and O'Donnell, P. (2005) Post-pubertal emergence of prefrontal cortical up states induced by D1-NMDA co-activation. Cereb Cortex. 15, 49–57.

Tseng, K.Y. and O'Donnell, P. (2007) Dopamine Modulation of Prefrontal Cortical Interneurons Changes during Adolescence. Cereb Cortex. 17, 1235–1240.

Tseng, K.Y., Mallet, N., Toreson, K.L., Le Moine, C., Gonon, F. and O'Donnell, P. (2006a) Excitatory response of prefrontal cortical fast-spiking interneurons to ventral tegmental area stimulation in vivo. Synapse. 59, 412–417.

Tseng, K.Y., Amin, F., Lewis, B.L. and O'Donnell, P. (2006b) Altered prefrontal cortical metabolic response to mesocortical activation in adult animals with a neonatal ventral hippocampal lesion. Biol Psychiatry. 60, 585–590.

Tseng, K.Y., Lewis, B.L. Lipska, B.K. and O'Donnell, P. (2007) Post-pubertal disruption of medical prefrontal cortical dopamine-glutamate interactions in a developmental animal model of schizophrenia. Biol. Psychiatry. Jan 2; [Epub ahead of print]

Volavka, J., Czobor, P., Cooper, T.B., Sheitman, B., Lindenmayer, J.P., Citrome, L., McEvoy, J.P. and Lieberman, J.A. (2004) Prolactin levels in schizophrenia and schizoaffective disorder patients treated with clozapine, olanzapine, risperidone, or haloperidol. J Clin Psychiatry. 65, 57–61.

Webster, M.J., Weickert, C.S., Herman, M.M. and Kleinman, J.E. (2002) BDNF mRNA expression during postnatal development, maturation and aging of the human prefrontal cortex. Brain Res Dev Brain Res. 139, 139–150.

Weinberger, D.R. (1987) Implications of normal brain development for the pathogenesis of schizophrenia. Arch Gen Psychiatry. 44, 660–669.

Weiser, M., Reichenberg, A., Rabinowitz, J., Kaplan, Z., Mark, M., Bodner, E., Nahon, D. and Davidson, M. (2001) Association between nonpsychotic psychiatric diagnoses in adolescent males and subsequent onset of schizophrenia. Arch Gen Psychiatry. 58, 959–964.

Wiersma, D., Wanderling, J., Dragomirecka, E., Ganev, K., Harrison, G., An Der Heiden, W., Nienhuis, F.J. and Walsh, D. (2000) Social disability in schizophrenia: its development and prediction over 15 years in incidence cohorts in six European centres. Psychol Med. 30, 1155–1167.

Williams, K., Russell, S.L., Shen, Y.M. and Molinoff, P.B. (1993) Developmental switch in the expression of NMDA receptors occurs in vivo and in vitro. Neuron. 10, 267–278.

Wyatt, R.J. (1991) Neuroleptics and the natural course of schizophrenia. Schizophr Bull. 17, 325–351.

Zahrt, J., Taylor, J.R., Mathew, R.G. and Arnsten, A.F. (1997) Supranormal stimulation of D1 dopamine receptors in the rodent prefrontal cortex impairs spatial working memory performance. J Neurosci. 17, 8528–8535.

Zecevic, N., Bourgeois, J.P. and Rakic, P. (1989) Changes in synaptic density in motor cortex of rhesus monkey during fetal and postnatal life. Brain Res Dev Brain Res. 50, 11–32.

Chapter 21

Prefrontal Cortical Circuits and Schizophrenia Pathophysiology

Patricio O'Donnell
Department of Anatomy and Neurobiology and Department of Psychiatry, University of
    Maryland School of Medicine, Baltimore, MD 21201, USA

21.1 Local Cortical Circuits as Functional Units

The prefrontal cortex (PFC) is a key brain region for a variety of high-order cognitive functions and catecholamines are critical for its proper functioning. The PFC sits at the highest end in the hierarchical organization of associative cortices, exerting control over broad sensorimotor processes This cortical region presents a complex synaptic organization, arranged so as to provide a tight control of the firing of its primary neurons, the pyramidal cells. Spatial and temporal aspects contribute to this control, with discrete clusters of PFC units perhaps encoding specific aspects of cognitive outcome.

Several different divisions comprise the PFC, and some important differences exist across species. In primates, the dorsolateral PFC (DL-PFC) is perhaps the most recently evolved brain region (Fuster, 1997). Other areas typically labeled as part of the PFC include orbitofrontal cortices, the medial PFC and the cingulate cortex. In rodents, although there is not an obvious DL-PFC, the most dorsal aspects of the medial PFC (which include the prelimbic cortex) are typically taken as a rudimentary DL-PFC. This is due to both areas receiving inputs from the mediodorsal nucleus of the thalamus and being involved in working memory and executive functions. Thus, the medial PFC does control the highest cognitive functions rodents can afford.

All PFC regions are heavily connected with the basal ganglia and limbic system. For example, the orbital and medial PFC receive important innervation from the hippocampus and amygdala (Carmichael and Price, 1995; Ishikawa and Nakamura, 2003). Also, all PFC areas project to specific striatal regions, and thalamocortical projections targeting the PFC are controlled by ventral basal ganglia loops (Groenewegen et al., 1990). This organization allows for the activity of a specific set of PFC ensembles to be fine-tuned by the feedback it receives from its corollary discharge through the basal ganglia (Fujii and Graybiel, 2005). Thus, prefrontal function is dependent on a complex pattern of afferent innervation that includes cortico-cortical projections as well as feedback from basal ganglia loops arriving via the mediodorsal thalamus. Monoamines and other modulators are likely to affect such synaptic influences, thereby shaping the level and pattern of PFC activity.

In addition to long loops via the basal ganglia, modulation of local synaptic circuitry may be critical to determine PFC function. Pyramidal neurons are the primary neural population in cortical regions, providing projections to target areas. They are influenced by the activity of other pyramidal cells and by local interneurons. Although interneurons are a minority among cortical neurons and there are several types, it is clear that they can tightly control the output of pyramidal neurons. Some types (i.e., chandelier cells and basket cells) provide GABA innervation to pyramidal neurons at proximal sites, including the cell body and the axon's initial segment (Lewis et al., 2005). As interneurons can be activated by pyramidal cells and by afferents to the PFC, they can be a substrate for feedback and feed forward inhibition. Their modulation by monoamines will therefore also shape the activity of PFC neural ensembles.

A unique characteristic of interneuron activity in all cortical regions studied is their synchronous firing. Interneurons do present gap junctions, allowing for a tight coupling of their electrical activity (Gibson et al., 1999; Galarreta et al., 2004). They also tend to fire more rapidly (Kawaguchi, 2001), and because of their high connectedness, this translates in synchronized activity at high frequencies. One pattern of activity that has received increasing attention is their contribution to cortical gamma oscillations (Buhl et al., 2003). Thus, interneurons are critical for high frequency patterns in the EEG, and may be involved in the binding phenomenon this synchronized activity can provide.

Given the complex organization of cortical circuits, any afferent that can modulate activity in pyramidal neurons and/or local inhibitory processes may have a strong impact in the sharpness of the representations determined by the activity of discrete neural ensembles. This modulation will ultimately affect executive functions such as goal setting, planning means to achieve the goals, and error monitoring.

21.2 Dopamine Actions in the Prefrontal Cortex

The dopamine (DA) projection to the PFC was identified by Anne Marie Thierry and colleagues (Thierry et al., 1973). The ventral tegmental area (VTA) has DA-containing neurons that project to the PFC and nucleus accumbens. Now it is known that many, if not most, VTA neurons are non-DA in nature, and many neurons projecting to the PFC release GABA (Steffensen et al., 1998). A small population of VTA neurons expresses glutamate vesicular transporter, and they do no coexpress GABA or DA synthesizing enzymes (Yamaguchi et al., 2007). Therefore, activation of mesocortical projections will involve release of DA and other transmitters, and their impact on PFC function in vivo will depend on a number of factors including the interactions among glutamate, GABA and DA receptors. VTA neurons become active in the presence of reward or reward-predicting stimuli, and they do so with a brief train of action potentials (Schultz, 1998), and this firing has been interpreted as a "teaching" signal that highlights relevance or saliency of external cues (Waelti et al., 2001).

A few in vivo studies assessed the effects of mesocortical activation, yielding quite variable results. With single-unit recordings, a mix of excitatory (increases in firing) and inhibitory (decreases in firing) effects have been reported, with anesthesia and the state of the animal likely contributing to the variability. With intracellular recordings, we have reported inconsistent effects of VTA stimulation on pyramidal neurons in the medial PFC with single-pulse stimulation, including excitatory post-synaptic potentials (EPSPs) (Mercuri et al., 1985; Lewis and O'Donnell, 2000), inhibitory postsynaptic potentials (IPSPs) and no changes (Lewis and O'Donnell, 2000). As single pulses likely mimic tonic DA cell firing and the behaviorally relevant firing seems to be a burst-like pattern (Schultz, 1998), we also tested the responses to VTA stimulation with trains of pulses. With this pattern, a persistent D_1-dependent depolarization resembling up states was evoked. However, along with the depolarization (that could be envisioned as an excitatory response) the neurons stopped firing (Lewis and O'Donnell, 2000). This was in fact a replication with endogenous DA release of what had been observed with local DA administration. Bernardi and colleagues had shown that intra PFC DA injection elicited a depolarization with firing suppression (Bernardi et al., 1982). In our experiments, the depolarization was blocked by D_1 antagonists, but the firing suppression was not. It is conceivable that the inhibition was the result of either mesocortical GABA projections or the activation of local PFC interneurons. We conducted some juxtacellular studies of fast-spiking interneurons in the PFC, assessing their response to VTA train stimulation. Consistent with the intracellular data, firing in pyramidal neurons was reduced by chemical VTA stimulation (Tseng and O'Donnell, 2006). Fast-spiking interneurons, on the other hand, were typically activated with a temporal course similar to the inhibition observed in pyramidal neurons (Tseng and O'Donnell, 2006). It is therefore likely that DA exerts its control over PFC function by modulating local circuits within this region.

The studies of the actions of PFC DA at a cellular level have been slow and mired with discrepancies and controversy. Electrophysiological studies provided different results depending on a number of factors (Seamans and Yang, 2004). Experiments in slices have revealed excitatory and inhibitory actions of D_1 receptors on pyramidal neurons (Gorelova and Yang, 2000; Henze et al., 2000; Wang and O'Donnell, 2001; Tseng and O'Donnell, 2004). A careful analysis of how the experiments were done reveals that when slices are incubated or recorded at room temperature the excitatory action of D_1 activation cannot be observed (Penit-Soria et al., 1987; Gulledge and Jaffe, 1998). It is possible that cooling the slices inactivates enzymes that are critical for the second messenger signaling activated by D_1 receptors. Thus, in conditions that preserve better physiological environment, D_1 activation has been consistently seen as excitatory. D_1 receptors can also activate interneurons (Gorelova and Yang, 2000; Gorelova et al., 2002) and may have presynaptic effects modulating glutamate release (Gao et al., 2001). The in vitro D_2 effects have also some degree of controversy. Some early studies indicated that D_2 agonists can reduce interneuron excitability in the PFC (Gorelova et al., 2002), but those recordings were conducted in slices from very young (preweaning) animals. More recently, we have reported that D_2 agonists can enhance fast-spiking interneuron excitability in PFC slices obtained from adult animals, suggesting that some component mediating the effects of D_2 activation on interneurons matures during adolescence (Tseng and O'Donnell, 2006). Species differences and cytoarchitectonical considerations may also contribute to some of the confusion. For example, recordings from adult monkeys have revealed that D_2 agonists do not excite interneurons in layers II–III (Gonzalez-Burgos et al., 2007), whereas in adult rats, D_2 agonists excite interneurons in layers V–VI (Tseng and O'Donnell, 2006). In short, the picture of cellular actions of DA within PFC circuits is far from clear. However, significant advances have been made at identifying the factors that make this modulation complex and extremely sensitive to small variations in the experimental conditions.

Although more is to be done to characterize DA actions in the PFC, the bulk of the evidence suggest a role as a filtering signal. The inhibitory effects of DA, primarily via D_2 receptors, seem to be aimed at suppressing weak inputs (O'Donnell, 2003). D_1 receptors, on the other hand, potentiate NMDA responses (Tseng and O'Donnell, 2004), thereby enhancing the activity of strongly activated units (O'Donnell, 2003). This combination could yield a contrast-enhancing function that would highlight behaviorally relevant ongoing activity, perhaps enabling NMDA-dependent plasticity mechanisms. In fact, although VTA stimulation attenuated responses to hippocampal afferents (Jay et al., 1995), it also facilitated hippocampal-dependent long-term potentiation (Gurden et al., 1999). Thus, the mesocortical modulation of PFC activity is dependent on the state of the system.

21.3 Late Developmental Changes: Adolescence as a Critical Period

As indicated above, some aspects of PFC physiology change during adolescence. Cortical GABA interneurons do exhibit a significant reorganization during adolescence. Chandelier cell synapses on pyramidal axons reach stable adult levels after adolescence (Anderson et al., 1995), and amygdala projections to the PFC also mature at that age (Cunningham et al., 2002). Electrophysiological evidence indicates that PFC interneurons acquire the ability of being excited by D_2 DA agonists after puberty in rats (Tseng and O'Donnell, 2004) and stimulants activate immediate early genes differently in the PFC from adolescent animals (Andersen et al., 2001). In monkeys, local PFC circuits including interneuron connections mature during adolescence (Lewis, 1997; Woo et al., 1997) and the DA innervation reaches its peak during that age (Rosenberg and Lewis, 1994; Rosenberg and Lewis, 1995). The DA modulation of NMDA responses also changes during adolescence, allowing the observation of up states in slices but only in those obtained from adult animals (Tseng and O'Donnell, 2005). Human imaging studies also revealed morphological changes in the PFC during adolescence (Giedd et al., 1999; Casey et al., 2000), and neurophysiological studies have shown that cognitive event-related potentials continue maturing into late adolescence (Segalowitz and Davies, 2004). Thus, it is possible that prefrontal inputs as well as local interneuronal function acquire adult functional levels during late adolescence.

The periadolescent maturation of the PFC may be of high importance for schizophrenia. Although there are both a clear genetic predisposition and developmental deficits in this disorder, symptoms do not appear until late adolescence or early adulthood. Indeed, imaging studies suggest abnormal postpubertal development as an important element in schizophrenia (Pantelis et al., 2005). It can be speculated that when DA fibers establish an adult level of contact with interneurons, just after adolescence, they may encounter an abnormal population of GABA neurons in a brain predisposed toward schizophrenia (Benes, 1997). This may yield insufficient interneuron activation, which may not be able to filter excessive cortical activity. Thus, the periadolescent maturation of these interneurons could be responsible for the delayed emergence of symptoms.

21.4 Schizophrenia: Lessons from Animal Models

Modeling schizophrenia has been an elusive goal, for obvious reasons. First, schizophrenia is a uniquely human condition, and it is certainly impossible to develop a model with 100% face validity (i.e., expressing schizophrenia symptoms). But the several decades of research in this area have provided several *useful* models, with various degrees of construct validity. Their usefulness has ranged from becoming tools to test potential new antipsychotics to ways to explore possible pathophysiological scenarios.

The earlier models were pharmacological in nature. As the DA hypothesis of schizophrenia gained strength, it was also obvious that amphetamine intake could cause psychotic episodes. Thus, psychostimulants became an early model (Sams-Dodd, 1998). Another compound capable of inducing hallucination is LSD. This may be due to activation of a number of modulators, including serotonin. However, the hallucinations generated by amphetamine or LSD intake do not resemble those seen in patients. Both drug types cause primarily visual hallucinations while schizophrenia patients present primarily auditory hallucinations. PCP and other noncompeting NMDA antagonists have been used more recently (Javitt and Zukin, 1991). Their use as pharmacological models is derived from the findings that PCP intake causes psychotic episodes in normal controls and a reappearance of symptoms in patients in remission (Luby et al., 1959; Rosenbaum et al., 1959). This agent also induces negative symptoms. Thus, NMDA antagonism offers the best validity among pharmacological models, and this has opened up the consideration of non-DA factors in at least some aspects of the disease. Perhaps the only weak aspect of this model is that it does not include a developmental factor.

Developmental models have emerged in the past couple of decades. They include experimenter-induced brain alterations and early effects of environmental manipulations. One of the most comprehensively studied developmental models is the neonatal ventral hippocampal lesion. This manipulation does not reproduce hippocampal changes in schizophrenia, as patients do not present a lesion. However, the absence of proper hippocampal innervation at a critical developmental period may have long-term consequences on PFC neural circuits. Rats with a NVHL exhibit a variety of behavioral anomalies that emerge after puberty, including enhanced locomotion (Sams-Dodd et al., 1997; Al-Amin et al., 2001), altered reaction to stress and amphetamines (Lipska et al., 1993; Kato et al., 2000), sensorimotor gating deficits (Lipska et al., 1995; Swerdlow et al., 1995), and deficits in social behavior in adult rats (Sams-Dodd et al., 1997). Lesioned animals show working memory deficits (Chambers et al., 1996; Moghaddam et al., 1999; Lipska et al., 2002) and enhanced liability for cocaine self-administration (Chambers and Self, 2002). Some studies have reproduced similar deficits in monkeys with a neonatal hippocampal lesion, which show abnormal social behaviors at adult ages (Bachevalier et al., 1999). The relevance of these findings to schizophrenia is highlighted by their reversal with antipsychotic drugs (Lipska and Weinberger, 1994; Le Pen and Moreau, 2002). Lesioned rats also exhibit anatomical and neurochemical changes. In the PFC, there have been reports of reduced BDNF levels (Ashe et al., 2002), decreased GAD67 mRNA (Lipska et al., 2003) and changes in spine density and spine length (Flores et al., 2005). Also, the DA response to stress is attenuated in the nucleus accumbens (Brake et al., 1999) and D_2 mRNA is decreased in the striatum (Lipska et al., 2003). Furthermore, some of the behavioral deficits were prevented with a PFC lesion (Lipska et al., 1998). These findings indicate that the PFC is critically affected in this model, and suggest that DA deficits may arise as a consequence.

Further characterization of DA alterations in these animals was obtained using electrophysiological methods. Electrical stimulation of the VTA, the source of DA innervation to the PFC, results in a suppression of firing in pyramidal neurons in normal animals (Lewis and O'Donnell, 2000). In lesioned rats, VTA stimulation evoked a dramatic increase in firing (O'Donnell et al., 2002), which can be interpreted as underlying a loss of filtering of weak or unwanted information that could disrupt goal-directed behavior and attention. VTA stimulation also produces an abnormally high cell firing in the nucleus accumbens in adult, not prepubertal rats (Goto and O'Donnell, 2002), which can also be prevented by haloperidol (Goto and O'Donnell, 2002) or a PFC lesion (Goto and O'Donnell, 2004). As all these abnormal responses are observed only after adolescence, this suggests that periadolescent maturation of prefrontal cortical circuits is affected by the absence of proper hippocampal innervation.

The increased PFC cell firing suggests a state of hyperexcitability in these animals. Indeed, fast-spiking interneurons in the medial PFC mature around adolescence (Tseng and O'Donnell, 2006) and fail to acquire the ability to become excited by D_2 dopamine receptors in lesioned animals (see preliminary data). The absence of proper hippocampal inputs to the PFC during early development could result, in addition to abnormal electrical activity, in a reduction of PFC BDNF (Ashe et al., 2002), a trophic factor that modulates GABA receptors (Jovanovic et al., 2004) and has receptors expressed at high levels in parvalbumin-positive interneurons (Gorba and Wahle, 1999). Reduced BDNF mRNA has also been observed in the PFC of schizophrenia patients (Hashimoto et al., 2005). In NVHL animals, PFC GABA transmission is impaired, as revealed by reduced GAD67 mRNA (Lipska et al., 2003) and increased levels of GABA-A receptor subunits (Mitchell et al., 2005). Thus, it is conceivable that abnormal interneurons are among the key elements underscoring behavioral, chemical and electrophysiological anomalies emerging during adolescence in animals with a neonatal hippocampal lesion.

Another model receiving increasing attention is the impairment of cell division during a very specific time window in gestation with an antimitotic agent. Injection of methylazoxymethanol acetate (MAM) at gestational day 17 causes deficits in the response of PFC neurons to VTA stimulation (Lavin et al., 2005), in the responses to stimulants (Flagstad et al., 2004), working memory (Gourevitch et al., 2004; Featherstone et al., 2007), and sensorimotor gating, which also emerged during adolescence (Le Pen et al., 2006). Other developmental models include raising animals in social isolation, immune challenges and prenatal stress. Isolation rearing causes neurochemical, behavioral and electrophysiological anomalies that are not very different from those caused by a NVHL (Jones et al., 1991; Geyer et al., 1993; Peters and O'Donnell, 2005), and many of these effects are only seen after puberty (Paulus et al., 1998) and can be reversed by antipsychotic drugs (Bakshi et al., 1998). Thus, many convergent findings are observed in NVHL, MAM-treated, and socially isolated rats.

All these manipulations cause deficits in sensorimotor gating, response to stress, social interactions, etc. Again, what appears to be a common pathophysiological mechanism are DA actions in corticolimbic systems. The specificity of these deficits regarding interneurons vs. pyramidal cells remains to be assessed. It is therefore encouraging that several different early manipulations may have their strongest eventual impact on the development of cortical microcircuits and acquire similar abnormal behavioral traits. Thus, it is possible that a combination of developmental deficits (resulting from a combination of developmental predisposition, perinatal and postnatal environmental factors) has a final common deficit in the assembly of PFC local circuitry.

21.5 Schizophrenia: A Cortical Pathophysiological Scenario

All the elements reviewed above could be synthesized into a unique, yet complex, set of pathophysiological conditions. Development- and synaptic transmission-related genes may predispose toward schizophrenia, most likely when combined with a perinatal environmental insult (Cannon et al., 2003). This could lead to widespread changes in cortical areas. As cortical circuits and their DA innervation do not mature until late adolescence (Rosenberg and Lewis, 1995), there are not obvious early physiological or behavioral consequences. However, the postadolescent period may be marked by the emergence of labile cortical circuits, with improper modulation of neural activity derived from abnormal interneurons. Cortical (and most importantly, prefrontal) circuits can cope with regular demands, but they may break down with increasing loads. It is possible that an improper periadolescent maturation of PFC circuitry can render the system into a state with reduced flexibility. In fact, GABA interneurons can be driven by DA (Gao and Goldman-Rakic, 2003; Tseng and O'Donnell, 2004), require network activity to develop, and NMDA blockade could critically affect them (de Lima et al., 2004). So, conditions that normally place intense demands on re-wiring and plasticity, coupled perhaps with stress, may extenuate interneuronal ability to keep pyramidal activity under control.

The glutamatergic hypothesis of schizophrenia can be reformulated taking interneurons into consideration (Coyle, 2004). There is evidence suggesting that noncompeting NMDA antagonists increase glutamate levels and pyramidal cell firing in the PFC (Moghaddam et al., 1997; Jackson et al., 2004). Imaging studies have also revealed that ketamine can increase PFC metabolic activity (Vollenweider et al., 1997). It is possible that PCP and other noncompetitive NMDA antagonists could be psychotomimetic because of their higher affinity for NMDA receptors located in interneurons. Cortical and hippocampal interneurons express NMDA receptors with NR2C and NR2D subunits, which are not seen in pyramidal neurons (Standaert et al., 1996; Scherzer et al., 1998). Different levels of NR2C and NR2D have been reported in brains from schizophrenia patients (Akbarian et al., 1996) and single-nucleotide polymorphisms in the NR2D gene have been associated with schizophrenia in a Japanese population (Makino et al., 2005). NR2D/NR1 NMDA receptors have a distinct pharmacological profile: they are less sensitive to competitive antagonists and highly sensitive to psychosis-prone noncompeting antagonists (Buller et al., 1994). They are also less sensitive to magnesium blockade (Kuner and Schoepfer, 1996), which makes them relatively easy to become activated (perhaps not needing AMPA-induced depolarization). NR2D receptors are also glycine-dependent (Williams, 1995), and this may contribute to the potential success of strategies enhancing NMDA function by blocking glycine transporter sites (Coyle and Tsai, 2004). In addition, NR2D receptors change during postnatal development. They are in high numbers during early development and decrease later. They also exhibit unique topographical distribution, being abundant in the adult thalamus and sparse in the cortex, measured either as mRNA (Wenzel et al., 1996) or protein (Dunah et al., 1996). NR2D receptors peak later than other NMDA receptor subunits during postnatal development (Monyer et al., 1994). Thus, it is possible to reconcile two major sets of findings in schizophrenia (i.e., that blockade of NMDA receptor causes psychosis and that GABA interneurons present anomalies in this disorder) if NMDA blockade were to take place in interneurons.

The functional consequences of this disorganization will be evident with the protracted periadolescent maturation of these circuits. Once the abnormal response of interneurons to DA emerges, DA is not able to strongly activate interneurons so there is not enough override of excessive pyramidal cell activity. This increased "noise" becomes transferred to the BG loops, and the system cannot be engaged so as to choose the appropriate behavioral response, nor it can switch strategies when the feedback systems report errors. Behaviorally relevant PFC ensembles cannot be readily distinguished from the noise. This overall scenario offers several lines of thought for new therapeutical approaches. Perhaps increasing GABA-A (alpha 2 subunit containing) activity could provide a way to boost impoverished interneuron activity. Alternatively, benzodiazepines could be beneficial during prodromal states. If this model is correct, then D_2 antagonists alone should not work. In fact, all effective antipsychotics are D_2 blockers but their selectivity is poor. However, there is some evidence of antipsychotics restoring electrophysiological deficits derived from poor interneuron activation in NVHL animals (Goto and O'Donnell, 2002). Indeed, we have shown that in NVHL animals, interneurons not only are not activated by D_2 agonists but in many cases there is an abnormal inhibition (Tseng and O'Donnell, 2006). It could be that the effect of neuroleptics is due to D_2 blockade of abnormal responses in interneurons that could allow D_1 receptors to engage some of them and partially counter excessive pyramidal cell firing.

In summary, DA actions in the PFC are critically dependent on how local circuits are modulated. Interactions among DA, GABA and glutamate do affect PFC function, shaping the behavior of distributed neural ensembles. Some aspects of these local interactions mature during adolescence, allowing the fine-tuning of complex cognitive functions during that age. In a brain predisposed toward schizophrenia or even with frontal circuitry anomalies, these will be translated into symptoms only after their normal periadolescent maturation allowed the deficits to become evident.

Acknowledgments

I thank several past lab members that contributed to this work (Kuei Yuan Tseng, Yukiori Goto, Barbara Lewis, and Yvette Peters). This was supported by NIH grants MH57683, MH60131, DA14020 and a NARSAD Independent Investigator Award.

References

Akbarian S, Sucher NJ, Bradley D, Tafazzoli A, Trinh D, Hetrick WP, Potkin SG, Sandman CA, Bunney Jr. WE, Jones EG (1996) Selective alterations in gene expression for NMDA receptor subunits in prefrontal cortex of schizophrenics. J Neurosci 16:19–30.

Al-Amin HA, Shannon Weickert C, Weinberger DR, Lipska BK (2001) Delayed onset of enhanced MK-801-induced motor hyperactivity after neonatal lesions of the rat ventral hippocampus. Biol Psychiatry 49:528–539.

Andersen SL, LeBlanc CJ, Lyss PJ (2001) Maturational increases in c-fos expression in the ascending dopamine systems. Synapse 41:345–350.

Anderson SA, Classey JD, Conde F, Lund JS, Lewis DA (1995) Synchronous development of pyramidal neuron dendritic spines and parvalbumin-immunoreactive chandelier neuron axon terminals in layer III of monkey prefrontal cortex. Neuroscience 67:7–22.

Ashe PC, Chlan-Fourney J, Juorio AV, Li XM (2002) Brain-derived neurotrophic factor (BDNF) mRNA in rats with neonatal ibotenic acid lesions of the ventral hippocampus. Brain Res 956:126–135.

Bachevalier J, Alvarado MC, Malkova L (1999) Memory and socioemotional behavior in monkeys after hippocampal damage incurred in infancy or in adulthood. Biol Psychiatry 46:329–339.

Bakshi VP, Swerdlow NR, Braff DL, Geyer MA (1998) Reversal of isolation rearing-induced deficits in prepulse inhibition by Seroquel and olanzapine. Biol Psychiatry 43:436–445.

Benes FM (1997) The role of stress and dopamine-GABA interactions in the vulnerability for schizophrenia. J Psychiatr Res 31:257–275.

Bernardi G, Cherubini E, Marciani MG, Mercuri N, Stanzione P (1982) Responses of intracellularly recorded cortical neurons to the iontophoretic application of dopamine. Brain Res 245:268–274.

Brake WG, Sullivan RM, Flores G, Srivastava L, Gratton A (1999) Neonatal ventral hippocampal lesions attenuate the nucleus accumbens dopamine response to stress: an electrochemical study in the adult rat. Brain Res 831:25–32.

Buhl DL, Harris KD, Hormuzdi SG, Monyer H, Buzsaki G (2003) Selective impairment of hippocampal gamma oscillations in connexin-36 knock-out mouse in vivo. J Neurosci 23:1013–1018.

Buller AL, Larson HC, Schneider BE, Beaton JA, Morrisett RA, Monaghan DT (1994) The molecular basis of NMDA receptor subtypes: native receptor diversity is predicted by subunit composition. J Neurosci 14:5471–5484.

Cannon TD, van Erp TG, Bearden CE, Loewy R, Thompson P, Toga AW, Huttunen MO, Keshavan MS, Seidman LJ, Tsuang MT (2003) Early and late neurodevelopmental influences in the prodrome to schizophrenia: contributions of genes, environment, and their interactions. Schizophr Bull 29:653–669.

Carmichael ST, Price JL (1995) Limbic connections of the orbital and medial prefrontal cortex in macaque monkeys. J Comp Neurol 363:615–641.

Casey BJ, Giedd JN, Thomas KM (2000) Structural and functional brain development and its relation to cognitive development. Biol Psychol 54:241–257.

Chambers RA, Self DW (2002) Motivational responses to natural and drug rewards in rats with neonatal ventral hippocampal lesions: an animal model of dual diagnosis schizophrenia. Neuropsychopharmacology 27:889–905.

Chambers RA, Moore J, McEvoy JP, Levin ED (1996) Cognitive effects of neonatal hippocampal lesions in a rat model of schizophrenia. Neuropsychopharmacology 15:587–594.

Coyle JT (2004) The GABA-glutamate connection in schizophrenia: which is the proximate cause? Biochem Pharmacol 68:1507–1514.

Coyle JT, Tsai G (2004) The NMDA receptor glycine modulatory site: a therapeutic target for improving cognition and reducing negative symptoms in schizophrenia. Psychopharmacology (Berl) 174:32–38.

Cunningham MG, Bhattacharyya S, Benes FM (2002) Amygdalo-cortical sprouting continues into early adulthood: implications for the development of normal and abnormal function during adolescence. J Comp Neurol 453:116–130.

de Lima AD, Opitz T, Voigt T (2004) Irreversible loss of a subpopulation of cortical interneurons in the absence of glutamatergic network activity. Eur J Neurosci 19:2931–2943.

Dunah AW, Yasuda RP, Wang YH, Luo J, Davila-Garcia M, Gbadegesin M, Vicini S, Wolfe BB (1996) Regional and ontogenic expression of the NMDA receptor subunit NR2D protein in rat brain using a subunit-specific antibody. J Neurochem 67:2335–2345.

Featherstone RE, Rizos Z, Nobrega JN, Kapur S, Fletcher PJ (2007) Gestational methylazoxymethanol acetate treatment impairs select cognitive functions: parallels to schizophrenia. Neuropsychopharmacology 32:483–492.

Flagstad P, Mork A, Glenthoj BY, van Beek J, Michael-Titus AT, Didriksen M (2004) Disruption of neurogenesis on gestational day 17 in the rat causes behavioral changes relevant to positive and negative schizophrenia symptoms and alters amphetamine-induced dopamine release in nucleus accumbens. Neuropsychopharmacology 29:2052–2064.

Flores G, Alquicer G, Silva-Gomez AB, Zaldivar G, Stewart J, Quirion R, Srivastava LK (2005) Alterations in dendritic morphology of prefrontal cortical and nucleus accumbens neurons in post-pubertal rats after neonatal excitotoxic lesions of the ventral hippocampus. Neuroscience 133:463–470.

Fujii N, Graybiel AM (2005) Time-varying covariance of neural activities recorded in striatum and frontal cortex as monkeys perform sequential-saccade tasks. Proc Natl Acad Sci U S A 102:9032–9037.

Fuster JM (1997) The Prefrontal Cortex. Anatomy, Physiology, and Neuropsychology of the Frontal Lobe. New York: Lippincott-Raven.

Galarreta M, Erdelyi F, Szabo G, Hestrin S (2004) Electrical coupling among irregular-spiking GABAergic interneurons expressing cannabinoid receptors. J Neurosci 24:9770–9778.

Gao WJ, Goldman-Rakic PS (2003) Selective modulation of excitatory and inhibitory micro-circuits by dopamine. Proc Natl Acad Sci U S A 100:2836–2841.

Gao WJ, Krimer LS, Goldman-Rakic PS (2001) Presynaptic regulation of recurrent excitation by D1 receptors in prefrontal circuits. Proc Natl Acad Sci U S A 98:295–300.

Geyer MA, Wilkinson LS, Humby T, Robbins TW (1993) Isolation rearing of rats produces a deficit in prepulse inhibition of acoustic startle similar to that in schizophrenia. Biol Psychiatry 34:361–372.

Gibson JR, Beierlein M, Connors BW (1999) Two networks of electrically coupled inhibitory neurons in neocortex. Nature 402:75–79.

Giedd JN, Blumenthal J, Jeffries NO, Castellanos FX, Liu H, Zijdenbos A, Paus T, A.C. E, Rapoport JL (1999) Brain development during childhood and adolescence: a longitudinal MRI study. Nat Neurosci 2:861–863.

Gonzalez-Burgos G, Hashimoto T, Lewis DA (2007) Inhibition and timing in cortical neural circuits. Am J Psychiatry 164:12.

Gorba T, Wahle P (1999) Expression of TrkB and TrkC but not BDNF mRNA in neurochemi-cally identified interneurons in rat visual cortex in vivo and in organotypic cultures. Eur J Neurosci 11:1179–1190.

Gorelova NA, Yang CR (2000) Dopamine D1/D5 receptor activation modulates a persistent sodium current in rat prefrontal cortical neurons in vitro. J Neurophysiol 84:75–87.

Gorelova N, Seamans JK, Yang CR (2002) Mechanisms of dopamine activation of fast-spiking interneurons that exert inhibition in rat prefrontal cortex. J Neurophysiol 88:3150–3166.

Goto Y, O'Donnell P (2002) Delayed mesolimbic system alteration in a developmental animal model of schizophrenia. J Neurosci 22:9070–9077.

Goto Y, O'Donnell P (2004) Prefrontal lesion reverses abnormal mesoaccumbens response in an animal model of schizophrenia. Biol Psychiatry 55:172–176.

Gourevitch R, Rocher C, Pen GL, Krebs MO, Jay TM (2004) Working memory deficits in adult rats after prenatal disruption of neurogenesis. Behav Pharmacol 15:287–292.

Groenewegen HJ, Berendse HW, Wolters JG, Lohman AHM (1990) The anatomical relation-ship of the prefrontal cortex with the striatopallidal system, the thalamus and the amygdala: evidence for a parallel organization. Prog Brain Res 85:95–118.

Gulledge AT, Jaffe DB (1998) Dopamine decreases the excitability of layer V pyramidal cells in the rat prefrontal cortex. J Neurosci 18:9139–9151.

Gurden H, Tassin J-P, Jay T (1999) Integrity of the mesocortical dopaminergic system is necessary for complete expression of in vivo hippocampal-prefrontal cortex long-term potentiation. Neuroscience 94:1019–1027.

Hashimoto T, Bergen SE, Nguyen QL, Xu B, Monteggia LM, Pierri JN, Sun Z, Sampson AR, Lewis DA (2005) Relationship of brain-derived neurotrophic factor and its receptor TrkB to altered inhibitory prefrontal circuitry in schizophrenia. J Neurosci 25:372–383.

Henze DA, Gonzalez-Burgos GR, Urban NN, Lewis DA, Barrionuevo G (2000) Dopamine increases excitability of pyramidal neurons in primate prefrontal cortex. J Neurophysiol 84:2799–2809.

Ishikawa A, Nakamura S (2003) Convergence and interaction of hippocampal and amygdalar projections within the prefrontal cortex in the rat. J Neurosci 23:9987–9995.

Jackson ME, Homayoun H, Moghaddam B (2004) NMDA receptor hypofunction produces concomitant firing rate potentiation and burst activity reduction in the prefrontal cortex. Proc Natl Acad Sci U S A 101:8467–8472.

Javitt DC, Zukin SR (1991) Recent advances in the phencyclidine model of schizophrenia. Am J Psychiatry 148:1301–1308.

Jay TM, Glowinski J, Thierry AM (1995) Inhibition of hippocampo-prefrontal cortex excita-tory responses by the mesocortical DA system. Neuroreport 6:1845–1848.

Jones GH, Marsden CA, Robbins TW (1991) Behavioural rigidity and rule-learning deficits following isolation-rearing in the rat: neurochemical correlates. Behav Brain Res 43:35–50.

Jovanovic JN, Thomas P, Kittler JT, Smart TG, Moss SJ (2004) Brain-derived neurotrophic factor modulates fast synaptic inhibition by regulating GABA(A) receptor phosphorylation, activity, and cell-surface stability. J Neurosci 24:522–530.

Kato K, Shishido T, Ono M, Shishido K, Kobayashi M, Suzuki H, Nabeshima T, Furukawa H, Niwa S (2000) Effects of phencyclidine on behavior and extracellular levels of dopamine and its metabolites in neonatal ventral hippocampal damaged rats. Psychopharmacology (Berl) 150:163–169.

Kawaguchi Y (2001) Distinct firing patterns of neuronal subtypes in cortical synchronized activities. J Neurosci 21:7261–7272.

Kuner T, Schoepfer R (1996) Multiple structural elements determine subunit specificity of Mg^{2+} block in NMDA receptor channels. J Neurosci 16:3549–3558.

Lavin A, Moore HM, Grace AA (2005) Prenatal disruption of neocortical development alters prefrontal cortical neuron responses to dopamine in adult rats. Neuropsychopharmacology 30:1426–1435.

Le Pen G, Moreau JL (2002) Disruption of prepulse inhibition of startle reflex in a neurodevelopmental model of schizophrenia: reversal by clozapine, olanzapine and risperidone but not by haloperidol. Neuropsychopharmacology 27:1–11.

Le Pen G, Gourevitch R, Hazane F, Hoareau C, Jay TM, Krebs MO (2006) Peri-pubertal maturation after developmental disturbance: a model for psychosis onset in the rat. Neuroscience 143:395–405.

Lewis BL, O'Donnell P (2000) Ventral tegmental area afferents to the prefrontal cortex maintain membrane potential 'up' states in pyramidal neurons via D_1 dopamine receptors. Cereb Cortex 10:1168–1175.

Lewis DA (1997) Development of the prefrontal cortex during adolescence. Neuropsychopharmacology 16:385–398.

Lewis DA, Hashimoto T, Volk DW (2005) Cortical inhibitory neurons and schizophrenia. Nat Rev Neurosci 6:312–324.

Lipska B, al-Amin H, Weinberger D (1998) Excitotoxic lesions of the rat medial prefrontal cortex. Effects on abnormal behaviors associated with neonatal hippocampal damage. Neuropsychopharmacology 19:451–464.

Lipska BK, Weinberger DR (1994) Subchronic treatment with haloperidol and clozapine in rats with neonatal excitotoxic hippocampal damage. Neuropsychopharmacology 10:199–205.

Lipska BK, Jaskiw GE, Weinberger DR (1993) Postpuberal emergence of hyperresponsiveness to stress and to amphetamine after neonatal excitotoxic hippocampal damage: a potential animal model of schizophrenia. Neuropsychopharmacology 90:67–75.

Lipska BK, Aultman JM, Verma A, Weinberger DR, Moghaddam B (2002) Neonatal damage of the ventral hippocampus impairs working memory in the rat. Neuropsychopharmacology 27:47–54.

Lipska BK, Swerdlow NR, Geyer MA, Jaskiw GE, Braff DL, Weinberger DR (1995) Neonatal excitotoxic hippocampal damage in rats cause post-pubertal changes in prepulse inhibition of startle and its disruption by apomorphine. Psychopharmacology (Berl) 132:303–310.

Lipska BK, Lerman DN, Khaing ZZ, Weickert CS, Weinberger DR (2003) Gene expression in dopamine and GABA systems in an animal model of schizophrenia: effects of antipsychotic drugs. Eur J Neurosci 18:391–402.

Luby ED, Cohen BD, Rosenbaum G, Gottlieb JS, Kelly R (1959) Study of a new schizophrenomimetic drug – Sernyl. Arch Neurol Psychiatry 81:363–369.

Makino C, Shibata H, Ninomiya H, Tashiro N, Fukumaki Y (2005) Identification of single-nucleotide polymorphisms in the human *N*-methyl-D-aspartate receptor subunit NR2D gene, GRIN2D, and association study with schizophrenia. Psychiatr Genet 15:215–221.

Mercuri N, Calabresi P, Stanzione P, Bernardi G (1985) Electrical stimulation of mesencephalic cell groups (A9-A10) produces monosynaptic excitatory potentials in rat frontal cortex. Brain Res 338:192–195.

Mitchell CP, Grayson DR, Goldman MB (2005) Neonatal lesions of the ventral hippocampal formation alter GABA-A receptor subunit mRNA expression in adult rat frontal pole. Biol Psychiatry 57:49–55.

Moghaddam B, Adams B, Verma A, Daly D (1997) Activation of glutamatergic neurotransmission by ketamine: a novel step in the pathway from NMDA receptor blockade to dopaminergic and cognitive disruptions associated with the prefrontal cortex. J Neurosci 17:2921–2927.

Moghaddam B, Aultman J, Weinberger D, Lipska B (1999) Neonatal damage of the rat ventral hippocampus impairs acquisition of a working memory task. Soc Neurosci Abstr 25:1891.

Monyer H, Burnashev N, Laurie DJ, Sakmann B, Seeburg PH (1994) Developmental and regional expression in the rat brain and functional properties of four NMDA receptors. Neuron 12:529–540.

O'Donnell P (2003) Dopamine gating of forebrain neural ensembles. Eur J Neurosci 17:429–435.

O'Donnell P, Lewis BL, Weinberger DR, Lipska BK (2002) Neonatal hippocampal damage alters electrophysiological properties of prefrontal cortical neurons in adult rats. Cereb Cortex 12:975–982.

Pantelis C, Yucel M, Wood SJ, Velakoulis D, Sun D, Berger G, Stuart GW, Yung A, Phillips L, McGorry PD (2005) Structural brain imaging evidence for multiple pathological processes at different stages of brain development in schizophrenia. Schizophr Bull 31:672–696.

Paulus MP, Bakshi V, Geyer MA (1998) Isolation rearing affects sequential organization of motor behavior in post-pubertal but not pre-pubertal Lister and Sprague–Dawley rats. Behav Brain Res 94:271–280.

Penit-Soria J, Audinat E, Crepel F (1987) Excitation of prefrontal cortical neurons by dopamine: an in vitro electrophysiological study. Brain Res 425:263–274.

Peters YM, O'Donnell P (2005) Social isolation rearing affects prefrontal cortical response to ventral tegmental area stimulation. Biol Psychiatry 57:1205–1208.

Rosenbaum G, Cohen BD, Luby JS, Gottlieb JS, Yelen D (1959) Comparison of sernyl with other drugs: stimulation of schizophrenic performance with sernyl, LSD-25, and amobarbital (amytal) sodium; I. Attention, motor function, and proprioception. Arch Gen Psychiatry 1:113–118.

Rosenberg DR, Lewis DA (1994) Changes in the dopaminergic innervation of monkey prefrontal cortex during late postnatal development: a tyrosine hydroxylase immunohistochemical study. Biol Psychiatry 36:272–277.

Rosenberg DR, Lewis DA (1995) Postnatal maturation of the dopaminergic innervation of monkey prefrontal and motor cortices: a tyrosine hydroxylase immunohistochemical analysis. J Comp Neurol 358:383–400.

Sams-Dodd F (1998) A test of the predictive validity of animal models of schizophrenia based on phencyclidine and D-amphetamine. Neuropsychopharmacology 18:293–304.

Sams-Dodd F, Lipska BK, Weinberger DR (1997) Neonatal lesions of the rat ventral hippocampus result in hyperlocomotion and deficits in social behavior in adulthood. Psychopharmacology (Berl) 132:303–310.

Scherzer CR, Landwehrmeyer GB, Kerner JA, Counihan TJ, Kosinski CM, Standaert DG, Daggett LP, Velicelebi G, Penney JB, Young AB (1998) Expression of *N*-methyl-D-aspartate receptor subunit mRNAs in the human brain: hippocampus and cortex. J Comp Neurol 390:75–90.

Schultz W (1998) Predictive reward signal of dopamine neurons. J Neurophysiol 80:1–27.

Seamans JK, Yang CR (2004) The principal features and mechanisms of dopamine modulation in the prefrontal cortex. Prog Neurobiol 74:1–58.

Segalowitz SJ, Davies PL (2004) Charting the maturation of the frontal lobe: an electrophysiological strategy. Brain Cogn 55:116–133.

Standaert DG, Landwehrmeyer GB, Kerner JA, Penney Jr. JB, Young AB (1996) Expression of NMDAR2D glutamate receptor subunit mRNA in neurochemically identified interneurons in the rat neostriatum, neocortex and hippocampus. Mol Brain Res 42:89–102.

Steffensen SC, Svingos AL, Pickel VM, Henriksen SJ (1998) Electrophysiological characterization of GABAergic neurons in the ventral tegmental area. J Neurosci 18:8003–8015.

Swerdlow NR, Lipska BK, Weinberger DR, Braff DL, Jaskiw GE, Geyer MA (1995) Increased sensitivity to the sensorimotor gating-disruptive effects of apomorphine after lesions of medial prefrontal cortex of ventral hippocampus in adult rats. Psychopharmacology (Berl) 122:27–34.

Thierry AM, Blanc M, Sobel A, Stinus L, Glowinski J (1973) Dopaminergic terminals in the rat cortex. Science 182:499–500.

Tseng KY, O'Donnell P (2004) Dopamine-glutamate interactions controlling prefrontal cortical pyramidal cell excitability involve multiple signaling mechanisms. J Neurosci 24:5131–5139.

Tseng KY, O'Donnell P (2005) Post-pubertal emergence of prefrontal cortical up states induced by D1-NMDA co-activation. Cereb Cortex 15:49–57.

Tseng KY, O'Donnell P (2007) Dopamine modulation of prefrontal cortical interneurons changes during adolescence. Cereb Cortex 17:1235–1240.

Vollenweider FX, Leenders KL, Scharfetter C, Antonini A, Maguire P, Missimer J, Angst J (1997) Metabolic hyperfrontality and psychopathology in the ketamine model of psychosis using emission tomography (PET) and [18F]fluorodeoxyglucose (FDG). Eur Neuropsychopharmacol 7:9–24.

Waelti P, Dickinson A, Schultz W (2001) Dopamine responses comply with basic assumptions of formal learning theory. Nature 412:43–48.

Wang J, O'Donnell P (2001) D_1 dopamine receptors potentiate NMDA-mediated excitability increase in rat prefrontal cortical pyramidal neurons. Cereb Cortex 11:452–462.

Wenzel A, Villa M, Mohler H, Benke D (1996) Developmental and regional expression of NMDA receptor subtypes containing the NR2D subunit in rat brain. J Neurochem 66:1240–1248.

Williams K (1995) Pharmacological properties of recombinant N-methyl-D-aspartate (NMDA) receptors containing the epsilon 4 (NR2D) subunit. Neurosci Lett 184:181–184.

Woo TU, Pucak ML, Kye CH, Matus CV, Lewis DA (1997) Peripubertal refinement of the intrinsic and associational circuitry in monkey prefrontal cortex. Neuroscience 80:1149–1158.

Yamaguchi T, Sheen W, Morales M (2007) Glutamatergic neurons are present in the rat ventral tegmental area. Eur J Neurosci 25:106–118.

Index

Printed in the United States of America.